Lightweight Magnesium Technology 2001-2005

PT-131

Edited by
Thomas Ruden

Published by
Society of Automotive Engineers, Inc.
400 Commonwealth Drive
Warrendale, PA 15096-0001
U.S.A.
Phone: (724) 776-4841
Fax: (724) 776-5760
www.sae.org
March 2006

For permission and licensing requests contact:

SAE Permissions
400 Commonwealth Drive
Warrendale, PA 15096-0001-USA
Email: permissions@sae.org
Fax: 724-776-3036
Tel: 724-772-4028

For multiple print copies contact:

SAE Customer Service
Tel: 877-606-7323 (inside USA and Canada)
Tel: 724-776-4970 (outside USA)
Fax: 724-776-0790
Email: CustomerService@sae.org

ISBN-10 0-7680-1781-5
ISBN-13 978-0-7680-1781-6
Library of Congress Catalog Card Number: 2006923048
SAE/PT-131

SAE Order No. PT-131

Printed in USA

Preface

Magnesium is the lightest of all structural metals. The compilation of magnesium papers in this book represents the "best of the best" from all of those published at the SAE World Congress from 2001 through 2005. Topics include development of new alloys, process technologies, mechanical and physical properties, and case studies of applications placed into production. The authors have played a key role toward increasing the use of magnesium by the global automotive industry.

The overall increase in the use of magnesium by the automotive industry, however, is due to the collective efforts of everyone who helps manufacture, design, educate, market, and promote its features and benefits. This increase is evident by the transformation of applications from aluminum, steel, and engineered plastics into magnesium. Examples include instrument panel structures, seat frames, center consoles, transmission cases, front-end and radiator support structures, and hybrid magnesium powertrains.

Advancements in magnesium continue toward developing even higher-performing alloys, improving processes, developing new applications, and identifying new technologies to further the competitive edge of magnesium as a lightweight, recyclable, and viable metal to meet global automotive needs.

I am honored to have the privilege of working with SAE and its authors to help promote these papers within the global automotive industry. Without their research, dedication, and enthusiasm toward sharing the advancements in magnesium, the magnesium story could not be told.

Magnesium—the metal having the lowest weight and the highest growth rate in the global automotive industry.

Thomas Ruden
Market Development
Lunt Manufacturing Co., Inc.

Table of Contents

Preface iii

Microstructure and Mechanical Properties of Squeeze Cast AZ91D Magnesium Alloy (2005-01-0330) 1
Rathindra DasGupta

Prediction of Distortion and Residual Stresses in a Magnesium High- Pressure Die Casting With Considerations to the Die-Casting Process (2005-01-0331) 9
Winston Sequeira

Use of Quality Mapping to Predict Performance of Thin-Walled Magnesium Die Castings (2005-01-0332) 17
Stian Sannes

Optimizing the Magnesium Die Casting Process to Achieve Reliability in Automotive Applications (2005-01-0333) 25
Haavard Gjestland

Effect of HPDC Parameters on the Performance of Creep Resistant Alloys MRI153M and MRI230D (2005-01-0334) 33
Germa Gertsberg

The Concept and Technology of Alloy Formation During Semisolid Injection Molding (2005-01-0336) 39
Frank Czerwinski

Magnesium Engine Cradle - The Uscar Structural Cast Magnesium Development Project (2005-01-0337) 45
Naiyi Li

The Application of Magnesium Die Casting to Vehicle Closures (2005-01-0338) 55
Patrick Blanchard

2005 Fuel Cell Vehicle and its Magnesium Power Distribution Unit (2005-01-0339) 63
Naiyi Li

Development of the 2006 Corvette Z06 Structural Cast Magnesium Crossmember (2005-01-0340) 69
Jonathan Aragones

2005 Ford GT Magnesium Instrument Panel Cross Car Beam (2005-01-0341) 77
Naiyi Li

Use of Cast Magnesium Back Frames in Automotive Seating (2005-01-0723) 85
Matthew Abate

Numerical Modeling of the Structural Behavior of Thin-Walled Cast Magnesium Components Using a Through-Process Approach (2005-01-0724) 93
Cato Dørum

Mechanical Properties of AM60B Die Castings: A Review of the AUTO21 Program on Magnesium Die-Casting (2005-01-0725) 101
J. T. Wood

Physical and Virtual Prototyping of Magnesium Instrument Panel Structures (2005-01-0726) 107
John Z. Lin

2004 Nissan 3.5L Cam Cover Material Study: Aluminum, Magnesium and Composite (2005-01-0727) 115
Eric Herrera

AJ (Mg-Al-Sr) Alloy Mechanical Properties: From Fatigue to Crack Propagation (2005-01-0729) 123
Pierre Labelle

Friction Stir Welding of Magnesium for Automotive Applications (2005-01-0730) 135
Kazutaka Okamoto

Corrosion Protection of Structural Magnesium Alloys: Recent Development (2005-01-0732) 141
Wenyue Zheng

Wrought Magnesium Alloys and Manufacturing Processes for Automotive Applications (2005-01-0734) 155
Alan A. Luo

Energy Absorption Capacity for HPDC Components (2004-01-0128) 167
Cato Døum

A Case Study of a Die-Cast Magnesium Structure Supporting Transmission Shifter Mechanisms and Interfaced With Other Structural Systems (2004-01-0130) 173
Thomas J. Ruden

Robustness of Large Thin Wall Magnesium Die Castings for Crash Applications (2004-01-0131) 183
Regis Alain

Reliable Light Weight Fastening of Magnesium Components in Automotive Applications (2004-01-0136) 191
C. Friedrich

Progress Toward a Magnesium-Intensive Engine: the USAMP Magnesium Powertrain Cast Components Project (2004-01-0654) 197
Bob R. Powell

Powertrain Components - Opportunities for the Die Cast AE Family of Alloys (2004-01-0655) 207
Per Bakke

High Temperature Mg Alloys for Sand and Permanent Mold Casting Applications (2004-01-0656) 213
Boris Bronfin

Evaluation of High Temperature, Cast Magnesium Alloy, Front Engine Covers (2004-01-0657) 219
Alan P. Druschitz

Comparative Evaluation of Automotive Oil Pans Fabricated by Creep Resistant Magnesium Alloy and Aluminum Alloy (2004-01-0658) 225
Peter Vert

AJ (Mg-Al-Sr) Alloy System Used for New Engine Block (2004-01-0659) 235
Pierre Labelle

Die Casting of Magnesium Alloys - The Importance of Controlling Die Filling and Solidification (2003-01-0183) 245
Sannes Stian

Bolt Load Compressive Stress Retention Testing of Magnesium Alloys (2003-01-0187) 255
Alan P. Druschitz

Die Cast Magnesium Alloys for High Temperature Applications (2003-01-0189)............................. 261
Per Bakke

Magnesium Die Casting Alloy AJ62x with Superior Creep Resistance, Ductility and Die Castability (2003-01-0190).. 269
Mihriban Pekguleryuz

A Comparative Study of New Magnesium Alloys Developed for Elevated Temperature Applications in Automotive Industry (2003-01-0191) 275
Eli Aghion

AMC-SC1: A New Magnesium Alloy Suitable for Powertrain Applications (2003-01-1365) 283
C. J. Bettles

Development in Lost Foam Casting of Magnesium (2003-01-0821)... 291
Michael Marlatt

Metal Quality - The Effects on Die Casters and End Users (2002-01-0078) 299
Gerry Wang

New Aspects of Temperature Behavior of AJ52x, Creep Resistant Magnesium Alloy (2002-01-0079).. 307
Pierre Labelle

Evaluation of Protective Coatings on Magnesium for Phosphate Process Compatibility and Galvanic Corrosion Prevention (2002-01-0081).. 317
Guilian Gao

Production and Die Casting of Semi-Solid Magnesium Alloy AZ91D (2002-01-0082) 325
M.T. Shehata

Elevated Temperature Forming of Sheet Magnesium Alloys (2001-01-3104).................................. 331
Paul E. Krajewski

Development of Creep-Resistant Magnesium Alloys for Powertrain Applications: Part 1 of 2 (2001-01-0422).. 337
Bob R. Powell

Tensile Creep and Microstructure of Magnesium-Aluminum-Calcium-Based Alloys for Powertrain Applications--Part 2 of 2(2001-01-0423) 345
Alan Luo

Bolt-Load Retention Behavior of a Die Cast Magnesium-Rare Earth Alloy (2001-01-0425)............ 355
Ian P. Moreno

About the Editor 365

2005-01-0330

Microstructure and Mechanical Properties of Squeeze Cast AZ91D Magnesium Alloy

Rathindra DasGupta, Phil Burton and Zach Brown
SPX Corporation, CONTECH Division

ABSTRACT

Squeeze casting is considered a "high integrity" casting process because it imparts qualities (higher tensile properties, in particular ductility due to reduced or absence of porosity in the matrix, and the ability to heat treat) to a metal that are difficult to achieve with conventional casting techniques including gravity permanent mold (GPM) and high pressure, high velocity (HPDC) die casting.

In recent years, the squeeze casting process has been widely used with various aluminum alloys to manufacture near-net shape automotive components requiring high strength, ductility or pressure tightness. However, with the emphasis on weight reduction, lower cost and improved performance of structural components, alternative lightweight materials including magnesium are now being seriously considered. Unfortunately, the use of magnesium as a structural material has been hindered by the lack of data on mechanical properties and the lack of new improved casting methods.

This paper, therefore, reviews the squeeze casting process and provides information on microstructures and mechanical properties of squeeze cast AZ91D magnesium alloy before and after heat treatment (T4 and T6 temper). The data obtained from this investigation are then compared with those from commonly used die casting aluminum alloys.

INTRODUCTION

With increasing emphasis on vehicle weight reduction, magnesium alloys (with the combination of low density, high specific strength, improved corrosion resistance and excellent castability) are gaining prominence as a structural material [1]. The magnesium alloys have been traditionally cast using the high pressure, high velocity (HPDC) casting process. Although the HPDC process is a well established and economical casting process [1], it is not usually used for manufacturing parts (particularly thick-walled components) requiring high structural integrity (fairly uniform microstructure across the wall thickness, reduced or absence of porosity in the matrix, pressure tightness, higher strength and ductility).

Improved mechanical properties and decreased porosity in thick-walled components can, however, be achieved via the use of squeeze casting process. In recent years, the squeeze casting process has been widely used with various aluminum alloys to manufacture near-net shape automotive components [2]. Mechanical properties and microstructures of these squeeze cast aluminum alloys are also well documented [2]. Unfortunately, there is little to no information on squeeze cast magnesium alloys.

The objective of this paper is to review the squeeze casting process in use today, and to assess the mechanical properties and microstructures of squeeze cast AZ91D magnesium alloy (AZ91D is the alloy most commonly chosen for general use applications), before and after heat treatment (T4 and T6 temper).

SQUEEZE CASTING PROCESS

The typical advantages of squeeze casting include:

1. Minimal turbulence
2. Minimal air entrapment
3. Reduced shrink porosity
4. Rapid solidification
5. Dimensional control similar to that of HPDC
6. Heat treatable (T6, T4)
7. Used for sections > 2.5 mm

A schematic of a Horizontal Vertical Squeeze Cast (HVSC) machine and its typical features are shown in Figure 1 and Figure 2, respectively [2]. The machine and tools are, in many respects, like those employed in a conventional die casting machine.

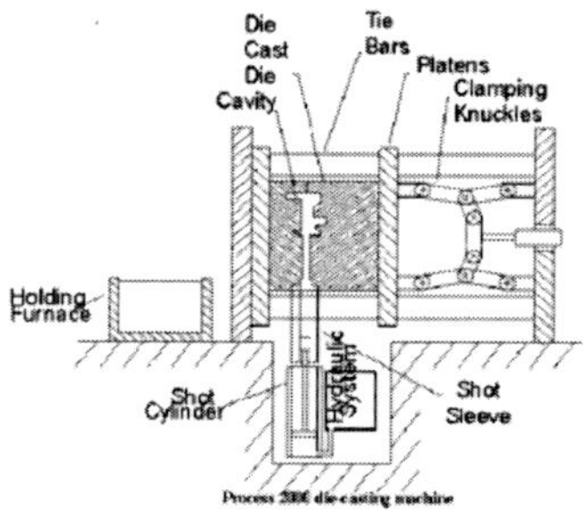

Figure 1: Schematic of HVSC machine [2]

1. Horizontal die clamping
2. Vertical, high pressure delivery system
3. Tilt-docking injection unit
4. Separated sleeve for cooling and transfer

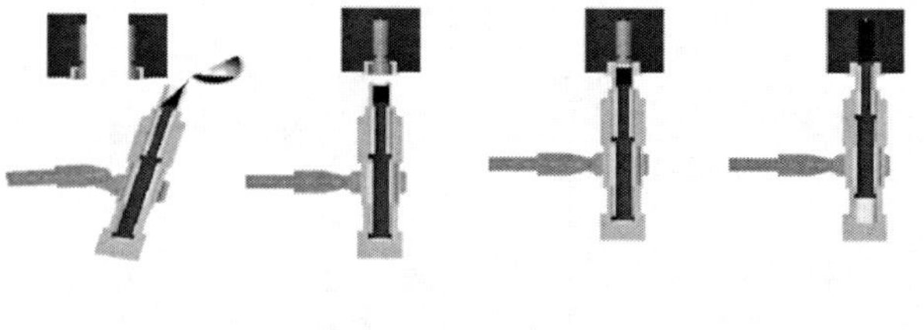

Figure 2: Typical features of HVSC machine [2]

EXPERIMENTAL PROCEDURE

Step plates (see Figure 3) were squeeze cast using the AZ91D magnesium alloy and a 350T HVSC machine. The step plates consisted of four sections with thicknesses of 2, 6, 10 and 14 mm. The dimensions of each step plate measured 152.4 mm in length and 101.6 mm in width.

Following squeeze casting, tensile properties were determined from tensile specimens machined from 6, 10 and 14 mm sections (see Figure 3) in the as-cast condition and after heat treatment (T4 and T6 temper). Fracture surface and "core" microstructures (as-cast and after T4, T6 temper) of select tensile specimens were subsequently examined and documented.

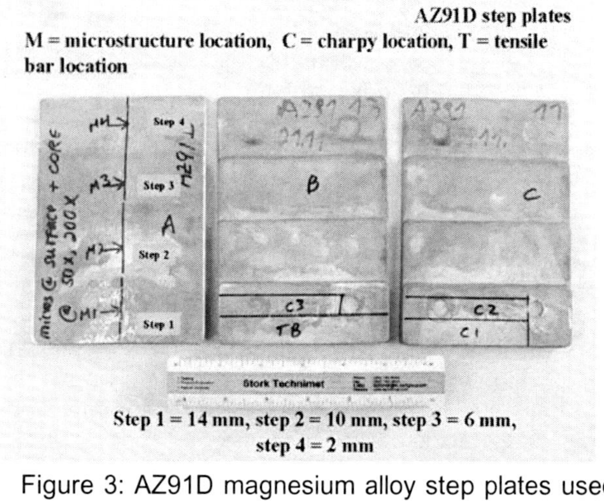

Figure 3: AZ91D magnesium alloy step plates used in this investigation

In addition, the various phases observed in the as - cast and heat treated microstructures were identified using the scanning electron microscope (SEM).

Finally, impact strength was obtained from unnotched charpy specimens machined from 14 mm sections in the as-cast condition.

RESULTS

A. As-Cast Tensile Properties:

Tensile properties of squeeze cast AZ91D magnesium alloy in the as-cast condition are summarized in Table 1. Tensile testing was conducted in accordance with ASTM E 8-01 and E 10-01. Also shown in Table 1, for comparison, are squeeze cast and semi-solid metal cast (SSM) data reported in literature [1].

Table 1: As-Cast Tensile Properties[1]

Sample ID	*Yield Strength (Mpa)*	*Tensile Strength (Mpa)*	*% Elongation*
Step 1, 14 mm	112 – 115	168 – 178	2.0 – 3.5
Step 2, 10 mm	115 – 117	178 – 185	2.0 – 3.5
Step 3, 6 mm	115 – 118	171 – 182	2.0 – 3.0
SSM [1]	105 +/- 6	171 +/- 10	3.4 +/- 1.0
Squeeze Cast [1]	104 +/- 4	183 +/- 12	4.5 +/- 1.2

Legend:
1 = Original gage dimensions were 6.35 mm x 6.07 mm; pretest gage length was 25 mm.

Key observations from Table 1 are:

1. The as-cast strength data obtained in this study compare favorably well with both squeeze cast and rheocast data documented in literature [1].

2. The lower percent elongation observed in this investigation (in comparison to that reported in literature) may be attributed to the presence of small inclusions on the fracture surfaces of the tensile specimens.

3. Wall thickness does not significantly influence tensile properties. This observation, however, does not agree with that reported by Dunlop et al [3].

In the study conducted by Dunlop et al (tensile tests were performed on flat HPDC tensile specimens ranging in thickness from 1 mm to 6 mm), yield strength of 185 Mpa was obtained for a 1 mm section as compared to 130 Mpa for a 6 mm thick section. The difference in yield strength between the two sections is attributed to the presence of a much higher volume fraction of a fine hard surface layer (comprised of small α-Mg grains and the β-$Mg_{17}Al_{12}$ phase) in the thin die cast sample [3].

B. As-Cast Microstructures and Phase Identification:

The typical microstructure (regardless of wall thickness) of squeeze cast AZ91D magnesium alloy is shown in Figure 4.

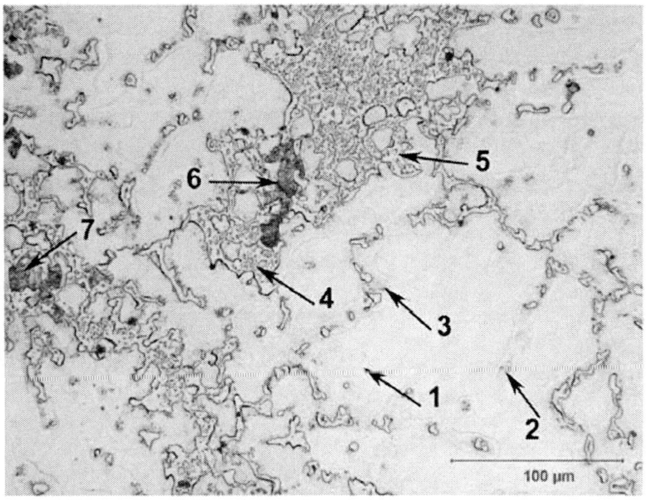

Figure 4: Optical micrograph of Step 1 (14 mm thickness); etched with 2% nital

The various microconstituents/phases identified with numbers & arrows in Figure 4 were subsequently examined with a scanning electron microscope (SEM) using the secondary electron mode. Chemical composition of these phases was obtained via the Energy Dispersive X-ray Spectrometry (EDS). At room temperature, squeeze cast AZ91D magnesium alloy reveals the following:

- Pro-eutectic α-Mg identified as "1" in Figure 4. SEM image of this phase (along with its chemical composition) is shown in Figure 5.

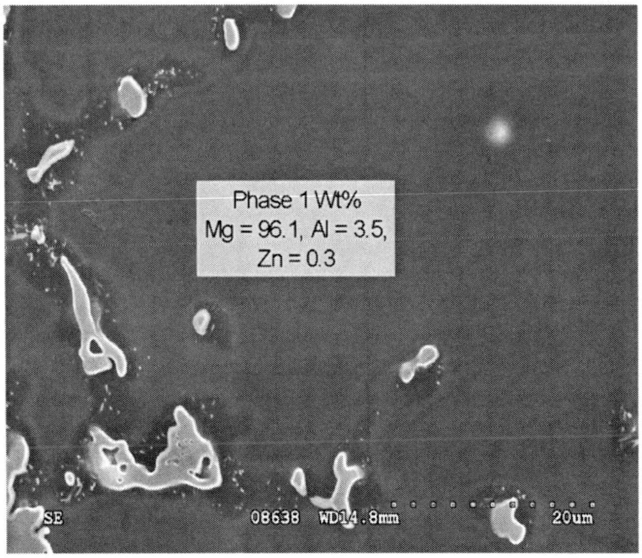

Figure 5: Scanning electron micrograph of phase "1" in Figure 4; etched with 2% nital

- Supersaturated solid solutions of Al in Mg designated as "2" and "3" in Figure 4. SEM image of phase "2" along with its chemical analysis is shown in Figure 6.

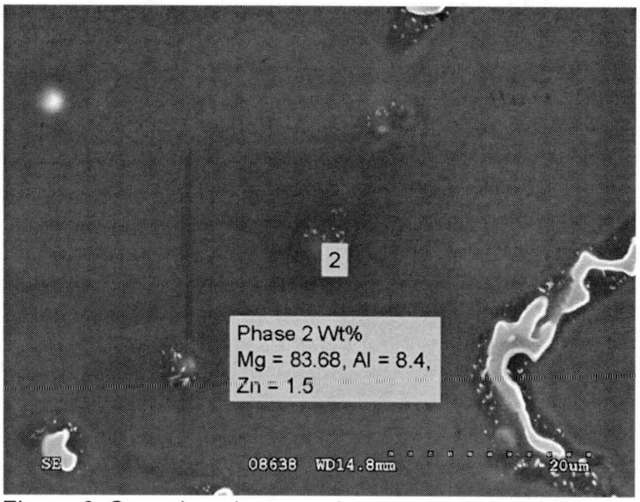

Figure 6: Scanning electron micrograph of phase "2" in Figure 4; etched with 2% nital

- The β-$Mg_{17}Al_{12}$ phase identified as "4" and"5" in Figure 4. Scanning electron micrograph of the β-$Mg_{17}Al_{12}$ phase (phase "5" in Figure 4) is shown in Figure 7. It should be noted that the β-$Mg_{17}Al_{12}$ phase (that is known to affect the mechanical properties of Mg-Al castings) does not form a continuous network. This observation, as noted later in this paper, influences the fracture behavior (crack path).

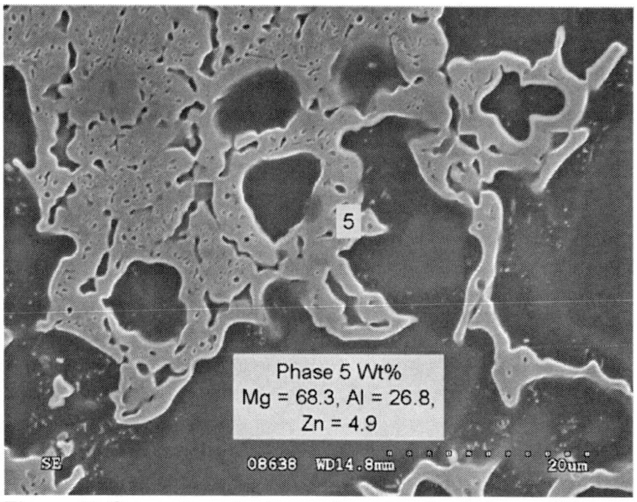

Figure 7: Scanning electron micrograph of phase "5" in Figure 4; etched with 2% nital

- Depending on the subsequent cooling rate in the solid state, the supersaturated solid solution of Al in Mg (phases 2 and 3 in Figure 4) can experience a discontinuous decomposition with formation of $Mg_{17}Al_{12}$ intermetallics in a lamellar (pearlite-like) form. The lamellar phases are designated as "6" and "7" in Figure 4. Scanning electron micrograph of a pearlite-like phase (phase “6” in Figure 4) is shown in Figure 8.

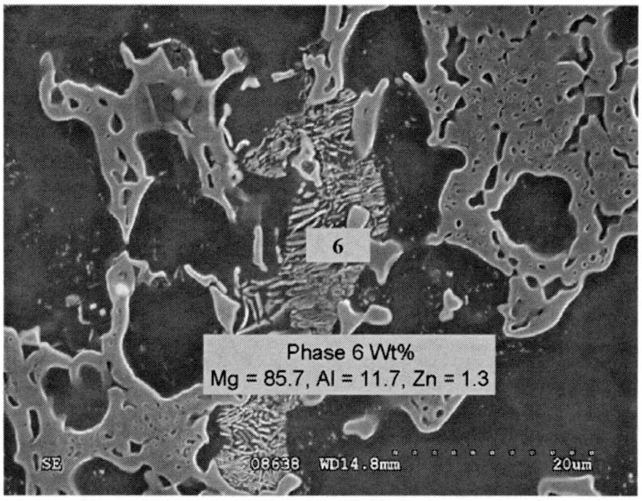

Figure 8: Scanning electron micrograph of phase “6” in Figure 4;etched with 2% nital

It should be noted that the lamellar structure was not evident in sections less than 6 mm. Furthermore, the lamellar structure was observed in the “core” of the thicker sections (slower solidification rates) than at the surface (faster solidification rates).

- During identification of the various phases in squeeze cast AZ91D alloy, scanning electron microscopic examination also revealed the presence of Al-Mn-Mg phase; see Figure 9.

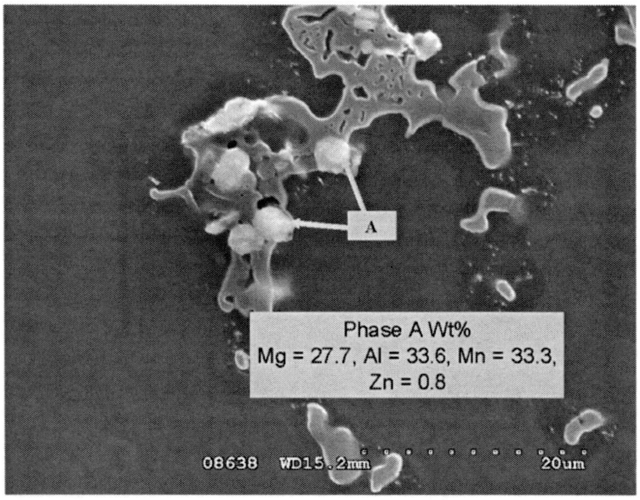

Figure 9: Scanning electron micrograph of Al-Mn-Mg in squeeze cast AZ91D magnesium alloy; etched with 2% nital

The effect of wall thickness on as-cast surface and “core” microstructures was also investigated in this study using the fractured tensile bars from each of the steps (14 mm, 10 mm and 6 mm). The results are as follows:

1. With decreasing section size, the microstructure exhibits a finer dispersion of the β-$Mg_{17}Al_{12}$ precipitates, both at the fracture surface and at the “core (see Figures 10a to 10c)).

2. For a given section size, fracture surface and the “core” do not yield a significant difference in the level of porosity; see Figures 11a and 11b. Furthermore, there is no significant difference in the size of the β-$Mg_{17}Al_{12}$ precipitates between the two locations.

3. The volume fraction of the β-$Mg_{17}Al_{12}$ precipitates does not appear to increase with decreasing section size.

Figure 10a: Sample #78; microstructure of fracture surface of tensile bar machined from 14 mm thick step (polarized light)

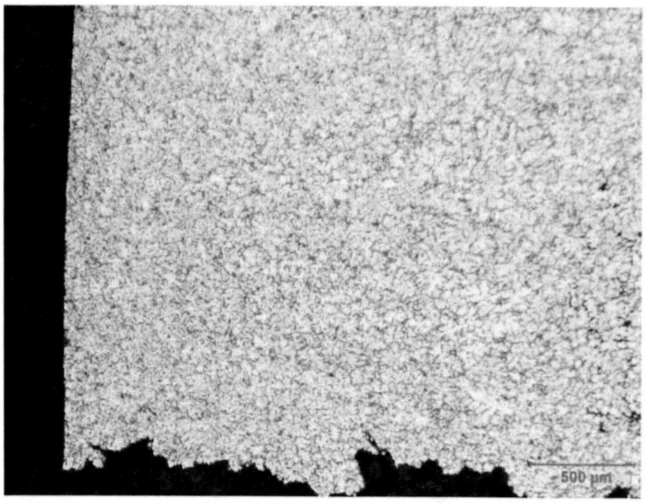

Figure 10b: Sample #78; microstructure of fracture surface of tensile bar machined from 10 mm thick step (polarized light); etched with 2% nital

Figure 10c: Sample #78; microstructure of fracture surface of tensile bar machined from 6 mm thick step (polarized light); etched with 2% nital

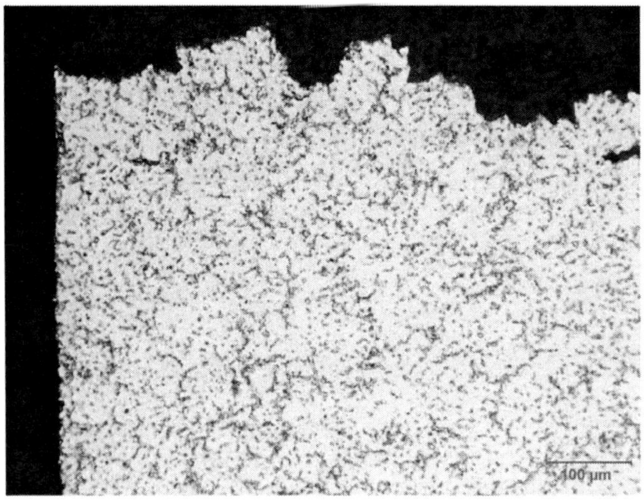

Figure 11a: Fracture surface microstructure for a 6 mm section (polarized light); etched with 2% nital

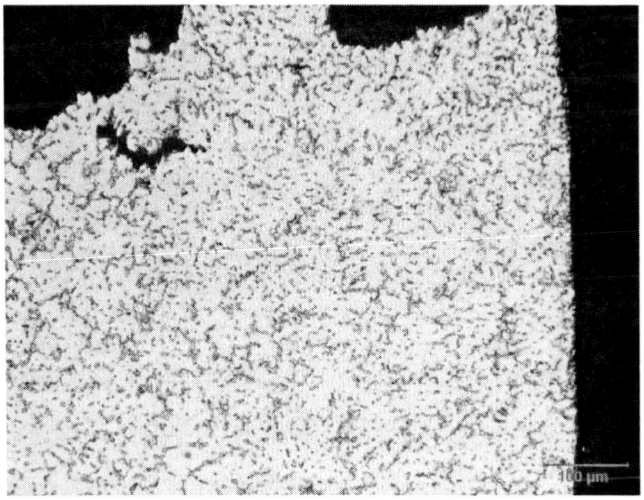

Figure 11b: "Core" microstructure for a 6 mm section (polarized light); etched with 2% nital

C. As-Cast Impact Properties:

The as-cast impact properties were determined at room temperature from un-notched specimens (10 mm x 10 mm section size) machined from the step plates (14 mm section size). The impact strength ranged from 7 J to 11 J.

Subsequent to impact testing, one of the charpy specimens was used to investigate the crack path in squeeze cast specimens. As shown in Figures 12a and 12b, the crack path is essentially intergranular and quite "tortuous" (due to resistance against crack extension since the β-$Mg_{17}Al_{12}$ phase network is not continuous). This observation is in agreement with that reported by Kleiner et al [1].

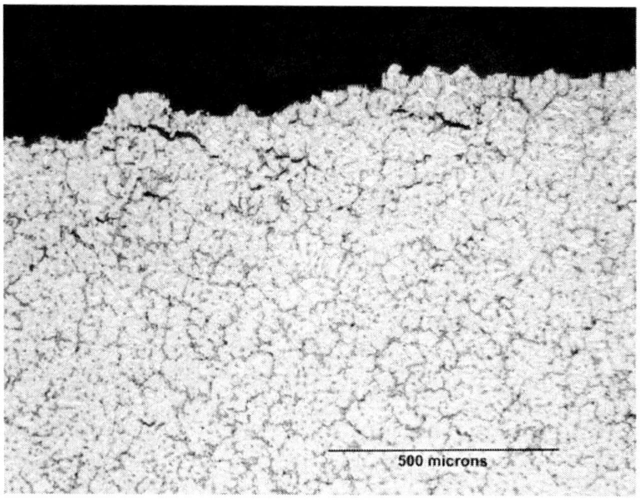

Figure 12a: Tortuous crack path in squeeze cast AZ91D magnesium alloy

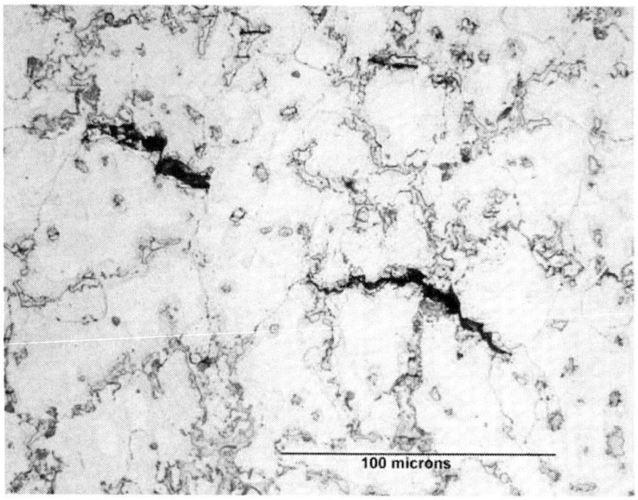

Figure 12b: Higher magnification view of Figure 12a

D. Tensile Properties of Heat Treated Squeeze Cast AZ91D Alloy:

Tensile properties of squeeze cast AZ91D magnesium alloy, after T4 and T6 temper, are summarized in Table 2. Tensile testing was conducted in accordance with ASTM E 8-01 and E 10-01.

The T4 temper consisted of solution treatment at 410°C for 6 hours followed by cooling with fan to room temperature. The T6 temper involved solution treatment at 410°C for 6 hours, fan cooling to room temperature, artificial aging at 215°C for 6 hours followed by air cooling to room temperature.

Also shown in Table 2, for comparison, are tensile properties of commonly used aluminum alloys (380 and 383) cast using various processes.

Table 2: Tensile Properties Comparison

Alloy & Temper	Process	Yield Strength (Mpa)	Tensile Strength (Mpa)	% Elongation
380-F	HPDC	130 - 155	172 - 185	1.2 – 1.4
	Squeeze cast	138 - 159	227 - 245	2 - 3
	SSM	131 - 152	241 - 255	2 - 4
383-F	HPDC	145 (avg)	200 (avg)	1.25 (avg)
AZ91D (F)	Squeeze cast	112 - 118	168 - 185	2.0 – 3.5
AZ91D (T4)	Squeeze cast	92 - 101	228 - 247	7.0 – 11.0
AZ91D (T4)	Rheocast [1]	94+/-5	241+/-19	8.3+/-1.6
AZ91D (T6)	Squeeze cast	135 - 140	196 - 210	2.0 – 2.5
AZ91D (T6)	Sand cast	118+/-2.8	202+/-10	3.5+/-1.5

Key observations from Table 2 are:

1. In comparison to the F temper, large improvements of tensile strength and elongation are observed after solution heat treatment (T4 temper); however, the yield strength is reduced.

2. The increase in ductility observed after a T4 temper is due to the dissolution of the brittle β-$Mg_{17}Al_{12}$ phase and of the discontinuous (lamellar) structure at elevated temperatures. Figure 13 shows the typical microstructure of squeeze cast AZ91D-T4 magnesium alloy.

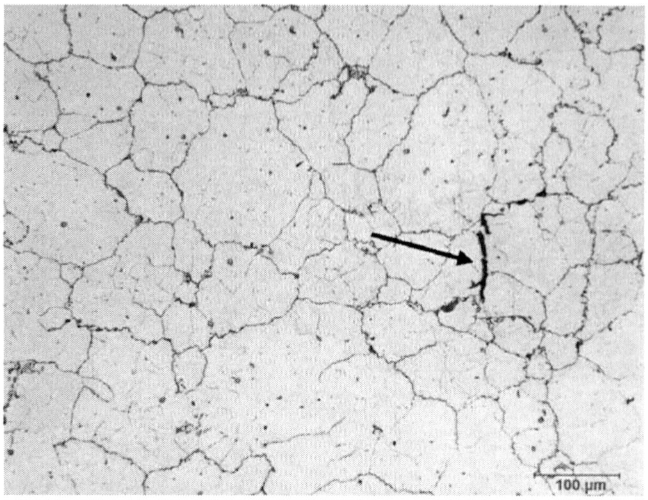

Figure 13: Microstructure of squeeze cast AZ91D-T4 magnesium alloy; some undissolved β-$Mg_{17}Al_{12}$ phase (arrow) is evident

3. Tensile strength of squeeze cast AZ91D-T4 is comparable to that obtained from tensile specimens machined from squeeze cast or semi-solid metal cast 380-F aluminum alloy castings.

4. Squeeze cast AZ91D-T4 reveals higher elongation than that achieved from squeeze cast or semi-solid metal cast 380-F aluminum alloy castings.

5. The tensile properties of squeeze cast AZ91D-T4 obtained in this study are comparable to those of rheocast AZ91D-T4 reported by Kleiner [1].

6. Yield strength of squeeze cast AZ91D-T6 is comparable to that obtained from tensile specimens machined from 380-F aluminum alloy HPDC castings; however, the tensile strength of squeeze cast AZ91D-T6 is higher than that reported for 380-F HPDC.

8. Yield strength and percent elongation of squeeze cast AZ91D-T6 are comparable to those obtained from tensile specimens machined from squeeze cast or SSM 380-F aluminum alloy castings.

9. The tensile properties observed in this study indicate that squeeze cast AZ91D-T6 may be a possible substitute for thick-walled aluminum alloy HPDC, particularly for 380 aluminum alloy castings.

10. Squeeze cast AZ91D-T6 exhibits better yield strength than do sand cast specimens. The higher strength of squeeze castings may be attributed to the presence of smaller (refined) grains.

11. Squeeze cast AZ91D-T6 reveals lower yield strength than 383-F aluminum alloy castings.

E. Microstructures of Heat Treated Squeeze Cast AZ91D Alloy:

Figures 14 and 15 show the typical microstructures of squeeze cast AZ91D after heat treatment (T6).

The various microconstituents (see Figure 16a) present after heat treatment were subsequently examined with a SEM. The various phases observed are summarized below:

- Undissolved β-$Mg_{17}Al_{12}$ identified as "1" in Figure 16a. SEM image of phase "1" along with its chemical composition is shown in Figure 16b.
- Undissolved lamellar intermetallics identified as "2" and "4" in Figure 16a. SEM image of the lamellar phase along with its chemical make-up is shown in Figure 16c.
- Supersaturated solid solution of Al in Mg designated as "3" in Figure 16a. SEM image of phase "3" in Figure 16a along with its chemical analysis is shown in Figure 16d.

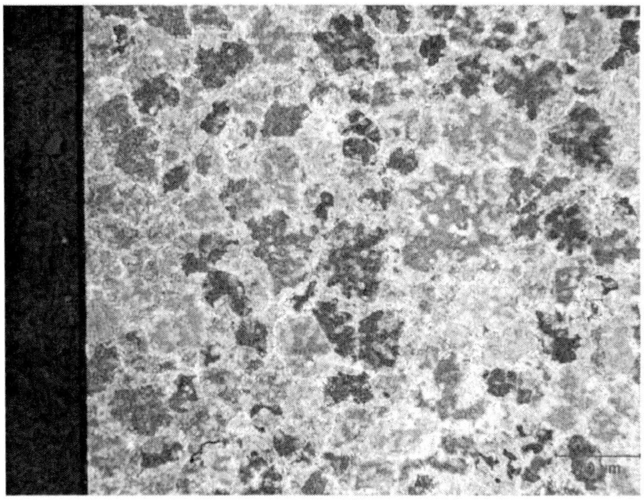

Figure 14: Surface microstructure of squeeze cast AZ91D-T6 magnesium alloy (14 mm section); etched with 2% nital

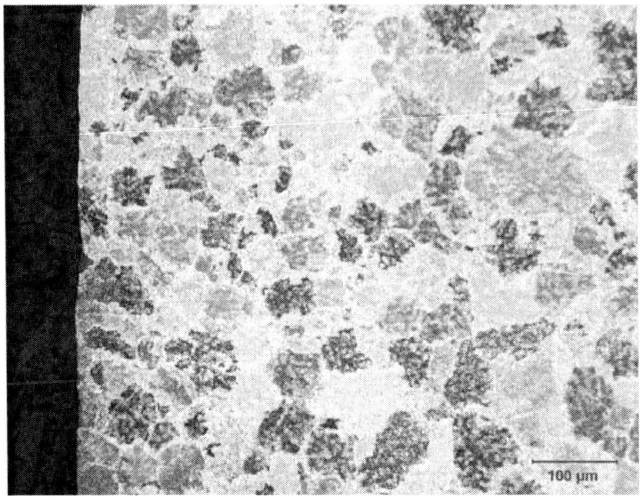

Figure 15: "Core" microstructure of squeeze cast AZ91D-T6 magnesium alloy (14 mm section); etched with 2% nital

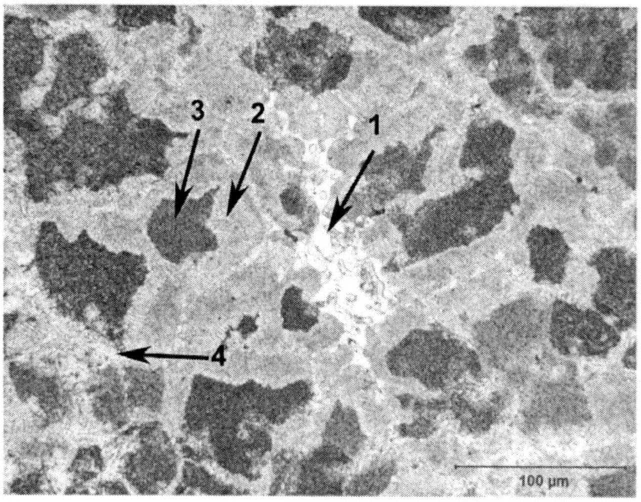

Figure 16a: Optical micrograph of squeeze cast AZ91D-T6 magnesium alloy; etched with 2% nital

Figure 16b: Scanning electron micrograph of phase "1" in Figure 16a; etched with 2% nital

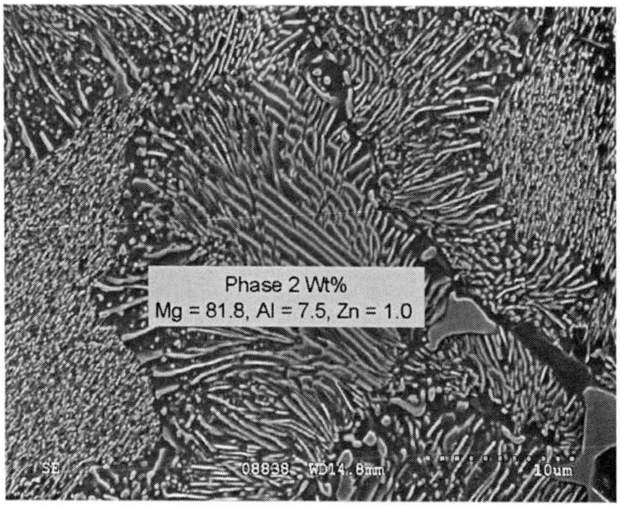

Figure 16c: Scanning electron micrograph of phase "2" in Figure 16a; etched with 2% nital

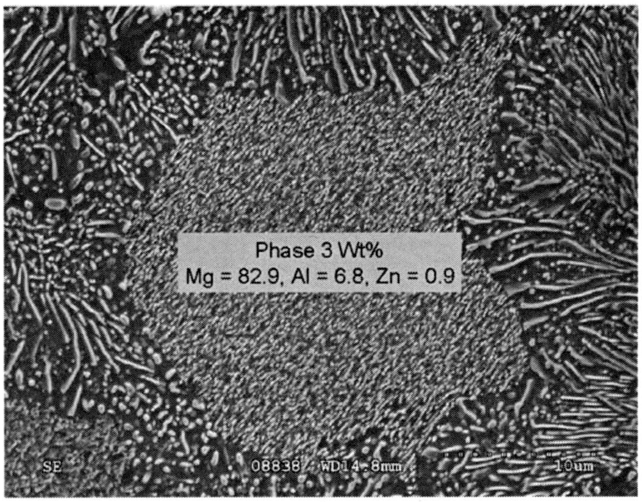

Figure 16d: Scanning electron micrograph of phase"3" in Figure 16a; etched with 2% nital

CONCLUSIONS

1. Wall thickness of squeeze cast AZ91D magnesium alloy does not significantly affect tensile properties.

2. The as-cast strength data obtained in this study compare favorably with semi-solid metal casting (UBE NRC) data reported in literature [1].

3. The β-$Mg_{17}Al_{12}$ phase, in as-cast microstructures, does not form a continuous network; thus, the crack path observed is quite tortuous.

4. Squeeze cast AZ91D-T6 yields properties that are comparable to those obtained from tensile specimens machined from 380-F HPDC castings.

5. Squeeze cast AZ91D-T6 exhibits better tensile properties (yield and tensile strengths) than sand castings. The higher strength of squeeze castings may be attributed to the presence of smaller (refined) grains.

6. In comparison to the F-temper, large improvements of tensile strength and elongation are observed after solution heat treatment (T4 temper) of squeeze cast AZ91D alloy; however, the yield strength is reduced.

7. The increase in ductility observed is due to the dissolution of the brittle β-$Mg_{17}Al_{12}$ phase and of the discontinuous (pearlite-like) structure at elevated temperatures.

ACKNOWLEDGMENTS

Special thanks are extended to LKR (ARC Leichtmetallkompetenzzentrum Ranshofen GMBH) for preparation of the step plates.

REFERENCES

1. S. Kleiner et al, "Microstructure and Mechanical Properties of Squeeze Cast and Semi-Solid Cast Mg-Al Alloys," Journal of Light Metals 2, 2002, pp 277-280

2. S. Corbit and R. DasGupta, "Squeeze Cast Automotive Applications and Squeeze Cast Aluminum Alloy Properties," SAE International Congress and Exposition, 1999, paper # 1999-01-0343

3. G. L. Dunlop et al, "Microstructure and Properties of Magnesium Alloy High Pressure Die Castings," Proceedings of the First Israeli International Conference on Magnesium Science and Technology," Magnesium 97, 1998, pp 108-115

2005-01-0331

Prediction of Distortion and Residual Stresses in a Magnesium High Pressure Diecasting with Considerations to the Diecasting Process

Winston Sequeira, Steve Sikorski and Soren Andersen
MAGMA Foundry Technologies Inc.

Charles Barnes
SPX Corporation

ABSTRACT

Magnesium cold chamber diecastings are steadily making inroads into the automotive industry. Parts that are routinely diecast include instrument panels, valve covers, steering columns, transmission case covers, seat pans, and door frames to name a few. In a bid to take full advantage of weight savings, the majority of the parts are optimized during the design process. This means wall thickness are reduced, and ribs are used to reinforce the part for stiffness. As a part of the design and optimization process, generic load analysis is carried out on the part without taking into account the residual stress developed during the casting process. The iterative process is continued until the design responsible engineer is satisfied with the outcome of the analysis. However, during the diecasting process, variations in the thermal patterns are obtained in the die mainly due to section thickness changes of the part, placement of cooling lines and spray patterns. On ejection from the die, the part will inherit a thermal history, which upon cooling to room temperature in air or in a quenching medium will result in part distortion and development of residual stress. Part warpage, which is a consequence of distortion, results in problems during trimming, machining, and assembly. On the other hand, residual stress beyond certain limits might lead to cracking of the part after production or during service. This paper discusses some of these aspects for a high pressure diecast magnesium automotive part.

INTRODUCTION

Near net shape, fast production rates, high volume requirements, excellent surface finish and dimensional repeatability are the most important factors that make high pressure diecasting an attractive route for manufacture of nonferrous castings. The casting purchaser likes flexibility of sourcing the product from different locations and evaluates the total process capability of the product. When the part is designed, standard shrinkage and draft tolerances are applied [1], as the designer does not have total control over the process variables that cause dimensional variations. This lack of advance knowledge of the process variables and its affects on dimensional variations has the potential to result in a diecast part that can significantly vary in terms of *dimensional repeatability.* In the diecasting process, variations in the thermal patterns are obtained in the die mainly due to direct casting variables such as part geometry, placement of cooling lines, spray patterns, and cycle time. On ejection from the die, the part will inherit a thermal history, which upon cooling to room temperature in air or in a quenching medium will lead to shrinkage, development of residual stress and associated part distortion. Simulation tools that carry out residual stress and distortion analysis are increasingly used during the early part optimization process, and these tools are now considered an integral part of a suite of casting optimization tools such as filling and solidification analysis. This paper discusses the development of residual stress and distortion in the diecastings, and illustrates typical pitfalls that could be avoided during manufacturing.

DISTORTION, STRESSES AND STRAINS IN DIECASTINGS

DISTORTION IN DIECASTINGS

Even though the diecasting process achieves high dimensional accuracy, deviations from dimensions could result due to one or more of the following reasons.

1. Shrinkage in diecastings is a function of wall thickness and shape variations such as

protrusions and contours. Shrinkage can be divided into two groups: (a) solidification shrinkage, which is caused as the metal cools from Liquidus to Solidus temperature, (b) thermal shrinkage caused by the cooling of the casting from ejection to room temperature. The illustration in Figure 1 shows how the thermal shrinkage in the diecasting process significantly influences the final shape of the component [2].

2. Thermal expansion of the diecasting die causes variations in dimensions of the casting. The variation in die dimensions due to thermal expansion is a function of the alloy that is cast, the shape of the casting, the cycle time and die cooling. Areas in the die that have a higher thermal mass to die-surface-area ratio will have higher temperatures as opposed to areas with lower ratio. Further, die temperature differentials within the die, and die spray have the potential to influence the thermal balance of the die.
3. The position of different die elements with respect to each other during die clamping have the potential to alter the dimensions of the casting. Cores, slides, die halves, guide pins, clearance between the slides all affect the dimensions of the casting during die lockup. Further, trapped flashing and debris within the die and in between the moving die elements have the potential to vary part dimensions.
4. Usually higher temperatures prevail in the vicinity of the gating system and the sprue regions of the die. Consequently, the thermal expansion of the die and the temperature of the gating system will differ thereby indirectly influencing the distortion and stress development in the gating system. Also, vacuum runners attached to the casting will have a constraining effect on the casting after ejection and subsequent cooling.
5. The cooling media will influence the development of stress and distortion in diecastings. Residual stress may be different if a casting is quenched.
6. Deliberate corrective actions on the die to increase die life (shot peening) or die polishing to remove hairline cracks will cause variations in casting dimensions.

The simulation tool illustrated in this paper, MAGMAstress, has the ability to consider deviations in casting dimensions caused by all the above factors, except due to die lock-up and shot peening. In addition, the simulation tool also has the ability to predict a variety of stresses and strains as a result of the overall diecasting process variables. These stresses and strains not only have a direct influence on part distortion, but also have the ability to warn the designer of any impending problems in the part such as cracking/fracture. These problems can either be solidification related (hot tears) or are a virtue of large unrelieved stresses within the part as a result of uneven cooling after ejection.

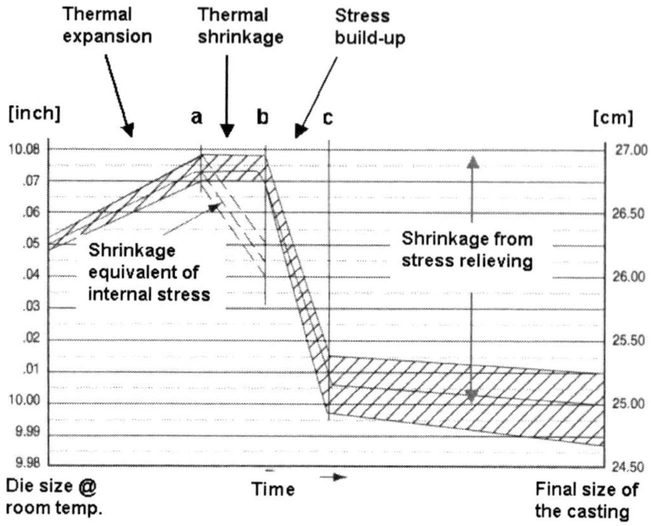

Figure 1: Dimensional variations in a typical diecasting during the casting cycle. 'a' indicates die size at operating temperature, 'b' casting size at ejection, 'c' casting size after quenching. The size variations shown are linear [2].

STRESS, STRAINS AND THEIR EFFECTS ON DIECASTINGS

Internal stresses in a casting develop because of restrictions to contraction. This internal stress can develop due to two reasons: (a) hindrance to contraction by the die or mold during solidification of the casting (b) hindrance to contraction because of casting geometry after the casting is ejected from the die. The following promote the occurrence of stress and strains due to the latter:

1. Non-uniform solidification occurs due to unequal cross sections or variation in wall thickness.
2. Large plane surfaces will warp without proper ribs for stiffness.
3. Inadequate shrinkage conditions can cause the casting to shrink on a certain part of the die. This will causes high stresses and casting deformation during ejection.
4. Sections that transition abruptly are regions with high stress build-up resulting in cracking of such sections.
5. Protrusions from the casting such as small bosses or standing features (fins) are unable to withstand shrinkage stress.

DISTORTION & GEOMETRIC TOLERANCES

Distortions or shape variations of diecastings have a strong influence on subsequent operations that are carried out on the part and the final assembly. Many diecasting shops machine castings in-house, and quite a few of these castings will undergo complete assembly and leak testing before shipment. Ascast surfaces of diecastings are milled flat for one or more of the following reasons:

(a) The drawing calls-out for a flatness that cannot be achieved by the casting process.
(b) The surface finish requirements are very stringent and warrant a milling operation.
(c) General machining processes often steer process plans to the machining of datum surfaces. The machined datum provides a known reference for fixturing, and in turn, the machined features reference the datum.

It is a well-known fact that diecastings have a high integrity 'skin' that extends a few hundred microns into the cast surface. Typically 0.5-1 mm (0.02-0.04 inch) of machining stock is recommended on flat surfaces so that during milling operation opening-up of subsurface porosity is avoided. In large machining-intensive castings, the machining of the datum is a common practice and leaving the surface ascast is an exception to the rule. If the diecasting has a flat surface, which forms a machined datum, and if the casting has a relatively large projected area, the distortion of the surface may lead to problems during machining if the machining stock is on the low side. This results in non-clean up of the flat surface thereby rejects. An example of this is shown in Figure 2.

Figure 2: Illustrates a diecasting with a flat-machined (milled) mounting face. This flat surface is used for mounting the casting on the machining fixture and becomes a datum for subsequent machining. This involves operations such as drilling, tapping, counter bore and facing (shown in inset).

Another typical problem, caused by residual stress in diecastings, is illustrated in an example shown in Figure 3. This part has two distinctly different geometric shapes connecting each other. Typical examples of diecastings with these shapes include covers used in automotive applications, double bowl type filter adapters, motorbike engine covers etc. Poor die design could lead to (a) preferential cooling or heating of one side of the geometry with respect to the other or, (b) unfavorable cooling of a poorly designed center section. This may result in bending or warpage of the part after the diecasting process in the directions indicated by large white arrows.

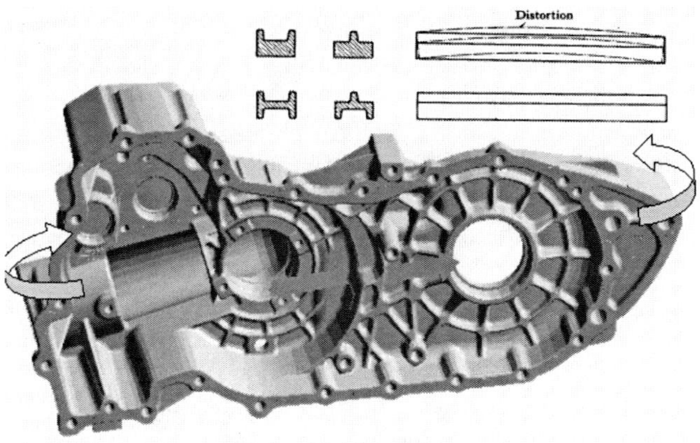

Figure 3: Shows a cold chamber aluminum diecasting with sections on either side connected by a center section (near red arrows). Due to unfavorable solidification, the diecasting may distort backwards as shown by the large white arrows. The schematic shown in the inset illustrates this for a simple shape with an unequal cross section. The distortion leads to warpage of the sealing face, which is normally machined.

There are no easy solutions to correct distorted surfaces that are to be machined or are used as datums after machining. Mechanically straightening the surface appears to be the only way to solve this problem, and this has been tried with limited success by using shims under trip tools. Of course, additional stock can be factored into the design if the surface is used as machining datum, but this increases the potential of opening-up of subsurface porosity and the chances of leakers during pressure testing.

DISTORTION AND LOCATING FEATURES

Misalignment of locating features of castings during assembly is a problem commonly experienced by the diecasting industry. Two common problems are:
1. Thin-walled, rib-stiffed, aluminum or magnesium diecastings having large projected areas pose many distortion related challenges during the diecasting process. Magnesium alloy diecastings currently in production or in development, such as instrument panels, large brackets, door/floor panels and window frames fall into this category. Warpage due to distortion is a major problem in these diecastings, and attention must be paid during the design stage to correct this issue before the part goes into production.
2. Large castings may have locating features that are stack-dimensioned from one end of the casting (datum). Due to the practice of applying linear shrinkage allowances to the casting, combined with a die design that may involve a series of inserts in the die, the casting

after ejection may have cored holes that are offset or unable to hold the specified positional tolerances. An example of this is shown in Figure 4 wherein a standard shrink factor, estimated to be 0.005 in/in or 0.006 in/in for aluminum diecastings, is applied to the part during die design. This shrink factor is not only supposed to take into account the shrinkage of the casting, it also factors in the thermal expansion of the H13 mold. Thermal expansion in the multiple-insert die assembly may vary due to varying temperature profiles. Residual stress analysis of the diecasting after ejection with its temperature field and the associated die/insert dimensional variations, may lead to the cast product drifting away from specified tolerances at room temperature. It is critical to understand that the final dimensional variations in the casting, as measured by using a CMM (coordinate measuring machine), are a summation of thermal shrinkage and distortion caused by the combined affect of residual stresses and strains. In situations wherein the casting tolerances drift, early residual stress analysis on the part and the die elements will provide the designer with key information on relative insert movements and resultant casting distortion. However, interpretation of results will require die design experience and ingenuity; core competencies developed by accumulating relevant knowledge over time.

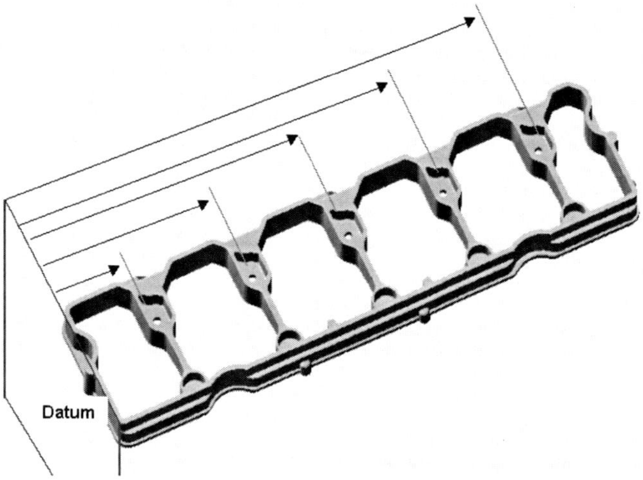

Figure 4: Shows a large casting, ~86 cm (34 inch) in length, with holes along the length of the casting. The holes are stack dimensioned from one end of the casting (datum). The die is made from a series of inserts.

One practical process-based solution commonly used to tackle both of the above mentioned adversities are to increase the dwell time of the casting in the die before ejection. The die constrains the casting for a longer time period, forcing the casting to better conform to the die features. This practice has been found to reduce warpage and to improve tolerances. The diecaster often pays a severe penalty for this practice, often increasing the cycle time by 15% or higher. Another solution, specific to problem #2 discussed earlier, is to correct the tool by reverse engineering. This may not be always possible, as the tool must render itself amenable for such a correction. Carrying out stress analysis early on during the design stage can provide valuable information to the designer, as any subsequent reverse engineering related tooling costs are typically absorbed by the diecaster.

COLD CHAMBER DIECAST AZ91D STEERING COLUMN

Illustrated in this section is a case study of a cold chamber diecast steering bracket made from magnesium alloy AZ91D. This alloy is a workhorse alloy used in non-load bearing type automotive applications. Typically, non-load bearing automotive applications require ductility values of ~3%, and magnesium alloy AZ91D, along with other mechanical property data, fulfills these requirements. More recently, AM series of alloys (AM50 and AM60) have been preferred because of their higher ductility even though they have somewhat lower tensile strength than AZ91D. Nevertheless, residual stress analysis discussed here is applicable to all diecast alloys as long as a compilation of accurate thermophysical and mechanical property data is available.

Before a residual stress analysis is performed on the steering bracket, the die was warmed up to the quasi-steady state. The warming up process included accurate cycle time, water lines, oil lines, and spray pattern. Doing this allows the die to achieve realistic thermal gradients, which in turn significantly influences the thermal gradient of the casting and the resultant residual stress calculations. In the final cycle, the casting was ejected at the appropriate time after die open, and the casting was allowed to cool in air for a time of 1000 s (~17 minutes). Experience for similar thin walled castings showed that after 1000 s, the casting reaches a temperature of around ~100°C.

Residual stress analysis starts as iterations beginning with the overall temperature field of the casting just after ejection (i.e., solidification result at die open). The capture of this thermal signature or the temperature field as a starting point for the stress analysis is considered extremely valuable for an accurate residual stress analysis. This temperature field is a function of die warmup, cycle time, water/oil circuitry, die spray, and die open time. Complex yet accurate thermal patterns in the casting at ejection are a virtue of these parameters. After capturing the temperature field of the casting just after ejection, the solver carries out thermal stress iterations over the time period (1000 s) as the casting cools in air. The solver used for carrying out stress calculations is described in detail elsewhere [3], and the application of the tool to other diecasting related problems has been illustrated by other authors [4,5,6]. If the casting is quenched in a cooling media such as water, the residual stress analysis may be performed from casting ejection through the end of the quenching process. Further iterations of the redistribution of distortion and stress may be carried out after trimming the casting (degating). For the steering bracket the simulation results presented are after trimming the gates, vents, and overflows.

For practical purposes, results captured when the casting reaches room temperature are of the highest significance. Termed as *Residual Stress* result files, these are stress results that are obtained when the casting reaches 20°C (68°F). Residual values of displacements, principal stress and Von Mises stress are listed in Table 1 along with their significance.

The results obtained for the steering bracket are shown in Figures 5-9. A summary of these results is given in Table 2. Final displacements that are obtained in the casting are shown in Figures 5-7. The overall displacements may be manipulated to obtain absolute values between two features in the casting. Also, a distortion factor (magnification factor) may be applied to the result file, which exaggerates the displacement values for viewing. Figure 6 illustrates this for the steering bracket, which shows the standing features (mounting faces) distorting towards the inside.

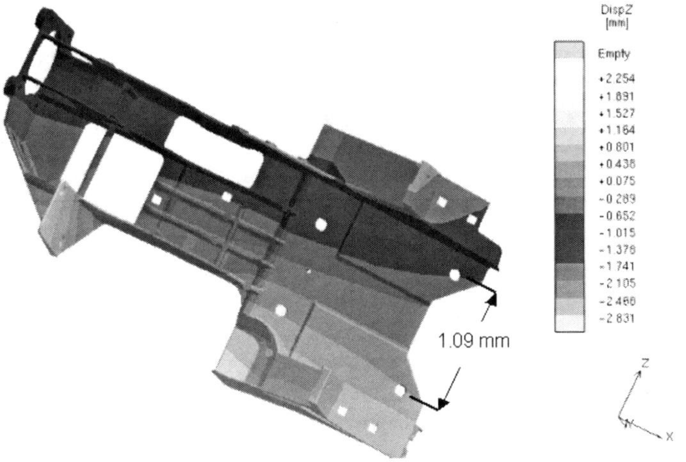

Figure 5: Displacement of the steering bracket in the z-direction. As an example, it can be seen that the distance between the holes marked in the figure has been reduced by 1.09 mm over a true distance of 147.08 mm. The value of 1.09 mm is the absolute displacement obtained after a residual stress analysis on the AZ91D steering bracket.

Figure 5 shows absolute displacements between two features in the diecasting that are approximately 147.08 mm (5.787 inch) apart. After residual stress analysis it was found that the absolute displacement over this distance was 1.09 mm. This is a displacement (shrinkage factor) of 0.007 inch/inch. Figure 7 shows two sets of features, which originate from the same datum plane. For the minor dimension of 373.90 mm it was found that the absolute displacement value was 3.026 mm and for the major dimension of 508.25 mm the corresponding value was found to be 4.323 mm. Calculations indicate a displacement (shrinkage factor) of 0.0081 inch/inch and 0.0085 inch/inch, respectively, for the two dimensions. From the above calculations, and as summarized in Table 3, it was clear that the final displacements in the steering column were nonlinear. This was expected and the information along with relative displacement of the part, as illustrated in Figure 6, provides the designer with early information on the dimensional variability of the part. For a part that warps, or for a part with critical location tolerance, this information is invaluable at the design review stage of the diecasting.

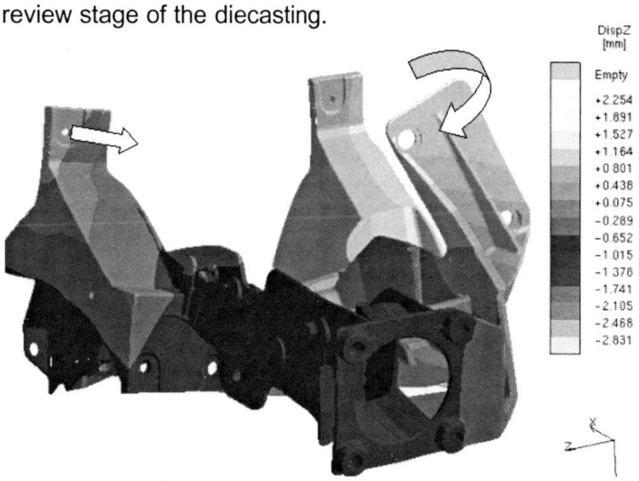

Figure 6: Displacement of the steering bracket in the z-direction magnified 20 times shows relative movements of the casting features due to the combined effect of shrinkage and distortion.

In terms of stress results there are two that are of primary importance – Principal stress and Von Mises stress. The magnitude of these values is of greater interest than their range. Higher value areas indicate areas in the casting that contain locked-in unrelieved residual stresses, which have the potential to cause early failure in the casting if it experiences loading conditions during service. For the steering column this was not the case and the loading conditions, if any, in the casting were a virtue of other components that were mounted on the part and casting assembly conditions.

The maximum and minimum principal stress values that were obtained were ~100 MPa and ~-100 MPa, respectively. The maximum principal stress (Figure 8) was much lower than the nominal yield stress (in tension), which is reported for diecast AZ91D to be 154 MPa at 25°C [7]. In analyzing the principal stress values, their practical significance must be realized. The values of maximum stress, evaluated in comparison to yield stress, suggest that the areas can withstand 54 MPa of additional stress to reach yielding and cracking. In contrast, areas in the casting with high minimum principal stress values are actually beneficial if the part experiences loading conditions (i.e., applying tension to areas that are in compression).

Residual stress	Significance
Displacements (x, y, z directions)	➢ Shows displacement or movement of the casting in x, y, z direction. ➢ If required, incremental displacements may be visualized as the casting cools to room temperature after ejection. ➢ A 3D composite picture of displacements can be built for correlation of results with CMM.
Von Mises stress	➢ *Von Mises stress* is computed from the stress tensor as calculated by the stress solver. ➢ Von Mises is commonly used to predict plastic deformation. ➢ In analyzing results, Von Mises is usually compared with the material's yield stress at room temperature. For practical purposes, one could develop a criteria based on a ratio of Von Mises stress to the materials nominal yield stress. Areas in the casting with a ratio close to one (or greater) are areas that are likely to result in plastic deformation.
Maximum Principal stress and Minimum Principal stress	➢ The software computes *Maximum and Minimum Principal stress* from normal stress and shear stress. ➢ Positive values indicate local areas in the casting that are in tension and negative values indicate areas that are in compression. ➢ High principle stress values are areas that are prone to cracking if the casting is subjected to additional loading conditions.

Table 1: Significance of results obtained from residual stress analysis of a casting.

Displacements in the casting		
	Values	Significance
x-direction	+2.77 mm to –3.27 mm	In the x-direction the casting distorted a total of 6.04 mm over a length of ~580 mm. This distorting includes both, the shrinkage of the casting and the distortion due to stress/strains in the casting. Figure 5 illustrates the distortion in the z-direction between two features (holes) in the casting.
y-direction	+1.97 mm to –1.80 mm	
z-direction	+2.25 mm to –2.83 mm	
Maximum Principal stress		
	Values	Comment on results
Maximum values	51 to 110 MPa	Maximum and minimum values for principal stress are computed from Normal and Shear stresses that are not discussed.
Minimum values	-75 to –100 MPa	
Yielding criteria		
Von Mises Criteria	75 to 115 MPa (general range) Some areas exceed 115 MPa, but are lower than 159 MPa.	Yield stress of diecast AZ91D at room temperature is ~159 MPa. Areas in the casting above this stress show permanent deformation.

Table 2: Values of displacements, maximum principal stress and Von Mises stress obtained after residual stress analysis on the AZ91D steering column.

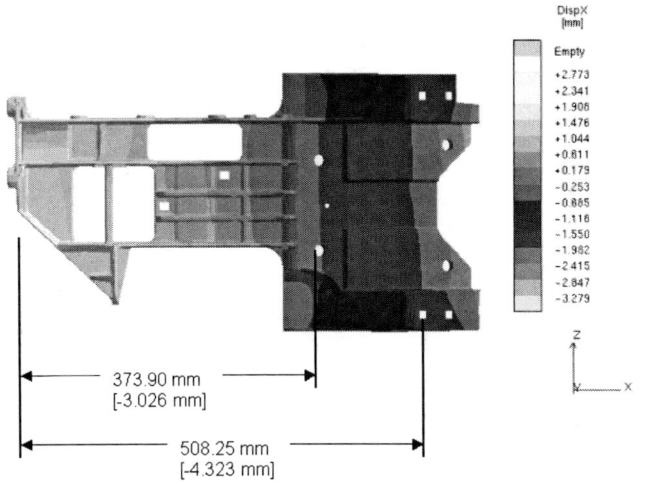

Figure 7: Displacement of the steering bracket in the x-direction for two dimensions. The minor dimension showed an absolute displacement of 3.026 mm over a true distance of 373.90 mm and the major dimension showed an absolute displacement value of 4.323 mm over 508.25 mm. The displacements were calculated to be 0.0081 inch/inch and 0.0085 inch/inch for the two dimensions respectively.

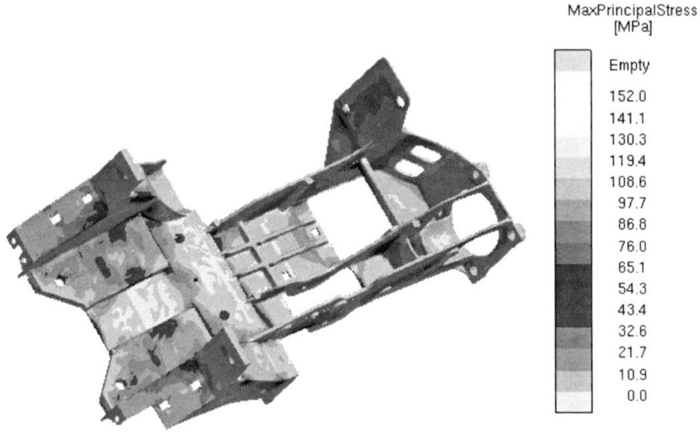

Figure 8: Maximum Principal stress shows sharp corners in the casting as areas with highest Maximum Principal stress values.

Von Mises stress (Figure 9) is of somewhat relatively lower significance than the principal stress. This stress, which may be considered a criteria condition to predict plastic deformation, is again benchmarked against yield stress. Yield stress of a part is a value that is obtained in a uniaxial tension test. But Von Mises stress for an isotropic material is a stress that is independent of the choice of axes, i.e. it is an invariant function providing the designer with a true indicator of plastic deformation criteria in a 3-D state of stress. In such a state of stress the Von Mises stress is always found to be lower than the material's nominal yield stress. Some areas of the steering column showed stress values in the range 115-130 MPa, well below the safety critical stress limit of 159 MPa. However, lower practical significance of Von Mises stress values stems from the fact that these values do not indicate the magnitude of stress in terms of sign (positive or negative). This makes it difficult to analyze them with reference to, say for example, the yield stress. In other words, the highest value of Von Mises stress may be areas in the casting that are in compression, which for most load bearing applications (generally in tension) is a safe proposition.

Feature	True distance between features	Displacement over the true distance	Displacement in terms of inch/inch (Shrinkage factor)
Casting width (z-direction)	147.08 mm	1.09 mm	0.0070
Casting length -- (x-direction)	373.90 mm	3.026 mm	0.0081
Casting length -- (x-direction)	508.25 mm	4.323 mm	0.0085

Table 3: Displacements after residual stress analysis between different true distances in the steering column.

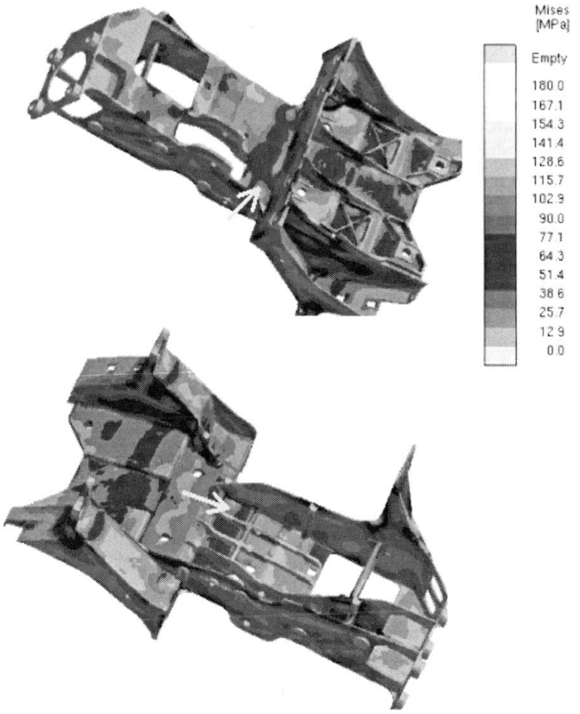

Figure 9: Von Mises stress showing areas of high Mises stress values. The areas highlighted are typical areas with high values of Mises stress.

There was an excellent correlation between the CMM measurements of the steering column and the results obtained from residual stress analysis of the part. The part shape was not designed after the residual stress of the casting was known. Rather, the comparisons were part of a research program that benchmarked residual stress results with actual CMM measurements. The software was then integrated within the product development group of the company, and is currently used to complement other product design tools.

CONCLUSION

This paper illustrates with examples the potential for using residual stress analysis for a variety of design related issues in the diecasting industry. The general principles discussed are not limited to diecastings alone, but are also applicable to casting made from other routes. Residual stress analysis must be considered an integral part of any casting and die design program. To this end, the tool is an add-on capability to previously existing casting optimization tools such as filling and solidification simulation. The wealth of information generated is expected to augment existing strategic product development tools that will raise the standards of an industry that is becoming increasingly competitive.

REFERENCES

1. NADCA Product Specification Standards for Die Castings (1994), North American Diecasting Association, Rosemont, IL, 1994.
2. Magnesium Diecasting Handbook, NADCA Publication #201, North American Diecasting Association, Rosemont, IL, p32, 1998.
3. J.Hattel, J.Thorborg, S.Andersen: Stress/Strain Modeling of Casting Processes in the Framework of the Control-Volume Method, Proceedings: Modeling of Casting, Welding and Advanced Solidification Processes VIII, Edited by B.G.Thomas and C.Beckermann, The Minerals, Metals and Materials Society (TMS), Warrendale, PA, p763-770, 1998
4. S.Andersen, J.Hattel, P.N.Hansen: Modeling of Thermal Induced Stresses in High Pressure Die Casting Dies, Trans.17th International Diecasting Congress and Exposition, Cleveland, OH, p99-103, 1993.
5. C.Rosbrook, R.Kind, R.Seefeldt: Casting Process Simulation of Stresses and Distortion in Castings and Dies, Die Casting Engineer, Sept/Oct 1999, p72-76.
6. R.Kind, S. Andersen, W. Sequeira, P. Cheng, Y.-L.Chu, M.C.Kite: Reverse Engineering for Surface Reproduction of Engine Block Water Jacket Insert and Thermal Stress Analysis for Water Jacket Insert Using Computer Stress Simulation, Trans. 21th International Diecasting Congress and Exposition, Cincinnati, OH, p433-437, 2001.
7. W.Sequeira and G.L.Dunlop: Microstructure, Mechanical Properties and Fractography of High Pressure Diecast Magnesium Alloy AZ91D, Trans. 22nd International Diecasting Congress and Exposition, Indianapolis, IN, p213-228, 2003.
8. L.W.Foster: Geo-metrics III, The application of Geometric Dimensioning and Tolerancing Techniques, Addison-Wesley Publishing Company, 1994.
9. Dimensioning and Tolerancing, ANSI Y145M-1982, American Society of Mechanical Engineers, 1982.

CONTACT

For further information please contact authors Winston Sequeira, Soren Andersen or Steve Sikorski at:

MAGMA Foundry Technologies Inc.
2340 S. Arlington Heights Rd,
Arlington Heights,
IL 60005
Tel: 847.427.1001

E-mail: info@magmasoft.com
Web site: www.magmasoft.com

2005-01-0332

The Use of Quality Mapping to Predict Performance of Thin-Walled Magnesium Die Castings

Stian Sannes, Haavard Gjestland and Håkon Westengen
Competence Centre Magnesium, Hydro Magnesium

Cato Dørum
NTNU - Structural Impact Laboratory, Dept. of Struc. Eng.

Otto Lohne
NTNU - Inst. of Material Science

ABSTRACT

A methodology that includes the processing history of the metal in the High Pressure Die Casting (HPDC) process in the simulations of the structural behavior of magnesium components has been established. In this methodology the results from the HPDC process simulations are used to modify the material model and the fracture criterion in the Finite Element Analysis (FEA). This paper focuses on the simulation of the HPDC process for thin-walled magnesium components. The close connection between the processing history and the mechanical properties of the casting mandates a careful analysis of the key factors influencing the final part performance. The definition of the boundary and initial conditions will strongly influence the ability to predict important features in the microstructure of the casting and consequently the final mechanical properties of the casting. In this context, improvements in HPDC process simulations are important in order to improve the predictability of the FEA of the final part performance.

INTRODUCTION

Virtual design has become a standard procedure in the product development process in the automotive industry. The use of virtual design is expected to become more and more important in the future, increasing the need for precise predictions of the component behavior under different service conditions. In this scenario the ability to make precise predictions of the part performance can actually become an important factor in the material choice for automotive applications.

The final design of an automotive application depends strongly on the load cases the application needs to serve. In order to evaluate this, FEA is used extensively. To improve the prognosis reliability of FEA from the current level, it has become necessary to take into account the local mechanical properties of the component. In order to do so, the material processing history needs to be incorporated. The challenge is, however, to extract and implement the key information in the FEA in an efficient way.

HPDC PROCESSING AND SIMULATION

The key parameter to focus on in the HPDC processing of magnesium is the thermal history of the metal. In the following the main steps in the HPDC process will be discussed, and some of challenges from a simulation point of view will be analyzed.

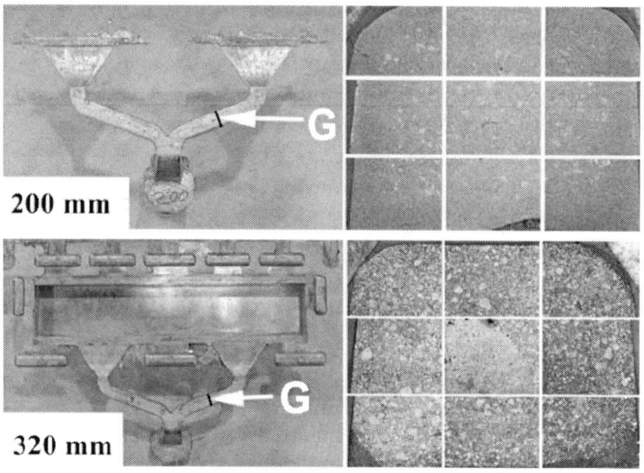

Figure 1 Formation of ESC in the shot sleeve. Fill test castings early and late (stroke=200 mm and 320 mm respectively) in the filling sequence. The micrographs to the right show the microstructure over the cross section G in the gating. Illustration used with kind permission from Hans Ivar Laukli /1/.

SHOT SLEEVE CONDITIONS

Significant heat loss takes place in the shot sleeve of a cold chamber die casting machine /1, 2, 3/. Many of the commercial HPDC simulation codes do not consider the shot sleeve where the temperature of the metal before injection is significantly modified compared to the initial pouring temperature. The heat loss causes coarse, pre-solidified crystals to be formed in the metal prior to the filling phase of the cavity. In the following discussion these pre-solidified crystals will be termed externally solidified crystals, ESC. ESC can be seen from the micrographs shown in Fig. 1. Early in the filling of the cavity only few ESC can be observed, while at the end of filling the fraction of ESC has become significant.

In the simulation model shown in Fig. 2 the shot sleeve was divided into 13 different sub-volumes; three volumes (11-13) covered the metal close to the shot sleeve wall, while the rest (1-10) covered the metal in the interior. The basis for choosing these volumes was a fraction solid criterion. As long as the fraction of solid in the metal is below a critical level it can be assumed to be free to follow the forced convection set up by the piston movement. As the fraction of solid exceeds the critical level, the ESC will start to interact. This metal is then assumed not to follow the flow and will accumulate in front of the piston. These ESC will then be injected into the die cavity at the end of the filling phase. A fraction of solid of 30% was applied as the critical level for free flow in this work for AM60. An initial simulation was therefore done without dividing the shot sleeve volume into sub-volumes. The thickness of the outer shell was defined by this fraction solid criterion at a time corresponding to the end of the 1st phase of injection.

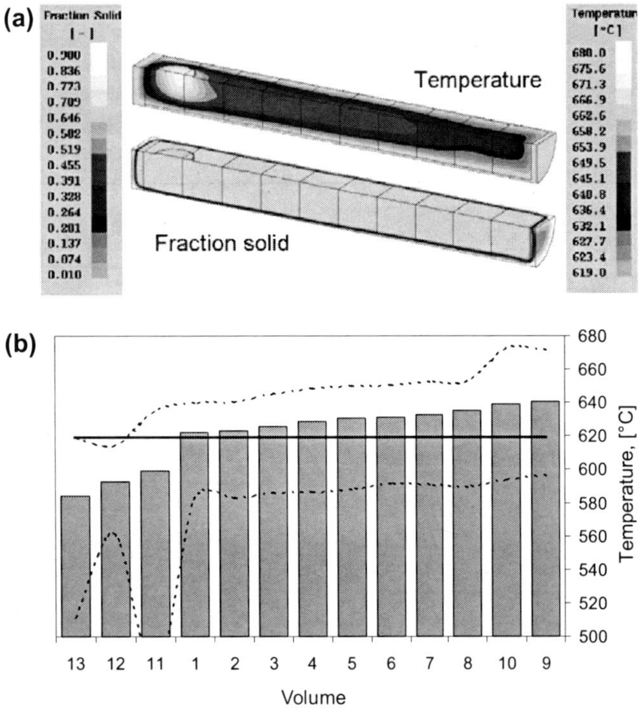

Figure 2 Shot sleeve modeling. (a) Temperature and fraction solid distribution. The pour hole is to the left while the gate end is to the right in the illustration. (b) Average temperature in the volumes. Volume no. 1 was originally the one closest to the gate end, while no. 10 was closest to the pour hole.

The temperature distribution and the corresponding fraction solid distribution after 2.4 s are shown in Fig. 2a. The bar chart in Fig. 2b shows the average temperature in the 13 volumes, while the upper and lower broken lines show the maximum and minimum temperatures, respectively. The solid line indicates the liquidus temperature of the alloy. The simulation was done by including the filling of the shot sleeve within 1.25 s and including a subsequent delay of 1.15 s corresponding to the 1st phase. No piston movement was simulated. Initial metal temperature was 680°C. The simulation shows an average fraction of solid in the complete volume of 14% at this stage. As a simple approach to model the temperature distribution of the metal flowing into the die cavity, a flow sequence of the sub-volumes with decreasing temperature was assumed, as illustrated by the bar chart in Fig. 2b. This is in agreement with simulations with moving mesh, assuming Newtonian fluid behavior, where it has been reported that the hottest metal leaves the shot sleeve first /2/. This distribution is also in qualitative agreement with the experimental observations in Fig. 1, where a progressively increasing fraction of ESC was observed in the fill test castings. As illustrated by the minimum temperature curve in Fig. 2 this simple approach does, however, not capture that ESC are distributed in the first metal that enters the die even though the average temperature is above the liquidus temperature.

FLOW PATTERN

Based on the above discussion it can be concluded that the metal that is to be injected into the die cavity has an inhomogeneous temperature distribution, along with a significant volume fraction of inhomogeneously distributed ESC. The next step is then for this metal to enter into the die cavity. Experimentally this can be studied using short shot experiments as shown in Fig. 3.

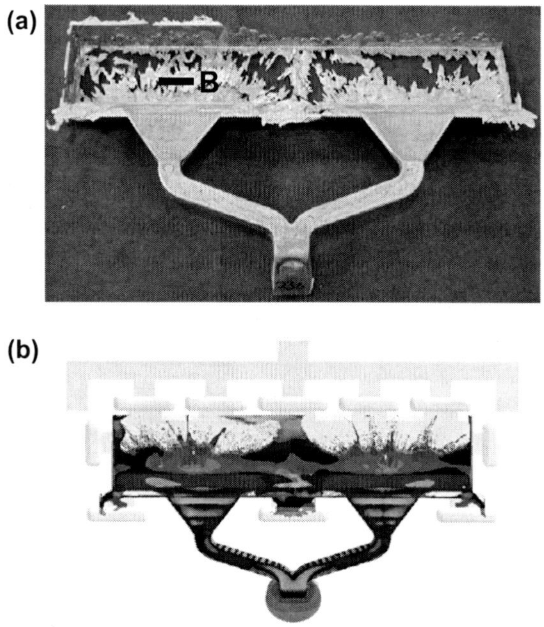

Figure 3 Comparison between the flow pattern from a fill test casting (a) and (b) the corresponding stage in the flow simulations.

The flow pattern revealed by the fill test showed a good correlation with the simulated flow pattern. This observation shows that the velocity distribution in the flow pattern must be qualitative in agreement with the real flow pattern.

The microstructure investigations in the gating system revealed that already early in the fill test the first coarse dendrite fragments appeared, with a progressively increasing fraction towards the end of filling. This increasing fraction of ESC in the flow is also recognized inside the casting, as illustrated in Fig. 4. In the completely filled casting, a surface layer of typical thickness 500 μm shows very few coarse dendrite fragments. Towards the center, the concentration of ESC increases progressively.

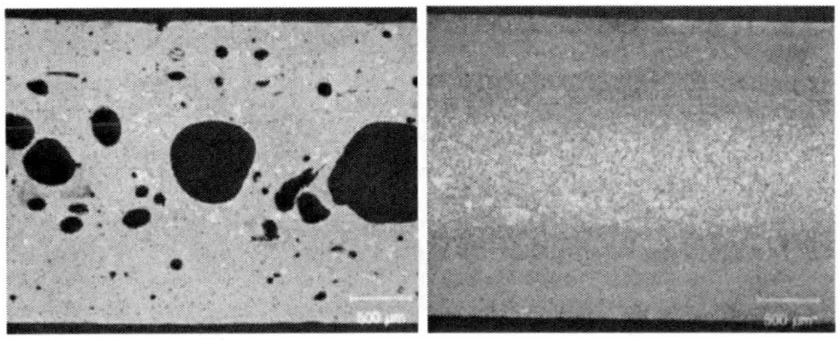

Figure 4 The microstructure at position B of the U-profile illustrated in Fig. 3. The first metal that enters the die cavity early (left) show few ESC compared to the metal at the end of filling (right).

SOLIDIFICATION DURING FILLING

In addition to the solidification of ESC that takes place before filling of the die cavity, solidification can also take place during filling of the die cavity /3, 4/. To illustrate solidification on the die wall during filling, Fig. 5 shows a paper clip that can be put in between the surface skins in a fill test casting. When the piston suddenly becomes arrested, the metal will continue to move unless it already has become stagnant due to solidification.

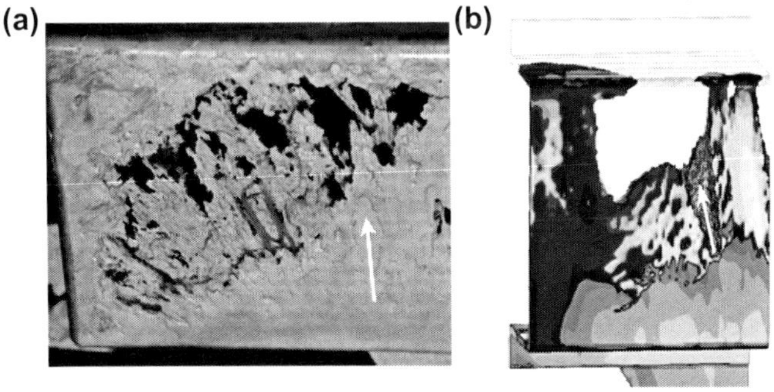

Figure 5 Solidification on the die wall during filling of the U-profile casting /4/. (a) A paper clip can be put in between the surface skins in a fill test casting. Notice also the longitudinal surface defect following the flow direction as indicated by the white arrow. (b) The corresponding simulated flow pattern.

In Fig. 5a the core is empty allowing the paper clip to be put in between the two surface skins. Along the flow fronts and toward the gating, surface patterns with the appearance of long tongues can be seen as indicated by the white arrow, Fig. 5a. Just below the start of the long tongues the surface appears as smooth. In this smooth area a solid skin was not formed during filling. These observations are captured in the simulated flow pattern, Fig. 5b. The white arrow indicates an area that is not yet filled, while on both sides the light areas indicate metal that has already solidified on the die surface. This means that when this area is filled a knit line will form between the flow fronts due to the difference in thermal history. Just beneath the solidified skin the fraction of solid is defined by the incoming metal flux. Below the white arrow in the simulation an area with no solidified skin is recognized. Due to the high impact from the metal flow that hits perpendicular to the die surface in this area, the surface skin is not formed. For more details about the simulation and the results, the reader is referred to Sannes et al. /4/.

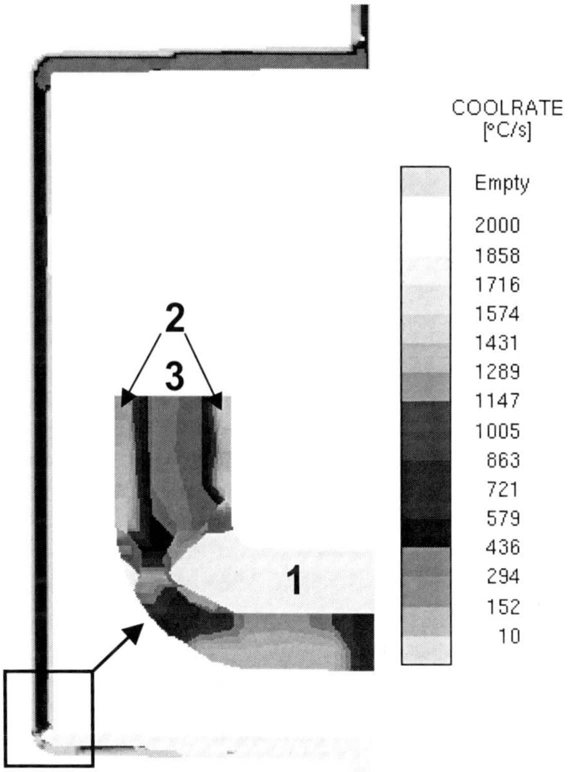

Figure 6 The simulated cooling rate over the cross section of the U-profile casting showed in Fig. 5.

The impact of this mixed thermal history on the mechanical properties can now be analyzed. Fig. 6 shows the simulated cooling rate of the flow pattern in Fig. 5b. The cooling rate correlates to the grain size, which again, has a strong correlation to the yield stress of the material /5/. In Fig. 6, three characteristic thermal histories are recognized in the cooling rate distribution.

Area 1, which appears empty in the magnified view of the lower left corner, has no defined cooling rate in the simulation. The reason is that it started to solidify in the shot sleeve. The cooling rate, defined as the cooling rate just above the liquidus, is therefore given by the cooling rate in the shot sleeve. With reference to the shot sleeve modeling shown in Fig. 2, the cooling rate in the shot sleeve is of the order of 10^1 °C/s. Area 2 of Fig. 6 is the skin area. In the simulation shown in Fig. 6, the cooling rate is of the order of 10^3 °C/s. An even more detailed model with higher mesh resolution has, however, shown that the cooling rate is as high as 10^5 °C/s for a skin of thickness of 100 μm /4/. The core, Area 3, between the surface skins has a metal history strongly influenced by the shot sleeve conditions. It may contain ESC, but this depends on the position in the casting as well as the cooling conditions in the shot sleeve. With the shot sleeve model illustrated by the bar chart in Fig. 2b, the core section should contain few ESC if it is far from the gate, since the hottest metal is leaving the gate first. On the other hand, if the core section is close to the gate, a high fraction of ESC should be expected. If the metal starts to solidify in situ in the core, the cooling rate is of the order of 10^2 °C/s, while the ESC formed in the shot sleeve has experienced a cooling rate about 10^1 °C/s.

HPDC PROCESS MODELLING AS INPUT TO THE FEA OF THE FINAL PART PERFORMANCE

From the above discussion it is expected that the local mechanical properties will vary over the casting. By simulating the thermal history carefully, the main characteristics in the filling and solidification pattern should be captured by the process simulations. Hence, the HPDC process simulation should contain the key parameters that influence the mechanical properties of the casting. The challenge is to extract the correct information from the HPDC simulation results and then to link the process simulation with the FEA of the final part performance in different load cases.

MECHANICAL PROPERTIES AND FRACTURE IN HPDC MAGNESIUM

The potential mechanical properties that are possible to achieve in HPDC magnesium are first of all controlled by the local solidification conditions. The cooling rate has a very strong correlation to the dendrite arm spacing and the grain size. These issues are discussed in more detail by Gjestland et al. /5/ in this proceedings. The potential mechanical properties are, however, modified by the introduction of defects. The defects will in general reduce the mechanical properties. Fig. 7a shows a micrograph of a test bar that has fractured after tensile testing. By careful inspection we see that close to the fracture, voids have opened up in the core region. No voids can be seen in the skin areas. On the other hand, the micrograph in Fig. 7b shows the fracture area in 3-point bending. In this case, cracks can be seen opening up in the skin area from the surface before it propagates further into the core area. The observations illustrated in Fig. 6 and 7 have formed the basis for the material model proposed in this work. The surface skin has a fine microstructure due to a high cooling rate. Fracture in this skin is initiated in the surface due to surface defects. The core has a microstructure composed of a mix of ESC and microstructure solidified in situ at a lower cooling rate than the skin. Fracture in the core is initiated by porosity.

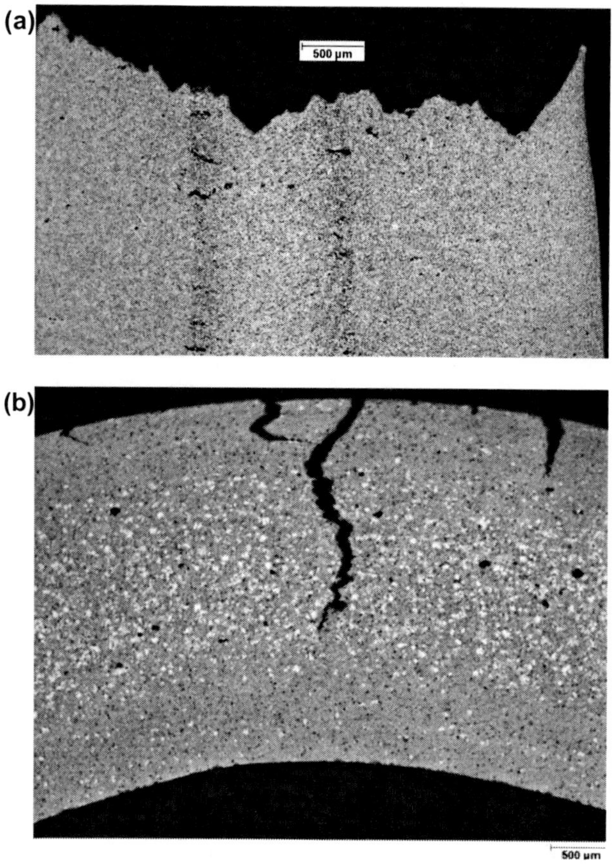

Figure 7 Fracture in (a) tensile test bar and (b) 3-point bending test specimens.

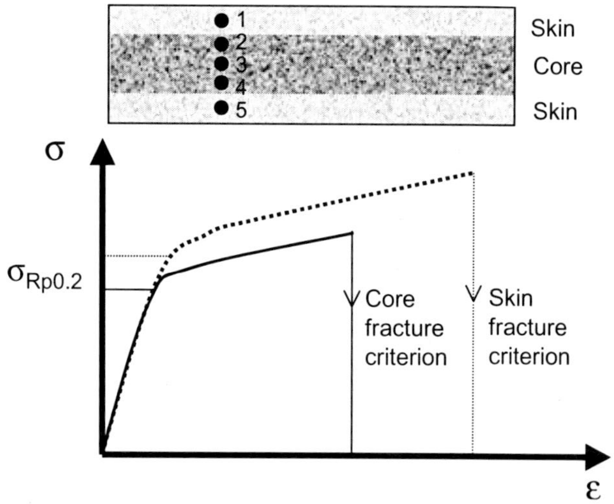

Figure 8 Schematic illustration of the model for implementation of HPDC process simulation to the constitutive model in FEA.

CONSTITUTIVE MODEL AND FAILURE CRITERIA

In order to implement the microstructure understanding described above a material model involving two skin layers separated by a core was used as illustrated in Fig. 8. Through the thickness the mesh has 5 nodes. Skin properties were assigned to the surface nodes no. 1 and 5, and the core properties to the three center nodes nos. 2, 3 and 4. In the model both core and skin properties are described by the same E-modulus and work hardening coefficient. To reflect the variation in cooling rate over the thickness, the skin and the core have different yield stress. For both skin and core the difference in tensile and compressive stress is taken into account. The main input from the process simulation is, however, used to modify the fracture criterion.

Due to the different behavior in compression and tension, the non-associated extended J2-flow theory was applied as the constitutive model /6, 7/. The yield surface in this model can be expressed as:

$$\Phi(\boldsymbol{\sigma},\varepsilon_e)=\sigma_e-\kappa(\mathrm{tr}\boldsymbol{\sigma}/\sigma_e,\varepsilon_e)=0$$

The material model is calibrated based on stress-strain curves obtained from uniaxial tensile and compression tests from the U-profile. Consequently, the porosity and void growth in the cast material is directly taken into account in the plastic domain. For simplicity, the elastic properties are assumed to be unaffected by the porosity. For more information about the constitutive model and the fracture criteria the reader is referred to Dørum et al. in this proceeding /7/. The main input from the HPDC process simulation will, however, be discussed here.

The fracture criterion is the same for both the skin and the core, but the criterion is modified by different input from the process simulation. The fracture criteria is the Cockcroft-Latham criterion that can be expressed as:

$$W=\int_0^{\varepsilon_f}\max(\sigma_1,0)\,d\varepsilon_e\le W_c$$

where σ_1 is the maximum principal stress and ε_f is the true plastic strain at fracture. This criterion implies that fracture is a function of the tensile stresses and equivalent plastic strain and has the dimensions of work per unit volume. If the maximum principal stress is compressive, fracture cannot occur. Furthermore, neither stresses nor strains alone are sufficient to cause fracture.

SURFACE DEFECTS

To modify the fracture criteria in the skin, the Quality map concept was introduced /8/. In the quality map for surface defects, the U-profile casting was characterized with detailed metallographic investigations with respect to cold flows and knit lines. This was done for different U-profile castings in a DOE matrix where melt varied. The quality map criterion with respect to surface defects was then developed based on inverse modeling between selected simulation results and the experimentally determined surface defect distribution using the MAGMA frontier module. The selected results were the flow length, the air contact time and the temperature distribution in the metal at the end of filling. The flow length of the melt is correlated to a higher potential for cold shuts with increasing flow length. The air contact time of the melt, i.e., the time a melt volume is in contact with the surrounding air at the flow front, is important since a longer contact time will increases the likelihood of forming oxides and porosity. A large drop in temperature and temperatures below the liquidus temperature will of course increase the probability for the formation of cold shuts and knit lines. Fig. 9 shows the calculated quality map with respect to surface defects for the U-profile.

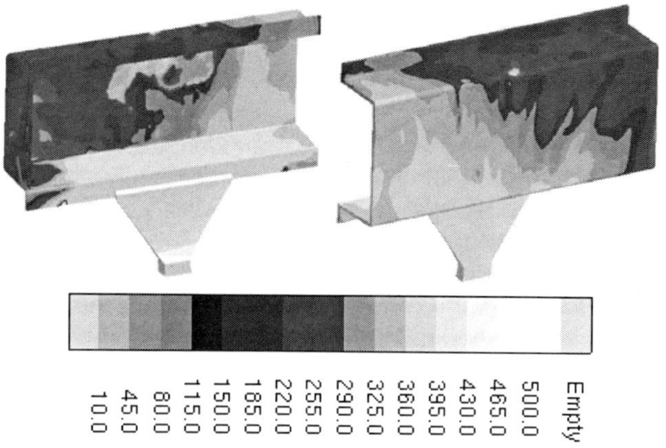

Figure 9 The Quality map distribution with respect to surface defects used to modify the fracture criterion in the skin.

The Quality map distribution with respect to surface defects must be interpreted as a probability map for cold shuts and knit lines, and it must be calibrated with the mechanical properties in the skin.

CORE DEFECTS

The core fracture criterion is modified by the simulated porosity distribution. During solidification, feeding takes place through an element in the mesh as long as the fraction of solid is less than a critical fraction of solid. Areas that become isolated with respect to feeding will accumulate solidification shrinkage in the remaining liquid. This solidification shrinkage is then converted to a percentage porosity based on the difference in density. In the simulation of the U-profile, a critical feeding fraction of 70% was applied and the resulting porosity distribution is shown in Fig. 10. The simulated porosity distribution shows porosity only in the interior of the cross sections, that is, in an area that corresponds to the core layer in the model framework, Fig. 8. In the transparent areas with light shading no porosity was predicted. The simulated porosity distribution is again interpreted as a probability map for porosity that needs to be calibrated with the mechanical properties in the core.

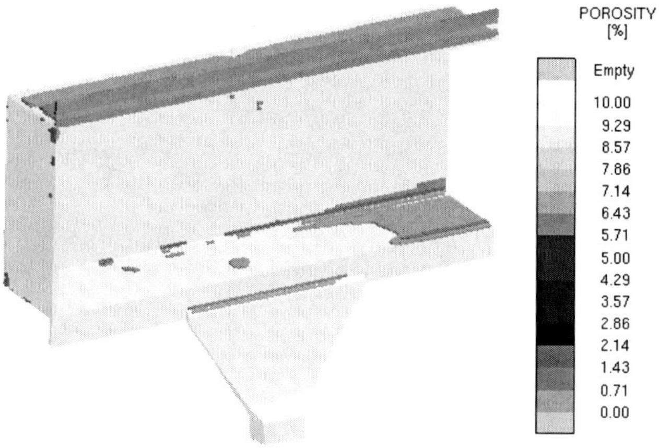

Figure 10 The simulated porosity distribution used to modify the fracture criterion in the core.

The probability distributions for surface defects and porosity were used to calibrate the local ductility /7/. The porosity distribution was used to calibrate the experimental results found from the uniaxial tensile tests. The local tensile ductility of the skin material, and the corresponding value for the "critical work", W_c, was calibrated using the experimental results from the 3- and 4-point plate bending experiments. The calibration was performed using an inverse modeling approach; the experiments are simulated numerically using LS-DYNA and choosing the fracture parameter W_c such that the numerical force-deformation predictions fit the experimental results.

DISCUSSION

The methodology presented in this work is a framework to integrate the information from the HPDC process simulation into FEA of the final part performance. As discussed above, assumptions and simplifications have been made and only what is believed to be 1st order effects are implemented in this work. The methodology is schematically illustrated in Fig. 11. The proposed framework is, however, an "add-on" approach and improvements and other effects and models can easily be implemented. This is, however, also pointing to the challenges that need to be solved. One should, however, bear in mind that if such a methodology should be of any practical value it must be easy to use and implement. This is putting restrictions on how sophisticated the material model can be.

In the HPDC processing simulation, thermal history is the key issue. Theoretically the HPDC process is extremely complex due to large temperature and velocity gradients. In the simple model of the filling and the 1st phase in the shot sleeve, a fraction of solid of 14% was estimated. The question is then how these ESC are distributed in the flow as they are injected into the die cavity. This is a complex task to solve. Numerically it involves a volume that is changing due to the piston stroke, and consequently techniques involving a moving mesh must be applied. The problem is also complex from a material modeling point of view. The metal changes its flow characteristics from being essentially a Newtonian fluid in the complete liquid state to locally becoming a fluid with thixotropic behavior as the fraction of solid increases. At high fractions of solid the metal will behave as a solid. Thus, with large temperature and velocity gradients, the metal will show large local variations in viscosity.

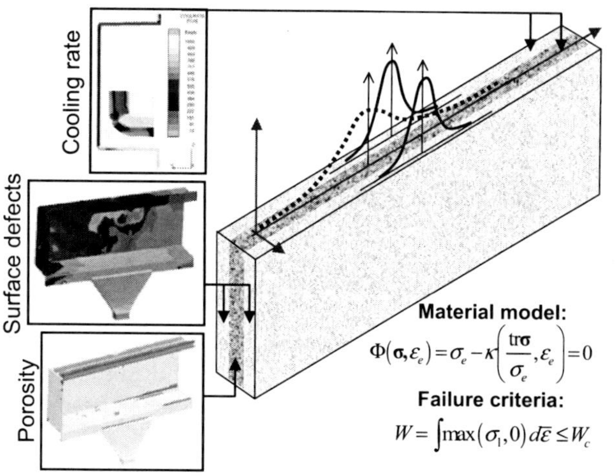

Figure 11 Schematic illustration of the material model and fracture criteria based on the HPDC process simulation.

When this mixture of complete liquid and ESC is entering the die cavity, the metal is accelerated and average metal velocities in the range 10-100 m/s are typical. This causes large local shear rates close to the surface of the die. The rheology of the ESC and liquid metal mixture, the high shear rates, the large temperature gradients and the simultaneously solidification, illustrate the complexity in the metal close to the die surface. Mesh resolution in the surface skin will then become a critical factor. When the grain size in the surface is of the order of 5–10 µm /2/ and the final skin thickness is in the order of 100-500 µm, it will be very difficult to capture the skin formation in the simulation with an element size of the order of 1000 µm (1 mm) /4/. As shown in Fig. 5, the thin skins formed at the die surface during filling are believed to be the origin for the knit lines and cold shuts. The quality criteria generating the quality map for surface defects has so far only been optimized with standard definitions for the mesh resolution /8/. Also for the process parameters and the initial and boundary conditions, only standard values were applied. Work is underway to incorporate both shot sleeve cooling as well as improved algorithms for finer resolution in the die-metal interface.

The implementation of the process and material understanding into the FE material model has been done with several simplifications. Besides the cooling rate that is used to modify the yield stress, the only input from the process simulation is the surface defect quality map and the predicted porosity distribution through the fracture criterion. This means that all fractures initiated in the surface are attributed to the surface defects, while all fractures initiated in the core are attributed to the porosity distribution. Also, the surface and core definitions in the FE mesh put limitations on the model. Since the FE mesh has a limited number of nodes over the thickness (in this work 5), the skin and core definitions will be restricted by the number of nodes.

In this work only input from the flow and solidification simulations have been applied as input to the FEA. But as already mentioned, the methodology is an "add-on" approach so input from thermal stress simulation should also be possible to implement. Residual stresses, hot tearing and geometrical imperfections like sinks and distortions are process related conditions that have a direct influence on the FEA. The performance of the material will be modified if it is pre-strained or if tears have been initiated due to thermal stresses. Sinks and distortions will influence the wall thickness and modify the geometry of the part. Consequently, this will have an impact on the initial part of the deformation when the deformation becomes localized.

In spite of all the simplifications made and the intentional omission of certain other effects, the methodology seems to give very promising results for the FEA of different load cases. The reader is referred to Dørum et al /7/ for further results on the FEA, where the methodology is compared with a standard approach using homogeneous material properties in 3 and 4 point bending tests.

SUMMARY

A methodology to integrate the information from the high-pressure die casting process (HPDC) simulation into finite element analysis (FEA) of the final part performance has been developed.

In the material model the cross section of a thin-walled magnesium die-casting is divided into two skin layers with a core layer in between. Different material properties are assigned to the different layers based on the input from the HPDC process simulations.

The main input from the processing simulation is used to modify the fracture criterion. The model for the fracture criterion is the same in the skins and in the core. In the fracture criterion the predicted surface quality with respect to knit lines and cold shuts is used to modify the skin fracture criterion, while the simulated porosity distribution is used to modify the fracture criterion of the core.

ACKNOWLEDGMENTS

MAGMA GmbH is acknowledged for the support and the co-operation with respect to the MAGMAsoft software and implementation of new quality criteria for magnesium high pressure die castings. SINTEF Material Technology is acknowledged for the support in developing new material models in LS-DYNA for Finite Element Analysis.

REFERENCES

1. Laukli H I, Lohne O, Sannes S, Gjestland H, Arnberg L, *The effect of variations in melt temperature on the grain structures in an AM60 die casting*, 6th Int. Conf on Magnesium Alloys and Their Applications, ed. K.U. Kainer, 2003, vol. 6, pp 182-189, Frankfurt, Germany
2. Laukli H., High pressure Die Casting of Aluminium and Magnesium Alloys–Grain Structure and Segregation Characteristics, PhD thesis, NTNU, Dept. of Materials Technology, ISBN 82-471-6324-1, 2004
3. Sannes S, Gjestland H, Westengen H, Laukli H I, Lohne O, *Die Casting of Magnesium Alloys – The importance of Controlling Die Filling and Solidification*, paper no. 2003-01-0183, SAE 2003, Detroit, USA
4. Sannes S, Gjestland H, Westengen H, Laukli H I, Lohne O, *Magnesium Die Casting for High Performance*, 6th Int. Conf on Magnesium Alloys and Their Applications, vol. 6, pp 725-731, ed. K.U. Kainer, 2003, Frankfurt, Germany
5. Gjestland H, Sannes S, Westengen H, Optimizing the Magnesium Die Casting Process to Achieve Reliability in Automotive Applications, paper no. 05M-224, SAE 2005, Detroit, USA
6. Dørum C, Hopperstad O S, Lademo O-G, Langseth M, Sannes S, *Behaviour and modelling of thin-walled cast components in magnesium and aluminium*, Int. Symposium on Crashworthiness of Light-Weight Automotive Structures, eds. M. Langseth and A. H. Klausen, 2004, Trondheim, Norway.
7. Dørum C, Hopperstad O S, Lademo O-G, Langseth M, Sannes S, Numerical modelling of the structural behaviour of thin-walled cast magnesium components using a through-process approach, paper no. 05M-215, SAE 2005, Detroit, USA
8. Hepp E, Lohne O, Sannes S, *Extended Casting Simulation for Improved Magnesium Die Casting*, Magnesium Alloys and Their Applications, ed. K.U. Kainer, 2003, vol. 6, pp 669-674, Frankfurt, Germany.

2005-01-0333

Optimizing the Magnesium Die Casting Process to Achieve Reliability in Automotive Applications

Haavard Gjestland, Stian Sannes, Jørild Svalestuen and Håkon Westengen
Competence Centre Magnesium, Hydro Magnesium

ABSTRACT

High pressure die casting is characterized by rapid die filling and subsequent rapid cooling of the molten metal in the die. These characteristics are favourable for magnesium die casting alloys. The high cooling rate favours the formation of a fine dendrite and grain structure, which in turn leads to substantial hardening; this refinement also provides improved ductility.

Since the cooling rate of the metal is highly dependent on both the process parameters and the geometry of the part, the three-dimensional flexibility associated with the latter factor means that the cooling rate cannot be uniform. This cooling rate difference in turn can lead to some variation in the mechanical properties between geometrically different portions of a die cast component. This variation is an inherent property of the material, in contrast to casting defects like microporosity, non-metallic inclusions, filling defects, and formation of hot cracks. The mechanical properties of the casting are also affected by the pre-solidification of metal in the shot sleeve.

In the present paper the correlation between the thermal conditions in the process and the resulting microstructure and mechanical properties in the casting is discussed.

INTRODUCTION

The growth in the application of magnesium die castings is made possible by technological improvements. Among several factors, the introductions of high purity alloys, fluxless melting, improved die casting technology and virtual design have been of great importance. The combined effect of these improvements offers possibilities for highly automated, cost competitive production of complex, light weight parts fulfilling the requirements of modern mass production.

Design and production of light weight structural applications are challenging tasks for die casters, as there is a continuous demand for improved material properties and robust manufacturing performance. The properties of a die cast part are first of all controlled by the local solidification rates in the component. Secondly, any defects formed during the casting process will reduce the theoretically optimal properties.

A key factor associated with prediction of the properties is the thermal conditions in the process from the melting furnace through the metering units and shot sleeve, and finally the filling of the die. Only by knowing the impact of the thermal history of all these elements on the properties is it possible to optimize the die casting process.

The current investigation firstly describes some basic correlations between the solidification rate of the metal and the resulting microstructure together with the mechanical properties. Such general correlations are valid for all alloys. The paper presents the results from investigations on the alloy AZ91D.

Secondly the work describes the general behaviour of a hot metal poured into shot sleeves with different thermal conditions. The paper presents the results from investigations on the alloy AM60B.

FACTORS THAT IMPACT THE MECHANICAL PROPERTIES

SOLIDIFICATION RATE

The grain size and the secondary dendrite arm spacing are both strongly dependent of the solidification rate. The faster the material is solidified, the smaller the grain size will be.

The heat content in the metal is extracted from the solidifying component through the massive steel die that is cooled by a circulating medium in channels in the die. Thin sections in the component and a cold die will enhance a high solidification rate. However, an excessively cold die may cause undesirable solidification during filling and lead to defects like cold shuts. Such defects can often be only cosmetic, but severe cold shuts could create problems in components requiring surface treatment, extremely tight dimensions, surface smoothness, etc. Die casters commonly use a die temperature in the range of 150 - 250 °C.

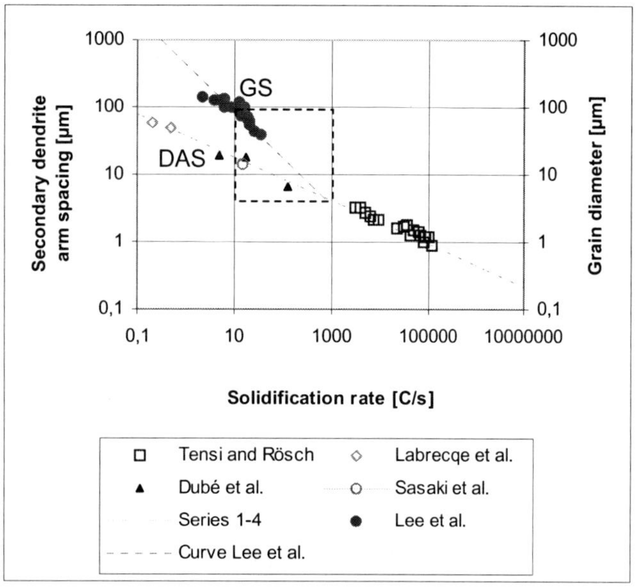

Figure 1. Grain size and secondary dendrite arm spacing as a function of the solidification rate for AZ91D. Operating window for die casting indicated by dashed rectangle [1]

Figure 1 shows the correlation between the grain size GS (and secondary dendrite arm spacing DAS) and the solidification rate for the alloy AZ91D [1]. In a die cast component of AZ91D, grain sizes in the range of 5 - 100 micrometers are typically found. This corresponds to local cooling rates in the range of 1000 - 10 °C/s. In order to have a more uniform grain size distribution in the product, die designers and manufacturers should focus on the local solidification conditions. Such uniformity can help to assure more homogeneity in the yield strength and the elongation throughout the part because of the relationship between these parameters.

GRAIN SIZE

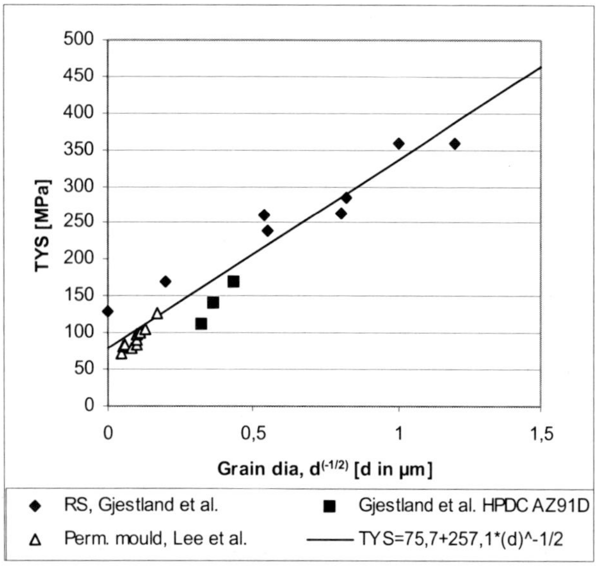

Figure 2. Tensile yield strength as a function of grain size in an AZ91D alloy [1]

A fine dendrite and grain structure provide improved tensile properties. In hexagonal structures such as magnesium, the tensile yield strength follows a Hall-Petch relationship as shown in Figure 2.

The tensile yield strength is proportional to the inverse square root of the grain size, and the hardening from the grain size is almost the only mechanisms for improved strength. The hexagonal crystal structure has also a limited number of slip systems at ambient temperatures. This causes inhomogeneous deformation, and fracture may occur at relatively low macroscopic deformations due to slip localization and build up of high local stresses at the grain boundaries.

With decreasing grain size the plastic deformation becomes more homogeneous, as more grains will have a favourable orientation relative to the deformation direction. This will lead to larger macroscopic strains before fracture, as illustrated for AZ91D and pure magnesium in Figure 3. Such strains are particularly valuable in applications where large post-accident deformation could be tolerated, but where fracture would be unacceptable.

Analytical observations from Figures 2 and 3 indicate that the grain size has to be near or below 10 µm in AZ91D to achieve a reasonable combination of mechanical properties. Other alloys follow a similar pattern.

These features will lead to variations in the properties of die cast parts with complex geometry. This is inevitable, and this kind of variation can only be controlled by careful design and selection of optimum casting parameters.

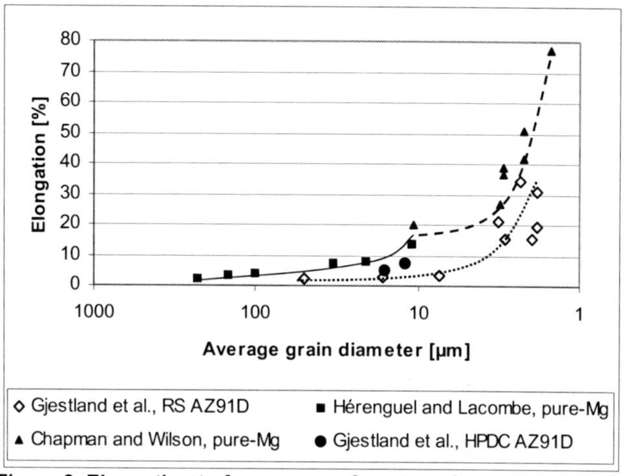

Figure 3. Elongation to fracture as a function of grain size in an AZ91D alloy and in pure-Mg [1]

SEGREGATION IN THE SHOT SLEEVE

Properties interpreted from diagrams like Figures 2 and 3 can be looked upon as potentially obtainable, providing no casting defects like cold shuts, knit lines, pores, alloy segregations etc. are present.

In the current presentation, focus will be put on the impact of alloy segregation caused by pre-solidification in the shot sleeve in high pressure die casting.

When hot (650 – 700 °C) metal is poured into the relatively cold (200 - 300 °C) shot sleeve, solidification starts on the sleeve wall. Figure 4 illustrates a computer simulation of a shot sleeve with pre-solidified material at the sleeve wall. As metal flows from the pour hole against the die, heat is extracted to the sleeve wall and the volume fraction of pre-solidified metal increases in the die-end of the sleeve.

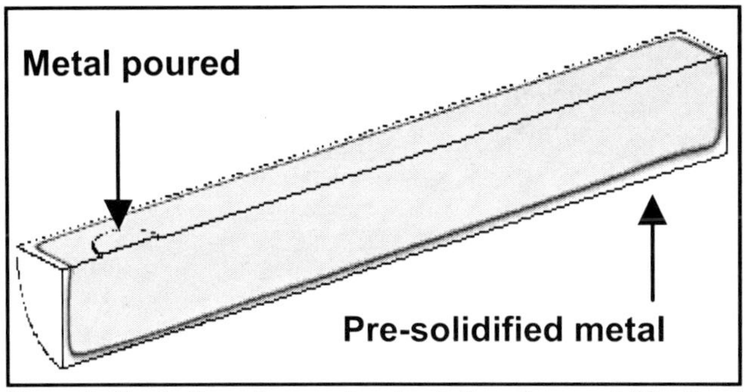

Figure 4. Illustration of computer simulation of pre-solidification in the shot sleeve

The fraction of metal that solidifies prior to the injection depends on the alloy, the thermal conditions in the sleeve and the retention period of the metal in the sleeve. When the plunger movement starts, 100 % liquid metal is supposed to flow firstly into the cavity. During injection, metal flow can tear off α-dendrite arms and the movement of the plunger will scrape off everything solidified on the wall. These fragments of the α-dendrites are injected as floating crystals in the liquid metal, into the cavity. The floating crystals, hereafter termed externally solidified crystals (ESC), are found throughout the casting. The highest volume fractions are typically located a) throughout the cross section close to the gates, and b) in the inner core of the casting wall away from the gates [2]. This is illustrated in a U-profile used to simulate box sections, see Figures 5a and b.

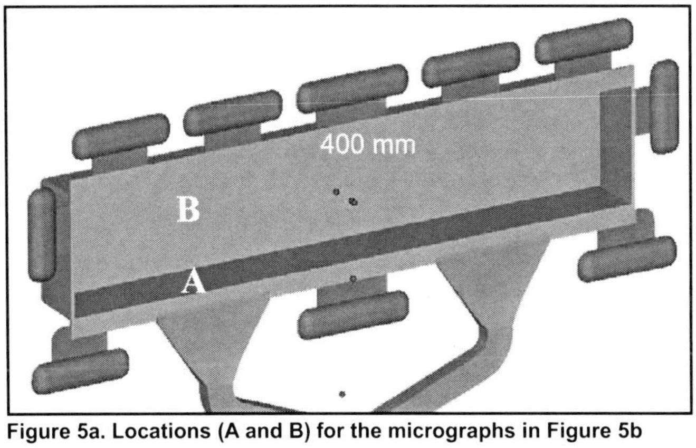

Figure 5a. Locations (A and B) for the micrographs in Figure 5b

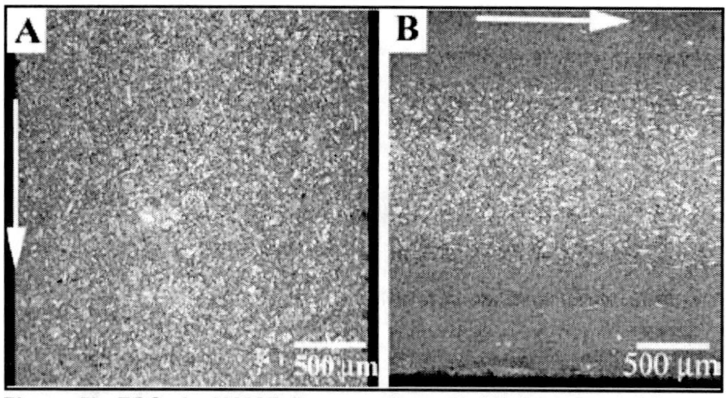

Figure 5b. ESCs in AM60B (large grains with bright contrast) in a U-profile type of casting. A) Close to the gate, B) Far from the gate. Arrows indicate flow direction [3]

In a cast tensile specimen, the ESCs are located in the centre zone of the specimens gauge length. The outer zone of the bar shows a fine grained microstructure almost free from ESCs. Between the two zones, pore bands can be formed, Figure 6.

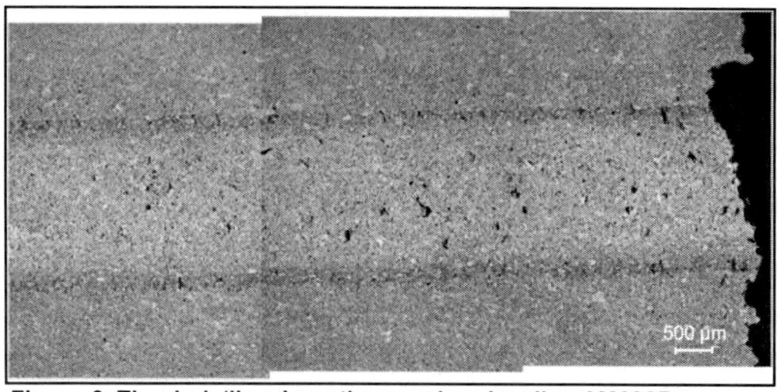

Figure 6. The dark 'lines' are the pore bands, alloy AM690B.

The ESCs and the porosity bands impact the elongation [4]. Figure 7 shows a longitudinal section of a tensile bar after tensile testing. During straining, pores close to each other have coalesced and caused fracture when the local stresses have exceeded the yield strength of the material. No cracks seem to be initiated in the fine grained surface zones.

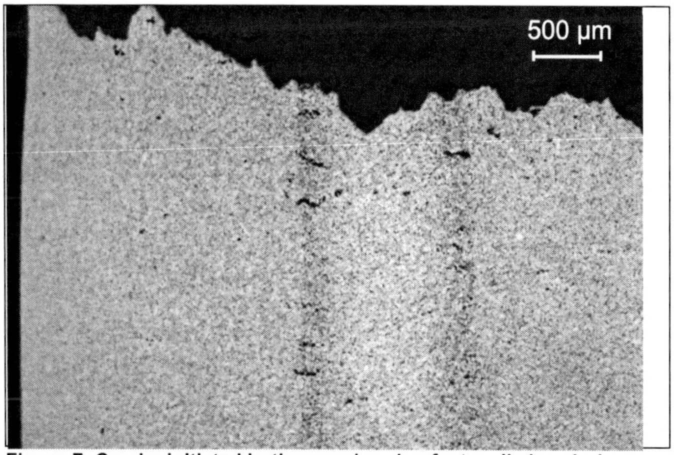

Figure 7. Cracks initiated in the pore bands of a tensile bar during tension, alloy AM60B.

To assure reliability in die cast automotive components, the presence of the ESCs must be taken into account. As the elongation of the casting seems to be strongly affected by the ESCs, their location and the distribution should be controlled by design and by gating of the die.

SIMULATION OF THE THERMAL CONDITIONS IN THE SHOT SLEEVE

The pre-solidification in the shot sleeve is affected by the heat transfer between the metal and the sleeve wall, the heat conductivity in the sleeve wall, the temperature of the sleeve and the contact time between the metal and the sleeve. Simulation was carried out to estimate the volume fraction of pre-solidified material formed under different thermal conditions in a sleeve with a diameter of Φ60 mm and a length of 365 mm.

In the simulation, a dosing time of 1.25 s and an additional time of 1.15 s from triggering of the plunger movement until the shot sleeve is 100 % full prior to acceleration of the metal in the 2nd phase are used. The fill fraction of the sleeve is 50 %.

Table 1. Computer calculation of fraction pre-solidified AM60B as a function of thermal conditions in the shot sleeve

	Sleeve condition	Vol. frac. pre-sol. metal
1	Heat transfer coefficient h(T)=5000 W/m^2 Heat conductivity in the sleeve λ= 35.4 W/m °C Temperature of the sleeve T= 200 °C	14 %
2	Heat transfer coefficient h(T)=600 W/m^2 Heat conductivity in the sleeve λ= 35.4 W/m °C Temperature of the sleeve T= 200 °C	0 %
3	Heat transfer coefficient h(T)=5000 W/m^2 Heat conductivity in the sleeve λ= 7.4 W/m °C Temperature of the sleeve T= 200 °C	5 %
4	Heat transfer coefficient h(T)=5000 W/m^2 Heat conductivity in the sleeve λ= 35.4 W/m °C Temperature of the sleeve T= 440 °C	2.5 %

Table 1 shows the results from the computer simulation of four different thermal conditions in the shot sleeve. The reference is a commonly used H13 tool steel sleeve where as much as 14 % of the metal is likely to solidify at the wall prior to injection. The three other conditions represent different ways of changing the thermal performance of the sleeve. All three conditions result in significant reduction in the pre-solidification.

In addition to the thermal conditions in the sleeve, the total time the metal is in contact with the sleeve and the fill fraction of the sleeve are important factors to reduce the pre-solidification. A die caster should therefore have in mind the impact the shot sleeve dimensions and the metering time have on the formation of ESCs when he is planning his process.

CASTING EXPERIMENTS WITH DIFFERENT THERMAL CONDITIONS IN THE SHOT SLEEVE

The four shot sleeve conditions computer simulated and presented in Table 1 were tested in practice in a 420 metric ton Bühler Evolution die casting machine. Standard ASTM tensile bars (gauge length 75 mm, gauge diameter 6 mm) were cast and tested according to ASTM 557B. Metallographic investigations were carried out to document variations in the microstructure.

Condition no 1

The shot sleeve was made from H13 tool steel. The heat conductivity of H13 tool steel is 35.4 W/m °C. The heat transfer coefficient between the metal and the sleeve wall was estimated to be 5000 W/m^2. The sleeve was kept at a temperature of 200 °C by circulating hot oil in the sleeve wall.

Condition no 2

Boron nitride was sprayed inside the H13 tool steel sleeve. The heat conductivity of H13 tool steel is 35.4 W/m °C. The heat transfer coefficient between the metal and the sleeve wall was now estimated to be 600 W/m^2. The sleeve was kept at a temperature of 200 °C by circulating hot oil in the sleeve wall.

Condition no 3

A H13 tool steel shot sleeve was equipped with a metal matrix composite (MMC) liner of Ti-SiC. The sleeve was kept at a temperature of 200 °C by circulating hot oil in the sleeve wall. The heat conductivity of the MMC liner is 7.4 W/m °C, about 20 % of the value of H13 tool steel. The heat transfer coefficient between the metal and the sleeve wall was estimated to be 5000 W/m^2.

Condition no 4

A H13 tool steel shot sleeve was equipped with heating cartridges in the sleeve wall. The sleeve was isolated outside to reduce heat conductivity to the die casting machine, and to reduce heat loss by radiation to the surroundings. The heat conductivity of H13 tool steel is 35.4 W/m °C. The heat transfer coefficient between the metal and the sleeve wall was estimated to be 5000 W/m^2. The sleeve was kept at a temperature of 440 °C by the heating cartridges.

RESULTS

MECHANICAL PROPERTIES

Figure 8 shows the mechanical properties of the four series of tensile test bars (200 samples of condition 1, 20 samples of condition 2, 93 samples of condition 3 and 39 samples of condition 4). The average values were improved when the volume fraction of pre-solidified material was reduced. The unmodified H13 tool steel sleeve gives the poorest properties. Any of the other three thermal conditions give an improvement in the balance of properties. However, when the difference in the volume fraction of ESCs is small, there are only minor differences in the properties, see also Table 1.

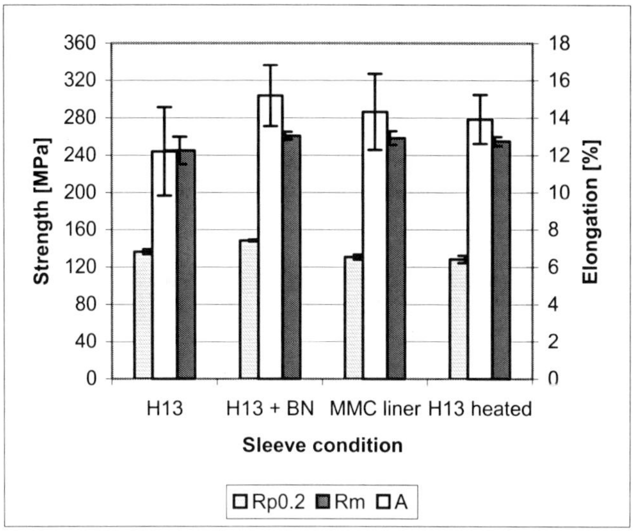

Figure 8. Bar graphs showing the average mechanical properties and one standard deviation in tensile bars cast with different thermal conditions in the shot sleeve, alloy AM60B

MICROSTRUCTURE

In Figures 9a-d the microstructures in a cross section of tensile bars cast with the four different shot sleeve conditions are shown. The appearance of the structures confirms the expected tendency in reduction of the volume fraction of pre-solidified material with improved shot sleeve thermal management, see Table 1.

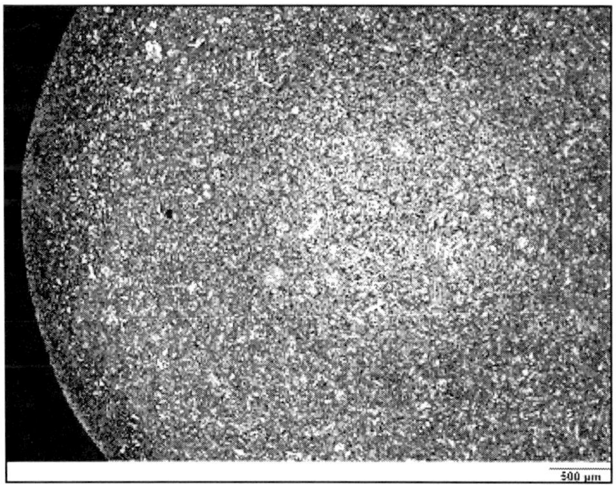

Figure 9a. Micrograph of tensile barin AM60B cast with a H13 shot sleeves. The 'white particles' are the ESCs

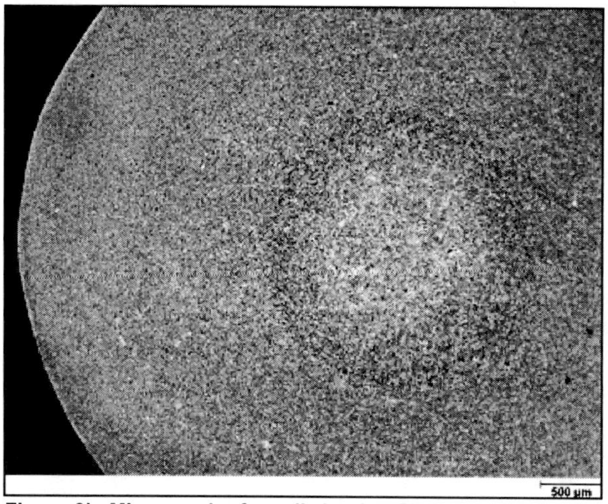

Figure 9b. Micrograph of tensile bar in AM60B cast with a H13 shot sleeve coated with boron nitride. The 'white particles' are the ESCs

Figure 9c. Micrograph of tensile bar in AM60B cast with a MMC-lined shot sleeve. The 'white particles' are the ESCs

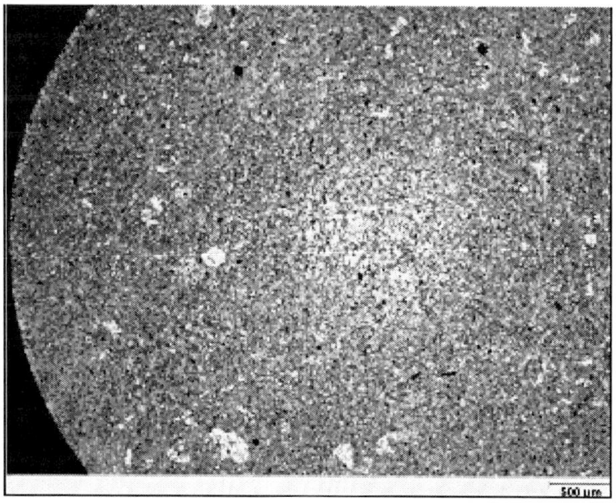

Figure 9d. Micrograph of tensile bar cast in AM60B with a heated H13 shot sleeve. The 'white particles' are the ESCs

The micrograph of a bar cast with the H13 sleeve, Figure 9a, shows ESCs distributed over the cross section but with a higher concentration in the centre, while there is almost no pre-solidification visible in the micrograph of the bar cast with boron nitride coating in the sleeve,

Figure 9b. Only a small volume fraction of ESCs can be seen in the bars cast with the MMC-lined and the heated sleeve, Figures 9c-d. There is no big difference between the two latter.

STATISTICAL EVALUATION OF THE RESULTS

Large series of tensile bars in the alloy AM60B were cast and tensile tested using the H13 sleeve, the MMC-lined sleeve and the heated H13 sleeve. Due to a too simple experimental setup with regard to the boron nitride coating, a scatter in the casting conditions was expected. A large series with boron nitride coated shot sleeve was therefore not produced.

The results were processed with statistical tools to evaluate the scatter in the results. Figures 10a-c show frequency bar graphs for the elongation values in the three series.

The H13 series, Figure 10a, shows a left skewed distribution with very low minimum elongation values. In the two other series, the 'tail' with the small values is removed, leaving a slight bimodal distribution with much less scatter than in the first series, Figures 10b-c.

The most important impact on the elongation value is the reduction in the scatter when pre-solidification is reduced. With the use of modified shot sleeve conditions represented by the MMC-lined sleeve or the heated sleeve, an end user will have more reliable date for his design.

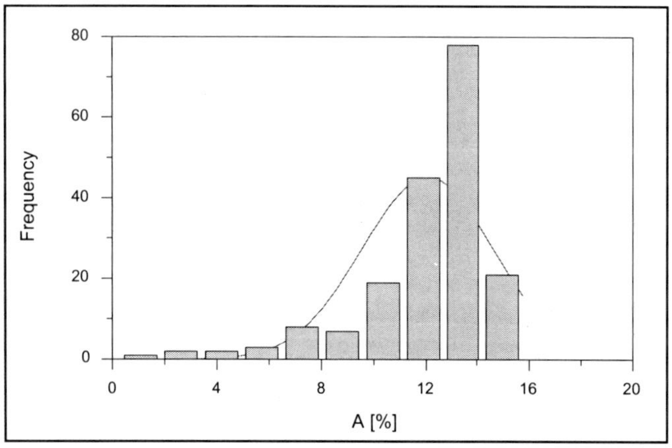

Figure 10a. Frequency bar graphs for the elongation in tensile specimens cast with a H13 shot sleeve (200 specimens of AM60B)

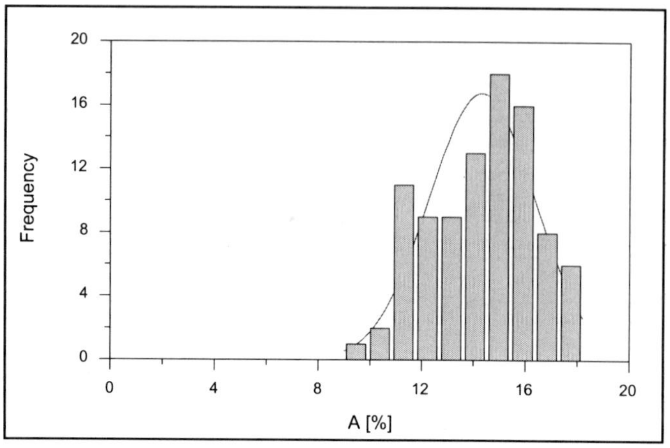

Figure 10b. Frequency bar graphs for the elongation in tensile specimens cast with a MMC-lined shot sleeve (93 specimens of AM60B)

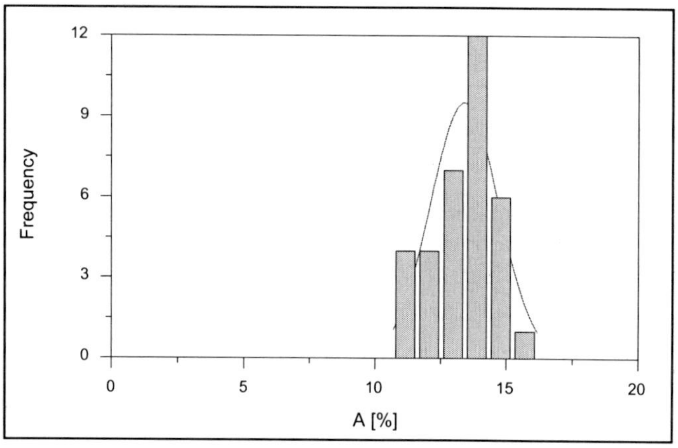

Figure 10c. Frequency bar graphs for the elongation in tensile specimens cast with a heated H13 shot sleeves (34 specimens of AM60B)

The results can also be presented in a 2-parameter Weibull analysis, Figure 11. The analysis clearly reveals that there is a significant difference in the scatter between the results from the H13 sleeve and the MMC-lined sleeve and the heated sleeve. The effect of the large reduction in volume fraction of ESCs, from 14 % to 5 % is significant. Lesser differences are indicated between the MMC-lined and heated sleeves; the scatter is less in the heated sleeve, but this is accompanied by lower maximum elongation. The reduction in ESCs here is only from 5 % to 2.5 %.

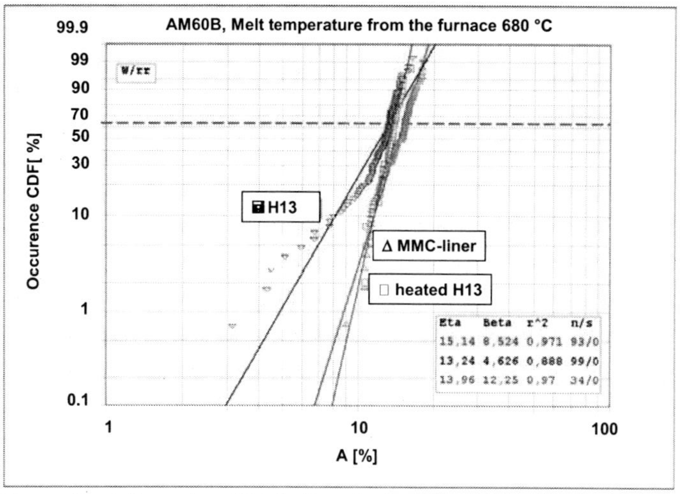

Figure 11. 2-dimensional Weibull analysis of the elongation in tensile bars cast with different shot sleeve conditions

CONCLUSIONS

- Use of conventional shot sleeves manufactured from of H13 tool steel results in a significant volume fraction of pre-solidified material in the sleeve due to heat loss in the metal prior to injection.

- Externally solidified crystals (ESCs) are α-magnesium dendrite fragments. When dispersed in the liquid metal and injected into the die cavity, these ESCs lead to reduction in the mechanical properties in the cast components.

- Pre-solidification in the shot sleeve can be decreased by a) reducing the heat transfer from the metal to the sleeve wall by insulating the inner sleeve wall, b) reducing the heat conductivity in the sleeve wall, c) reducing the heat transfer by maintaining a higher temperature in the sleeve wall. These efforts in turn will lead to improved mechanical properties in the casting.

- Heat loss reduction results in a higher temperature in the melt injected into the cavity. This enhances the fluidity for filling the given die, as well as potentially leading to better filling of more complex dies. A negative effect of enhancing fluidity is a larger tendency for flashing of a given die.

- A Best Practice for magnesium die casting to obtain most reliable properties in a component should include:

    - Component and die design for highest possible solidification rates in all sections of the component.

    - Use of shot sleeves and process parameters causing low heat loss in the metal prior to injection.

REFERENCES

1. H. Gjestland, S. Sannes, H. Westengen and D. Albright, Effects of Casting Temperature, Section Thickness, and Die Filling Sequence Concerning Microstructure and Mechanical Properties of High Pressure Die castings, NADCA conference, Indianapolis, September 2003.
2. H. I. Laukli, High Pressure Die Casting of Aluminium and Magnesium Alloys – Grain Structure and Segregation Characteristics, Doctoral Theses, Norwegian University of Science and Technology, 2004.
3. H. I. Laukli, O. Lohne, S. Sannes, H. Gjestland and L. Arnberg, Grain Size Distribution in a Complex AM60B Magnesium Alloy Die Casting, Int. J. of Cast Metals Research, 2003, vol. 16, no. 6, pp. 515-521.
4. Crack initiation and crack propagation in HPDC tensile bars, Hydro Magnesium Competence Centre, Internal report, August 2003.

2005-01-0334

Effect of HPDC Parameters on the Performance of Creep Resistant Alloys MRI153M and MRI230D

G. Gertsberg, N. Nagar, M. Lautzker, B. Bronfin and N. Moscovitch
Dead Sea Magnesium Ltd

S. Schumann
Volkswagen AG

ABSTRACT

The growing demand for the use of magnesium alloys in the production of automotive powertrain components led to the development of creep resistant diecasting alloys MRI153M and MRI230D.
The present paper addresses the main high-pressure die casting parameters, which significantly affect the performance of components, produced of these new alloys. A systematic study was carried out in order to correlate die-casting parameters to the performance of new alloys.
The results obtained clearly indicated that optimization of molten metal and die temperatures, injection profile parameters and lubrication mixtures allowed to improve the die castability and service properties of the new alloys and produce high performance components with intricate geometry. This was manifested by production of several practical demonstrators such as gearboxes, oil pans, oil pumps and crankcases.

INTRODUCTION

High pressure die casting (HPDC) is a high volume manufacturing process for components of different sizes and shapes. Mg alloys have become promising and reliable materials for die cast applications due to their low density and manufacturing advantages such as outstanding castability and machinability. Magnesium alloys can be cast into components with very low wall thickness and small draft angles having dimension tolerances 50% tighter than aluminum castings [1]. Magnesium alloys have a lower latent heat of fusion and reduced tendency to die pick-up and erosion compared to aluminum alloys. Magnesium die casting process requires reduced die casting machine cycle time and allows 2-4 times longer die life compared to their aluminum counterparts. Commercial magnesium alloys also exhibit good ambient temperature mechanical properties. However, poor creep properties of conventional magnesium alloys restrict their use for powertrain applications. In order to overcome the above limitations several new die casting alloys were recently developed [2-8].
The target to develop a HPDC alloy, which exhibits an adequate combination of die castability, creep performance, corrosion resistance, room temperature strength and affordable cost, was also addressed by DSM and Volkswagen AG [9,10]. This comprehensive work resulted in the development of two new magnesium alloys designated MRI153M and MRI230D.
MRI153M is a Beryllium free, low cost, creep resistant alloy with the capability of long-term operation at temperatures up to 150°C under high loads. This alloy exhibits die castability, corrosion resistance and mechanical properties similar to or better than those of AZ91D alloy with superior creep resistance at 130-150°C under stresses of 50-85 MPa.
This makes the alloy a superior alternative for applications such as gearbox housings, valve covers, intake manifolds and other high temperature applications for the automotive industry.
MRI230D is a die cast alloy that was developed to address the powertrain applications of the automotive industry such as engine blocks operating at temperatures up to 190°C. The alloy has excellent creep resistance combined with good castability, high strength and superior corrosion behavior. Automatic transmissions housings and bedplates are part of the applications, which can be produced by HPDC of MRI230D alloy.
The mechanical and physical properties of the above alloys are well known and documented [9,10]. It should be pointed out that the castability and the HPDC process characteristics of the newly developed alloys are

different from those of conventional Mg alloys. Hence, in order to achieve high quality sound die cast components process optimization and die modifications are required. The present work aims at optimization of injection parameters and gating system design at HPDC of newly developed alloys. Special attention is drawn to determination of the effect of molten metal and die temperatures as well as the lubricant data on the properties of the newly developed alloys.

EXPERIMENTAL PROCEDURE AND RESULTS

In order to address the above objectives comprehensive experiments were performed using IDRA OL320 cold chamber die casting machine with locking force of 345 tons. The trials consisted of 3 main subjects:

1. HPDC process optimization.
2. The effect of molten metal and die temperature on the filling time.
3. Optimization of gating system.

HPDC PROCESS OPTIMIZATION

An oil pump consisting of three parts, weighing 0.19, 0.18 and 0.35 kg, was selected as a demonstrator component (Figure 1) and cast in MRI153M and MRI230D as well as AZ91D and AM50A, in order to determine the significant HPDC process parameters

The aim of these trials was to minimize the component defects such as:

- Insufficient filling
- Gas and shrinkage porosity
- Surface defects (hot cracking, soldering and tearing).

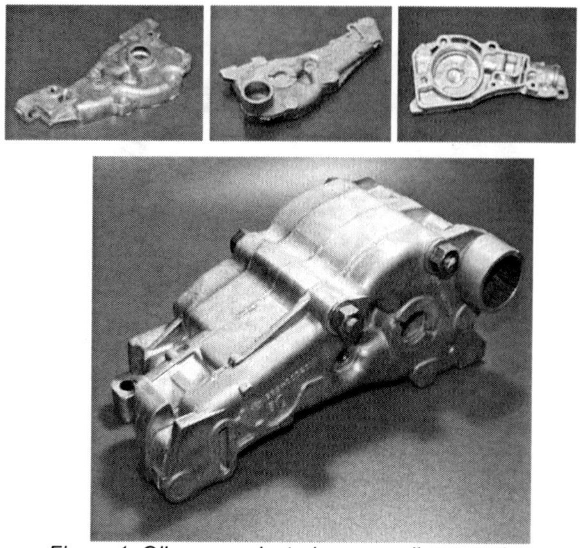

Figure 1. Oil pump selected as new alloys casting demonstrator

The parameters investigated were:

- Molten metal (alloy temp. in the furnace) and die temperature (Thermocouple measurement located 30mm from the die surface)
- Lubricant data (spraying time and mixing ratio).
- Dwell time.

The optimized parameters obtained in course of experiments are listed in Table 1. These parameters are very useful tools at die casting of actual components in MRI153M and MRI230D alloys.

Table 1. HPDC data for commercial and new alloys

Mg Alloy	Temperature data		Spraying and Lubrication		Dwell time (%)
	Melt (°C)	Die (°C)	Spraying time (%)	Mixing ratio	
AZ91D	650-670	150-200	100	1/40-1/80	100
AM50A	680-695	150-200	100	1/40-1/80	100
MRI153M	660-670	200-250	120-150	1/25-1/40	70-80
MRI230D	680-690	200-250	150-170	1/15-1/40	50-75

THE EFFECT OF MOLTEN METAL AND DIE TEMPERATURE ON THE FILLING TIME.

These trials were performed using the cavity presented in Figure 2.

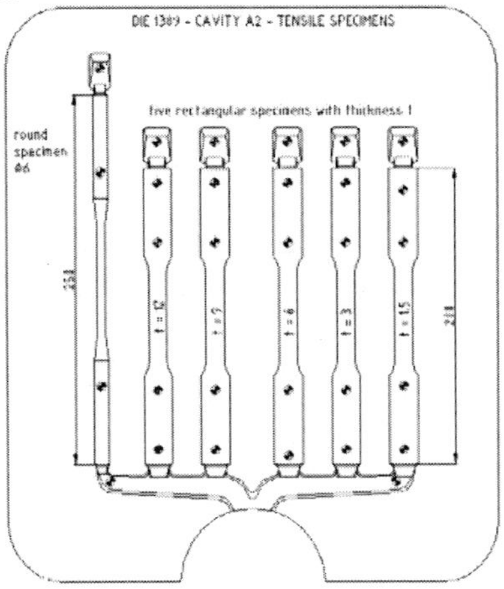

Figure 2. Cavity with different types of tensile samples

This cavity allows the production of six tensile specimens. One round specimen with 6 mm diameter and five rectangular specimens with thickness of 1.5, 3, 6, 9 and 12 mm.

In order to determine the optimal filling time for each thickness, several experiments were performed with a wide range of second phase velocities from 1 to 7m/s. for different melt and die temperatures (Table 2). The aim of these experiments was to determine the minimum second-phase velocity or maximum acceptable filling time, which would allow obtaining sound castings.

Table 2. Experimental parameters for evaluation of filling properties

Experiment No.	1	2	3	4	5	6
Casting temp. °C	710	690	690	690	670	670
Die temp. °C	200	200	160	140	160	130

For example, Figure 3 presents the map obtained for MRI230D alloy.

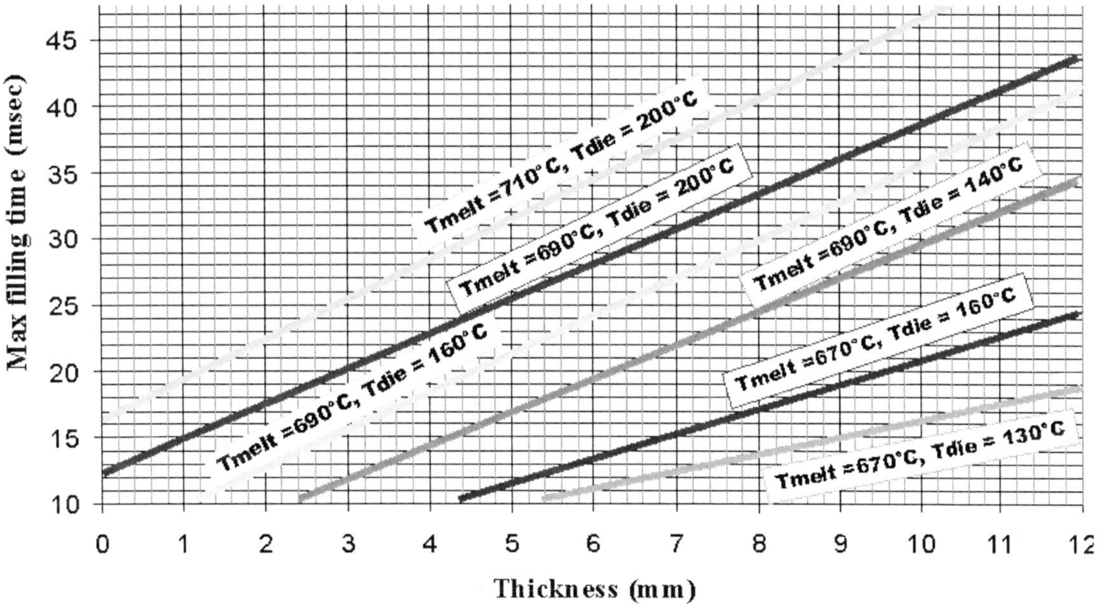

Figure 3. The correlations between molten metal and die temperatures, and the maximum filling time.

GATING SYSTEM OPTIMIZATION

In order to optimize the gating system for MRI153M and MRI230D alloys the following parameters were selected and studied:

- Gate velocity.
- Gate thickness.
- Die temperature.

The die and the part used during this study are presented in Figure 4. The part net weight was 0.83 kg. The gating system was machined during the trials in order to alter gate area from 1.17 cm^2 to 2.35 cm^2. The die temperature was varied from 80 to 160°C while second phase velocity was altered from 1 to 5 m/s for three different gate areas (Table 3). Molten metal temperature was 670°C for MRI153M and 685°C for MRI230D.

Figure 4. The die used for gating system optimization

Table 3: Experimental parameters

Gate area [cm²]	1.17			1.8			2.35		
T_{die} [°C]	80	120	160	80	120	160	80	120	160
Second phase velocity [m/sec] (Plunger vel.)	1, 2, 3, 4, 5								

The trials were stopped when soldering was detected. Figure 5 and Figure 6 present the correlations obtained for MRI153M and MRI230D, respectively.

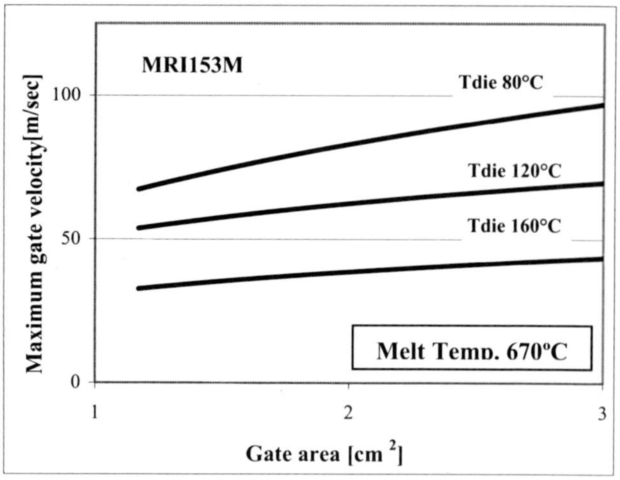

Figure 5. MRI153M gating system optimization.

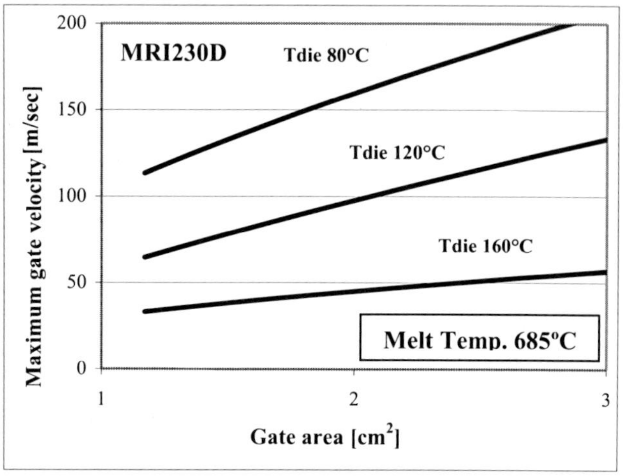

Figure 6. MRI230D gating system optimization.

DISCUSSION

The first series trial indicated that casting defects could be minimized by HPDC process optimization. It was found that sound parts could be obtained with increased spraying time, increased mixing ratio and decreased dwell time. (Table 1). Adherence to the recommended procedure would contribute to decreased cycle time and high productivity.

In addition, it is also recommended to consider the parameters map (Figure 3), which demonstrates correlation between the molten metal and die temperatures and the second phase velocity or the cavity filling time.

It is evident from the developed map that for a given wall thickness higher melt and die temperatures allow the use of lower injection velocities and consequently higher cavity filling time. On the other hand, it should be taken into consideration that too high second phase velocity may result in atomized flow pattern and a potential for increased gas porosity and decreased properties.

In order to minimize surface defects such as soldering, it is recommended to use for given gate area and die temperature the values of gate velocities lower than outlined by curves shown in Figs.5 and 6.

The presented procedures were used in several casting trials with both MRI153M and MRI230M alloys. Following their implementation, the productivity and the process characteristics have been improved and high quality components such as oil pans, gearbox housing and transfer cases were easily die cast. Recently Audi has successfully die cast hybrid magnesium aluminium engine blocks using MRI153M and MRI230D alloys [11,12].

CONCLUSIONS

1. Adequate HPDC process parameters related to melt and die temperatures, lubrication and injection parameters are available for MRI153M and MRI230D.
2. Based on various casting trials it was manifested that proper die design and HPDC process optimization result in high quality components cast in MRI153M and MRI230D.
3. MRI153M and MRI230D are promising and suitable materials for powertrain applications.

REFERENCES

1. G.S. Cole "Issues that Influence Magnesium's Use in the Automotive Industry", Material Science Forum, V.419-422, 2003, 43-50.
2. S. Koike et al., "Development of Lightweight Oil Pans Made of a Heat-resistant Magnesium Alloy for Hybrid Engines", SAE Technical Paper 2000-01-1117, Detroit, MI, 2000.
3. B.R. Powell et al., "The Die Castability of Calcium-Containing Magnesium Alloys: Thin-Wall Computer Case", Magnesium Technology 2002, Ed. H.I. Kaplan (Warendale, PA: TMS, 2002), 123-132.
4. K. Pettersen et al., "Creep-Resistant Magnesium Alloy Development", Magnesium Alloys and Their Applications, Ed. K.U. Kainer (Weinheim, Germany: Wiley-VCH ,Verlag GmbH, 2000), 29-34.
5. M.Pekguleryuz et al, "Magnesium Diecasting Alloy AJ62x with Superior Creep Resistance", Magnesium Technology2003, Ed. H.I. Kaplan, (Warrendale, PA: TMS, 2003), 201-206
6. P. Bakke, H.Westengen and K.Pettersen, "Die Cast Magnesium Alloys for High Temperature Applications", SAE Technical Paper 2003-01-189, Detroit, MI, 2003.
7. P. Lyon, J.F. King, and K. Nuttal "A New Magnesium HPDC Alloy for Elevated Temperature Use", Proceedings of the Third International Magnesium Conference, Ed. G. W. Lorimer, (Manchester, UK, 10-12 April 1996), 99-108.
8. F. King, "Development of Practical High Temperature Magnesium Casting Alloys", Magnesium alloys and their applications, Ed. K.U. Kainer (Weinheim, Germany: Wiley-VCH, Verlag GmbH, 2000), 14-22
9. B. Bronfin et al., "Die Casting Magnesium Alloys for Elevated Temperature Applications", Magnesium Technology 2001, Ed. J.N. Hryn (Warrendale, PA: TMS, 2001), 127-130.
10. E. Aghion et al., " Dead Sea Magnesium Alloys Newly Developed for High Temperature Applications", Magnesium Technology 2003, Ed. H.I. Kaplan, (Warrendale, PA: TMS, 2003), 177-183. .
11. J. Bohme et al., "Audi Hybrid Magnesium Technology – A new approach to the lightweight engine block", 61st Annual World Magnesium Conference, New Orleans, May 9-12, 2004 .
12. S. Schumann, "Future Use of Magnesium in Cars – Potential and Prerequisite", 12th Magnesium Automotive and User Seminar, Aalen, 13-14 September, 2004.
13. P. Vert et al., "Comparative Evaluation of Automotive Oil Pans Fabricated by Creep Resistant Magnesium Alloy and Aluminum Alloy", SAE Technical Paper 2004-01-0658, Detroit, MI, 2004

2005-01-0336

The Concept and Technology of Alloy Formation During Semisolid Injection Molding

Frank Czerwinski
Development Engineering, Husky Injection Molding Systems Ltd.

ABSTRACT

The application of semisolid technologies for processing of magnesium alloys is described. The benefits of using the semisolid state and processing capabilities of Husky's thixosystem are introduced. Our main attention is focused on exploring Thixomolding® for the generation of alloys by the mixing and partial melting of particulate precursors with different chemistries. Experiments with magnesium-based precursors revealed that the partition of alloying elements between the liquid matrix and remaining primary solid as well as the microstructure of created alloys were controlled by the processing temperature.

INTRODUCTION

After thirty years since its discovery, semisolid metal processing is establishing itself as a viable technique in producing high integrity components that compete with those made by conventional techniques. The advantages of processing metallic alloys in a semisolid state are attributed to the globular solid particles which control their thixotropic properties at high temperatures and reduce the content of dendritic forms after subsequent solidification.

Injection molding provides high flexibility in terms of the processed solid contents [1]. This feature is attributed to the fact that it combines the slurry making and component forming operations into one step. Moreover, the slurry is accumulated in the direct vicinity of the mold gate and does not loose its heat before reaching the mold. Technologies, which generate globular forms during the reheating of preformed billets, such as thixocasting or thixoforging, attempt to operate at high solid contents with typical values around 50%. Similarly, rheo-techniques, creating globular forms during the solidification of completely liquid alloys, implement agitation until relatively low temperatures of the solidus-liquidus range. This is achievable because, while dendritic features create a solid skeleton within the entire alloy volume roughly at 15% solid, globular forms allow free flow and agitation up to almost 50% solid. As opposed to major semisolid technologies, injection molding allows for the processing of slurries within a wide range of solid fractions, with present commercial Thixomolding® applications focused on the range from 5 to 30%.

In addition to the conventional route of net shape forming, the injection molding system offers a number of unique processing opportunities. First, the molded alloy is tightly closed in the machine barrel, filled with an inert gas of argon to protect it against reactive oxygen. Then, the barrel's content is precisely heated by external source and stirred by the rotating injection screw. All these features create a system capable of generating specific alloys directly before injection [2]. Selected aspects of the theory and technology of alloy generation in a semisolid state using injection molding are presented in this paper.

ADVANTAGES OF SEMISOLID PROCESSING

A distinction should be made between the several major ways how the semisolid state affects the process and component properties. The frequently quoted processing benefits, related to reduced operating temperature, are universal for all alloys: lower energy consumption, no handling of liquid metal, better yield from the raw material due to lower oxidation, evaporation or other losses related to melt overheating and longer die life. The second part of the benefits arises from an improvement in the product's properties. The usual assumption that the benefits of semisolid processing arise exclusively from the flow behaviour of the partially solidified metal is, therefore, a simplification.

The influence of the semisolid alloy on the component's properties is complex, and the major factors are listed in Fig. 1. First, the slurry affects product integrity throughout reduction of porosity. The turbulent flow of a liquid alloy into a mold can result in the entrapment of air and mold gases into the melt which, in turn, may translate into micro and macro porosity (Fig. 2a). The smooth flow of the semisolid slurry allows for minimizing these defects. Similarly, a reduced liquid content within

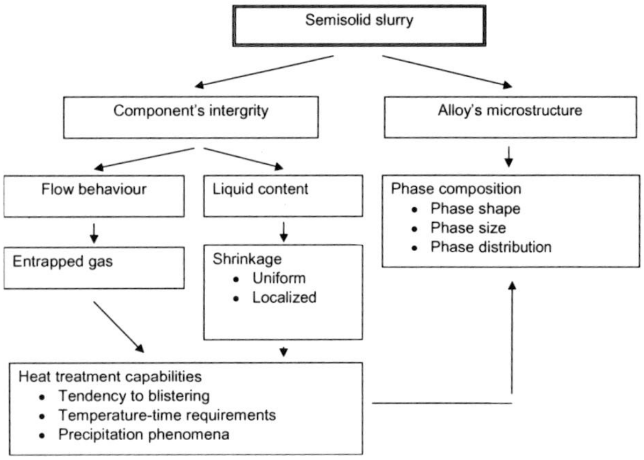

Figure 1. A role of the semisolid state in controlling the component's integrity and alloy's microstructure.

the semisolid slurry leads to lower shrinkage porosity, typical for solidification of superheated melts (Fig. 2b), which not only improves the material properties, but also allows achieving near-net shape capabilities. High internal integrity makes a component heat-treatable. As was the case for general benefits mentioned above, the enhanced component's integrity has a universally positive effect for all alloys.

In addition, a presence of globular solid particles within the alloy slurry essentially changes the alloy's microstructure after solidification. The typical morphological differences are portrayed in Fig. 2c. The new features, unique for semisolid processing, are unmelted globules with a size below 100 μm. The globules are surrounded by a solidification product of the liquid fraction. The primary solid leads to a segregation of chemical elements which translates to a new distribution of phases and microstructural components. It is clear that the presence of an α-Mg rich unmelted solid will affect the composition of the liquid and its tendency to the formation of the intermetallic phases, e.g. the $Mg_{17}Al_{12}$ compound in Mg-Al or Mg-Al-Zn alloys. This phase controls the properties of magnesium alloys, including strength, ductility and corrosion resistance [3,4]. The influence of globular solids on direction of changes in alloy properties is not obvious and should be evaluated for individual chemistries. Heat treatment capability allows for modifying the as-cast microstructure to further improve the alloy properties. The present literature does not clearly separate between the structural factor and the component-integrity factor and their contributions to the improvement in properties.

THIXOSYSTEM CHARACTERISTICS

The new generation unit, with a clamp force of 500 tons, and based on Hylectric and Quadlock machines, is shown in Fig. 3. It is a part of the system which includes, additionally, a high-speed top-entry robot for part removal and die lubricant application, material loading equipment, a part cooling conveyor and mist filtration, as

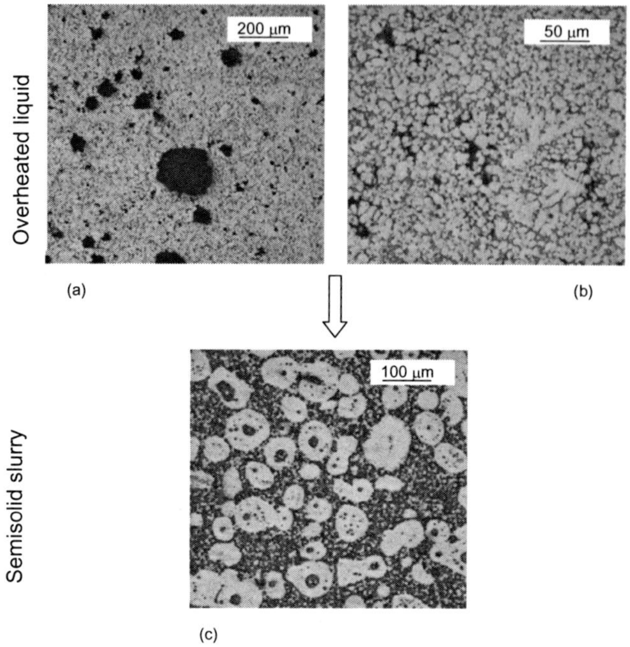

Figure 2. Microscopical images showing differences between die casting from a superheated melt (a,b) and injection molding from a semisolid state (c): (a) AM60B, entrapped gas porosity; (b) AZ91D, shrinkage porosity, dendritic structure; (c) AZ91D-globular structure.

well as mold lubricant mixing and distribution. The machine barrel has a diameter of 70 mm, a length of approximately 2 m and its temperature profile is maintained by coaxial style heating coils, imbedded into grooves, machined in the outer diameter of the barrel. The barrel heaters are grouped into several independently controlled zones. Additional heater zones are situated along the barrel head and nozzle adapter. Molds are preheated typically to 200°C and the alloy is injected at a screw velocity up to 6 m/s. As described previously [1,5], the hopper is filled with cold feedstock which is heated and conveyed within the machine barrel while protected against oxidation by argon gas. Upon accumulating the required shot size in the front of the non-return valve, the screw is accelerated forward, injecting the alloy through the sprue and gates into the mold cavity. The slurry velocity during injection depends on the size of the mold's gating system.

The major parameters controlling injection molding include: (i) cylinder barrel temperature; (ii) screw rotation speed; (iii) injection velocity; (iv) injection pressure and (v) mold temperature. Although all of them affect product quality, the barrel temperature, or more precisely its profile, is of critical importance since it affects the transformations in the semisolid or liquid state.

Figure 3. The general view of Husky thixosystem with a clamp force of 500 tons.

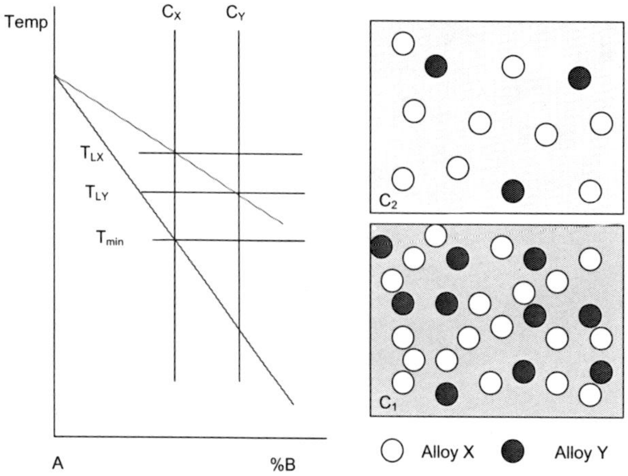

Figure 4. The concept of alloy generation by semisolid state mixing of two thixotropic slurries X and Y.

THE CONCEPT OF MIXING THIXOTROPIC SLURRIES

The simplified mechanism of alloys' formation during the semisolid mixing of two alloys of the same elements can be deduced from a comparison of the phase diagram and a selection of room temperature microstructures (Fig. 4). Since the alloy is conveyed within a barrel by an injection screw, there is an intense mixing and redistribution of chemical elements in a liquid state. During solidification/melting considerations it is generally assumed that, at the same time, there is negligible diffusion in a solid state. Another assumption is that the formation of an alloy is possible when both precursors coexist in a semisolid state. This condition was used during the selection of the lowest mixing temperature, marked as T_{min} in Fig. 4.

According to the hypothetical phase diagram (Fig. 4a), when alloy X entered the semisolid state, the melting of Y was already well advanced. As a consequence, the predominant portion of the liquid phase, which is rich in B, was derived from the alloy Y. For the two alloys considered here, at all molding temperatures, X contributed more solids while Y supplied more liquid. While increasing the preheating temperature, the B content in the liquid was reduced due to a solution of the solid phase rich in A. Assuming equilibrium conditions, represented by the phase diagram, at temperatures of the semisolid coexistence of both alloys, the compositions of solid and liquid derived from both alloys are the same. The short molding time, however, does not allow for substantial changes in the chemistry of the solid state, which remains close to its initial stage; typical for a completely solid alloy. At temperatures equal to or higher than the liquidus of Y, marked as T_{LY}, this precursor became completely liquid and all solids were derived from the X chips. Above temperature T_{LY}, the Y alloy contributes exclusively liquid with the same chemistry as that of the original solid state. The liquid is then diluted in B by mixing it with the A-rich liquid originating from the X source.

The above analysis is further complicated by the particulate morphology of the feedstock used here. Of key importance during semisolid mixing is the melting behaviour of mechanically comminuted chips. As documented previously [6, 7], the cold-worked chips, after exceeding the solidus, transform to globular solid particles suspended in the liquid, rich in lower-melting point metal. The created alloy is formed, therefore, by mixing the solid globules and the liquid phase originating from both grades of chips (Fig. 4, right side). The microstructure does not represent the homogeneous mixture of both precursors, but it is selectively influenced by either X or Y, depending on temperature. The lower the processing temperature, the higher the contribution of Y to the alloy matrix and X to the unmelted solid.

TECHNOLOGICAL VERIFICATION USING MG-AL-ZN AND MG-AL PRECURSORS

In order to verify the semisolid mixing concept, two commercial magnesium alloys, AZ91D and AM60B, in the form of mechanically comminuted chips, were used (Fig. 5, top). Both alloys contain Al as the major alloying element and the addition of Zn in AZ91D represents an advantage since it may be used as a marker to track the solid origin. After the mixing of two particular heats of AZ91D and AM60B grades in a volume ratio of 1:1, the alloy created had an overall chemistry of Mg-7.5%Al-0.35%Zn. Since it was suspected that the properties may not be entirely typical for this composition but may also be affected by the processing temperature, the alloy is referred to further as MIXTURE.

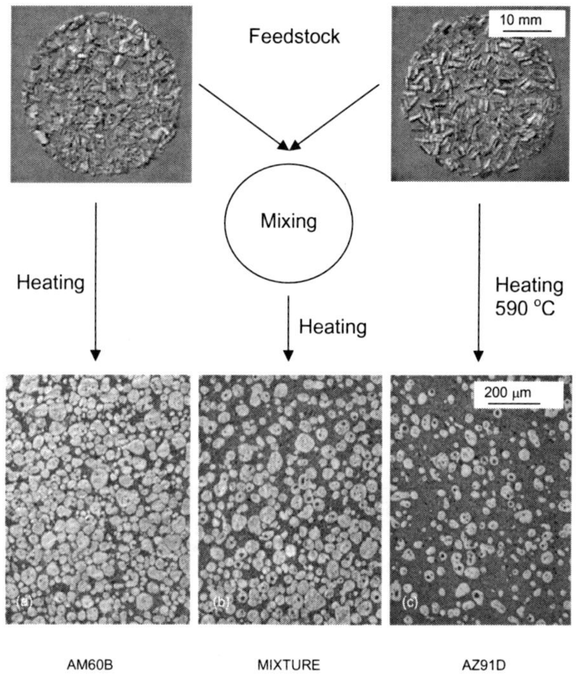

Figure 5. The experimental verification of the alloy generation concept using AZ91D and AM60B grades: the morphology of chipped particulates (top) and resultant microstructures after preheating to 590 °C (bottom).

MELTING DIFFERENCES OF AZ91D AND AM60B

Two precursor alloys, AZ91D and AM60B, have approximately 20 °C difference in liquidus temperatures. The solid content f_s was calculated according to Scheil's equation:

$$f_s = 1 - \{(T_m - T)/(T_m - T_L)\}^{-1/(1-Ko)} \quad (1)$$

where T_m is the melting temperature of the pure metal, T_L is the liquidus temperature of the alloy and K_o is the equilibrium distribution coefficient. The graphical representation of equation (1) for the AZ91 chemistry is discussed in Ref. 6.

MICROSTRUCTURAL DETAILS

Due to a deviation from the equilibrium state, both AZ91D and AM60B alloys, under typical solidification conditions, contain the $Mg_{17}Al_{12}$ phase. The phase forms by an eutectic reaction during sufficiently rapid cooling from the liquid as a result of coring. The same phenomenon is expected to take place within alloys obtained by AZ91D and AM60B mixing.

A selection of the microstructures of AZ91D, AM60B and the alloys generated by their blending are shown in Fig. 5. Under magnifications of optical microscopy, the microstructure of blended alloys was similar to those described previously for elemental precursors [1,5]. No obvious features within the matrix and globules were distinguished to link their origin with either the AZ91D or AM60B grade. While comparing the solid content in Fig. 5 with calculations based on Scheil's equation (1), some discrepancy may exist since the alloy melting within the injection molding system is essentially different from that within the furnace. The residency time of the alloy within the injection barrel is very short, of the order of 100-300 s, and the amount of heat transferred to the feedstock depends on the conveyance rate. As a consequence, there may be no temperature equilibrium between the barrel walls and its content. Short residence time, moreover, does not homogenize the microstructure of the magnesium feedstock, known for its segregation phenomena.

The microstructures in Fig. 5 emphasize the differences in the melting temperatures of AZ91D and AM60B precursors. During molding at the 590 °C cycle, both precursors were in a semisolid state and the created alloy exhibited 53% solid (Fig. 5b). The mixing scenario changed after raising the temperature to a 600 °C cycle, when the AZ91D alloy became practically molten. The AM60B alloy separately molded according to this cycle contained 49% solid and the alloy generated after their mixing reached 15% solid. A further increase in temperature led to mixing of fully liquid AZ91D with semisolid AM60B.

Some amounts of zinc are carried from AZ91D to the created alloy and its influence on phase composition should be assessed. Under equilibrium conditions up to 4% of Zn, the created alloy should not form any new phase beyond those present in AZ91D and AM60B. If zinc exceeds this amount, a three-phase region is entered involving the ternary intermetallic phase ϕ. This compound creates an eutectic reaction at a temperature of about 360 °C. According to X-ray measurements conducted on AZ91D thixomolded alloy, α-Mg and $Mg_{17}Al_{12}$ were identified as the only phases; the same should apply to the created alloy [2].

It also appears that, for an average granule size of the order of 60-80 μm, solid-state diffusion is difficult to neglect. Assuming that the diffusion coefficient of Al in Mg at 600 °C equals 3.33×10^{-12} m^2s^{-1} [7] and for the processing time of 180 s, the root mean square displacement of Al atoms equals 25 μm. Enrichment of the globule surface in Al, observed in fact during EDX analysis, would reduce the solidus location and cause melting. The practical implication should be solid-size reduction after an increase in processing temperature, which is not the case during injection molding. The ambiguity is created since the eventual melting, due to a diffusional change in chemistry, is superimposed on other size evolutions within thixotropic structures, caused by the interaction between particles leading to coagulation, spheroidization and Oswald ripening. As a consequence, particle size reduction may be compensated.

CONCLUSIONS

The mixing of partially melted particulate precursors with different chemistries leads to formation of new alloys with microstructure selectively affected by each precursor. According to this concept, a critical role is played by the thixotropic structure of the semisolid slurry which originated from the particulate nature of the feedstock. The experimental verification revealed that the injection molding technique is suitable for the purpose of alloy generation by semisolid state mixing.

REFERENCES

1. F. Czerwinski, Assessing Capabilities of Thixomolding System in Semisolid Processing of Magnesium Alloys, International Journal of Cast Metals Research, Vol. 16, 389-396, 2003.
2. F. Czerwinski, The Generation of Mg-Al-Zn Alloys by Semisolid State Mixing of Particulate Precursors, Acta Materialia, Vol. 52, 5057-5059, 2004.
3. F. Czerwinski, A. Zielinska-Lipiec, P.J. Pinet and J. Overbeeke, Correlating the Microstructure and Tensile Properties of a Thixomolded AZ91D Magnesium Alloy, Acta Materialia, Vol. 49, 1225-1235, 2001.
4. F. Czerwinski, The Oxidation Behaviour of an AZ91D Magnesium Alloy at High Temperatures, Acta Materialia, Vol. 50, 2639-2654, 2002.
5. F. Czerwinski, Injection Molding Magnesium Alloys, Advanced Materials and Processes, Vol. 160, issue 11, 31-33, 2002.
6. F. Czerwinski, On the Generation of Thixotropic Structures during Melting of Mg-9%Al-1%Zn Alloy, Acta Materialia, Vol. 50, 3265-3281, 2002.
7. F. Czerwinski, The Microstructural Development of Mg-9PctAl-1PctZn Alloy during Injection Molding, Metallurgical & Materials Transactions, Vol. 33A, 2963-2972, 2002.

CONTACT

Dr. Frank Czerwinski, Development Engineering, Husky Injection Molding Systems Ltd., Bolton, Ontario, L7E 5S5 Canada, tel. 905 951-5000 ext. 3263, fax 905 951-5365, email fczerwinski@husky.ca, web www.husky.ca

2005-01-0337

Magnesium Engine Cradle – The USCAR Structural Cast Magnesium Development Project

Naiyi Li
Ford Motor Company

Richard Osborne
General Motors Corporation

Bruce Cox
DaimlerChrylser Corporation

Donald Penrod
Manufacturing Services and Development Inc.

ABSTRACT

The Structural Cast Magnesium Development Project is a jointly sponsored effort by the US Department of Energy (DOE) and the US Council for Automotive Research (USCAR) Automotive Metals Division (AMD) to identify and resolve technical and manufacturing issues that limit the light weighting opportunities of applying large-scale structural cast magnesium automotive components. This project, which began in the end of year 2001, comprises General Motors, Ford, DaimlerChrysler and thirty-four other North America companies and organizations. The project has its overall objective set to determine the technical feasibility and practicality of producing and implementing a *one-piece front engine cradle casting*. This paper provides an overview of the project scopes and up-to-date accomplishments. The engine cradle design criteria are introduced in various aspects including magnesium alloys property database development, component design and analysis, corrosion protection and coating strategy, assembly constraints, crashworthiness assessments, noise vibration harshness performance, and durability. To assess its manufacturability of the engine cradle, a large-scale structural cast, the project adopts two casting processes, i.e. high pressure die cast and low pressure permanent mold. With the support of science and technology, the project activities also focus on developing "tools", 1) numerical simulation model that predicts mold cavity fill and casting solidification, 2) material simulation models that predict cast component monotonic and cyclic properties, and 3) non-destructive evaluation equipment, procedures and process sensors. The ultimate goals of the project are to enhance the casting processes and large-scale structural automotive products, and to promote the magnesium infrastructure in North America.

INTRODUCTION

In 1994, the US Department of Energy (DOE) and the United States Council for Automotive Research (USCAR) together created the Partnership for a Next Generation of Vehicles (PNGV) with the goal of developing a "Supercar" which could attain 80 mpg. It was quickly recognized, that weight reduction would be a necessary enabler for this aggressive increase in fuel efficiency from the current car CAFE levels of 27.5 mpg. The Automotive Materials Partnership (USAMP), a new partnership was created to manage this aspect of the Supercar program. With the USAMP support, a five-year project entitled "Design and Product Optimization of Cast Light Metals" (CLM) was initiated in 1995, with the goal of developing the knowledge and technology to significantly reduce the vehicle weight of chassis, body and interior components and subsystems through the application of cast light metal structural components. The CLM deliverables, which included few magnesium material property data, were documented in April 2000. [1]

With the merits of magnesium, it has been shown that a wider vehicle application of cast magnesium offers a promise of nearly 100 Kg of mass reduction [2, 3]. Currently, US automakers do not produce any of the cast magnesium components deployed in their vehicles. A relatively small industry base supplies them, which usually is not sufficient in size to conduct major R & D tasks necessary for cost-effective and aggressive magnesium applications. Further, aggressive vehicle weight saving applications require high to moderate strength, ductile magnesium castings having the ability to function in cyclic loading, elevated temperature and harsh corrosive environments. To widen the application of ductile magnesium castings it will be necessary to improve our scientific knowledge and provide comprehensive design input, optional casting processes, improved corrosion resistance, soft failure (fail-safe) mode analysis and joining technologies. The Structural Cast Magnesium Development Project was initiated and begun in the end of year 2001 as a jointly sponsored effort by DOE and USAMP Automotive Metals Division to identify and resolve technical and manufacturing issues that limit the light weighting opportunities of applying large-scale structural cast magnesium automotive components. This project team comprises General Motors, Ford, DaimlerChrysler and thirty-four other North America companies, universities and organizations.

The SCMD has been conducted in two parallel tracks with specific objectives, i.e. industry and science tracks, respectively. Along the industry track, this project is intended to significantly improve the Big 3 and supplier industry knowledge/expertise of how to manufacture and apply structural magnesium cast components and resolve critical technical and manufacturing issues that limit the large-scale application of structural cast magnesium automotive components. On the science track, this program aims to leverage research work to be developed by the National Laboratories. The project activities focus on developing the science and technology to implement a production intent one-piece front engine cradle casting with the following R&D activities:

- Develop enhanced magnesium alloy property database and cast component.
- Develop baseline cost model for production intent magnesium cradles.
- Compile and develop comprehensive cast component design guide.
- Improve numerical modeling to correlate die casting fill and solidification and its casting properties.
- Enhance process monitoring, feedback controls and non-destructive techniques that improve the casting quality assurance.
- Develop magnesium constitutive property and its numerical modeling techniques to predict mechanical properties throughout cast components.
- Provide cost effective solutions to atmospheric and galvanic corrosions.
- Investigate robust joining techniques and materials.
- Develop component and vehicle level testing to validate durability of cast magnesium cradle.
- Provide technology and commercial transfer throughout the automotive value chain.

Note that the production intent component ideally to be on pre-competitive basis was identified to be magnesium engine cradle or subframe shown in Figure 1 for the 2006 GM's Corvette high performance Z06 model. In essence, there is a "fast-track industry aspect" to the SCMD project that will convert an existing (production) aluminum engine cradle to magnesium. The reader may find additional product point of view in a separate paper. [4] The main purpose of this paper is to provide an overview of the project scopes and up-to-date accomplishments, particularly in the material technical aspects.

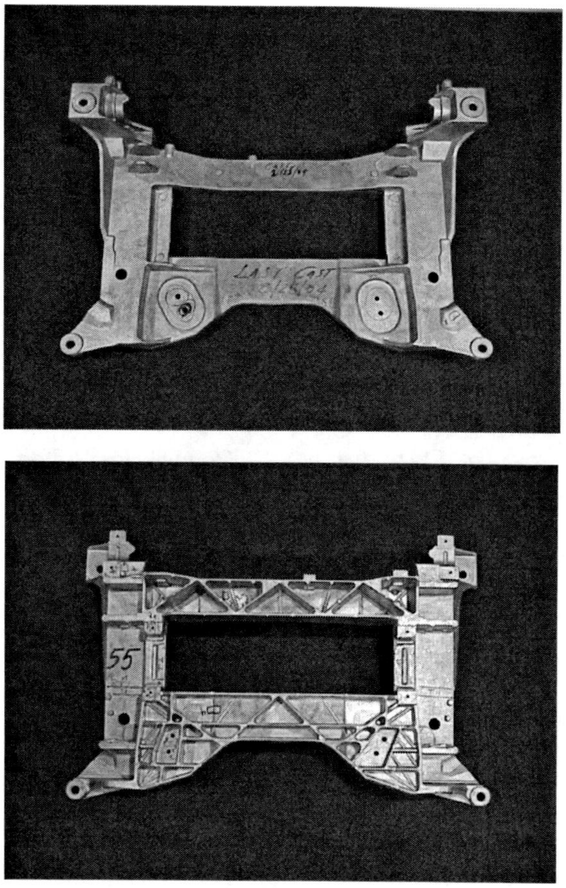

Figure 1. Corvette 2006 magnesium engine cradle

The scientific objectives of the SCMD project were three fold: (1) identify commercial magnesium alloys to meet functional requirements of the magnesium cradle, (2) assess casting processes to produce lowest cost, high integrity, ductile, structural castings that are cost competitive with aluminum, and (3) Investigate and understand the science of magnesium galvanic and atmospheric corrosion behavior by evaluating corrosion behavior of common alloys and processes and developing suitable low cost corrosion protection system to ensure durability of the cradle casting. These objectives are embodied in the tasks described in the following,

MAGNESIUM ALLOYS

Based on the statement of requirements, the magnesium cradle must be able to withstand normal operating temperatures ranging from –50 to 125 degree C. The commercial magnesium alloys considered for the converted magnesium cradle include AM60, AM50, AZ91 and even alternate creep resistant alloys such as MRI 202S, MRI 153M, AJ52X, AJ62X and AE1. With the scope of the SCMD program, two casting processes, namely high pressure die casting (HPDC) and low pressure permanent mold (LPPM), were selected to produce, high integrity, ductile, structural castings that are cost competitive with aluminum ones. The alloys and casting processes are listed in the following,

magnesium alloy	material supplier	casting process
AM60B	n.a.	HPDC
AM50A	n.a.	HPDC
MRI 153M	Dead Sea	HPDC
AJ52x	Noranda	HPDC
AJ62x	Noranda	HPDC
AE1	Hydro	**HPDC**
AZ91	n.a.	LLPM
MRI 202S	Dead Sea	**LLPM**

Among the magnesium alloys shown in the table above, the AE1 (MgAl4RE4) alloy, produced by Hydro Magnesium with 4% aluminum, 2% rare earth mischmetal, was determined to be the alloy for the engine cradle produced by the HPDC process. The selection criteria were based on factors such as the high temperature creep and strength properties, castability, room temperature ductility exceeding 8%, etc. The Rare Earth (RE) mischmetal portion of the alloy is typically comprised of Cerium (Ce), Lanthanum (La), Neodymium (Nd) and Praseodymium (Pr) elements. These four elements will be incorporated into the production spectrographic chemistry evaluation, in order to ensure a complete analysis of the MG-AL-RE alloy series. Initial Hydro Magnesium tests results indicate the potential for improvements, over AE42 (MgAl4RE2) alloy, in the areas of high temperature creep and bolt load retention, yield strength, fatigue performance and castability. Note AE42 alloy (MgAl4RE2), another Mg-Al-RE alloy, is primarily comprised of 4% aluminum, 2% rare earth mischmetal and the balance magnesium. Die castings were produced from the AE44 alloy for both the USCAR Structural Cast Magnesium Development (SCMD) Project and the GM Magnesium Front Engine Crossmember production programs.

The detailed material database covering various alloys including AE1 was developed in this program using test specimens exercised from both coupons and components. Its title page and typical working pages containing data are shown in Figure 2. The working pages show the range of properties and data that are contained in the tree architecture that was used to access data sets with users' preferences. Each test result is shown, specimen by specimen, as well as the statistical summary of the test set. All data and graphs are exportable to facilitate use by automotive engineers and researchers. The casting procedures, specimen geometries, and sample histories also contained in the material database.

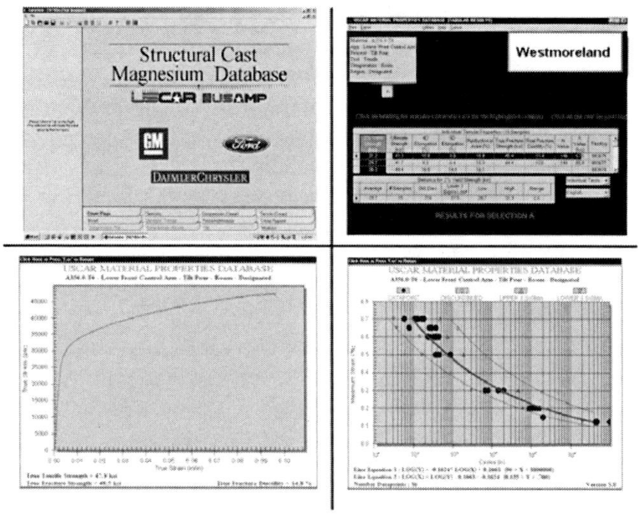

Figure 2. The SCMD material database of cast specimens.

CASTING PROCESSES AND ALLOYS

Magnesium alloy castings can be produced by almost all of the conventional gravity- and pressure-casting methods: sand, low pressure permanent and semi-permanent mold; investment casting; and high pressure die casting. [5] The choice of a casting method for a particular part depends on factors such as the part design geometry, the application, the properties required, the total number of castings required, and the castability of the alloy. Up to date, the most popular and volume-oriented casting process for magnesium is high pressure die casting (HPDC). For the HPDC, the mold, which is clamped together by hydraulic force, is rapidly filled at the level of 5 to 100 ms by injecting the molten metal through runners and narrow gates. The HPDC die designs always offer high cooling or solidification rate, which leads to a find grained casting. Typical advantages of HPDC magnesium casting include thin walls, net shapes, high casting rates, and long die life.

In addition to the SCMD program, other magnesium HPDC automotive research programs include the Mercedes-Benz 7-speed transmission case [6], the BMW Inline (I-6) engine block and Audi inline (I-4) engine block [7]. However, the lack of understanding of solidification behavior for magnesium alloys has become one of major obstacles to expand automotive casting applications. In fact, each alloy has to be readily castable where the term "castability" is defined as a measure of a metal's propensity to be transformed from a molten state into high quality solid casting [8].

A major research effort here in to establish a new magnesium high-pressure die casting technology to support the development of the SCMD cradle casting by aggressively using computer-aided simulation techniques. Although such technology has been successfully applied to the aluminum high-pressure die casting process, it has a number of shortcomings when applied to magnesium. In the cyclic HPDC process, solidification begins in the shot sleeve even before the melt enters the die cavity. This can result in banded defects that may be detrimental to mechanical properties [9]. For example, magnesium alloys are more susceptible to gas porosity than aluminum alloys due to the larger hydrogen gas solubility and to increased oxide inclusion defects during the filling of the die cavity. Other concerns include die sticking (i.e. an alloy's affinity for tool steel), hot tearing susceptibility, element segregation near casting surfaces and the effects of element range variation on different casting parameters (i.e. viscosity). With all these considerations, new advanced simulation technology specifically for magnesium high-pressure die casting is needed to accurately support simulation prior to any mold construction. In the SCMD, the effort supports our casting design and tool development so we have the best opportunity at producing a functional casting without extensive trials and tool modifications. The tasks to enhance magnesium casting simulation capabilities include the following,

- HPDC air entrapment from shot and cavity fill turbulence
- Shrinkage porosity formation
- Oxide inclusions
- Alloy phase for non-equilibrium HPDC process
- Estimate of mechanical properties
- Hot cracking prediction

Integration of a HPDC shot sleeve into the die casting flow simulation as an example shown in Figure 3 provides more accurate and realistic results than by using a velocity boundary condition where the shot sleeve simulation predicted 35%-50% more trapped gas than the velocity boundary simulation while the flow patterns of the simulations exhibit no significant difference. [10]

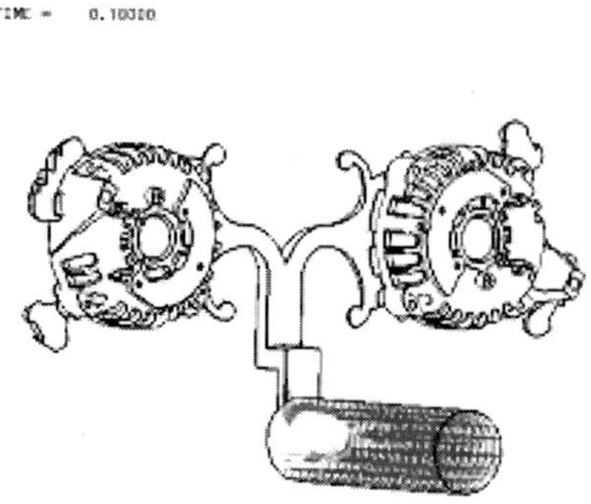

Figure 3. Example of integration of shot sleeve motion in fill analysis

With trapped gas being a large contributor to gas porosity and casting defects, it is important to be able to recognize these problems through accurate simulations. The newly developed trapped-air algorithm [11] can be applicable to large complex thin-wall high-pressure magnesium die casting. For the magnesium high-pressure die castings are thin and its heat capacities are low, the amount and location of the trapped-air has a direct impact upon the quality of the casting. One example of Ford GOR casting shown in Figure 4 compares the percentage of the porosity measured using the combination of metallography and digital image analysis [12], with the X-ray grading, and the predictions of the simulations for the high-porosity regions (>3%) and low-porosity regions. It has shown that there is a good agreement between the experimentally measured values and the porosity distribution trends predicted by the computer simulations. [13]

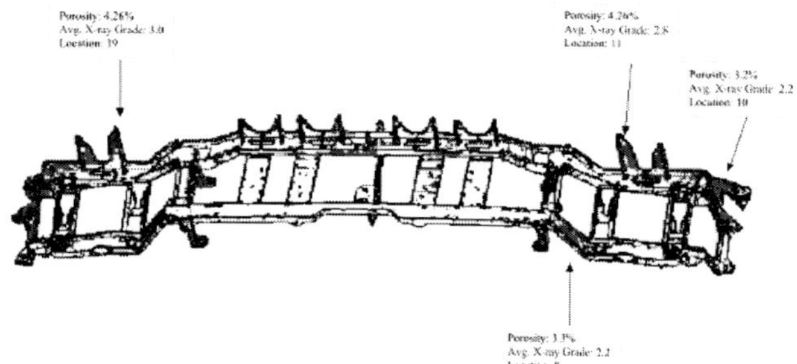

Figure 4. Casting porosity measured vs. computed

A better understanding of the rheological behavior of partially solidified alloys is required in order to reduce the negative effects of banding defects on mechanical performance while optimizing the filling behavior of magnesium alloys during HPDC. Figure 5 shows a typical fill stage of the HPDC simulation compared with as-cast magnesium cradle where the aforementioned research results were incorporated within the finite element based EKK software.

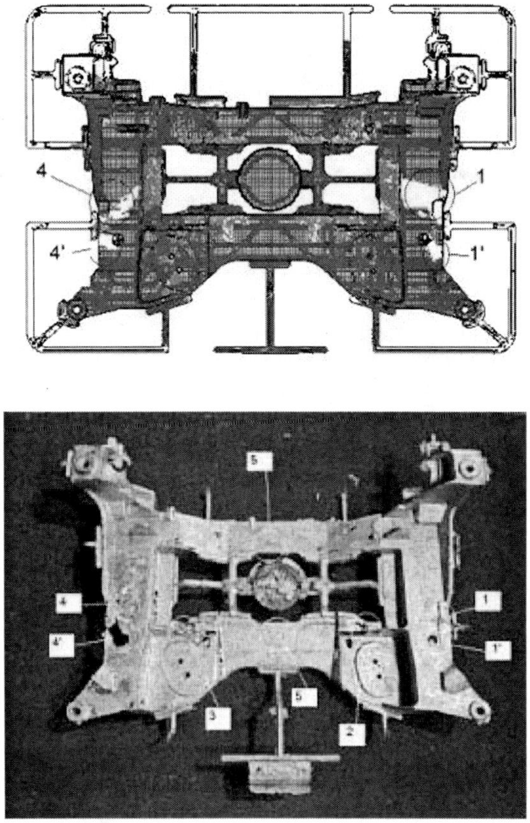

Figure 5. Casting fill analysis vs. HPDC as-cast cradle

In addition to the HPDC process, the SCMD program took a dual-path approach to conduct research to produce and a structural engine cradle via the low-pressure permanent mold (LPPM) process. There is a substantial minority of the program participants that believes low-pressure technology offers better chance of success at comparable to slightly higher costs. This effort is to help develop critical process and production variables for magnesium LPPM casting and determine the potential of the LPPM as a process option in the production of magnesium structural components. The major tasks include,

- Component modeling for fluid flow and solidification
- Production of cast iron low-pressure permanent mold
- Commission of magnesium enabled low-pressure casting machine
- Necessary equipment modifications and supplies required for the production of castings
- Research in the development of magnesium permanent mold coating
- Evaluation of environmental friendly cover gases
- Alloy development and grain refining study
- Castability and production of magnesium alloy cradle castings from alloy AZ91 to MRI 202S
- Non-destructive casting evaluation to determine suitability of product
- Destructive testing of casting for determination of mechanical properties

The LPPM requires cores to be made in chemically bonded sand process using magnesium sand inhibitors. Cradle components will be produced using magnesium alloy AZ91 to monitor tooling modification, alloy development (grain refining) issues, and mold coating development. The Initial LPPM magnesium cradle casting is shown in Figure 6. Pending verification on the AZ91 production, additional production run will be made with the MRI 202-S alloy. As opposed to the HPDC, the relatively slow cooling rates encountered in the low pressure casting process tend to produce coarse microstructures with low ductility. It is believed that casting ductility can be improved by decreasing the grain size with additions of grain refiners. Therefore, the casting research and development is directed in studying the mechanics of existing techniques (including the de-gassing aspects of the additives) and developing more efficient methods of grain refining rather than the traditional zirconium-based approach.

Figure 6. Initial LPPM magnesium (AZ91) cradle casting

Utilization of up-front computer modeling (used for both HPDC and LPPM processes) resulted in significant savings (time and costs) and eliminating casting defects in the initial castings. Based on the prototype castings produced by both process, there is an estimated 35% weight saving in the magnesium cradle vs. the current aluminum production part. This weight reduction (vehicle mass savings) will result in a reduction in fuel consumption and emissions. The significance to casting industry is that the investigation of producing the same casting by the two different processes (HPDC and LPPM) has a great potential for the utilization of existing aluminum casting companies to expand their operations into magnesium, without huge capital and facility expenditures.

MECHANICAL PROPERTY AND MICROSTRUCTURE

Cast Magnesium alloys often exhibit large variability in fracture related mechanical properties such as ductility (elongation) and strength. Significant specimen-to-specimen variations in the ductility of AM50 were observed at all temperatures. These samples have been successfully correlated to the presence of defects through careful qualitative and quantitative fractographic observations and quantitative microstructural data obtained through digital image analysis and stereological techniques. [12] Quantitative correlation between tensile elongation and total area fraction of defects (oxides, pores, flux residues, etc.,) of the fracture surface has already been demonstrated. [1] It was found the area fraction of porosity measured in the fracture surfaces of the tensile test specimens is much larger than the average volume fraction of the porosity in the corresponding bulk microstructure. Therefore, the fracture path preferentially goes through the regions of highly localized clusters of large pores.

To support the modeling of mechanical properties, the program has research efforts made by personnel from Georgia Institute of Technology, Alfred University, Oak Ridge National Laboratory (ORNL) and Sandia National Laboratory (SNL). The goal was to investigate and develop a mathematical model to describe the casting/properties correlation and hopefully to support the mechanical property simulation model work performed by ORNL and SNL to link cast magnesium microstructure to fracture, and monotonic/cyclic mechanical properties. SNL is planning to expand the mathematical multi-scale "*failure model*" developed in the cast light metals (CLM) project to be applicable to magnesium alloys. [1,14] To improve understanding of the crack nucleation and porosity generated in magnesium castings, academia participants has conducted (a) the characterization of crack nucleation and growth (Alfred University) and (b) the investigation of causes of casting defects and the technology to separate gas and shrinkage porosities (Georgia Institute of Technology). With all these efforts, the program expects to develop scientific tools for robust magnesium casting manufacturing.

Radiographic analysis of production parts and test samples will determine discontinuity types and grades at predicted high-stress locations. Lawrence Livermore National Laboratory (LLNL) has investigated the possibility of using fiber optics in-mold thermal monitoring system for HPDC in a high-volume production facility. Investigation and possible elimination of impurities in the metal charged into the furnace will reduce defects. Further, LLNL is investigating to either digitize the ASTM E505 [15] Reference Quality Indicators' radiograph film or actual cast standards to develop new RQIs (Reference Quality Indicators) for cast magnesium as shown in Figure 7. As a result, it provides a quantitative method for component x-ray radioscopic inspection to develop more consistent and measurable standards required for magnesium casting quality assurance.

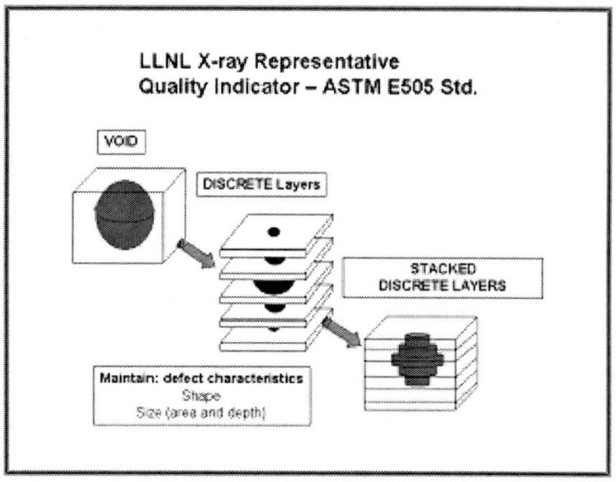

Figure 7. New RQI (Reference Quality Indicators) for cast magnesium

CORROSION SOLUTIONS

In general, one of the main issues impeding the large-scale applications of magnesium alloys in automotive industry is the less than desirable corrosion resistance where galvanic corrosion is often the major concern. Magnesium, often considered as anodic metallic material in the electrochemical activity, is the least noble of all the structural metallic materials as shown in Figure 8.

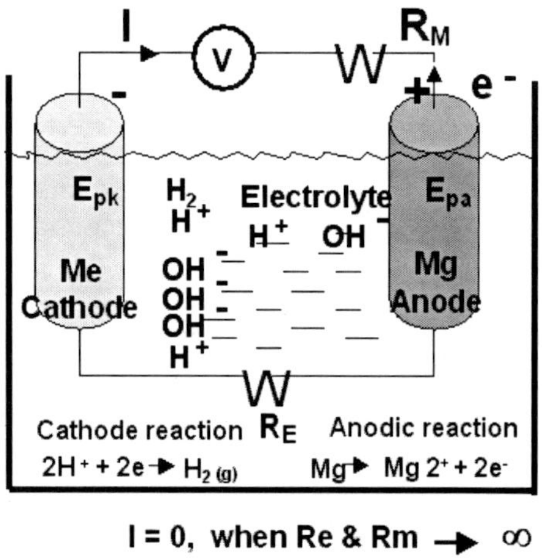

Figure 8. Mechanism of magnesium galvanic corrosion

The corrosion mechanism is usually attributed to micro-galvanic corrosion between the matrix and the more noble or cathodic intermetallic materials. Reducing the electro-potential difference or increasing the ionic resistance between the metals can all effectively enhance the galvanic corrosion resistance. Many types of fasteners and fastener coating have been developed to minimize galvanic corrosion of magnesium components. [16,17,18] The SCMD program has initiated a research project to investigate and understand the science of magnesium galvanic and atmospheric corrosion behavior. [19] With a research partnership with the Materials Technology Laboratory of Natural Resources Canada (CANMET), an effort was made to aim at defining and developing magnesium/fastener galvanic corrosion and corrosion prevention technology as pertaining to magnesium automotive structures, in particular the front cradle. The project focuses on the fastener material and environmentally friendly coatings, magnesium coatings, and the reduction of galvanic corrosion between the fastener and the magnesium structure where the major tasks include the following,

- Identify, select and evaluate commercially available magnesium coating systems and fastener assemblies based on economics and performance via erosion-corrosion testing
- Evaluate coated magnesium panels and fastener/magnesium panel assemblies under various salt spray and submerged salt water conditions.
- Conduct Stress Corrosion Cracking (SSC) susceptibility study
- Optimize coating and fastener assembly combination(s) in terms of durability and economics
- Conduct field test to attain lab test confirmation.

The external corrosion was appraised using the ASTM B117 [20] salt spray test as the coating screen test. Pending the coating result, GM 9540P test standard was chosen as the method for evaluating the galvanic corrosion susceptibility of Mg (coated and uncoated) fastened with various fasteners, M8, M10, and M14. Similar to the front radiator support [17], some intermediate results of magnesium alloy AM60 is shown in Figures 9 and 10. Commercially available coating for Mg alloys that can satisfy aggressive testing conditions established by the industry include chromate-free conversion coatings, such as Alodine 5200 [21] and MAGPASS [22]. Alodine 5200 is an organometallic titanium based primer to be used as a chromate-replacement conversion coating for light metals. This primer is usually followed with one or more topcoats. For this work, an epoxy powder coat was used as the topcoat. New Cr-free conversion coating products such as MAGPASS coating are currently reviewed for their cost-effectiveness. The Alodine 5200 treatment with a powder coat selected for this study has demonstrated effective protection against galvanic corrosion in the testing conditions applied.

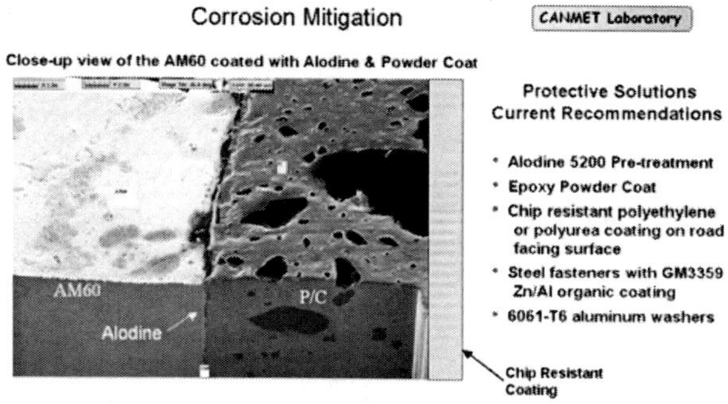

Figure 9. AM60 corrosion mitigation

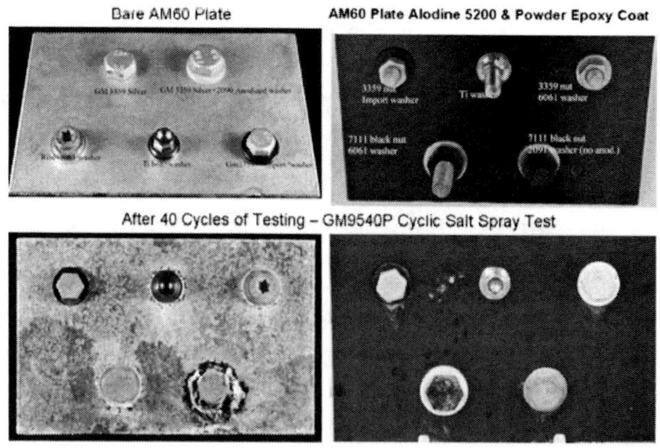

Figure 10. Corrosion salt spray test per GM9540P

The long-term corrosion rate of alternative fastener materials such as Al 6XXX and its compatibility with Mg alloys are also assessed. Galvanic corrosion between fasteners and Mg alloys can be effectively mitigated though the use of appropriate types of washers as long as the alkali generated by the corroding magnesium did not attack the aluminum under immersion or spray conditions. The application of sealants, glues or lubricants can also create an insulating gap and keep water out of the fastener crevices. The application of an epoxy coating or other types of spray coating to cover the fastener and extend onto the surrounding Mg surface can also be envisaged. In this work, washers made of AA 6061 and anodized 6061 can all produce significant protective effects.

JOINING TECHNOLOGIES

Tensile, creep resistance and stress relaxation characteristics of magnesium alloys at elevated temperatures are important for automotive applications [23]. In the past decade, several new creep-resistant magnesium alloys have developed by utilizing alloying additions of Ca, Sr, Re, Zr and Si. The advantages of these newly developed alloys in terms of creep properties are evident compared to the conventional creep-resistant magnesium alloys such as AS and AE alloys. Although strong research activities on developing creep-resistant magnesium alloys have been carried out, very few automotive magnesium alloys with excellent high temperature properties are developed and implemented for commercial vehicles. [24] The SCMD program is committed to gain scientific knowledge of creep theory and its effect of various processes of manufacturing magnesium automotive components, particularly for the AE1 or MRI 202S creep resistant alloys developed recently.

There are two types of testing developed up to date to evaluate compression creep relaxation of Mg alloys. The first one is the compression creep relaxation testing described in ASTM E328 [25]. It provides quantitative information on uniform stress relaxation under constant strain. The second one is the bolt-load retention (BLR) testing under different temperatures and preloads. [26, 27] Though not standardized yet, it is practical and engineering friendly in quantifying material performance. A research effort has been made at CANMET to comprehensively review the stress relaxation and the bolt load retention behavior of Mg alloys. [28] The aim is to provide a basis for a standard BLR test procedure of Mg alloys with common specimen and test fixture.

It is believed that the bolt load retention behavior of Mg alloys depends on both compressive creep property and the compliance of the bolt joints where in general the stress in Mg components at the bolt joints is not uniform. In terms of BLR as shown in Figure 11, the rear attachment joints of the cradle to body structure are the most critical ones in the front cradle application because they are the closest to high temperature regions in the cradle and they are required to be stiffer than that of the front attachment joints. It's worthy to note that the temperature at both the rear and front bolt joint attachments here is about 102°C while the maximum cradle temperature is 125°C at about 8 inches away from the rear bolt joint.

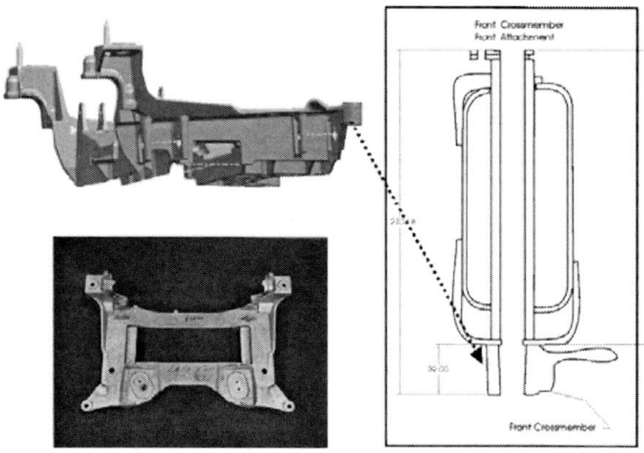

Figure 11. Sections of the critical bolt joint in a front cradle

Aligned with the SCMD program, the research effort has been ongoing to develop joining methodology to ensure absence of galvanic corrosion when attaching magnesium components to dissimilar steel and aluminum components, while ensuring robust mechanical fastened couple at operating temperatures and environments. Recommendations are made for a cylindrical common BLR test specimen based on the results from the literature review and from a survey of typical bolt joints as shown in Figure 12. [28]

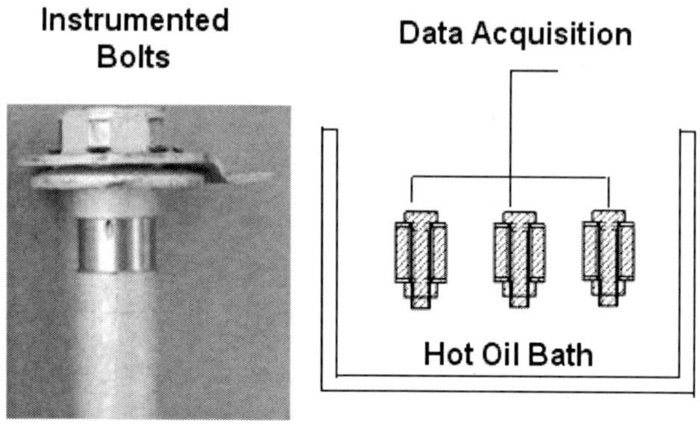

Figure 12. Setup of bolt load retention test

The compressive yield strength is an important material property in castings in terms of bolt-load retention for automotive applications. Tensile-compressive yield asymmetry is used to describe the physical responses of the yield strength and flow stress of a material when it is loaded under tension and compression, respectively. The tensile-compressive yield asymmetry in wrought Mg alloys is a well-established phenomenon due to texture formation and twinning asymmetry [29]. However, cast polycrystalline Mg alloys are generally considered to have random texture and therefore, the yield asymmetry mechanism is not quite the same as that for the wrought Mg alloys. Among the tasks of joining research and technologies, one topic is to study the tensile-compressive yield asymmetry of cast Mg alloys based on their microstructural characterization after deformation. Ultimately the result can be applied to identify the effect of mechanical twinning occurring in cast Mg alloys and help the fastener design and development.

In regarding to the casting process and its effect on microstructures, a study with the AM50 magnesium alloy showed that LPDC process produces much coarser microstructures than HPDC does where the LPDC average grain size is almost one order of magnitude larger, lower strength but better creep resistance than HPDC AM50. Non-equilibrium segregation of Al appearing during both processes affixes to dendritic microstructure in the LPDC samples and to casting cells in the HPDC samples. Tensile-compressive yield and creep asymmetries were shown in AM50, i.e. better deformation resistance in tension than in compression. Microstructural examinations on tested samples show that the deformation mechanisms in AM50 were slip plus twining at temperatures up to 125°C. The examinations indicate the important role of twining in the asymmetry phenomenon. [30] Similar work on AE1 and MRI 202S will be conducted and reported later.

SUMMARY

This paper introduces the USCAR/USAMP/AMD/SCMD program and summarizes the up to date achievements in terms of both scientific research and production intent demonstration and validation, i.e. the Corvette Z06 front engine cradle to help achieve its performance and fuel economy targets. The program efforts not only support science and technology development, but also identify and resolve technical and manufacturing issues that limit the light weighting opportunities of applying large-scale structural cast magnesium automotive components, which provide a weight reduction potential of 35 percent with respect to aluminum. With numerical and analytical tools developed, the project leads to enhance the casting processes and large-scale structural automotive products, and to promote the magnesium infrastructure in North America.

ACKNOWLEDGMENTS

The authors would like to thank their colleagues at Daimler Chrysler, Ford, General Motors for their support in the development and implementation of this program. These colleagues include Paul Bujalski of Daimler Chrysler, Carl Johnson and Jim deVries of Ford, Larry Ouimet, Kasi Goundan and James Quinn of General Motors, Joe Carpenter of the US Department of Energy. The active participation and commitment of all the non-OEM companies and organizations is greatly acknowledged.

The authors acknowledge that this research was supported in part, by Department of Energy Cooperative Agreement No. DE-FC05-02OR22910 (and/or for work conducted before 12-31-01, DE-FC05-95OR22363). Such support does not constitute an endorsement by the Department of Energy of the views expressed herein.

REFERENCES

1. Design and Product Optimization for Cast Light Metals, published and distributed by American Foundry Society, 2001, ISBN 0-87433-246.
2. H. Friedrich and S. Schumann, The Second Age of Magnesium – Research Strategies to Bring the Automotive Industry's Vision to Reality, Proceedings of the 2nd Israeli International Conference on Magnesium Science and Technology, pp 9-26, Dead Sea, Israel, February, 2000.
3. G. Cole, Magnesium Applications in Ford Motor Company's PNGV Project P2000 Contour, Proceedings of the 54th International Magnesium Association, Toronto, Canada, June 1997.
4. J. Aragones, K. Goundan, S. Kolp, R. Osborne, L. Ouimet, and W. Pinch, Development of the 2006 Corvette Z06 Structural Cast Magnesium Crossmember, SAE paper 2005-01-0340, Michigan, April 2004.
5. Casting, volume 15, ASM Metals Handbook, ASM International.
6. A. Barth, Mg-Application in the 7G Tronic Gear, Proceedings of the 61st International Magnesium Association, New Orleans, May 2004.
7. J. Bohme, Audi Hybrid Magnesium Technology, Proceedings of the 61st International Magnesium Association, pp 108-116, New Orleans, May 2004.
8. R. Beals, et al., Fundamental Research Needs for the Magnesium Powertrain Cast Components Project, Magnesium Technology, pp 11-17, TMS, Raleigh, March, 2004.
9. L. Backerud, E. Krol, and J. Tamminen, Solidification Characteristics of Aluminum Alloys, v.2 Foundry Alloys, AFS, 1990.
10. S. Elmore, M. McLaughlin, C.W. Kim and G. Backer, Integration of Shot Sleeve Motion into the Die Casting Filing Analysis, pp 11-14, NADCA Transactions, Indianapolis, September 2003.
11. G. Backer, M. Ranganathan, J. Heimsch, M. McLaughlin and C.W. Kim, Simulation of Flow-Induced Gas Entrapment and its Effect on Porosity in Aluminum Die Castings, 10-16, NADCA Transactions, Cincinnati, Oct 2001.
12. A. Balasundaram and A.M. Gokhale, Digital Image Analysis Technique for Characterization of Shrinkage and Gas (Air) Porosity in Cast Magnesium Alloys, Magnesium Technology, TMS, New Orleans, Feb. 2001.
13. M. McLaughlin, C.W. Kim, G.P. Backer, J.E. Prindiville, A.M. Gokhale and S.G. Lee, An Application of Trapped Air Analysis to Large, Complex High Pressure Magnesium Casting, Die Casting Engineer, NADCA, Sept. 20 04.
14. A.M. Waters, H.E. Martz, K.W. Dolan, M.F. Horstemeyer and R.E. Green Jr., Statistical Void Analysis from CT Imagery with Applications to Damage Evolution in an AM60B Magnesium Alloy, 15th World Conference on NonDestructive Testing, Rome, Italy, October 2004.
15. ASTM-E505, Standard Reference Radiographs for Inspection of Aluminum and Magnesium Die Castings.
16. D. Hawke and T. Ruden, Galvanic Compatibility of Coated Steel Fasteners with Magnesium, J. of

Materials & Manufacturing, vol.104, pp 358-362, 1995.
17. G. Gao, G. Cole, M. Ricketts, J. Balzer and P. Frantzeskakis, Effects of Fastener Surface on Galvanic Corrosion of Automotive Magnesium Components, Proceedings of the 2nd Israeli International Conference on Magnesium Science and Technology, pp 321-338, Dead Sea, Israel, February, 2000.
18. J. Balzer, P. Dellock, M. Maj, G. Cole, D. Reed, T. Davis, T. Lawson and G. Simonds, Structural Magnesium Front End Support Assembly, SAE paper 2003-01-0186, Michigan, March 2003.
19. W. Zheng, C. Derushie, R. Zhang and J. Lo, Protection of Mg Alloys for Structural Applications in Automobiles, SAE paper 2004-01-0133, Michigan, March 2004.
20. ASTM-B117, Standard Practice for Operating Salt Spray (Fog) Apparatus.
21. D. Davis, Industrial Paint & Powder (USA), vol. 74, no. 6, pp 34-36, June 1998.
22. P. Kurze, T. Singe and J. Diesing, Corrosion Protection of Magnesium Materials, Metalloberf-lache [Metalloberflache], vol. 54, no. 9, pp 22-24, Sept. 2000.
23. T.K. Aune and T.J. Ruden, High Temperature Properties of Magnesium Die Casting Alloys, SAE paper 920070, pp 1-7, Michigan, Feb 1992.
24. H. Hu, A. Yu, N. Li and J. Allison, Potential Magnesium Alloys for High Temperature Die Cast Automotive Applications: A Review, J. Materials and Manufacturing Processes, vol. 18, no.5, pp 687-717, 2003.
25. ASTM-E328, Standard Test Methods for Stress Relaxation for Materials and Structures,
26. F.C. Chen, J.W. Jones, T.A. McGinn, J.E. Kearns, A.J. Nielsen and J.E. Allison, Bolt-Load Retention and Creep of Die-Cast Magnesium alloys, SAE paper 970325, pp 13-21, Michigan, Feb 1997.
27. S. Loike, K. Washizu, S. Tanaka, T. Baba and K. Kikwas, Development of Lightweight Oil Pans Made of a Heat-Resistant Magnesium Alloy for Hybrid Engines, SAE paper 2000-01-117.
28. S. Xu, J.P. Thomson and M. Sahoo, A Review on Stress Relaxation and Bolt Load Retention of Magnesium Alloys for Automotive Application, Canadian Metallurgical Quarterly, vol. 43, no. 4, pp 489-505, 2004.
29. E.A. Ball and P.B. Prangnell, Tensile-Compressive Yield Asymmetries in High Strength Wrought Magnesium Alloys, Scripta Metall. Mater., vol. 31, pp. 111-116, 1994.
30. S. Xu, J. Li, V.Y. Gertsman, J.P. Thomson and M. Sahoo, Tensile-compressive Yield and Creep Asymmetries of Cast Magnesium Alloys, 16th Canadian Materials Science Conference, Ottawa, June, 2004.

CONTACT

Naiyi Li, nli1@ford.com

Richard Osborne, richard.osborne@gm.com

Bruce Cox, bmc8@daimlerchrysler.com

Donald Penrod, dep@gls3c.com

2005-01-0338

The Application of Magnesium Die Casting to Vehicle Closures

Patrick J. Blanchard, Gregory T. Bretz, Saravanan Subramanian and James E. deVries
Ford Motor Company

Andrew Syvret, Andrew MacDonald and Peter Jolley
Aston Martin

ABSTRACT

During the last decade, advances in magnesium die casting technology have enabled the production of large lightweight thin walled die castings that offer new approaches for low investment body construction techniques. As a result, many OEMs have expressed an interest in magnesium door closure systems due to investment reduction opportunities, coupled with potential weight savings of up to 50%. However, for such applications, product engineers are faced with the challenge of designing for stiffness and strength in crash critical applications with a material of lower modulus and ductility compared to wrought sheet product. Concept designs for side door systems have been presented in the literature, and indicate that structural performance targets can be achieved. However, to date, series production designs feature a multitude of supplementary sheet metal reinforcements, attached to die castings, to handle structural loads. While this approach can still offer performance benefits, the additional cost of tooling and assembly has a negative impact on both overall weight and the business rationale. On the contrary, the magnesium door concepts presented in this paper describe the development of side door systems designed to replace the bulk of sheet metal stampings by a single magnesium die casting. A summary of the design, analysis, prototyping and testing stages is reported, in addition to the development of a series production door system for a 2004 model year vehicle. A review of manufacturing and test results demonstrate how magnesium can be used effectively in the manufacture of low investment, lightweight vehicle closures.

INTRODUCTION

The impact of vehicle mass on fuel economy is well understood and is considered an important factor in the drive to increase corporate average fuel economy. In fact, significant gains were reported in the early 1980s as vehicle curb weights declined on average by 20% [1]. However, weight targets are now under increased pressure as the trend to add new features and functionality to vehicle models continues. Consequently OEMs have examined a wide range of materials and processes in an attempt to satisfy consumer needs while controlling costs. In some cases, simple substitutions of steel for aluminum have been made [2]. However, forming limitations of aluminum alloys often restricts their application. To overcome this, hybrid door systems combining stamping, extrusions and castings have been developed. However, final weight benefits can easily be offset by increases in assembly complexity and variable cost.

Magnesium, on the other hand, has been used successfully for many automotive applications. Properties including low density and high strength-to-weight ratio, make magnesium an attractive material for reducing component mass. Consequently, several structural castings have been reported over the last decade, with the scope of application increasing as larger, more complex die castings are demonstrated [3,4]. One particular application that has been the focus of several programs is the development of a die cast vehicle closure panel [5,6,7]. Many of these proposals have remained as concepts, however, production versions of liftgates and side doors have emerged [6]. Perhaps the most notable of these have been those implemented by Mercedes for their CL and SL models. However, the general trend remains, where by loading of the magnesium casting is limited, and supplementary reinforcements are fastened to the door assembly to meet structural requirements. Once again, this can have a negative impact on the overall weight and assembly cost.

This paper focuses on the development of a new magnesium door inner panel designed to eliminate supplementary stampings through integration of key reinforcements into the die-casting. Hence, the magnesium panel becomes an integral part of the door structure and contributes to the static and dynamic crash performance of the vehicle. A summary of the design, analysis and testing for an early concept is presented, followed by the development of a production intent door system, implemented on the MY2004 Aston Martin DB9. A final weight assessment is provided for both designs in addition to an economic analysis to identify technology costs relative to conventional steel stamping.

CONCEPT DEVELOPMENT

Development of the magnesium door architecture was divided into two distinct phases. Initially, concept prove-out and static durability testing was performed using designs based upon a fully framed side door assembly from the MY2000 Ford Contour. Prototypes were fabricated using gravity sand casting, and a combination of CAE and prototype testing was used for technology evaluation. At this point in the program, the business rationale was also developed for several production volumes using high pressure die casting as the assumed processing route.

Following concept evaluation of the Ford Contour prototypes, a production intent door design was developed at Aston Martin for implementation on the MY2004 DB9. This second phase, or implementation phase of the program, focused on two key areas: crash performance and corrosion protection. For the former, a production die casting tool was machined to manufacture components of representative part quality. For the latter, cyclic corrosion testing was performed to evaluate suitable protection systems for closure applications.

FORD CONTOUR MAGNESIUM DOOR ASSEMBLY

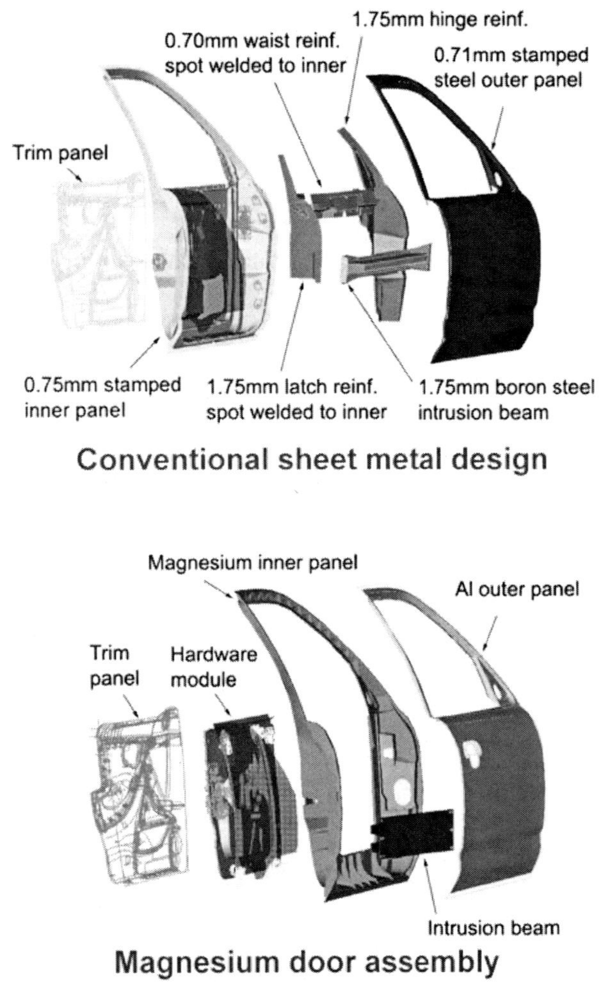

Figure 1. Lightweight magnesium door architecture.

Figure 1 compares the magnesium door architecture to the conventional sheet metal design used in the production version of the Ford Contour. In this early concept, an emphasis was placed on the adoption of a modular architecture. This enabled off line testing of internal door hardware and simplified assembly. Also, the open architecture of the inner panel reduced assembly complexity facilitating loading of the window glass. Latch and hinge reinforcements were integrated into the die casting, while the inner belt line reinforcement was integrated into the thermoplastic hardware module.

As with many automotive applications, closure design is driven primarily by stiffness requirements rather than strength. Consequently, a full complement of design standards such as torsion, drop off, sag and lateral rigidity were used to determine final wall thickness requirements for the die casting. Ultimately, nominal thicknesses between 2mm to 5mm were specified to meet the design criteria. Variations in section thickness over the part are shown in Figure 2.

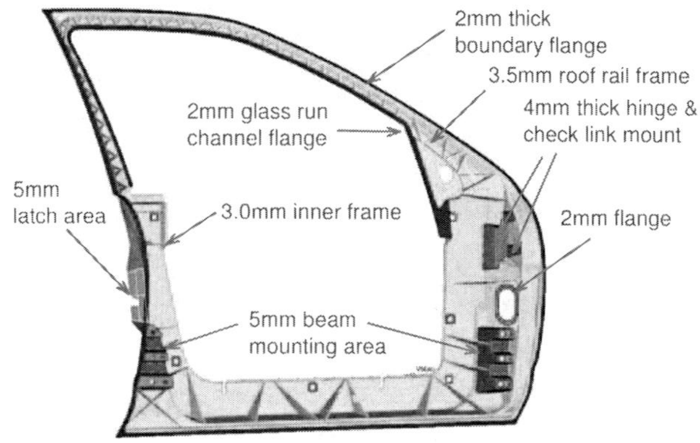

Figure 2. Casting section thickness variations magnesium door inner panel.

The lower portion of the frame was specified as 3.0mm with the exception of the hinge and latch region, which were increased to 5mm. To assist in seal mounting and hemming operations, both the perimeter flange and inner flange of the window frame were set to 2.0mm. The former, directly influencing the final thickness of the hem joint between inner and outer panels. Stiffness of the upper frame was achieved through an increase in wall thickness to 3.5mm combined with liberal placement of ribbing along the entire length of the frame.

	Ford Contour Door-In White Weight Summary				
	Conventional Door Assy		Mg Door Assy		
	Gage (mm)	Mass (kg)	Material	Gage (mm)	Mass (kg)
Door inner	0.75	4.81	AM60	2.00 - 5.00	4.32
Module	-	-	PP/Glass	2.50	1.50
Door outer	0.71	5.47	Al laminate	1.20	1.80
Waist reinf.	0.70	0.40	-	-	-
Hinge reinf.	1.75	1.86	-	-	-
Latch reinf.	1.75	1.02	-	-	-
Beam	1.75	2.00	Composite	3.00	2.00
	TOTAL	**15.56**		**TOTAL**	**9.62**

Table 1. Weight comparison of conventional and magnesium door assemblies.

A final weight comparison of the conventional and new door in white assembly is shown in Table 1. At 9.62kg, the magnesium door assembly offers a weight reduction of 5.94kg (38%) relative to the stamped steel design. Further examination of the data shows the steel inner panel mass, including all reinforcements, to be 8.09kg. This compares with 5.83kg for magnesium inner and hardware module. At 28%, the weight reduction offered by the inner assembly was less than anticipated. However, in practice, the stiffness of the magnesium inner panel was greater than the stamped steel version to enable transfer of the inner waist rail reinforcement to the thermoplastic hardware module. This resulted in some design inefficiencies. Furthermore, the stiffness offered by the magnesium inner panel enabled the use of a low density aluminum/plastic laminate for the outer panel. This offered a considerable weight reduction opportunity of 3.67kg compared to the steel outer panel.

PROTOTYPING AND EXPERIMENTAL VALIDATION

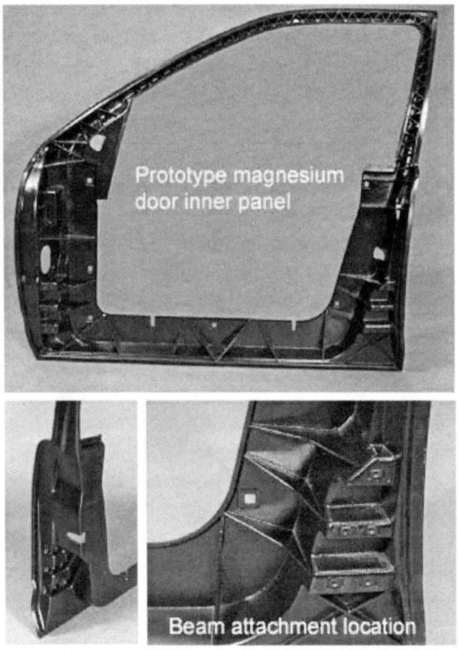

Figure 3. Magnesium door inner panel prototype based upon the MY2000 Ford Contour

A series of door inner panels, as shown in Figure 3, were manufactured for the purposes of prototype testing. Due to lead time and investment constraints, parts were manufactured using gravity sand casting as opposed to high pressure die casting. Differences in metal solidification rates between die casting and sand casting prevented any assessment of ductility or corrosion. However, fully functional door assemblies were assembled for an evaluation of stiffness based criteria. An additional door assembly was subjected to key life testing to prove out the durability of hardware systems to stressed environmental exposure. Note that the inner panel shown in Figure 3 is coated with an epoxy powder coat for corrosion protection.

Test Descriptions	Requirements	CAE in mm	Exp. in mm
Door Torsional Rigidity	<4 mm	2.4	5.7
Frame Lateral Rigidity	<5 mm	3.1	3.4
Door dropoff	<1.25 mm	0.7	0.7
Door sag	<38 mm	5.4	13.3
Door Slam	84000 cycles	14 milln	stopped at 84K

Table 2. Performance predictions vs. experimental testing for the Ford Contour magnesium door assembly.

For production purposes, compliance with over 90 design specifications was required for the door assembly. However, for concept prove-out, the majority of performance evaluations were based upon finite element models to simulate system response for key loading conditions. Subsequent experimental testing was performed for a limited number of tests. This provided an opportunity to highlight any discrepancies in theoretical predictions and provide further insight into the performance of the magnesium door assembly. A review of the results in Table 2 shows that CAE predictions indicate that all five tests meet specification. Experimental testing, on the other hand, did highlight a test failure for torsional stiffness, in addition to an inconsistency between the prediction and experimental result for door sag. In both cases the measured value was over double that predicted. This prompted a closer examination of the modeling and experimental test data to identify the source of the discrepancy. In conclusion, the difference between predicted and experimental data was attributed to a combination of hinge pin clearance and plastic deformation of the prototype hinges used on the magnesium door assemblies. Therefore, it was anticipated that a reduction in both torsional displacement and sag would be realized if production hinges were fitted.

Aside from the torsion result, all of the modeling and experimental data demonstrated the capabilities of the magnesium design, while working within the same design envelope as the steel counterpart. Slam open durability testing also showed no signs of damage to the frame or hardware at three times life (84,000 cycles).

BUSINESS CASE

An economic evaluation was performed to understand the cost implications and impact to future vehicle program teams. The analysis compared the fully accounted cost of the magnesium door assembly for two separate scenarios. The first scenario assumed limited production volumes, hence low automation and capital investment. At the other extreme, the second analysis assumed a high level of automation and investment to identify the suitability of the technology to high volume programs. A summary of the analysis is contained in Figure 4. For low volume applications, the cost benefit is clearly visible. For example, at 10K vehicles per year the total cost of the magnesium door assembly is 13% less expensive than it's steel counterpart. The majority of this advantage being due to part integration and associated reduction in tooling investment. However, as production volumes increase, the overall cost difference between the steel and magnesium door assembly becomes marginal. Indeed, as volumes approach 30k/annum, any investment savings offered by the magnesium door assembly are offset by penalties associated with increased variable cost. Beyond 30K vehicle per year, the most cost effective manufacturing route is offered by conventional sheet metal stamping.

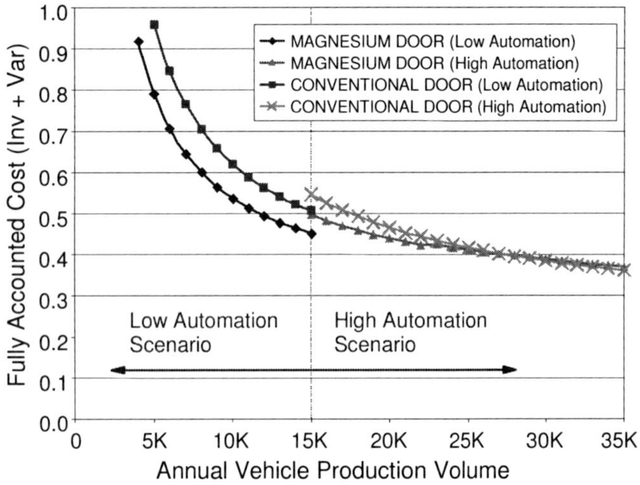

Figure 4. Economic comparison of conventional and magnesium door assemblies for the MY2000 Ford Contour.

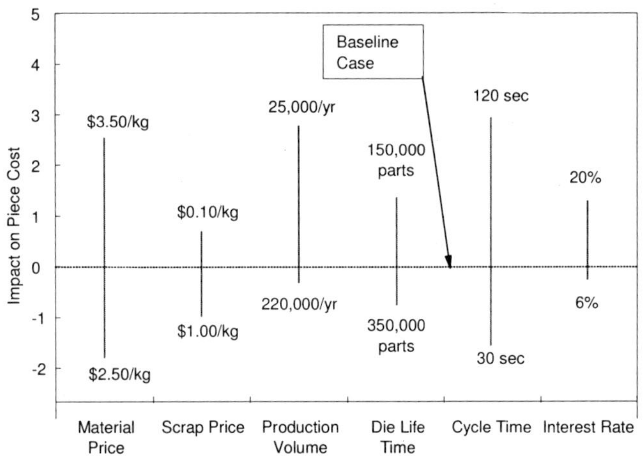

Figure 5. Sensitivity analysis of piece cost to model input data.

The precise breakeven point for the cost analysis will of course depend largely on the sensitivity of the model output to the input data. Figure 5 highlights this point by indicating the impact of key input data on the variable cost of the magnesium assembly. Not surprisingly, material price has a large impact relative to the other variables. It is also worth noting that at the time of this analysis, the baseline assumption for material price was $2.90/kg ($1.32/lb). Therefore, a shift in the breakeven point towards the higher volumes would be expected using today's materials costs. Once, at the higher volume, further economies of scale are minimal. Instead, the single most important input to have a major impact on variable cost is machine utilization. Hence reductions in process cycle time can still offer significant cost savings for higher volume applications.

In terms of technology implementation, the mutual economic benefits offered by competing manufacturing technologies will be influenced by the target application and the extent to which component integration is possible. Hence certain vehicle closures such as liftgates may present a more or less appealing business case compared to side door applications. Other applications of interest would also include rear super cab doors on trucks, in addition to other access panels. Even so, the current analysis would suggest that the front side door assembly still offers potential for significant weight savings at minimal to zero cost, particularly for low to medium volume programs.

TECHNOLOGY IMPLEMENTATION

The magnesium door assembly, based upon the Ford Contour, provided an opportunity to prove out the initial concept and manufacturing feasibility. However, open issues including crashworthiness, and validation of corrosion protection systems, remained as key obstacles to be overcome prior to technology implementation. To address these concerns, a secondary phase of the development program was initiated to complete prove out of the technology using data gathered from components manufactured by high pressure die casting. In practice, this phase of the program also became the implementation phase as the tooling was machined to be production ready assuming a successful outcome to the project.

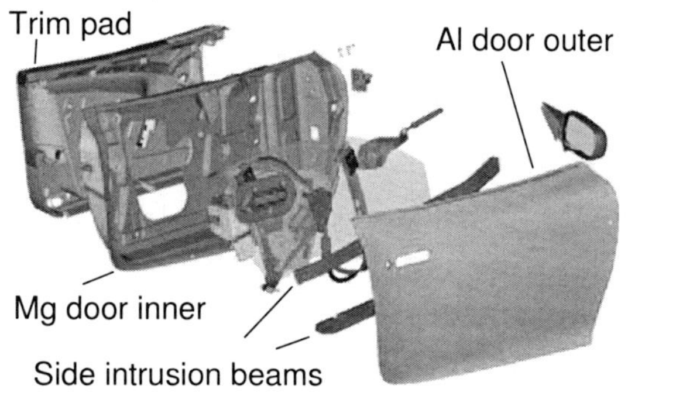

Figure 6. Aston Martin DB9 magnesium door assembly.

Figure 6 shows an exploded view of the magnesium door assembly designed within phase two of the project. As part of the technology transfer process, design responsibility for the production intent door inner was passed over to the engineering team at Aston Martin. Consequently, there were some departures from the original door designed for the Ford Contour. These included the specification of a frameless door system to protect for convertible, and a non- modular architecture. Aside from these points, the feature integration was similar. All major reinforcements were integrated into the die casting. Even the outer belt reinforcement was integrated into the inner panel. However, this did necessitate machining of a slot along the waist rail for the purpose of a glass drop. A photograph of the final door inner castings is shown in Figure 7.

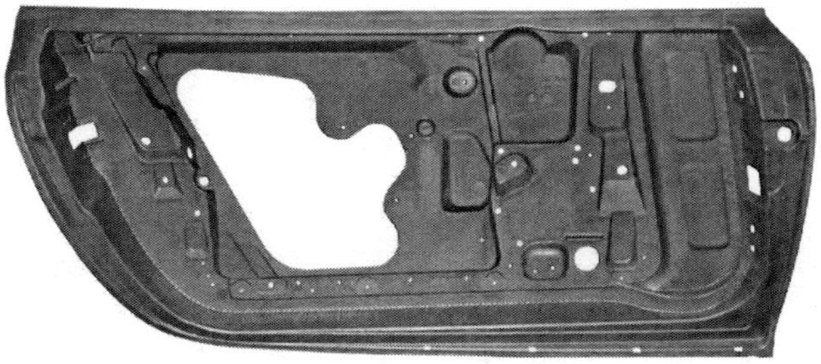

Figure 7. Aston Martin DB9 magnesium door inner panel.

STATIC DURABILITY ANALYSIS

Local variations in casting thickness were determined based upon specifications for hardware attachment. However, global variations were established by dividing the inner panel into sections such as those shown in Figure 8. The relationship between wall thickness and stiffness was then established for a complete series of static durability tests. The example analysis, shown in Figure 8, examines the effect of variable wall thickness on door drop deflection. In this case, the upper hinge attachment area, section 5, appears to have the greatest effect. A 20% increase in wall thickness results in a 36.4% increase in stiffness. Therefore, combined with results for other load cases, weight savings were possible by increasing wall thickness in only the critical areas whilst maintaining a minimum thickness elsewhere. Of course, some limitations were imposed as results from die casting fill simulation and solidification analyses became available.

SECTION	Change in Door Drop For 20% Gauge Increase (mm)	% Change in Door Drop Relative To Door Only System Contribution (+ve Stiffer)
SEC 1	-0.013	-1.7%
SEC 2	0.046	7.9%
SEC 3	-0.003	-0.6%
SEC 4	-0.009	-1.5%
SEC 5	0.215	36.4%
SEC 6	0.016	2.8%
SEC 7	-0.003	-0.6%
SEC 8	-0.039	-6.7%
SEC 9	0.112	19%

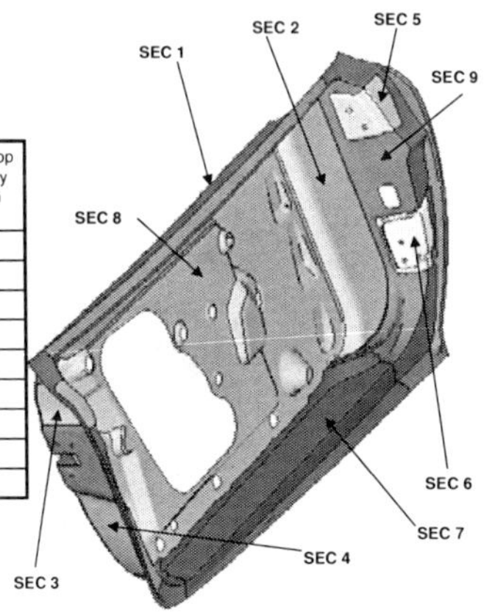

Figure 8. The effect of wall thickness on door drop off.

WEIGHT SAVINGS

A weight breakdown for a prototype version of the Aston Martin DB9 door in white assembly is shown in Table 3. At 3.7kg, the inner panel offers a further reduction of 1.1kg compared to the prototype based upon the modular design of the Contour. However after further analysis, localized increases in section thickness were made to assist in die filling. This resulted in a final part weight of 4.32kg for magnesium inner panel alone. Since a steel version of the DB9 door was not conceived, a precise weight saving for this component cannot be determined. However, based upon frameless door systems of similar dimensions, the final design of the DB9 door assembly was projected to offer a 43% weight save relative to an equivalent stamped steel assembly.

Door In White Weight Summary	Mass (kg)
Door Side Hinge * 2	0.144
M10 Bolt * 4	0.120
M8 Bolt * 3	0.045
Lower Door Beam	1.296
Upper Door Beam	2.079
Door Inner Panel	3.711
Door Outer Panel	1.808
Beam Extension Plate	0.166
Check Arm Casting	0.254
Door Check Arm Adhesive	0.042
Lower Beam Folding	0.049
Door Outer Adhesive	0.035
TOTAL DOOR IN WHITE	**9.749**

Table 3. Weight breakdown of an Aston Martin DB9 prototype door assembly.

CRASH ANALYSIS

Once the stiffness criteria had been met for the static load conditions, non-linear crash analyses were performed to predict the behavior of the door system during side impact for a series of US and European safety standards. Due to the direct load transfer from the barrier into the door, the load-deflection predictions for the FMVSS-214 [8] were perhaps the most challenging to meet. Figure 9 shows a CAE rendering of the test setup. During the test a 305mm (12") diameter rigid pole is pushed 457mm (18") into the side of the vehicle. Since contact with the rocker panel and other structural members is minimal, the loads must be transferred into the vehicle structure by the door beams and casting. Hence the deformation characteristics of the casting and overall ductility were extremely important.

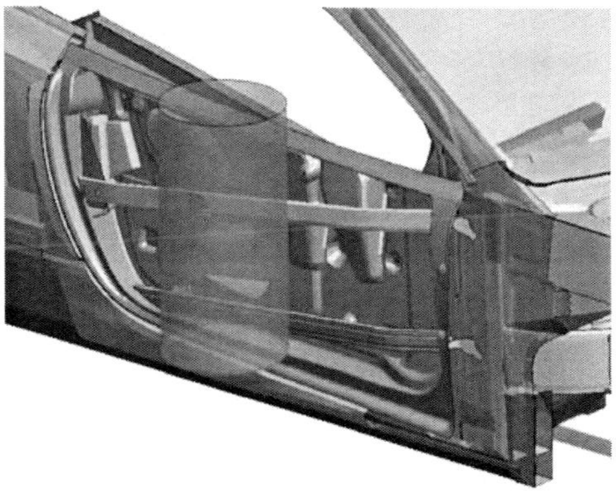

Figure 9. FMVSS 214 test setup.

To promote maximum elongation AM50 was selected as the magnesium alloy. While values of 10% have been reported and demonstrated for crash critical applications [9,10], a lower strain limit was assumed due to the safety critical nature of the component. This is also more consistent with property data reported for other large magnesium die casting such as instrument panels [11]. By virtue of the CAE methodology, the upper strain level was set for each element to this predefined limit. During the numerical simulation, elements reaching this strain value were deleted from the mesh and therefore unable to transfer further load during the analysis. In practice, this is an extreme case, as this has the effect of concentrating the loads over a smaller area of the structure resulting in premature failure of the part. However, a conservative approach was adopted to minimize any chance of failure in the final design. Areas of the door inner casting affected most by the side impact test can be seen in Figure 10. While a few localized high stress regions were observed, maximum strain levels, for the most part, were limited to the rocker region (location3) and mid way along the waist rail (location 4). A final review of load deflection data indicated that the door assembly would far exceed requirements for this test.

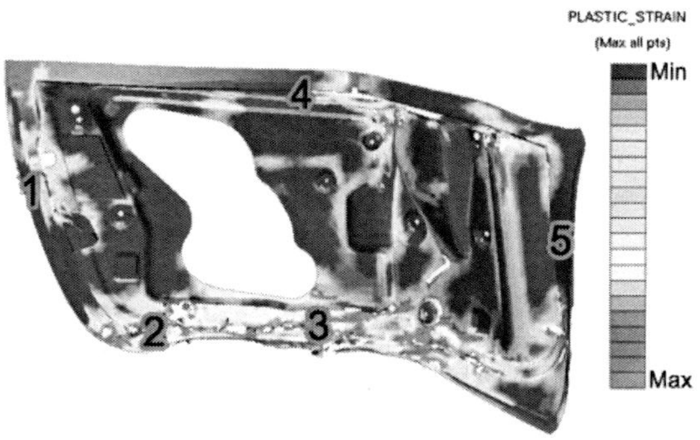

Figure 10. Strain map of a preliminary magnesium door inner panel design following static FMVSS-214 testing.

EXPERIMENTAL TESTING

Experimental validation of the non-linear crash analysis was achieved by performing full vehicle crash tests. Door assemblies using die cast door inner panels were mounted to vehicles to verify the performance of the lightweight door design. However, prior to full vehicle testing, local strain data were obtained for an early batch of die cast door inner panels to ensure parts met the specified strain criteria. Based upon strain prediction from crash simulations, the areas marked as 1 to 5 in Figure 10 were of particular interest. Hence, flat sub-size tensile test coupons, measuring 4" by ¼" gauge width were cut from these areas and tested according to ASTM E8. Figure 11 shows the results based upon a set of 135 samples (3 samples from each of 5 locations, for 9 doors).

In general, the mean strain to failure far exceeded both test requirements and typical values reported in the literature. Results for locations 1, 2 and 4 (latch, beam attachment, and waist rail) exceeded 15%. Furthermore, the hinge area reported strain values up to 28%. The rocker region, however, performed to a lesser extent with a mean value of 8%. Realistically, some decline in ductility was expected in this region due to the gating configuration of the tooling. Even so, further modifications to the process and tooling were instituted to improve the overall ductility in this region.

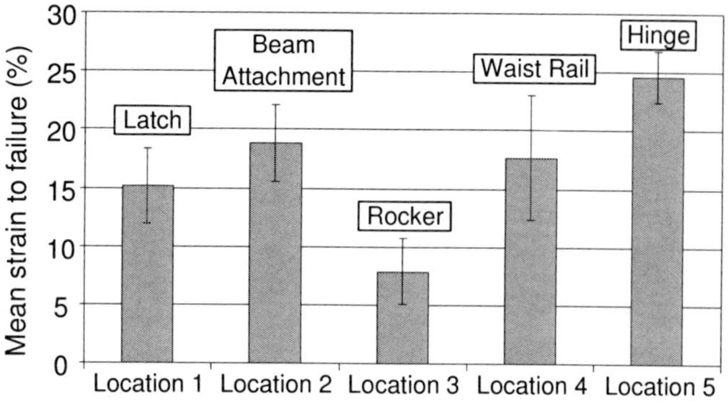

Figure 11. Mean tensile failure strains of castings by location (see Figure 10).

Based upon a final assessment of inner panel ductility, the full vehicle side impact crash testing proceeded as planned. Figure 12 shows a summary of results for the static FMVSS 214 test. The photograph shows the deformation of the door assembly, following removal of the outer panel. As expected, the door system performance exceeded requirements for each stage of the test by a comfortable margin. At 6" of ram displacement, the door performance exceeded requirements by 97%. Similarly, performance at 12" and 18" exceeded requirements by 87% and 60% respectively. In summary, ductile behavior was observed throughout the door inner casting, resulting in high strength and good load transfer into the surrounding vehicle structure. Some cracks were observed in the area of the rocker panel. However these remained localized mid way along the lower edge of the casting.

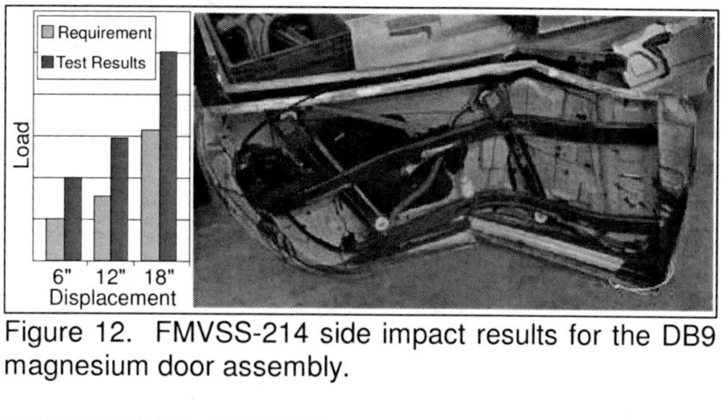

Figure 12. FMVSS-214 side impact results for the DB9 magnesium door assembly.

DIMENSIONAL CONTROL

Aside from weight considerations, integration of major reinforcements into the casting provided an opportunity to reduce build variance for the door assembly. The stiffness of the inner panel alone was also considered a benefit from a handling perspective. Final edge of part dimensions were achieved through trimming operations. The same applied to the majority of holes and apertures, almost eliminating the need for machining operations. Since the outer belt reinforcement was part of the die casting, machining was required to create the glass drop channel.

In general global tolerances were held to +/- 0.5mm over the component. Dimensional data from a die casting produced during process development is shown in Figure 13. In this case, tolerance specifications are met for 21 out of 22 points measured. However, for this pre-production door inner, one position (location 2) does fall outside the specification by 0.039mm. In practice, further adjustments to the die casting process were made to bring this area of the door back into compliance.

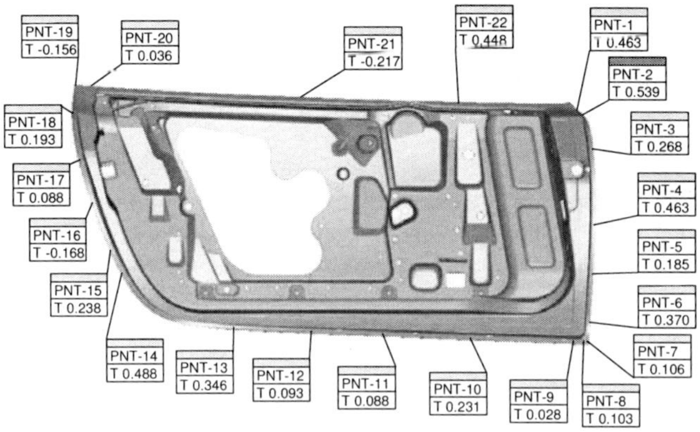

Figure 13. Dimensional data for a pre-production DB9 magnesium die cast door inner panel.

CORROSION PROTECTION

To satisfy the corrosion performance requirements for the door assembly, a strategy was adopted where the inner panel was isolated from the remainder of the sub assembly. Given the availability of a large number of surface pretreatments and coatings, an extensive series of screening trials was performed to identify suitable protection systems. Corrosion testing focused on a selection of chrome based conversion coatings, chrome free conversion coatings, anodized coatings, plus a variety of powder coat and electro-coat systems.

Test coupons were made consisting of cast magnesium samples bonded to aluminum coupons to simulate the hem flange in the door assembly. This was following by cyclic corrosion testing according to Ford test method BI123-01 [12]. In total, each of the coating systems was subjected to 100 cycles of corrosion testing. This resulted in a clear differentiation between suitable and poor performing systems. A detailed review of the test procedures and results is the subject of two separate publications [13,14]. However, a summary of the corrosion testing is included in the paragraphs below.

Of the conversion coatings tested, the chrome based system appeared to the most robust. Results from chrome free conversion coatings based upon Ti/Zr were mixed. Some performed as well as the chrome system, while others offered considerably less protection. The large inconsistency in results was attributed to poor control of the cleaning stages prior to conversion coating application. Special attention is therefore required at this stage of processing to maintain performance comparable with the chrome system. Powder coat choice, on the other hand, appeared to be only a secondary effect. Of those tested, any differences in performance were further diminished by the addition of the paint base and clear coat.

Anodizing was considered an alternative to the application of a conversion coating. Therefore, coupons were subjected to a variety of anodizing processes to evaluate the performance benefits. In general, all of the anodized systems performed exceptionally well when combined with a top coat of either powder coat or electro-coat. Furthermore, the performance exceeded that of any of the conversion coatings, regardless of composition.

In summary, of the systems tested, coatings using an anodized base layer offered the most effective protection. The chrome conversion coating performed almost as well, once combined with an epoxy-polyester powder coat. Some chrome free alternatives were also worth considering, although others produced largely inconsistent results. From the perspective of vehicle requirements, satisfactory performance was achieved using the system based upon the chrome conversion coating. Therefore, since the anodized coatings were accompanied by a significant cost penalty, the chrome conversion coating was selected for use ahead of a planned transition to chrome free systems in line with European End-Of-Life vehicle directive, 2000/53/EC.

CONCLUSION

The results from a two stage development program demonstrate how a magnesium die casting can be used effectively as a fully structural member within a side door assembly. Major portions of the door structure that are typically produced from multiple metal stampings, were replaced by a single magnesium casting that also integrated features that would normally require secondary machining operations. The part count reduction simplified the sub assembly process, reduced tooling investment in addition to providing the opportunity to optimize the design for both performance and weight. Projected weight savings for the Aston Martin DB9 door in white assembly were 43%.

Aside from a demonstrated capability to meet engineering standards, an economic assessment was performed to identify the true cost of technology implementation. Based upon a fully framed door design, derivative programs at low to medium volumes would benefit from the technology at no cost penalty. For higher volume vehicles, the stamped steel assembly offers the most cost effective route. However, a detailed analysis of the target application would be required to determine the breakeven volume for the two manufacturing technologies.

ACKNOWLEDGMENTS

The authors would like to acknowledge the contribution made by all members of the project team. In addition, throughout the door development program, staff at Ford and Aston Martin received extensive support from a wide range of sources. A special thanks goes to the following people: Carl Johnson, Jim Franko, Greg Dobieralski, Wayne Simonds, Chris Kirkum, Mark Salmonowicz, Ken Kendall, Kieron Carvell, Roland Snell, John Hill, Bob McCune, Dan Houston, Mike Scanlon, Arne Jurgens, Steve Buckley, Dave Wagner, Xiaoming Chen, Manfred Dunst, Gottfried Peer, Jake Zindel, Mei Li, and Duane Becknell.

REFERENCES

1. http://www.nhtsa.dot.gov/cars/rules/cafe/NewPassengerCarFleet.htm.
2. Seksaria, D.C. and R.K. Patel. "An Ultra-Light Thin Sliding Door Design - A Multi Product Multi Material Solution." SAE Technical Paper Series, 2002-01-0391.
3. Balzer, J., et al. "Structural Magnesium Front End Support Assembly." SAE Technical Paper Series, 2003-01-0186.
4. Friedrich, H. and S. Schumann, "Research for a "new age in magnesium" in the automotive industry." Journal of Materials Processing Technology, 2001. 117(3): p. 276-281.
5. Jonason, P., et al. "MAGDOOR- Magnesium in Structural Application." SAE Technical Paper Series,1999-01-3198.
6. Schreckenberger, H., M. Papke, and S. Eisenberg. "The Magnesium Hatchback of the 3-Liter Car: Processing and Corrosion Protection." SAE Technical Paper Series, 2000-01-1123.
7. "Design of a magnesium/aluminum door frame", Automotive Engineering, 1993, Vol 101, 5,pp 57-60.
8. Federal Motor Vehicle Safety Standard 214, Side Impact Protection.
9. Ames, W. and W. Altenhof. "Observations of the Relative Performance of Magnesium and Aluminum Steering Wheel Skeletons with Identical Geometry." SAE Technical Paper Series, 2000-01-0784.
10. Aune, T. and L. Riopelle. "An Evaluation of the Mechanical Properties Critical to the Design of Interior Magnesium Components." SAE Technical Paper Series, 960415.
11. Alain, R. et al. "Robustness of Large Thin Wall Magnesium Die Castings for Crash Applications." SAE Technical Paper Series, 2004-01-0131.
12. Ford Laboratory Test Method BI 123-01. "Painted Sheet Metal Corrosion Test (Laboratory Simulated Arizona Proving Ground Test)", 2001.
13. Bretz, G.T., et al. "Adhesive Bonding and Corrosion Protection of a Die Cast Magnesium Automotive Door." The Minerals, Metals and Materials Society. 2004: Magnesium Technology 2004.
14. Blanchard, P.J., et al. "Evaluation of Corrosion Protection Methods for Magnesium Alloys in Automotive Applications." The Minerals, Metals and Materials Society, Magnesium Technology 2005.

2005-01-0339

2005 Fuel Cell Vehicle and its Magnesium Power Distribution Unit

Naiyi Li, Edson Chagas and Kelly Zechel
Ford Motor Company

ABSTRACT

The High Voltage Power Distribution Unit (PDU) is constructed of magnesium in support of Fuel Cell Electric Vehicle (FCEV) weight reduction efforts. The PDU distributes and controls a nominal 75 kilowatts of power generated by the Fuel Cell, the primary source of High Voltage power, to all the vehicle loads and accessories. The constraints imposed on the design of the PDU resulted in a component highly susceptible to general and galvanic corrosion. Corrosion abatement was the focus of the PDU redesign. This paper describes the redesign efforts undertaken by Ford personnel to improve the part robustness and corrosion resistance.

INTRODUCTION

Fuel cell vehicles (FCV) are relatively new in the automotive industry. When compared with a conventional internal combustion engine, the fuel cell engine is cleaner, quieter and potentially more energy efficient. With the strong merits of fuel cell technology, Ford Motor Company and other major automakers have been conducting various research and development programs to determine the technical feasibility and practicality of producing fuel cell vehicles and developing its associated infrastructure.

One of the efforts in Ford Motor Company is to deliver a demonstration fleet of 30 fuel cell vehicles to California, Florida, Michigan, Vancouver and Germany beginning in 2005. The Fuel Cell Electric Vehicle (FCEV) is based on the Focus platform. The vehicle is powered by a combination of hydrogen fuel cells and electric batteries that are recharged through the braking system. The FECV includes High Voltage Nickel Metal Hydrate battery pack that provides back up and boost power to the Fuel Cell. The battery is recharged as necessary by the Fuel Cell and the recuperated braking energy from the electro hydraulic regenerative braking system. A 5000 pound per square inch hydrogen storage tank provides the hydrogen for the Fuel Cell. The Focus FECV speed and acceleration performance is uncompromised with a driving range of up to 200 miles.

FCEV weight constraints have demanded intensive use of lightweight materials throughout the vehicle. This paper provides an overview of the demo vehicle program and up-to-date accomplishments focusing on lightweight material applications. Magnesium, about 30% lighter than aluminum, is chosen to play an important role in these fuel cell vehicles. One of the magnesium applications in the vehicle program is for a power distribution unit (PDU) housing. Due to the low volume FCEV production, the part was sand cast with AZ91E.

The combination of material selection, casting process, part geometry and vehicle packaging fosters corrosion susceptibility. Although corrosion was taken into consideration in the initial design and the component passed a validation chamber test, the part failed the Total Vehicle Accelerated Corrosion Test. As a result, the PDU had to be redesigned. This paper describes the PDU design criteria and functional requirements, component design, magnesium alloy property, magnesium casting process and quality, corrosion protection and coating strategy, assembly constraints, etc. More attention is paid to improve the PDU's casting quality and corrosion protection based on the vehicle corrosion test results. Consequently the PDU's performance targets were achieved in a timely fashion.

PRODUCT DESCRIPTION

The Power Distribution Unit provides the interface from the Fuel Cell System (FCS) to the vehicle electrical loads. The main function of the PDU is to switch and distribute the electrical output of the Fuel Cell to the vehicle High Voltage (HV) buss. It houses the following major components and associated functions:

- HV Contactors: Switches the power On and Off from the FCS to the vehicle.
- HV Fuses and Associated Busses: Provides over-current protection to the distributed circuits.
- Fuel Cell Measurement (FCM) Unit: Monitors various parameters associated with the vehicle HV buss and communicates over HS-CAN with the FCS. The aluminum FCM heat sink also serves as the PDU cover.

Of specific interest to this paper, the PDU/FCM monitors the isolation of the HV buss bars to chassis and between HV positive and HV negative. The PDU/FCM shuts off power to the vehicle should this isolation fall below accepted safety levels. Water intrusion in the PDU, due to corrosion, can trigger an isolation fault.

Magnesium Alloy Properties

The standard practice for magnesium alloy designation identifies the main compositional elements and the associated percentages in the alloy [1]. The AZ91E magnesium alloy's two principle elements are aluminum (A) and zinc (Z) in the corresponding percentages of 8.1-9.3% and 0.4-1%. The designation (E) represents a slight alloy variation from (A-D) compositions. (Table 1)

Notably, higher iron content than specified accelerates the corrosion rate due to its ability to act as cathodic sites in a galvanic cell in the presence of a salt-water electrolyte. However, the overall magnesium alloy iron tolerance depends upon the manganese content found within the alloy. If iron exceeds 0.0005%, the iron-to-manganese ratio shall not exceed 0.032. [2] The overall impact of iron contamination in a magnesium alloy affects the corrosion rate in an exponential fashion. Elevated levels of nickel, copper and cobalt also increase susceptibility to corrosion in harsh environments, but at varying levels dependent upon the cast form, which influences grain size. [3]

Sources of contamination can be introduced in a several ways. The purity of the incoming magnesium alloy ingots, particularly recycled material, offers a potential source of elemental contamination. In addition, the manufacturing process could elevate the magnesium alloy's iron content through leaching of a steel chill plate. Material residue found in crucibles, heat-treatment solutions or water quench solutions also provide opportunity for increased levels of iron, nickel or copper.

Chemical Composition		
Base Element	Mg (Magnesium)	
Al (Aluminum)	8.1 to 9.3%	% by wt
Cu (Copper)	0.015% maximum	% by wt
Fe (Iron)	0.005% maximum	% by wt
Mg (Magnesium)	* 89.129 to 91.33%	% by wt
Mn (Manganese)	0.17 to 0.35%	% by wt
Ni (Nickel)	0.0010% maximum	% by wt
Si (Silicone)	0.20% maximum	% by wt
Zn (Zinc)	0.40 to 1.0	% by wt
Balance	Mg (Magnesium)	
Others Total	0.30% maximum	% by wt
Others Each	0.01% maximum	% by wt
Composition Note	Mg is listed as a Balance element for this material, but its % value is not listed in the specification. The value shown is a calculated one.	

Table 1, AZ91E Magnesium Alloy Composition Limits [2]

Sand Cast Magnesium Alloys

Sand cast molds are commonly used for low-volume production in lieu of die casting. Magnesium-aluminum-zinc alloys are generally easy to cast, but are limited in certain respects. Sand casting AZ91E magnesium alloy results in micro shrinkage and high creep potential. As a result, AZ91E is not suitable for applications where temperatures exceed 95ºC (200ºF). In addition, the large grain size promotes micro cracks along grain boundary where the sand cast process results in larger grain size products than die casting. Consequently casting surface defects increase the component's susceptibility to corrosion unless protected with pre-treatments, sealants and coatings in harsh environments.

Geometric Properties

The PDU physical package dimensions are 340mm (L) x 240mm (H) x 240mm (W) and its weight is of 3.6 Kg shown in Figure 1. The PDU is mounted under the vehicle in the transmission tunnel and consequently exposed to extensive road moisture and debris. It is secured to the vehicle by four attachment points; in the front it attaches to the sled runner welded studs and in the rear it attaches to the Fuel Cell Stack bracket. In addition to these four mounting points, the part is also fastened to the center of the tunnel by a welded stud to a Part In Assembly (PIA) steel bracket. This last attachment point, referred to as the Safety Strap, has no structural function; its purpose is to prevent the part to be removed while still energized. In order to unfasten this bracket from the vehicle, the HV connectors must be removed which actuates the mechanical Emergency Power Off (EPO) circuit, causing the PDU to open the main contactors and shut-off power to the vehicle HV buss.

A carbon fiber splash shield mounted to the bottom of the PDU and the vehicle floor pan provides some protection from stone pecking at the PDU front face. The HV connections to the vehicle are located in the front face of the PDU. In the back of the PDU, two self-aligning stack connectors (positive and negative) are coupled to their mating sides in the Fuel Cell Stack. The vertical and horizontal alignments of the PDU are crucial to the performance of the stack connectors. The tight tolerances demanded by the electrical and physical interfaces of the PDU could not be achieved with the sand casting alone and necessitated some machining operations. The machining process results in various sharp corners and other geometric characteristics, which negatively impact the corrosion resistance of the part due to moisture retention and poor coating adhesion.

The horizontal surfaces are flat allowing for moisture and road debris accumulation. The mounting points to the vehicle are also unfavorable from a drainage standpoint. Actions to improve these conditions were limited due to the negative impact it would have on the established vehicle packaging and PDU tooling. The magnesium enclosure interfaces were also analyzed for their potential contribution to galvanic corrosion and general corrosion through moisture and debris accumulation on/near the enclosure. They were classified as subsystem or vehicle-level interfaces.

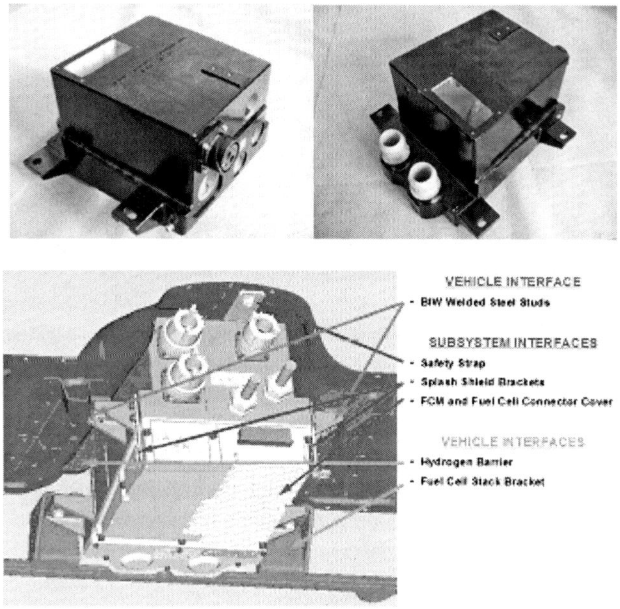

Figure 1 – PDU Geometry and Interfaces

Subsystem Interfaces

The following interfaces are present in the PDU:

- FCM and stack connector covers, made of 6000 series aluminum, did not contribute discernibly to galvanic corrosion of the magnesium housing.
- Splash shields and the safety bracket, made of cold rolled steel and powder coated, contributed to galvanic corrosion of the magnesium housing.
- Vehicle connectors, made of anodized aluminum, did not contribute discernibly to galvanic corrosion of the magnesium housing.

Vehicle Interfaces

The following interfaces are present in the overall vehicle package:

- Rear attachments to the FCS bracket, made of Al 356, secured with black organic corrosion protection coated bolts (S306), contributed to galvanic corrosion of the magnesium housing at the bolt interface.
- Front attachments to uncoated steel welded studs fastened with corrosion protection coated nuts (S427) contributed to galvanic corrosion of the magnesium housing.
- The hydrogen barrier, a reinforced foam barrier installed between the top and sides of the PDU and the transmission tunnel, rests upon the top surface of the PDU. This barrier prevents any potential hydrogen leaks from the FCS to migrate up the cowl and into the cabin. However, the hydrogen barrier negatively impacts the PDU as it accumulates moisture at the PDU's surface and prevents airflow around the part. Paint bubbles underneath the hydrogen barrier contributed to general corrosion of the magnesium housing.

Corrosion Problem Statement

All Ford Motor Company vehicles are subjected to the corporate engineering test procedure, Total Vehicle Accelerated Corrosion Test, executed at the Arizona Proving Grounds (APG). The test determines the corrosion resistance of the assembled Fuel Cell Electric Vehicle. The test consists of 30 cycles of mileage accumulation over various road conditions including dust, gravel, cobblestone, potholes and salt spray/baths augmented with controlled humidity soaking and drying replicating worst-case customer environments. The test procedure specifically limits the iron content of all water and solutions to 1 mg/L maximum.

The Attribute Prototype Fuel Cell Electric Vehicle (FCEV) was subjected to the corrosion test. The vehicle was towed through the test cycles and, as a result, the vehicle teardown revealed component failures at the end of 30 testing cycles. The most notable concerns included extensive corrosion in the mounting attachments points as shown in Fig 2 and exposure of the high voltage copper buss bar. Also, some issues were identified at the interface of the aluminum Fuel Cell Measurement (FCM) unit cover and the magnesium PDU housing. Initially it was thought that the FCM aluminum cover had induced galvanic corrosion, however the edge of the PDU, where it mates with the FCM, was masked during the coating application leaving the magnesium exposed in a high splash area and susceptible to general corrosion. Finally, the removal of the steel mounting bracket exposed significant material loss in the portion of the housing in direct contact with the steel bracket as shown in Fig 3.

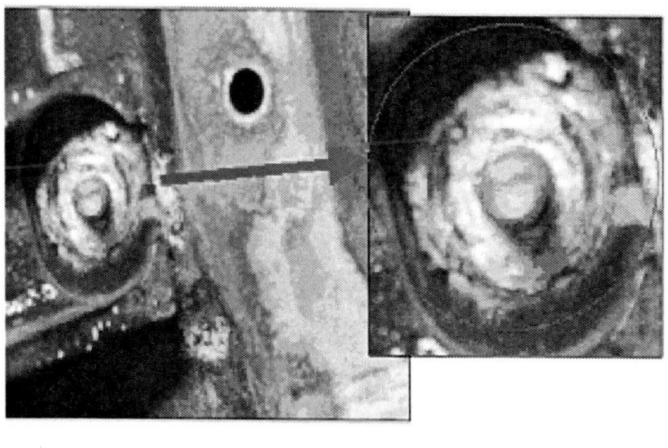

Figure 2 – Galvanic Corrosion at Welded Studs

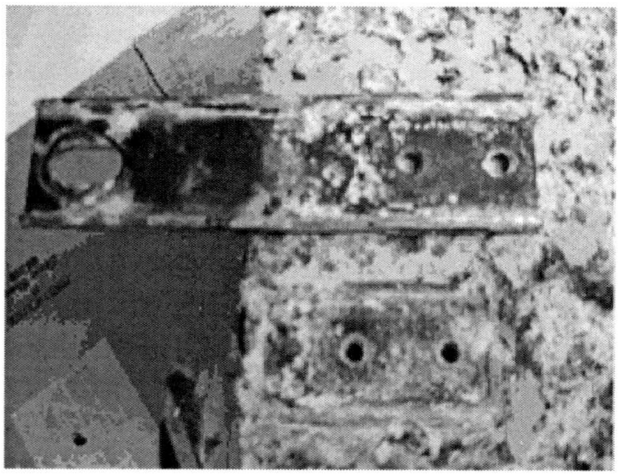

Figure 3, Galvanic Corrosion at the steel Safety Strap

General Corrosion Description

General corrosion is commonly characterized by uniform surface erosion. Typically, general corrosion does not exhibit any macroscopic area of the metal surface distinguishable as solely anodic or cathodic. Temperature and pressure govern the solubility of corrosive elements found in an alloy and consequently influence the general corrosion rate. So an alloy's general corrosion rate follows the chemical reaction rate, which generally is temperature dependent.

Motor vehicles have the added challenge of foreign materials being introduced at a velocity, i.e. erosion. Water, wind, stones, mud and salt introduced at a velocity often remove protective material treatments and coatings exposing more metal to corrosion over time. In addition, the presence of chlorides, sulfates and foreign materials that hold moisture on the surface can promote corrosion and pitting of unprotected magnesium material.

General Corrosion Issues

The primary failure mechanism of the PDU housing was general corrosion. The physical location of the PDU in the vehicle exposes it to harsh environmental factors such as water, wind, stones, mud and salt. Uniform corrosion across the housing surface results in a thinning of the entire casting. Since the powder-coat exterior finish does not adhere evenly across the sharp radii in the casting's geometry, it exposes the magnesium in these areas to environmental factors earlier than the rest of the casting surface. In addition, the sharp radii capture greater amounts of foreign material and therefore retain greater amount of moisture leading to more extensive general corrosion. The hydrogen barrier introduces a similar effect by capturing debris on the top of the flat housing surface.

Furthermore, the PDU's general corrosion rate was accelerated by iron contamination found in the Confirmation Prototype (CP) magnesium alloy housing. An early APG sled test, used to evaluate coating alternatives, resulted in better performance of the Attribute Prototype (AP) control part, over the CP experimental part. The confounding test results prompted elemental analysis of the magnesium castings from two different production batch runs. The analysis revealed 0.008% iron versus the 0.005% allowable. However, as noted earlier, higher iron content is permitted as long as the iron-to-manganese ratio does not exceed 0.032. [3] In the case of the CP PDU housing material, 0.018% manganese was present equating to an iron-to-manganese ratio of 0.044.

Finally, the heat treatment initially applied to the AZ91E magnesium alloy casting was a simple solution treatment (T4) instead of a solution treatment which was then aged (T6). The corrosion rate improves 13X for AZ91E alloy castings which receive a T6 solution treatment and aging over a simple T4 solution treatment. [3] Specific reference to the AZ91E ASTM B80 standard explicitly specifies the use of T6 solution treatment and aging in the casting process.

Galvanic Corrosion Description

The secondary failure mode of the magnesium PDU housing was galvanic corrosion. Galvanic corrosion damage occurs when two dissimilar materials are coupled in a corrosive electrolyte like salt water. The driving force for galvanic corrosion is the potential difference between the dissimilar metals. In a bimetallic couple, the less noble material becomes the anode of the corrosion cell and tends to corrode at an accelerated rate, compared with the uncoupled condition.

A galvanic series implies a particular electrolyte solution; hence for each specific solution a different order or series will ensue. The U.S. Army MIL-STD-889 outlines the "Practical Galvanic Series" for materials exposed to sea water [4],

Active (Anodic)
Magnesium
Mg alloy (AZ – 31B)
Mg ally HK – 31A
Zinc
Al 6061
Stainless steel 430 (active)
Lead
Iron (cast)
Copper (plated, cast or wrought)
Nickel (plated)
Titanium
Silver
Gold
Graphite
Noble (Cathodic)

Table 2 – Practical Galvanic Series Excerpt [4]

Galvanic Corrosion Issues

The PDU attaches to uncoated steel studs with S427 (electrolytic zinc base coat + aluminum organic) finished nuts in the front attachment points. Salt water creates a galvanic cell between the magnesium attachment points (anodic) and the steel studs (cathodic). The rear attachments use a S306 black organic coated steel fastener however scratches on the protective finish expose the magnesium housing directly to steel creating the conditions for galvanic corrosion. The steel brackets have an electrically conductive powder coat finish that supports galvanic interaction with the magnesium housing if scratches on the protective finish expose the steel to the magnesium housing, particularly after the onset of general corrosion.

Corrosion Mitigation Actions

With the results of the APG corrosion test, the PDU team initially recommended that the part be cast in aluminum. However at a weight penalty of 3 Kg, the program rejected this alternative. As mentioned above, galvanic interaction between the aluminum-to-magnesium interfaces were believed to be the main reason for the failure. Corrosion on areas distant from galvanic interface indicated that general corrosion was also a significant factor. After 30 cycles the degradation of the magnesium surfaces were already advanced to the point where little could be learned from the returned part in terms of how the corrosion started and progressed. A PDU subjected to the less aggressive chamber test, provided cause for challenging of this assumption.

A design of experiments was created to verify the effect of the FCM aluminum 6000 series on the magnesium interface, whereby, through a combination of coating applications on the FCM and PDU, four possibilities of treatment for this interface were present on the test part:

1. Aluminum FCM cover coated and magnesium housing coated.
2. Aluminum FCM cover uncoated and magnesium housing coated.
3. Aluminum FCM cover coated and magnesium housing uncoated.
4. Aluminum FCM cover uncoated and magnesium housing uncoated.

At 5 cycles it was evident that the corrosion was more pronounced in quadrants (3) and (4) without discernible difference between them. Ultimately, the galvanic interaction between the aluminum FCM cover and the magnesium housing, although existent, was negligible. Design efforts were then directed to address the robustness of the casting process, pretreatment and coating of the part to improve robustness against general corrosion.

General Corrosion Corrective Actions

Casting Process: The iron and copper contamination of the alloy was addressed by changing the chill plate material from Al 319 to Al 356. Only virgin ingots were acceptable as raw material versus the recycled ingots used for AP and CP parts. The heat treatment also changed from T4 to T6.

Pre-treatment: For CP parts the pretreatment after machining was Anomag™, this thin coating (μm) was deemed insufficient to provide surface protection of the magnesium casting [5, 6]. The following alternatives were considered: Tagnite™, Keronite™ and Anomag™ followed by impregnation.

Coatings: The CP level part was coated with a conductive powder coat, this choice was made to improve EMC performance. However, it is believed that the conductive element in the coating (carbon) would react with the magnesium in the presence of electrolyte (after peeling or chipping). For the redesign Tagnite™ and Keronite™ suppliers recommended their own coatings, respectively Rockhard™ and powder coating. Also, Anomag™/impregnated samples were coated with PPG's Frame Coat for comparison.

Coating Coverage: The edge of the PDU extends 2-3mm (sand cast tolerances) beyond the FCM cover. As pointed out earlier this area was left uncoated for CP. Monitoring of subsequent tests and prototype fleet vehicles showed that corrosion started earlier in this area. For the redesign all exposed magnesium surfaces were fully coated.

With the various pretreatment and coatings subjected to 30 cycles of the corrosion chamber test, this test results identified the application of Tagnite™ and Rockhard™ to the redesign PDU.

Galvanic Corrosion Corrective Actions

Vehicle Attachments: Nylon sleeves were introduced to isolate the front welded studs and the rear bolts. Aluminum 6000 series washers were also added.

PIA Steel Brackets: All three brackets are nylon coated. The Safety Strap had to be masked so that a ground path for the PDU housing could be established.

Figure 4 shows the improvement of corrosion protection on the PDU at the end of 20 test cycles at APG with the results of this study.

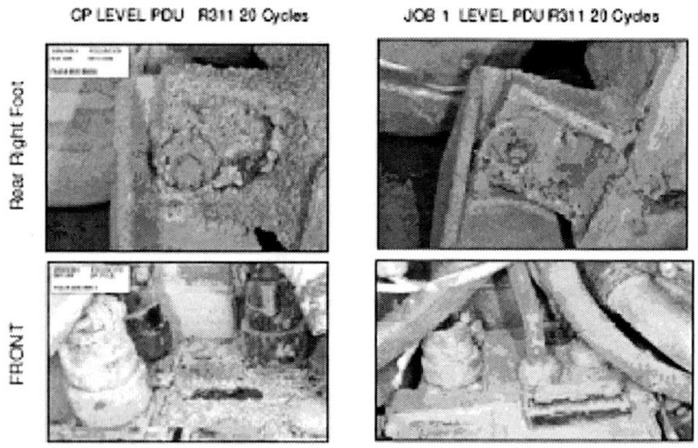

Figure 4, CP and the redesign PDUs at the end of 20 APG Test Cycles

SUMMARY

The High Voltage Power Distribution Unit (PDU) is constructed of magnesium AZ91E in support of Fuel Cell Electric Vehicle (FCEV) weight reduction efforts. This paper describes the redesign efforts to improve the part robustness and corrosion resistance. The PDU tooling costs, lightweight requirements and vehicle packaging constraints offered a formidable design challenge. The redesigned part successfully completed the APG corrosion test in August 2004.

ACKNOWLEDGMENTS

The authors would like to acknowledge Don Jordan, Andrew Gonce, Bob McCune, Greg Bretz, John Heinholz, Dan Schumacher from Ford Motor Company and Dave Robson and Luke Belfie from Sparton Electronics for their contributions throughout the investigative process.

REFERENCES

1. ASTM Standard B275, Standard Practice for Codification of Certain Nonferrous Metals and Alloys, Cast and Wrought.
2. ASTM Standard B80, Standard Specification for Magnesium-Alloy Sand Castings.
3. Barbara Shaw, Corrosion Resistance of Magnesium Alloys, ASM Handbook Volume 13A: Corrosion: Fundamentals, Testing and Protection, ASM International, Materials Park, OH, 1993.
4. U.S. Military Standard Specification MIL-STD-889.
5. Gerry G. Wang, Kevin Stewart, Richard Berkmortel, and Jan Ivar Skar, Corrosion Prevention for External Magnesium Automotive Components, SAE 2001-01-0421, Michigan, March 2001.
6. Guilian Gao and Mark Ricketts, Evaluation of Protective Coatings on Magnesium for Phosphate Process Compatibility and Galvanic Corrosion Prevention, SAE 2002-01-0081, Michigan, March 2002.

CONTACT

Naiyi Li, nli1@ford.com

Edson Chagas, echagas1@ford.com

Kelly Zechel, kzechel@ford.com

2005-01-0340

Development of the 2006 Corvette Z06 Structural Cast Magnesium Crossmember

Jonathon Aragones, Kasi Goundan, Scott Kolp, Richard Osborne, Larry Ouimet and William Pinch
General Motors Corporation

ABSTRACT

Since its very beginning in 1953, Corvette has been a pioneer in light weight material applications. The new 6th generation corvette high performance Z06 model required aggressive weight savings to achieve its performance and fuel economy targets. In addition to aluminum body structure and some carbon fiber components, the decision to use a magnesium front crossmember was identified to help achieve the targets. An overview of the Structural Cast Magnesium Development (SCMD) project will be presented which will provide information on key project tasks. Project focus was to develop the science and technical expertise to manufacture and validate large structural magnesium castings, which provide a weight reduction potential of 35 percent with respect to aluminum. The die cast magnesium cradle is being produced from a Mg-Al-RE alloy, designated AE44, for high temperature creep and strength performance as well as casting ductility requirements. Extensive validation and post-test evaluations were performed on the Magnesium Crossmember.

BACKGROUND

United States Council for Automotive Research (USCAR) and the Department of Energy (DOE) support materials research necessary to overcome the technical and manufacturing issues that limit the light weighting application of cast magnesium components. Vehicle weight reduction is a significant enabler to meet future competitive and mandated fuel economy pressure. A wider application of cast magnesium in vehicles offers a promise of nearly 100 kg of weight reduction. There are current body and chassis applications that utilize multi-piece fabrications to meet complex geometry and strength requirements. These designs could be integrated into single piece magnesium castings providing lower mass, improved dimensional control and lower cost through part consolidation. North American (NA) cast magnesium applications include; Instrument Panels, Steering Wheel Armatures, Seats, Brackets and Powertrain Transfer Cases. All of these applications except for Transfer Cases are used in non-cyclic load situations in friendly non-corrosive environments. Aggressive weight saving applications will require high strength, ductile castings having the ability to function in harsh environments under cyclic loading conditions.

Domestic automakers will be under increasing pressure to reduce CO_2 emissions and improve vehicle fuel economy. Cast magnesium structures have the potential to reduce vehicle mass, which is directly linked to improving CAFÉ. The NA auto industry currently uses approximately 60,000 metric tons per year of magnesium, which is equivalent to some 4.5 kg per vehicle. Significant increases in magnesium usage will assist automakers to meet competitive fuel economy targets. Without good scientific knowledge and solutions to difficult issues, our magnesium usage will be limited along with future options for weight reduction.

Currently, US automakers do not manufacture any of the cast magnesium components used in their vehicles. A relatively small industry supply base produces them, which is not sufficient in size to conduct all the R & D necessary for further magnesium applications. This USCAR/DOE sponsored project has brought together DOE, domestic automakers (Ford, General Motors & Daimler Chrysler), national laboratories, universities, suppliers and industry associations to provide project funding, science and manufacturing resources. The active materials consortium within USCAR is the United States Automotive Material Partnership (USAMP) with a role of conducting material research.

SCMD PROJECT

Magnesium is the lightest of all structural metals and can provide a 65-70% weight reduction compared to steel and a 35% reduction over aluminum. High Pressure Die Casting is the most often used method for manufacturing magnesium components. Magnesium castings can provide small or large complex parts which are dimensionally more accurate than fabricated and joined components. A typical wall thickness range between 1–5mm and a variety of alloy options are available to provide a range in properties for applications requiring good ductility and elevated temperature capabilities. The magnesium industry is less mature than both the steel and aluminum industries, and that is a key reason why USAMP is focusing significant research effort to overcome technical and manufacturing issues that limit the light weighting application of cast magnesium components.
A current active magnesium project within the USAMP portfolio is the "Structural Cast Magnesium Development" (SCMD), which has a strategic focus of enabling

magnesium castings to meet mass, performance and cost targets. Aggressive magnesium structural applications will support USAMP and DOE vehicle mass reduction objectives.

SCMD OBJECTIVES

The SCMD core team developed a dual approach to resolving and implementing large, complex structural cast magnesium components. Gaps in the manufacturing and vehicle application technology were identified and a plan was developed that would address each key issue. A demonstration component was also selected which would be a technical challenge. A concurrent dual activity path provided the team with the best synergy to understand and address the science and implementation issues. Significant resources were assembled with principle investigators from National Laboratories, Universities, OEMs and the supplier industry. The key science challenges focused on the following topics:

1. Development of an accurate magnesium alloy material database and design guide.
2. Investigation of alternative casting processes.
3. Enhancement of modeling software that predicts high pressure mold fill and solidification
4. Corrosion mitigation and joining technologies.
5. Quantitative fractographic and microstructual characterization parameters, and develop relationships between material properties and the component cast structure.
6. Develop a simulation model that integrates material microstructure / properties with Finite Element Modeling (FEM) in order to predict failure accurately.
7. Improve Non-Destructive Evaluation (NDE) methods for inspecting safety critical structural components.

To prove our technical knowledge, an aggressive chassis component was selected as the demonstration casting (Figure 1).

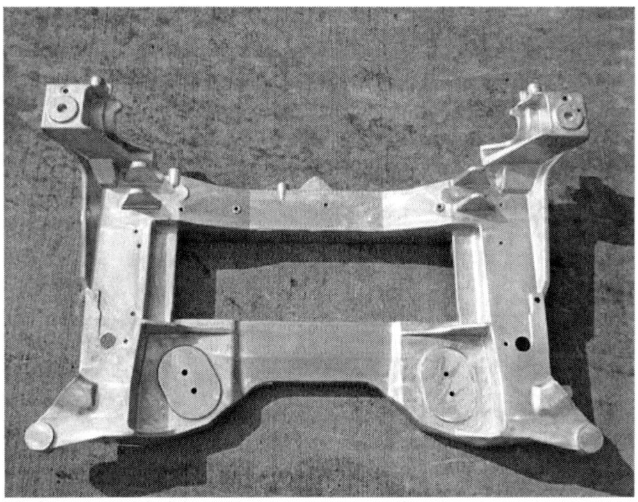

Figure 1: Z06 Corvette magnesium front crossmember casting.

DEMO SELECTION

The demonstration component required a part that would be available for vehicle validation in the later half of the SCMD project. Proposals were gathered from the Big 3 domestic OEMs and it was determined that the 2006 Corvette C6 Front Engine Cradle provided that necessary challenges and met the C6 validation timing initiating in mid 2004. The Corvette Front Cradle proved to be a great demonstration candidate for showcasing a magnesium chassis application with high service loads, temperatures reaching 125°C and under vehicle corrosion challenges of road debris, salt and water spray. A significant limitation with regard to cradle perimeter dimensions was encountered during the initial redesign phase. A sports car like Corvette is densely packaged with essentially no opportunity to acquire needed space to grow the size of the redesigned cradle. Product engineers insisted that the redesigned magnesium cradle have the same attachment points and overall exterior size of the production cast aluminum cradle. So the only opportunity to meet component stiffness and strength requirements was through appropriate application of ribbing and section wall thickness. Finite Element Modeling (FEM) was used to ensure that the redesigned magnesium component satisfactorily met all service load requirements. Advanced fill and solidification simulation tools were also used to develop the cast tooling, component geometry and process parameters for optimum part weight. The production cast aluminum cradle has a weight of 15.8 kg and the redesigned magnesium cradle weight of 10.3 kg, yielding a weight saving of 35%.

MAGNESIUM ALLOY

The two (2) rear cradle mounts were located adjacent to the catalytic converters, which resulted in a localized operating environment of 125^{0}C. The AE44 (MgAl4RE4) alloy, produced by Hydro Magnesium, was selected to enable meeting the high temperature creep and strength performance, as well as room temperature casting ductility requirements. The Rare Earth (RE) mischmetal portion of the alloy is typically comprised of Cerium (Ce), Lanthanum (La), Neodymium (Nd) and Praseodymium (Pr) elements. These four elements will be incorporated into the production spectrographic chemistry evaluation, in order to ensure a complete analysis of the MG-Al-RE alloy series. Initial Hydro Magnesium tests results indicate the potential for improvements, over AE42 alloy, in the areas of:

- High temperature creep and bolt load retention
- Yield strength
- Fatigue performance
- Castability

AE42 alloy (MgAl4RE2), another Mg-Al-RE alloy, is primarily comprised of 4% aluminum, 2% rare earth mischmetal and the balance magnesium. The AE42 alloy is the only currently listed Mg-Al-RE alloy in GM's Material Specifications for Pressure Die Cast Magnesium Alloys (GMW3028).

Die castings were produced from the AE44 alloy for both the USCAR Structural Cast Magnesium Development (SCMD) Project and the GM Magnesium Front Engine Crossmember production programs. As of November 2004, the testing is still ongoing for both programs.

MATERIAL TESTING

The integrity of components used in Chassis load bearing applications is of great importance contributing to vehicle performance and durability. With numerous casting process variables, it is possible to experience inconsistent product quality from part to part. Hence a certain number of tests are recommended for material compliance and product acceptability.

The recommended tests are:

1. Mechanical properties
2. Internal integrity
3. External integrity

Excising test specimens at various locations in casting based upon the FEA stress analysis usually assesses the mechanical strength of a part. At high stress locations based on a specific load case the expected strength is invariably high. There are usually 25 or so load cases to be considered and some of them produce high stresses in the component at specific locations. These high stress areas must be scrutinized very carefully.

EXTERNAL INTEGRITY

The external integrity is common for all casting processes. The parts must be assured of absence of cracks and crack-like anomalies detrimental to the performance and durability of the components. This is usually done during prototype stages and early parts of production. Once the process capability is demonstrated it may be eliminated during the production. The castings are tested for surface cracks by Liquid Penetrant Inspection (LPI) methods.

PROCESS SPECIFIC TESTING

The mechanical properties specimen excising and preparation method, and internal integrity test reference comparison are specific to the high-pressure die-casting selected for this component.

Mechanical Properties: The basic mechanical properties usually monitored in monotonic tensile testing are ultimate tensile strength, yield strength, and elongation. There are specific industrial standards such as ASTM[1] and ISO[2] available for test procedures. The die-casting acquires its strength by its skin and hence it is of high importance how the test specimen is selected and prepared for testing. To keep the skin intact as much as possible, flat and thin areas of the selected locations of the casting are used. The excised specimens are machined flat by removing a minimal amount of the skin and tested. Since tensile testing from excised specimens is not a recommended test for high-pressure die casting (due to partial and variable amount of removal of skin) this test is used for reference only. When on-going component level casting performance testing is adopted during regular production, this test may be limited to early stages of production.

Internal Integrity: The internal integrity is tested by Radiographic/Radioscopic inspection methods. ASTM-E 505[3] Standard Reference Radiographs specific to high-pressure die-castings are used for grading categories and severity levels of discontinuities. The types of anomalies normally encountered are porosity, cold fill, shrinkage and foreign material. The acceptable level of these anomalies is solely dependent on the durability of the components for the entire service life. The levels may also vary by location based on the stress levels within a component. The internal integrity evaluation is usually an on-going test for safety critical components such as the crossmember.

CASTING PERFORMANCE TESTING

The ideal way to test the performance of the casting is by a casting performance test (Figure 2). In the casting performance test the whole casting is left intact. A load vs. deformation test is selected for this purpose. The casting is supported at locations where it is attached to the frame. A load is applied by a hydraulic ram at the left rear motor mount, and the deformation is measured at this location by monitoring the ram movement. The ultimate load and total deformation are recorded and compared to castings from daily lots (Figure 3).

This assures the quality of parts is maintained due to day-to-day casting process variation when received at the customer's location.

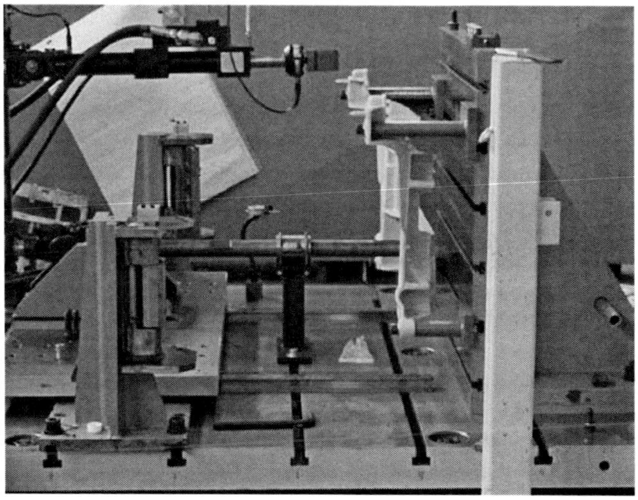

Figure 2: View of casting performance test set up

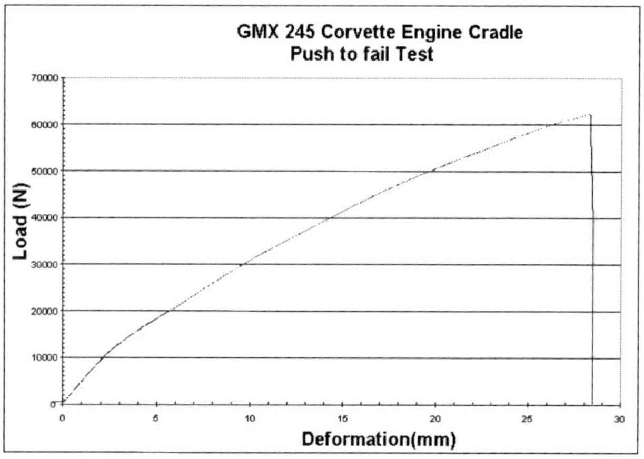

Figure 3: Typical Load Vs Deformation Curve

DESIGN STRATEGY

One of the main issues with the design of the Magnesium Crossmember was addressing its lower yield strength (140 MPa) and lower UTS (240 MPa), than its aluminum counterpart (Yield strength: 195 MPa; UTS: 265 MPa). This matter was handled by adding more interior ribs to the part, to strengthen it in areas of concern. The Magnesium Crossmember has considerably more ribbing designed into it in order to increase its overall stiffness and strength, to make it more comparable to the Aluminum Component. In order to prevent high porosity, a maximum wall thickness of 5mm was used in most regions. Areas of high stress and low stiffness received additional ribbing, rather than increasing the casting's wall thickness. However, some areas of the casting are thicker due to the need for additional bolt hole thread depth, or to close off the end of previously designed through holes, to prevent corrosion between the Magnesium Crossmember and Steel bolts.

The following figures show a comparison of the underneath ribs between the Magnesium Crossmember (top part) and Aluminum Crossmember (Figure 4).

Minimum draft angles of 2 degrees, and minimum radii of 3.0 mm for corners and 4–6mm for fillets were used to improve the casting process. A higher taper angle on the walls and a higher radius at corners and fillets prevents the tool from locking up during use. Some of the radii on the Crossmember are considerably higher than 6.0 mm, in order to prevent stress risers on the component. An example can be seen in the figure below. Increasing the radius at this corner reduced the initial high stress from 217 to 175 Mpa (Figure 5).

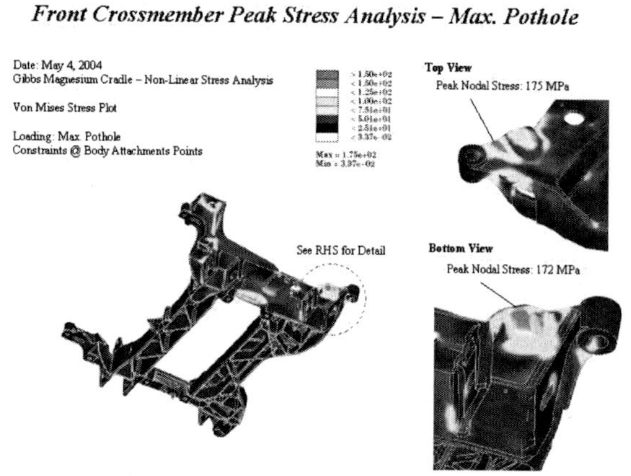

Figure 5: High Stress Area at Corner

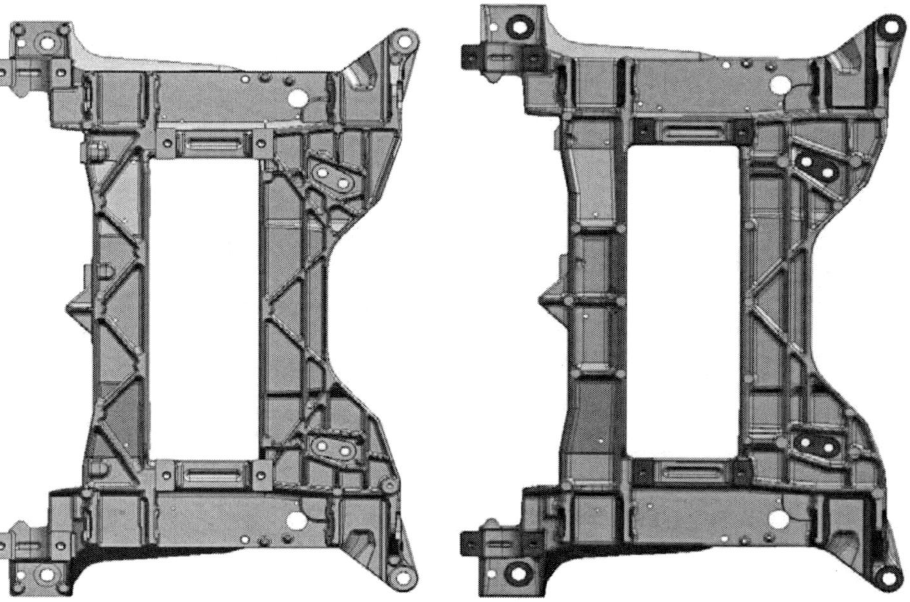

Figure 4: View of Ribs Under Magnesium (top part) and Aluminum Crossmembers

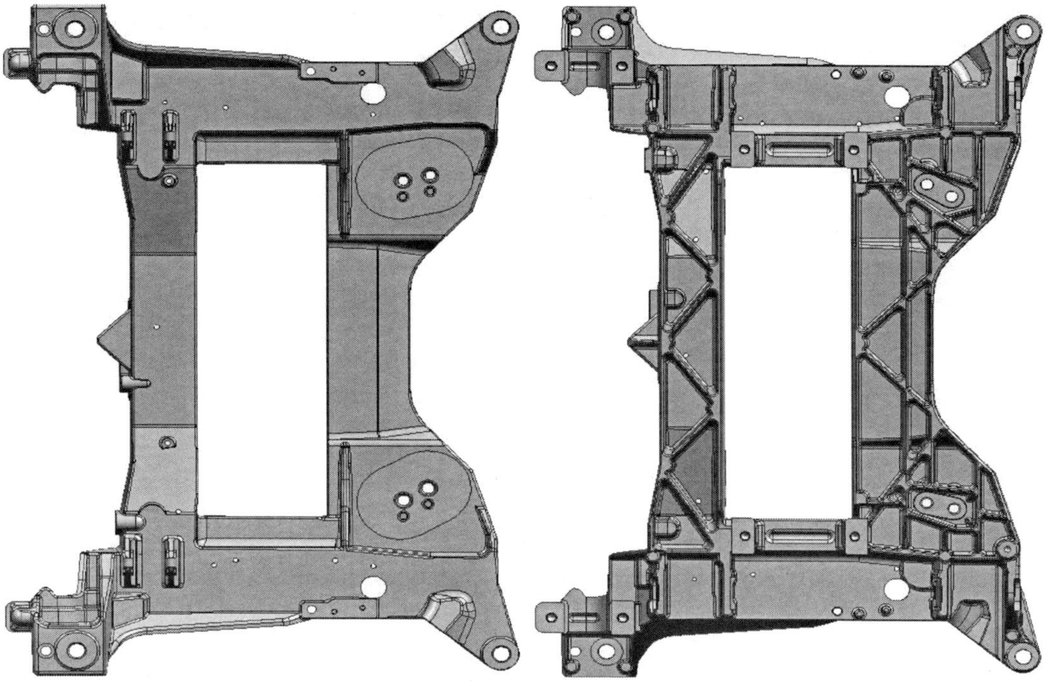

Figure 6: Aluminum Isolator locations (Bottom & Top Views)

Another design issue with the Magnesium Crossmember was corrosion. In order to prevent accelerated corrosion at the joints (1) all Magnesium casting surfaces were designed to be a minimum of 5.0mm away from any interfacing Steel components, and (2) an aluminum isolator was placed between the crossmember and every steel component (fasteners, brackets, etc.), and (3) all through holes were closed off to prevent corrosion in the threads. Figure 6 shows all the locations of the added aluminum isolators (highlighted in orange).

Due to issues with strength and packaging, not all of the casting surfaces could be designed with the minimum 5.0mm clearance. If this was the case, the interfacing components were modified in various ways in order to prevent the accelerated corrosion. If both the Crossmember and part in question could not be modified, the parts were validated on a combination of corrosion tests and durability cars, to determine if the close condition was acceptable.

The aluminum isolators were designed out of Al6061-T6. This material was chosen because of its low Fe content to prevent corrosion, as well as its high strength and hardness characteristics so the joints would be unaffected.

A major requirement of the assembly plant was that all of the Aluminum isolators be sub-assembled to the Crossmember, so that their assembly line would be unaffected. Due to packaging restraints, high stresses in certain concentrated areas, and the design status of the interfacing components' already being completed, none of the isolators could be attached to the Crossmember through mechanical methods. A suggestion of using aluminum fasteners and nuts was raised, but was deemed unacceptable due to the loss in strength going from a steel fastener to Aluminum. Cupped aluminum washers sub-assembled to the bolts and nuts was also proposed, but was not possible due to the high increase in piece price per fastener. Also, this solution would not be feasible for any of the interfacing steel components. With the constraints provided to the team, the group determined that the best way to attach the isolators was to use an adhesive to sub-assemble them to the Crossmember.

The material chosen was a two-part acrylic adhesive ("Lord" brand 403/19 equivalent). Multiple clamp load retention tests were run by Meridian Technologies, Inc., the casting supplier, in order to determine the best adhesive. The main sets of criteria for picking this adhesive were (1) fairly similar performance to the aluminum Crossmember on the clamp load retention tests, (2) no negative effects to the corrosion of the parts, and (3) its ability to keep the isolators attached during delivery to the assembly plant.

The change in material also requires a design change to the amount of threads for every threaded hole on the Crossmember. Usually for Aluminum components, the standard for amount of threads is 2 times the hole diameter. However, since Magnesium has a lower YS and UTS, the amount of threads was increased to 3 times its hole size. This caused the addition of material in certain regions, in order to allow for the extra threads.

PART VALIDATION

The Magnesium Crossmember validation is performed at three separate levels, vehicle, subsystem, and component level. The durability requirement is to pass the equivalent of one vehicle customer life. A customer life is correlated to collected vehicle road load data or a number of test cycles and developed into a test requirement. After the testing is completed, a review and evaluation of the test parts are performed.

VEHICLE LEVEL

Vehicle level validation consists of on road durability, lab simulation, and crash worthiness. One full vehicle lab simulation on a 17-channel multi axis test stand was performed for durability evaluation. The multi axis test was driven from edited road load data. Two vehicles ran durability schedules at the GM proving grounds equivalent to one durability life and ten years corrosion. Additional vehicles were run through 40 potholes and a rough road and abuse schedule for extensive evaluation at the suspension peak load cases. No issues were found with the Magnesium Crossmembers. Several barrier vehicles were evaluated according to MVSS certification and crashworthiness without detriment to the vehicle safety standards.

SUBSYSTEM LEVEL

The subsystem level validation consisted of the full front suspension on an 8-channel multi axis test stand (Figure 7). The test applies loads at the wheel simultaneously in all three directions. Three prototype and three production samples were run to three customer lives without incident to the Magnesium Crossmember. An additional sample was run after 10 years of lab corrosion without incident. Additional steering gear and engine mount tests were performed with specific load cases to further evaluate joint integrity and crossmember performance.

Figure 7 Multi Axis Test Stand

COMPONENT LEVEL

Extensive fastener joint testing, lab accelerated corrosion, and material properties evaluations were performed on the Magnesium Crossmember. A material casting process level test was developed to validate the material mechanical properties (Figure 2). The interface and functional durability requirements were validated at the vehicle and subsystem level.

FASTENER TESTING

Due to the lower strength of magnesium, there was a concern that there would be a negative impact to all of the joints on the component. Because of this, all joints were put through a torque-to-failure test and high and low temperature cyclic relaxation tests. Also, all of the joints on durability cars and validation lab tests are torqued down to the low end of their spec, in order to see if the joint will survive under actual loads.

FEA RESULTS

Due to the similarity between the Aluminum and Magnesium Crossmembers, the predicted stressed of the two parts are fairly similar, since the loads placed on them are the same. However, since the YS and UTS of the Magnesium Crossmember are lower, these high stresses are more of a concern, and must be addressed within the design of the component. The two primary areas of concern are during bolt up of the steering gear and pothole loads (figures 8 and figure 5).

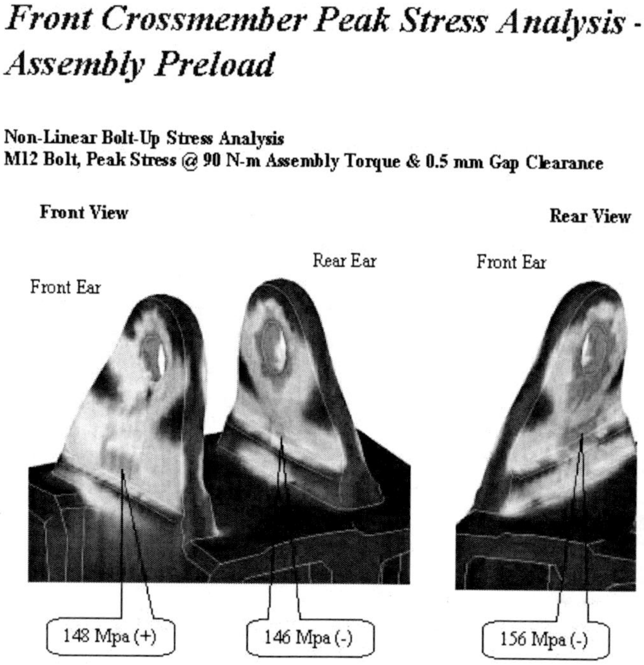

Figure 8 Steering Gear Clevis Stress due to Bolt-Up load

CORROSION TESTING

Corrosion is a major issue when dealing with magnesium. Because of this, extensive testing was performed on different scenarios. Both coated and uncoated crossmembers with production and prototype aluminum components were put through a static accelerated corrosion test in order to see how the parts would perform. Also, two sets of tests were run, with modified sets of aluminum isolators. The isolators were modified in order to provide better corrosion resistance in those areas. Additional corrosion testing is also being run on durability cars.

NOISE & VIBRATION

Analysis between the Magnesium and Aluminum Crossmembers were performed from a NVH perspective. The following graphs (Figures 10, 11, and 12) show that the Aluminum Crossmember performs better. Because of the nature of the Z06, this slight degradation in performance has been deemed acceptable.

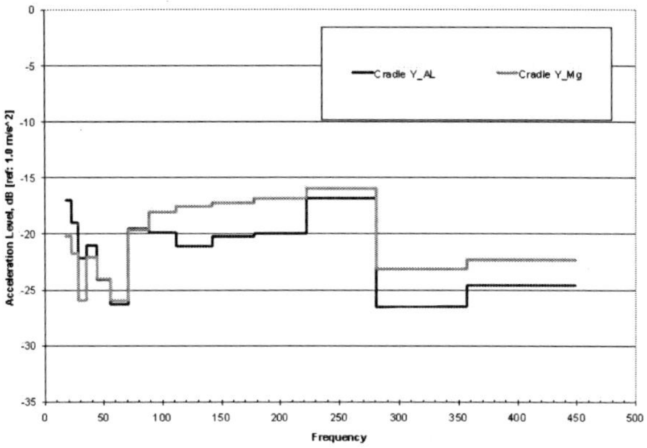

Figure 10: Mg vs. Al Vibration – Course Road 60kph

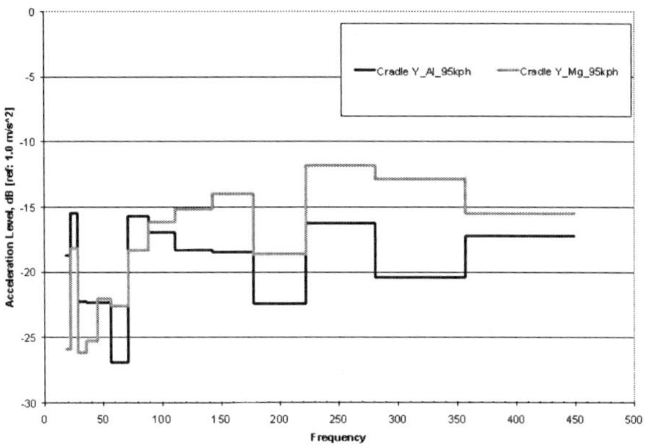

Figure 11: Mg vs. Al Vibration – Course Road 95k

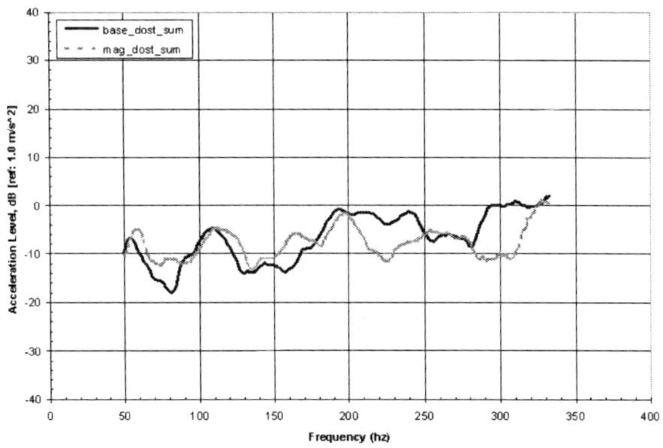

Figure 12: Driver Seat Track Vibration – Power sum of Triax. Low Gear Sweep

CONCLUSION

The Magnesium Crossmember is one of the enablers for a lighter weight vehicle. Development of the Corvette Z06 Crossmember required overcoming key challenges for magnesium as a structural component. The inherent differences of magnesium properties with other structural materials required special consideration when developing and validating the component and interfaces. The magnesium alloy, crossmember design, and validation were focused on these areas to meet the usage requirements. The material properties which had prevented magnesium alloys from being used in chassis applications have been identified to allow rapid application to future products.

ACKNOWLEDGMENTS

The authors would like to acknowledge Dave Hill and Mike Meloeny for championing the Magnesium Crossmember into a production program. Dave Zimmerman for program application contributions. John Kingsley for initial design work. Meridian Technologies and Hydro Magnesium for casting and alloy expertise.

REFERENCES

1. ASTM-B 557. Standard Test Methods of Tension Testing Wrought and Cast Aluminum- and Magnesium-alloy Products
2. ISO 6892. Metallic Materials – Tensile testing
3. ASTM-E 505. Standard Reference Radiographs for Inspection of Aluminum and Magnesium Die Casting

2005-01-0341

2005 Ford GT Magnesium Instrument Panel Cross Car Beam

Naiyi Li and Xiaoming Chen
Ford Motor Company

Tim Hubbert
Lear Corporation

Richard Berkmortel
Meridian Technologies Inc.

ABSTRACT

Ford GT 2005 vehicle was designed for performance, timing, cost, and styling to preserve Ford GT40 vintage look. In this vehicle program, many advanced manufacturing processes and light materials were deployed including aluminum and magnesium. This paper briefly explains one unique design concept for a Ford GT instrument panel comprised of a structural magnesium cross-car beam and other components, i.e. radio box and console top, which is believed to be the industry's first structural I/P from vehicle crash load and path perspectives. The magnesium I/P design criteria include magnesium casting properties, cost, corrosion protection, crashworthiness assessments, noise vibration harshness performance, and durability. Magnesium die casting requirements include high pressure die cast process with low casting porosity and sound quality, casting dimensional stability, corrosion protection and coating strategy, joining and assembly constraints. Magnesium alloy AM60B is used here for its good castability and ductility required for crashworthiness. The castings produced with different casting cavity and fill conditions, i.e. vacuum and aerobic, respectively, will be presented. The resulting magnesium cross-car beam structure enables the Ford GT to reach vehicle performance targets in a timely fashion and budget.

INTRODUCTION

Magnesium has been used as a structural material for automotive components with proven ability to reduce automotive weight at acceptable cost in many automotive applications [1,2]. Magnesium is increasingly becoming an attractive alternative to steel, aluminum, and polymer composites for vehicle weight reduction due to its ability to simultaneously meet vehicle integrity requirements while significantly reducing component weight. Three major attributes of magnesium are as follows,

- Magnesium has a lower density. At 1.8 kg/L, magnesium alloys are 20 percent lighter than polymer composites, 30 percent lighter than aluminum alloys and 75 percent lighter than steels and zinc alloys.

- Magnesium has outstanding castability. Components can be cast at 1.0 to 1.5 mm wall thickness and 1 to 2 degree draft angles, which are typically 1/2 that of aluminum. Magnesium's extensive fluid flow capability allows a large casting to replace a multitude of steel fabrications (such as radiator supports, instrument panels, and steering wheels) where ductility, energy absorption and impact resistance are important; e.g., a steel cross-car beam instrument panel (I/P) has upwards of 35 different parts while a comparable die cast magnesium I/P contains only 5.

- The surface "skin" of die cast magnesium has better mechanical properties over the bulk [3]. Consequently, thinner (ribbed) and lighter die-castings of magnesium are less costly, and can have sufficiently high strength per unit area to compete with heavier aluminum and plastic sections.

2005 Ford GT, space-framed vehicle as shown in Figure 1 was designed for performance, timing, cost, and styling to preserve Ford GT40 vintage look. Many of the key Ford GT program directives revolved around system level specifications for vehicle dynamics and how this car would "feel" when compared to the best supercars in the world, for example, the Ferrari 360 Modena. The use of magnesium for the instrument panel structure was identified early in the concept stage of the program as a viable way to reduce vehicle weight and improve vehicle dynamics.

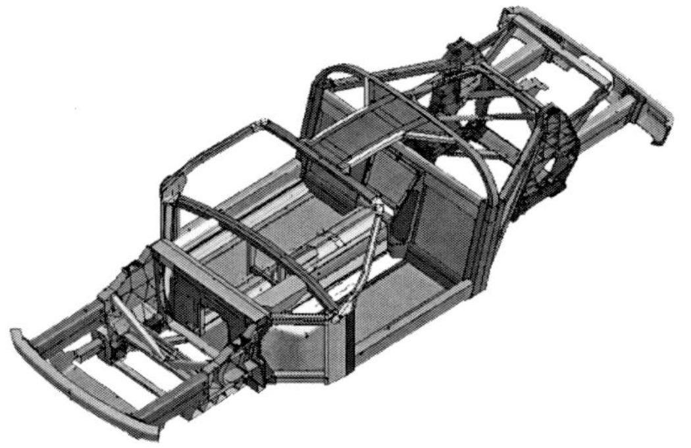

Figure 1. Ford GT 2005 space-frame vehicle

In terms of producing magnesium components, the main challenge, however, is encountered when performing die-casting of larger structural components that are characterized by thin walls. In such cases, molten magnesium penetrating long flow lengths in thin-wall mold cavities tends to freeze in the runner system during the turbulent flow and high-heat transfer rates characteristic of high pressure die-casting machines, producing porosities and unacceptably high variability of ductility, say between 3 to 10 percent in the cast part.

Other than casting design and manufacturing, several component performance characteristics were identified prior to the adoption of magnesium for Ford GT structural crosscar beam. First, design and FEA efforts were conducted in the vehicle model to assure the magnesium I/P beam would contribute significantly to meet the torsional stiffness of the vehicle. Also, due to magnesium's damping capacity, the magnesium beam was considered critical to the support structure vibration characteristics of the steering column and would provide a specified natural frequency. This paper describes the details of Ford Motor Company's 2005 MY FORD GT supercar which has implemented an industry first structural I/P from vehicle crash load and path perspectives.

OVERALL DESIGN PROCESS

To execute the instrument panel design, the interior team of the Ford GT was assembled quickly and comprised a number of experienced design engineers, safety experts, manufacturing and material specialists, and CAE analysts. The design of the magnesium I/P beam and console top was driven by the functionality, package constraints, NVH, durability and safety requirements. The team completed the preliminary design shown in Figure 2 through brainstorming, considering all the requirements. With the main objective to reduce vehicle weight and improve vehicle dynamics, the design objectives and performance measures cab be categorized in the following,

- Full frontal rigid barrier impact (FMVSS 208)
- Dynamic side impact (FMVSS 214)
- Stiffness and modal frequency analysis
- Weight to cost benefit

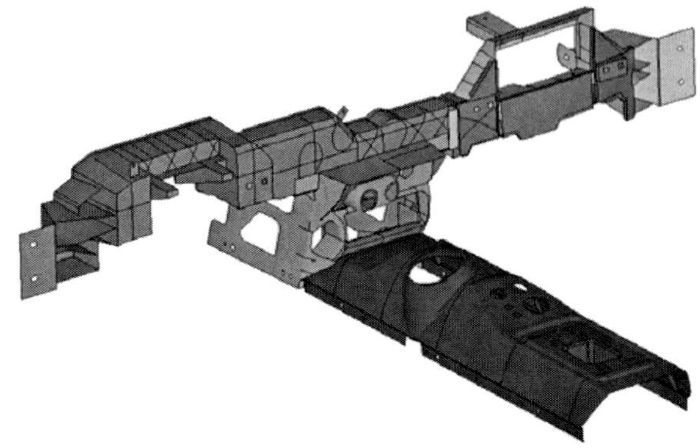

Figure 2. Design of the Magnesium I/P crosscar beam and console top

For design iterations depicted in Figure 3, the team applied finite element analysis (FEA) as a major tool. The finite element models of the magnesium I/P beam were analyzed using LS-DYNA® (version 960) for dynamic analysis where the properties of magnesium were prescribed by the 'simplified Johnson Cook' model, i.e. MAT 98 of LS-DYNA®. FEA gave directions and suggestions to modify the I/P beam design to meet requirements. Note the two different levels of finite element models were built, i.e. sub-vehicle model and full-vehicle model for better simulation and correlations to meet the integrity/stiffness requirement. Also, this analysis flow enables us to apply appropriate loading and boundary conditions during both sub- and full-scale crash events with numerical efficiency.

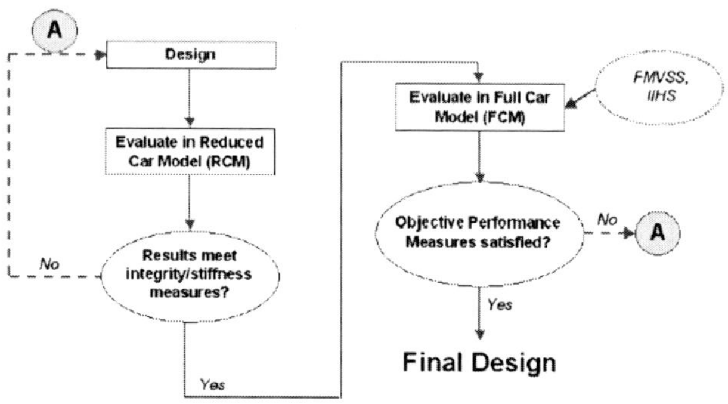

Figure 3. Flow Diagram of Mg I/P Design Development

Given that the class A surface of the I/P was essentially established by the Ford GT concept car, the structure necessary to achieve the NVH and safety requirements needed to be contained under the pre-defined skin. In order to maximize the safety and NVH performance of the I/P structure, targets for both systems were established and fed into FEA models for optimization.

Note that the space frame components are mainly made of aluminum extrusions. The magnesium structural beam and its connections to the aluminum extrusion cowl become a significant part of the load paths during a frontal crash event if any. In addition, the I/P beam and its connections to the hinge pillar and cowl provide significant support to the crush forces induced during impact. Safety targets include the amount of load from a particular frontal crash pulse that would need to be transferred from the front structure, into the cowl, outward from the I/P structure, up through the A-pillars and toward the rear of the vehicle. Following the vehicle chassis design, the distance between the cowl cross member and the I/P beam is set to 400 mm or 16 inches. In order to provide additional load paths for frontal impact crashes and support for better NVH performance, two aluminum struts in longitudinal direction are employed to connect the cowl and the I/P where the conceptual 'truss' design is shown in Figure 4.

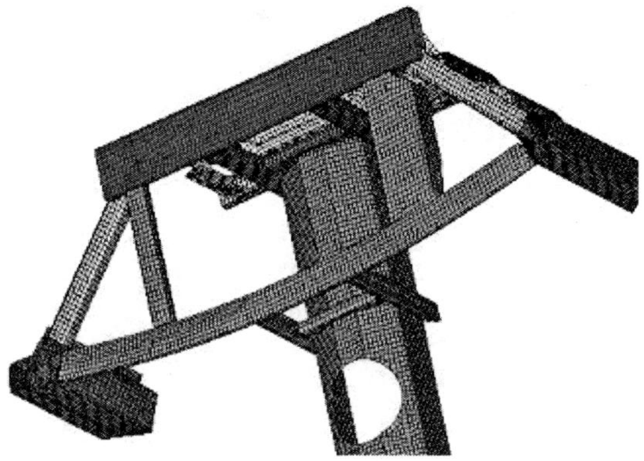

Figure 4. I/P design concept with two struts

The design is to enhance occupant protection upon frontal crash. The center radio stack structure, to house the radio, start button, passenger airbag deactivation switch and 12V power supply is bolt fastened to the vehicle, and transfers any load induced in the I/P structure down into the tunnel. In essence, the occupant safety is secured but not compromised during a crash event. The resultant structure was designed to be an effective load path and a visible, stylish design element. Working closely with the vehicle design studio personnel, we have the final center magnesium casting which provides an improvement in safety and NVH as well as a quality-integrated look.

Typically I/P beam is not designed as a load-carrying component in vehicle. In Ford GT, the I/P beam is a load-carrying component. It will take the load from the struts and then transfer it to the tunnel of the vehicle. The load target for each strut is at 25 KN (5,600 lbs.) based on the energy management of the front structure. In another words, the I/P beam needs to be able to take total load of 50 KN (11,000 lbs.) while maintains its integrity during a frontal crash event. The I/P beam also needs to be designed by considering the material property variations [4, 5].

One FEA sub-vehicle model shown in Figure 5 was developed to evaluate the I/P beam design. It has about 100,000 shell elements. With the boundary and loading conditions, i.e. the two struts subjected to the target load of 25KN each while the hinge pillar, roof and the back of the tunnel totally constrained, the resultant plastic strains experienced in the I/P beam were calculated by using LS-DYNA®.

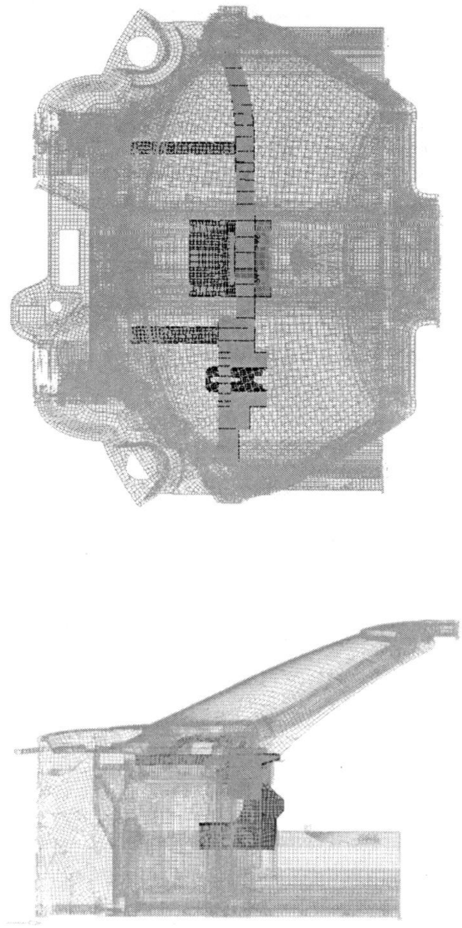

Figure 5. FEA sub-vehicle model

Ultimately the I/P beam design is to be complete based on the iterated FEA results, and I/P beam section geometries, casting ribs, wall thickness and I/P beam to body attachments defined throughout the process. A serious of FEA simulations using the full-vehicle model validated the magnesium I/P beam designed starting with the sub-vehicle model. It was found that the loading on each strut does not exceed 25KN shown in Figure 6 and the magnesium I/P beam could carry the desired load while maintaining its integrity. The simulation results also provide a viewgraph based on the plastic strain contour of the I/P beam, which may be used as valuable information for casting process development.

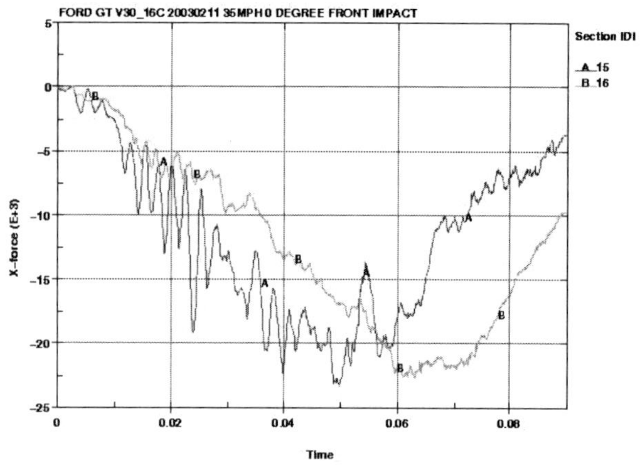

Figure 6. Loads go through the struts in full vehicle model

To gain confidence on the design, material properties and FEA results, one component-level testing was developed to evaluate only the driver side beam subject to a quasi-static pull loading as shown in Figure 7. Care was taken during the pull test to eliminate any dynamic effects. The test results of both load vs. displacement and failure areas compare favorably with the simulated one done by LS-DYNA [6].

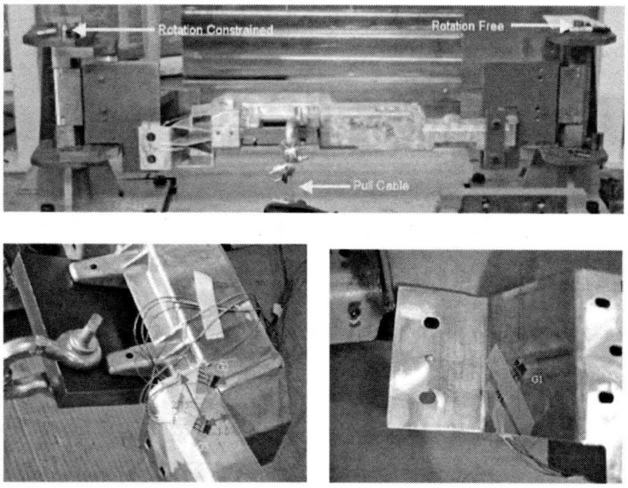

Figure 7. Instrumented beam and pull test setup

NVH CONSIDERATIONS

The NVH targets, established for the I/P structure by the program were to ensure the driver receives dynamic response from the steering column, steering wheel, I/P beam and pedal package without undesired system noise. The targets of resonance bending frequency (Hz) for both vertical and lateral bending modes are as follows,

	Vertical Bending	Lateral Bending
steering wheel & column	45	48
steering wheel, column & I/P beam	38	40
in-vehicle	35	35

The vertical bending stiffness target for the steering column & I/P beam fixed to ground is set to 1200 (N/mm).

In general, bending frequencies drop as the system boundary conditions migrate from bedplate to crosscar beam to in-vehicle. The reason is that each step adds compliance to the overall system, which lowers the bending mode frequency. The testing was conducted with the GT steering wheel/airbag as shown in Figure 8. From NVH design considerations, reducing the steering wheel/airbag mass is the most effective way to achieve the target bending frequencies. Note here increasing I/P beam stiffness and/or body stiffness will not significantly increase the bending frequencies.

Figure 8. NVH test setup of the magnesium I/P beam

In part of the efforts, the team built the NVH models using HYPERMESH® software. With the OPTISTRUCT solver of HYPERMESH®, normal mode analysis was conducted to check natural frequencies and stiffness of the steering column and the I/P beam. The simulations evaluated the design concepts, beam geometry and attachments. It was revealed that both the beam stiffness and the beam attachment to the tunnel were critical for NVH performance. A number of design changes were made based on the FEA analysis. As a result, the responses of resonance bending frequency of the I/P beam become acceptable, and the design lead the body attachment stiffness and the I/P beam stiffness to be acceptable which provides adequate structural integrity.

ENERGY ABSORPTION BRACKET PACKAGE REQUIREMENT

It is necessary to accommodate the travel distance for the energy absorption (knee bolster) brackets to meet the femur load requirement. A die-cast part can be designed to provide the EA brackets with two inches of crush distance by chamfering the rear bottom of the I/P beam as shown in Figure 9. This design not only maintains the beam cross-section for stiffness and load carrying capacity but also allows the EA brackets to achieve its design objective to absorb energy expected in crash event.

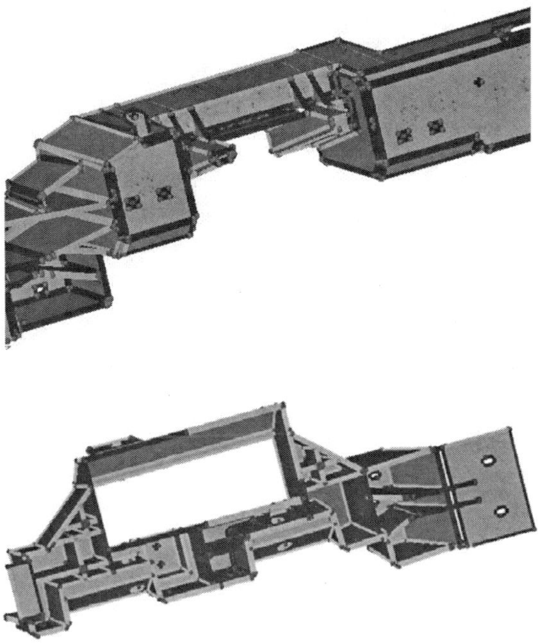

Figure 9. I/P beam design for EA (knee bolster) distance

BEAM – COVER CLEARANCE REQUIREMENT

Another design consideration was to meet head impact requirements by making sufficient clearance between the I/P beam and the panel cover. The impact zone to be checked is shown in Figure 10. A series of impact simulations were performed with LS-DYNA®. The simulation results indicate that the panel cover and the I/P beam will not be in contact with the designed clearance between them.

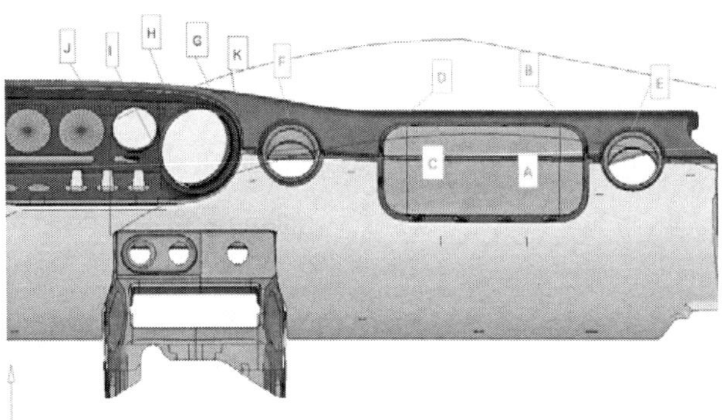

Figure 10. Impact zone definition

MANUFACTURING PROCESS

The Ford GT I/P beam contains seven magnesium castings. There are driver beam, passenger beam, three cover plates, center stack radio box, and console top as shown in Figure 11. There were 6 prototype beams made of sand casting process. Once the preliminary design was confirmed, these castings were produced with the high pressure die casting cold-chamber process with caster size up to 1200 ton. Different from AZ91D, the magnesium alloy used for this application is the AM60B alloy per ASTM-B94 [7] (magnesium alloyed with 6% aluminum and <1% manganese), which offers higher elongation, and is extensively used for crash structural components like instrument panels, seats frames or steering columns. This ductile magnesium alloy belongs to the AM family (magnesium alloyed with aluminum and manganese).

Figure 11. Interior view

The design and manufacturing sequence are as follows. The design of these castings was completed using SDRC-IDEAS® and FEA analysis utilizing LS-DYNA® later validated the casting design to meet its functional requirements. The die tools were prepared with gating and runners based on the tooling supplier's own experience. Some results of fill and solidification analysis on the driver side beam were conducted using finite-element based code EKK® as shown in Figure 12 where potential porosity may be predicted in the castings. Going through the casting simulation analysis, we may presume this will happen prior to production of the car and help to minimize casting defects caused by premature design.

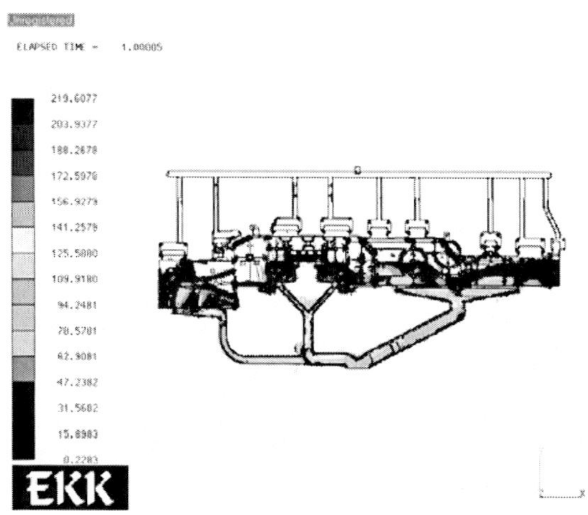

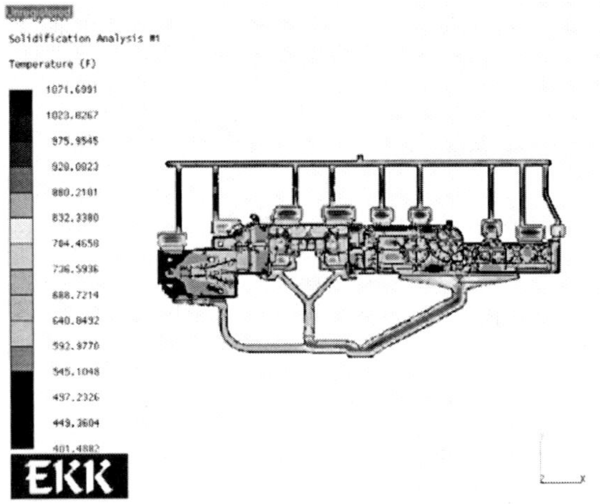

Figure 12. Fill and solidification analysis of the driver beam

The properties of magnesium alloy AM60 do vary in a large high pressure die casting component. The I/P beam has to be designed considering the material property variations. We verified the mechanical properties of some sub-size tensile specimens extracted from the castings, particularly the driver and passenger beams per ASTM-B557 standard specifications [8]. Among the properties, the elongation is worthy to note that the beam was expected to have the averaged value in the range of 3 to 8%. However, when testing properties on tensile coupons of the I/P beam, where elongation properties are sometimes not in the range of 6 to 8% as reported in the literature, the scatter encountered exists. It seems that coupon tensile values may mislead and be biased by both the process and the intrinsic potential of the alloy. [5]

Investigations have been carried out on coupons to try to explain the difference in elongation scatter, and to get a better understanding of the local properties and their correlation to the microstructure. By taking efforts along metal quality to meet the AM60 chemical composition specification and the die tools modifications for better casting filling, the improved results of the average elongation of 10 beams at 4 specific locations are shown in the following,

Sample location	Before	After
2	1.05	6.72
5	1.70	5.97
6	2.23	3.24
7	1.53	4.34

Further, in addition to tensile coupons, a number of physical tests have been extended to build reliable design data from the I/P beams at component level rather than coupon level. The efforts include both the quasi-static pull test and the dynamic pendulum test described earlier. Based on the tests at the component level, we have confirmed the IP beam meet the design requirements.

As for casting quality assurance with NDE techniques, we have used the high energy of 9 MeV x-ray computed tomography system at Ford to evaluate the casting per ASTM-E505 specifications [9] at 1mm scan increment. A typical scanned result of the computed tomography is shown in Figure 13. These findings provided the directions to optimize the casting process window and the design to minimize casting defects including porosity and cracks.

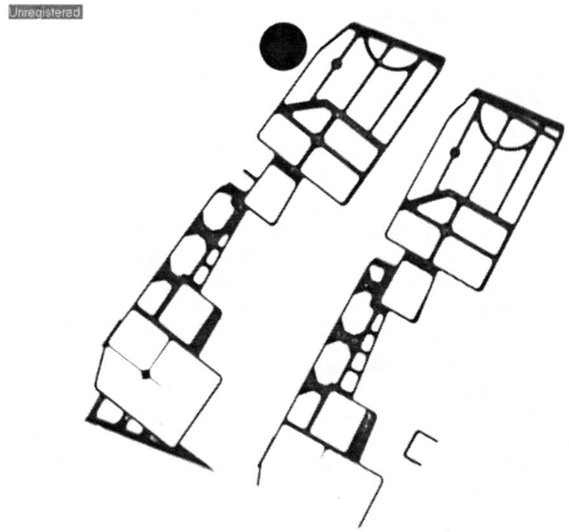

Figure 13. X-ray scan image of the driver beam (one layer)

DIMENSIONAL CONTROL / BUILD QUALITY

During the vehicle's development cycle, DVA's (dimensional variation analysis) results showed that the dimensional capabilities of the magnesium die casting components and casting process, and of the GT I/P in general, would improve I/P fit-and-finish, and decrease undesired gaps. We achieved this in several ways, including post-casting part handling, careful selection of datum structures for assembly methods, fixtures and design considerations.

CORROSION PROTECTION AND I/P BEAM ASSEMBLY

In general, there are two corrosion concerns, i.e. general and galvanic. There is no general corrosion concern here with the I/P beam castings because they are protected interior pieces. They are not likely to be exposed to foreign materials such as water, wind, stones, mud and salt introduced at a velocity to remove protective material treatments and coatings which lead metal to corrosion over time. However, the presence of chlorides, sulfates and foreign materials that hold moisture on the surface can promote corrosion and pitting of unprotected magnesium material. As for the galvanic corrosion concern, caused the electrical potential difference between the dissimilar metals, the less noble material, magnesium in this case, becomes the anode of the corrosion cell and tends to corrode at an accelerated rate, compared with the uncoupled condition. To prevent both corrosion concerns, we have the following procedures for the casting surface finish,

- Henkle Alodyne 5200 pre-treatment coating
- Protech powder coat
- Dow Betamate 1480, a one-component epoxy adhesive for joining the two beams and cover plates
- Mechanical fasteners to aid during the adhesive curing process.

WEIGHT BENEFIT

The total weight of the I/P beam castings assembled is 13 lbs. Regarding weight saving with magnesium I/P beam, there is no comparator as Ford GT does not have a carry-over I/P beam. Typically, 30% weight saving from an equivalent aluminum version is common.

CONCLUSION

Ford GT 2005 vehicle in Figure 14 was designed for performance, timing, cost, and styling to preserve Ford GT40 vintage look with the program starting in year 2002. Ford's commitment to the use of lightweight materials throughout its entire range of vehicles has continued to increase over the last two decades. Magnesium substitution in the automotive industry has increased steadily. Ford Motor Company and Lear Corporation jointly worked on the program to design and implement the magnesium instrument panel's crosscar beam on the Ford GT. The resulting magnesium cross-car beam structure enables the Ford GT to reach vehicle performance and safety targets in a timely fashion and budget.

Figure 14. 2005 Ford GT

ACKNOWLEDGMENTS

The authors would like to thank the magnesium I/P beam team personnel from FORD, LEAR, MERIDIAN, DOW AUTOMOTIVE, ALTAIR, ARUP and MAYFLOWER who worked so hard to ensure that all the bases are covered, especially personnel of LEAR who developed the initial I/P beam durability & stiffness CAD models. We would also like to acknowledge John Hill, Ron Cooper, Tom Teknos, Jeff Webb from Ford Motor Company for their technical contributions and the generous support from Ford GT program and R&AE management throughout the program.

REFERENCES

1. D. Albright, 'Global Automotive Applications Review', the 13th Automotive Magnesium Seminar, International Magnesium Association, Plymouth, MI, 10 April, 2002.
2. J. Balzer, P. Dellock, M. Maj, G. Cole, D. Reed, T. Davis, T. Lawson and G. Simonds, 'Structural Magnesium Front End Support Assembly', SAE paper 2003-01-0186.
3. W. Sequeira and G. Dunlop, 'Microstructure, Mechanical Properties and Fractography of High Pressure Diecast Magnesium Alloy AZ91D,' p213-221, NADCA Congress 2003, Indianapolis, IN.
4. X. Chen, T. May, J. Webb and D. Wagner, Magnesium Automotive Structure Component Design, 59th Annual World Magnesium Conference, International Magnesium Association, pp101-106, 2002.
5. R. Alain, T. Lawson, P. Katool, G. Wang, J. Jekl, R. Berkmortel, L. Miller, J. Svalestuen and H. Westengen, Robustness of Large Thin Wall Magnesium Die Castings for Crash Applications, SAE paper 2004-01-0131.
6. Kumar, E. Nelson and T. Hubbert, Use of Magnesium in Strustural Crashworthiness Applicatioins, 61st Annual World Magnesium Conference, International Magnesium Association, pp130-137, 2004.
7. ASTM-B94, Standard Specification for Magnesium-Alloy Die Castings.
8. ASTM-B557, Standard Test Methods of Tension Testing Wrought and Cast Aluminum- and Magnesium-Alloy Products.
9. ASTM-E505, Standard Reference Radiographs for Inspection of Aluminum and Magnesium Die Castings.

CONTACT

Naiyi Li, nli1@ford.com

Xiaoming Chen, xchen@ford.com

Tim Hubbert, thubbert@lear.com

Richard Berkmortel, rberkmortel@meridian-mag.com

2005-01-0723

Use of Cast Magnesium Back Frames in Automotive Seating

Matthew Abate
Intier Automotive Seating

Mike Willman
Meridian Technologies

ABSTRACT

The seat frame can play an important part in improving the overall craftsmanship and safety of the seat. This paper discusses the design and research done to a cast magnesium seat back frame to affect seat craftsmanship and safety.

INTRODUCTION

The seat back frame can play an important role in overall seat craftsmanship, safety, and optimized seat mass. In contrast to traditional steel frames, which consist of multi-piece stampings, bent tubing and wires, which are welded or swedged together a magnesium frame is a single part casting. These joining processes cause issues with tolerances and craftsmanship. Cast magnesium frames have no welding which means less manufacturing variations. With improved dimensional control, a cast magnesium frame will provide better craftsmanship to the overall seat.

Craftsmanship can be defined as overall appearance of the seat, with particular focus on intuitiveness operation and effort to actuate items such as armrests, head restraints and recliners. The magnesium back frame can help achieve the high level of craftsmanship if designed properly.

The magnesium back frame can be designed to define the STO (seat trim outline) and help support the trim cover/foam pad. Typically, the upper corners of a welded frame do not support the upper corners of a seat back. This could lead to trim/foam break down, especially in 2nd or 3rd row seating, where the seat back could be used as an assist handle to get out of the seat or vehicle.

The magnesium back frame has the option for more complex load paths and optimized design due to the manufacturing process.

OVERVIEW

Steel frames were originally designed for this program. To save mass, the customer requested the use of magnesium back frames. As the magnesium frame project progressed additional, benefits came forward. To compare how a magnesium back frame can improve the craftsmanship of the seat over a tradition welded steel frame, first look at part integration. Welded frames can be comprised of many different parts: different material thicknesses, material types and size. The 60% split bench seat steel back frame consists of 14 parts (side members, weld nuts, tubes and stampings). The 40% split back frame consists of 10 parts. The masses of these steel frames are 2.67 kg (5.88 lbs) (60%) and 1.95 kg (4.3 lbs) (40%). The masses of the cast magnesium frames are 1.32 kg (2.92 lbs) and 1.03 kg (2.26 lbs). A combined weight savings of 2.27 kg (5.0 lbs) was achieved using a cast frame versus a weld steel frame.

The one-piece cast magnesium frame saves mass and allows specific attachment features to be integrated into the back frame for head restraint sleeves, torsion bars or clock springs, plastic bezels, hard board panels, recliner bracket, pivot points and soft trim/foam support.

HEAD RESTRAINT SLEEVE ATTACHMENT

One of the more common methods of head restraint sleeve attachment is using a pair of independent steel tubes that are welded to the back frame. A common problem with this method is controlling the parallelism of each tube. If one tube is more than 1.0mm out of parallelism head restraint efforts can increase 15%. Comparing a typical steel frame design (steel tubes welded to the frame) and a cast magnesium c-section frame, the head restraint efforts decreased 14.7%(up) and 57.5% (down). The steel frame head restraint up effort is 70.3 N and the down effort is 48.1 N. The magnesium frame head restraint up effort is 60.0 N and down effort is 20.4 N.

To poke-yoke the sleeve attachment a design may have different tubes, one tube having one notch and the other two notches. These notches can also have sharp edges that can affect handling and sleeve installation. The magnesium frame can be designed with the poke-yoke feature built in.

A magnesium back frame can accommodate several types of designs for sleeve attachment, such as a tube or c-section. See figure 1 for an example of the c-section used for this project.

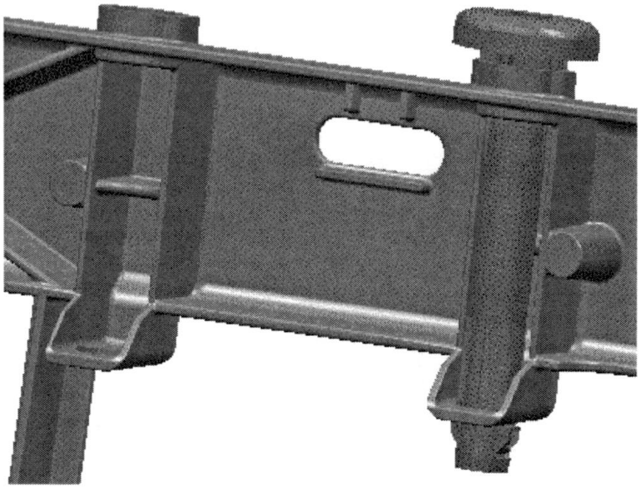

Figure 1 – Back frame c-section.

If using a tube design, relief pockets must be incorporated to avoid thick cross sections. Thick wall cross sections are subject to porosity and shrinkage. Ribs can be added to the relief pocket, if needed. A c-section design allows for top and bottom points of contact between the sleeve and frame. It also allows for a smooth flat surface. The flat surface can face either forward or rearward. Having the flat surface face rearward allows the back of the seat to have a very clean and smooth appearance. The relief pockets will show through the trim cover, thus affecting the appearance of the seat back. The final size of the attachment features are machined or drilled. This machining process is done in pairs; being machined in pairs the sleeve-to-sleeve positional tolerance and parallelism is better than a welded frame. If one of the tubes is a datum referenced back to datum's ABC, the true position of the secondary tube can be +/- 0.25mm, see figures 2 and 3. With a typical weld frame, that same tolerance would be +/- 0.75mm. In addition, during the drilling process a chamfer can be added to the top of the attachment to aid handling and installation.

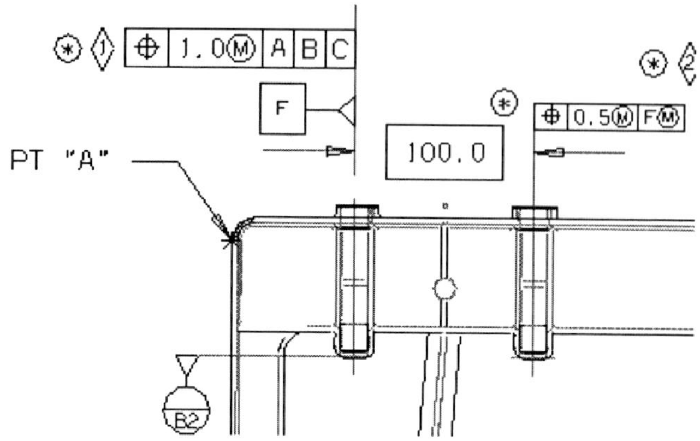

Figure 2 - Head restraint GD&T.

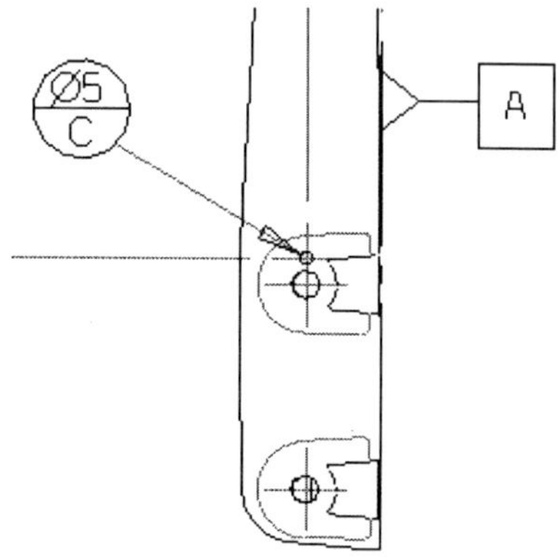

Figure 3 - Datum A and C locations.

BACK PANEL ATTACHMENTS

A magnesium back frame can have back panel attachments designed in without special machining. If the hard board back panel uses plastic "x-mas tree" fasteners, the attachment holes can be built into the die. Hole diameter can be as tight as ±0.15mm. This is the preferred method of attaching a hardboard panel. A cored hole requires no secondary machining, which keeps the piece price down. A cored hole can be added to almost any location on the back frame. Core pins must be in the line of draw.

MECHANISM ATTACHMENTS

Mechanism attachments can be designed into the magnesium back frame using cast embossments for the attachment fasteners. The minimum thread engagement should be designed to accommodate up to 2 times the nominal bolt/screw diameter. To optimize mass and service of each joint, 1.5 times nominal bolt/screw diameter was used, but joint testing is recommended. The joint should be designed for a minimum of five bolt rundowns to the maximum torque. The necessary ID hole size in the embossment is processed by drilling or by coring it as part of the casting. Thread rolling screws were found to provide greater thread strength than a standard machine type thread with a tapped hole, mainly because the thread rolling screw joint can account for 100% radial thread engagement verses 65%, which is typical for a machine type thread with a 6g/6H tolerance class. Even though there are special thread forms made for magnesium, an adequate joint design can be accomplished using a standard Type TT (trilobular) thread form. The only real advantage of using the special magnesium thread forms is a slight reduction in debris generated in the thread forming process and slightly increased rundown speeds. Using a M10 Type TT screw, a typical structural joint would require an emboss height of 20.0mm with a hole ID of 8.94+\-0.15mm. A typical torque range would be 50.0 Nm +\- 5.0 Nm with a property class 9.8 or 10.9 screw.

A torque to failure study was performed using a standard M10 Type TT screw; magnesium plates 20mm thick and pre-drilled holes (Ø8.94 mm). A 2.0mm thick washer was used to simulate an attachment bracket. See Table 1 for results. All torque results are in N-m. Sample 5 reached the tool capacity and shut down. All other samples, the bolt striped out. See figure 4 for a torque to failure rundown trace.

SAMPLE	Max drive torque	Max Torque
1	17.6	97.6
2	19.4	106.1
3	18.7	99.9
4	18.1	90.0
5	21.3	125.9
6	19.1	104.0
7	17.9	100.9
8	20.0	97.9
9	18.4	101.1
10	19.0	89.6

Table 1 – Torque to failure study.

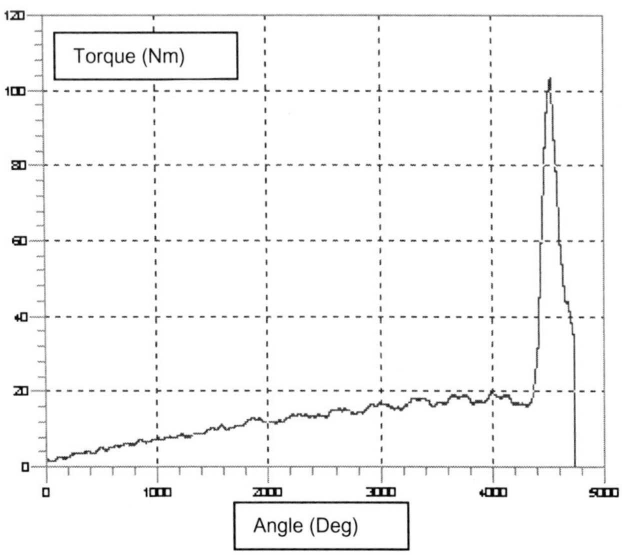

Figure 4 – Torque vs. angle trace sample 6.

Samples 11 to 14 had additional run-downs to verify that at least five are achievable using a type TT thread. Samples 11 and 12 were only tested for one and two rundowns. Samples 13 and 14 were rundown 5 times before torque to failure trial. The drive torque is the torque reading from the final additional rundown. The initial drive torque would be similar to table 1. All trials were able to meet initial torque setting. See table 2 for results. All torque results are in N-m. The torque to failure rundown failure in all samples was bolt strip out.

SAMPLE	Max drive torque	Max torque to failure	Initial torque setting	Rundowns before failure
11	5.0	90.3	50	1
12	8.0	84.8	50	2
13	2.0	72.1	55	5
14	2.0	97.8	60	5

Table 2 – Multiple bolt rundowns.

Based on the above data, a Type TT thread forming screw will work in magnesium; the performance remains adequate after five run-downs. The torque to failure results after five run-downs is still in the 60–90 Nm range, which will still support the initial torque setting of 50N-m ± 5.0 N-m. The draft on the attachment area can be reduced from 2 degrees to 0.5 degrees.

By contrast a typical steel frame would use a weld nut, clinch nut or extrusions plate. The cast magnesium frame integrates the attachment feature into the frame itself.

Since the embossment is a large volume of magnesium (cross section) one concern is porosity, this is minimized with gas venting. Each embossment should have its own vent to allow the gases to escape and liquid magnesium to fill the cavity.

TRIM AND FOAM SUPPORT

A magnesium back frame is not limited to steels minimum bend radii for tubing or stampings. A casted part can be designed to follow the STO of the seat back to help support the trim cover and foam pads. A thin profile seat can benefit from the versatility of a magnesium back frame. The frame can be designed to follow the STO within 10mm. The frame can be designed to support the upper corners of the seat; this proves helpful especially in second and third row seats where passenger may grab the seat corner in front of them to assist with ingress and egress. Without this support, the foam pad will wear over time and have a negative affect on the appearance of the seat.

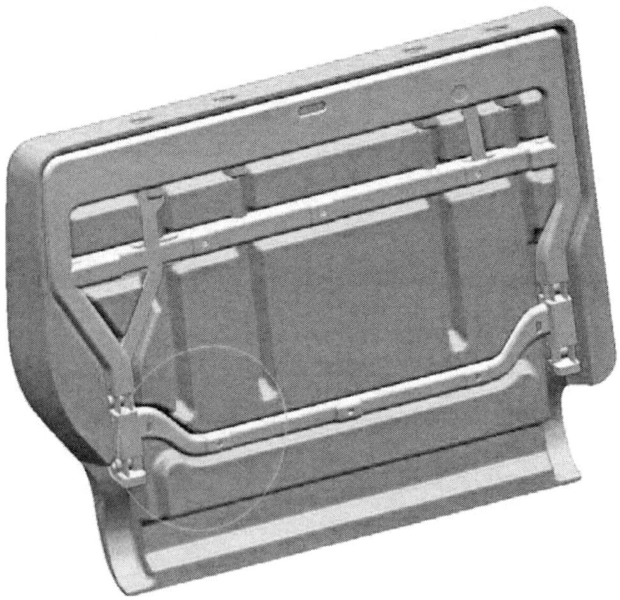

Figure 5 – Frame to foam nesting.

Figure 5 shows a magnesium back frame to foam pad nesting. The close frame to foam relationship is critical for freestanding split bench seats to achieve straight edges and minimal gaps between seats.

The rough edges of overflows and gates are minimized with the use of trim dies. Trim dies break off the excess material at the overflow and gates. Any rough edges that remain are less than 0.8mm for feed areas or gates (figure 6) and less then 0.6mm for overflow areas (Figure 7). Parting line flash height is nominally 0.38mm. If extra attention is required for certain areas, hand filing is an option. All wall sections can be designed with a radius to reduce damage to the trim/foam interface.

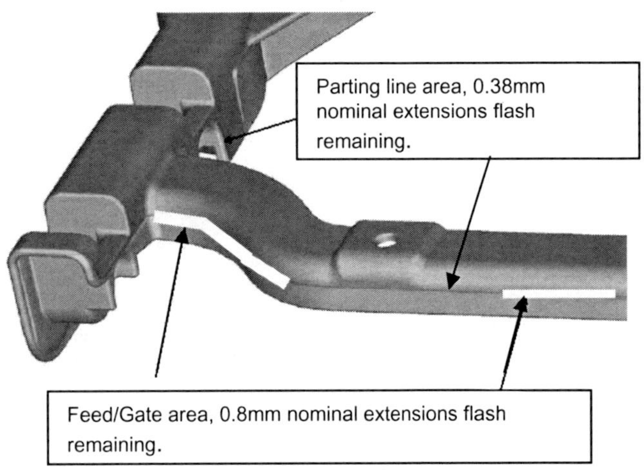

Figure 6 – Gate and parting line flash height.

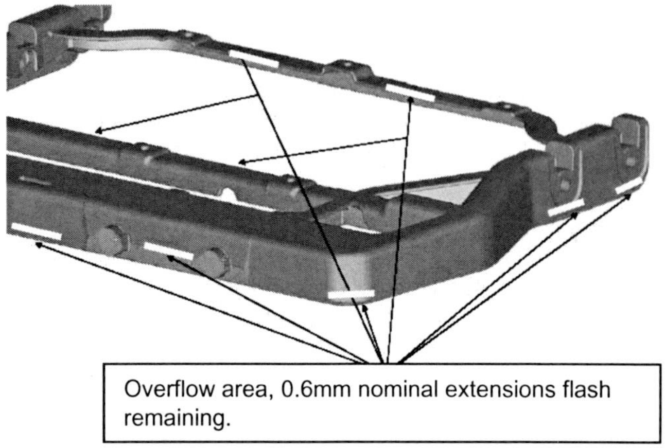

Figure 7 – Overflow nominal flash height.

DEVELOPMENT

Extensive Finite Element Analysis (FEA) was completed before tooling was started. This allowed both prototype (P20) and production (H13) tools to start within 3 weeks of one another. A frame cross-section that can meet both government regulations and customer craftsmanship requirements is an L-shape cross section. A c-section was used along the top to incorporate the head restraint sleeve attachment and provide additional strength for FMVSS 201 and 225 testing. See figure 8.

Although magnesium is a ductile material, the design needed to be comparable with a steel frame. FEA was run to compare back frame stiffness. See table 3 for results. Back frame stiffness can affect the performance of other mating parts. The magnesium back frame deflects more than the steel structure, but the torsional stiffness ratios from o/b corner to i/b corner are relatively close. Table 3 represents due care testing.

Load Applied	Displacement @ loading location (mm)					
	Magnesium Frame			Steel Frame		
	O/B	Ctr	I/B	O/B	Ctr	I/B
89 N @ O/B corner	0.6	0.38	0.07	0.16	0.13	0.03
89 N @ center	0.38	0.64	0.32	0.16	0.51	0.15
89 N @ I/B corner	0.07	0.32	0.49	0.03	0.13	0.16
890 N @ O/B corner	5.91	3.76	0.71	1.65	1.33	0.3
890 N @ center	3.73	6.26	3.14	1.64	5.21	1.55
890 N @ I/B corner	0.68	3.2	4.83	0.3	1.28	1.64

Table 3 – Magnesium vs. steel back frame torsional stiffness comparison.

Asides from FEA, the magnesium back frames were verified for safety by actual structural testing. Static and dynamic tests were used. Static testing included the following: rearward load fatigue (20,000 cycles), occupant restraint (447.4 N-m moment about h-point), latch strength (24 x seat back weight through C.G. forward horizontal pull), head restraint strength test (372.8 N-m moment about the h-point through the head restraints) and rearward load strength (1.668 kN load 279mm above h-point rearward horizontal pull). Dynamic testing included the following: Acceleration (forward and rearward, unoccupied) and occupied due care sled test using three 50% mannequins (forward and rearward). The frames met all customer and federal requirements.

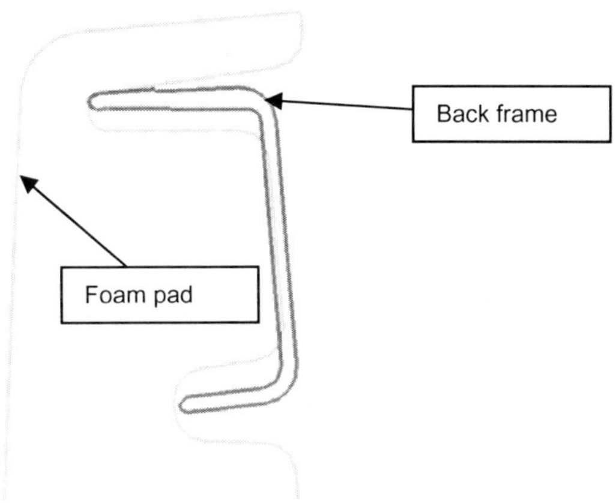

Figure 8 – C-section of back frame.

The L-shape wall section also helps meet mass targets and provides a positive locating feature for the foam pad. See figure 9. The average wall thickness used was 3.0mm. The draft of the walls used was 2 degrees, except for bolt attachment areas, which was reduced to 0.5 degrees.

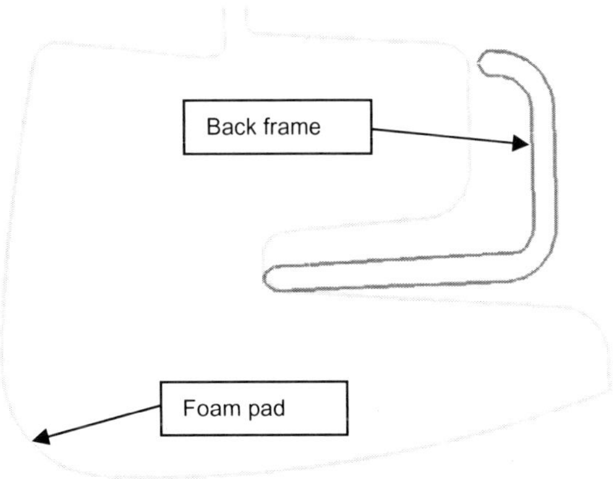

Figure 9 – L-section of back frame.

Vertical ribs can be added based on FEA to help carry load. Getting good correlation between FEA and actual testing during the DV (prototype) stage is helpful prior to PV (production) release. This can help eliminate unnecessary testing on future changes by using FEA. Several rounds of FEA runs may be needed, but once the correlation is met, all future changes can be validated by FEA before any tooling modifications. Actual testing can be reduced or if the engineering team agrees not needed. The die cast process allows local structure, such as thickness, to achieve performance with minimal weight impact. If the FEA failure mode represents the physical testing failure mode, then there is good correlation. See figures 10 and 11 for examples of correlation between FEA and actual test results. The load case for figure 10 was the head restraint strength test. The FEA prediction and the actual test results are the ultimate strength required of the customer specification. The required load of 1068 N is applied through the head restraints, after applying a 372.8 N-m moment through the head restraint. The requirement for the ultimate load is no structure failure. The load achieved in figure 11 was 2845 N. The design target of 6% plastic strain predicted the facture at approximately 2500 N.

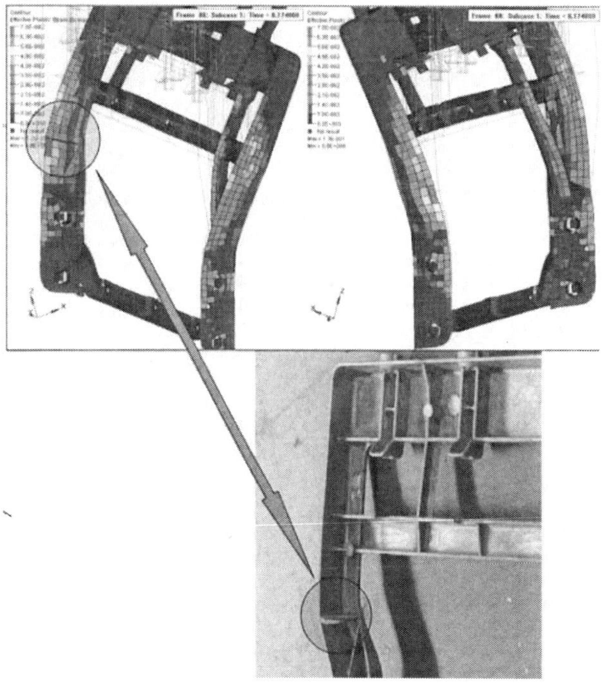

Figure 10 – FEA correlation to physical testing.

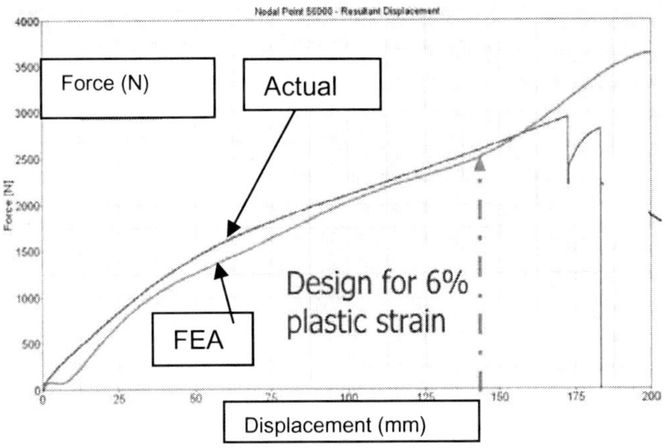

Figure 11 – Head Restraint Strength Test FEA vs. physical testing.

To keep tooling costs at a minimum all attachment and strength features were kept in the line of draw.

Magnesium is one of the lightest structural materials in the world (1.8 g/cm^3). The magnesium alloy used for this project was AM60B. AM60B was selected for its excellent energy absorption characteristics and outstanding ductility. Although AM60B has slightly lower strength (241 MPa vs. 230 MPa) than AZ91 alloy its improved ductility (6% elongation at break vs. 3%) and resistance to crack propagation makes it a versatile material.

Adding the parting line to the CAD model can prove useful to control flash height in critical areas. Where the frame edge and foam pad come in contact, minimal flash is allowed. If the flash height is not controlled, the edge of the frame will cut through the foam pad during a jounce/squirm test. Flash heights need to be agreed upon before the production tooling is kicked off. Critical areas on the frame also need to be established. This will aid the design of the trim dies.

Establishing porosity zones prior to any prototype testing is a best practice technique. All load carrying members, bolt attachment and sections that are thicker than the adjacent wall need to be classified as a potential porosity zone. Along with establishing zones, each zone should have a level of porosity associated with it. Refer to ASTM E505-01 for porosity levels. See Figure 12 for an example of porosity in a M10 bolt attachment embossment.

Casting 3-2 (3.5mm dia. Porosity)

Figure 12 – An example of porosity in an embossment.

All DV (prototype) parts need to be inspected for porosity (x-ray) with the worst case parts tagged for specific structural testing. See figure 13 for an example of an x-ray with porosity ranking. In addition, adding the porosity zones for critical areas to the drawing is a good practice.

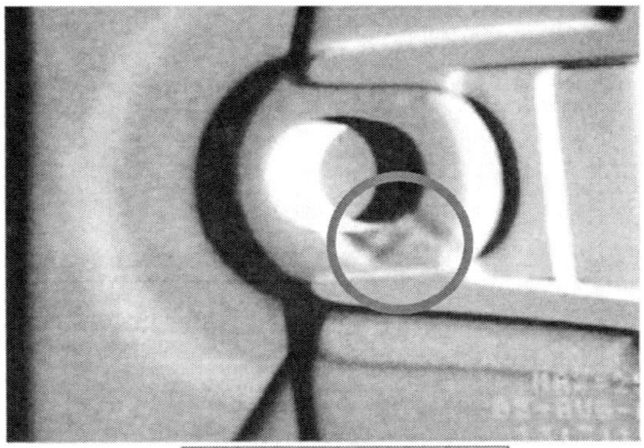

Part 21 – Location 003
Ranking 8

Figure 13 – X-ray showing porosity in an attachment feature.

Additional porosity studies were completed to determine what the size of the porosity had on peak torque readings. See table 4 for porosity versus torque study. The test was completed using prototype back frames (P20 tool). Each frame was x-rayed and porosity size (diameter) recorded. Refer back to figure 12 for an actual part used in this study. Even with a porosity diameter in the 3.5mm range, peak torque reading was still above the 60 Nm requirements. The torque readout for this sample was 100.8 Nm.

Casting Hole Number	Porosity Size Diameter (mm)	Peak Torque (Nm)
1-1	1	107.9
1-2	2	102.9
1-3	None	87.4
1-4	2	71.3
2-1	None	90.5
2-2	3	117.9
2-3	None	101.3
2-4	None	132
3-1	1.5	112.9
3-2	3.5	100.8
3-3	2.5	88.8
3-4	None	94.5

Table 4 – Porosity size vs. peak torque.

CONTROLLING POROSITY

Porosity is most always present in die-castings. Prevention is the best defense to minimize porosity. Parameters can be built into a control window to establish optimum die-castings. Unfortunately, for proprietary reasons not all parameters can be listed. Such parameters include, but are not limited to:

1. Gate areas and positions of entry into the casting die.
2. Runner systems geometry.
3. Injection speeds: first, second and third phases.
4. Injection pressure.

Current die cast technology offers electronic surveillance solutions to monitoring the process, unlike past die casting. Modern die cast cells are equipped with features such as electronic temperature controllers, linear displacement transducers and pressure transducers (Jackson, 2003). These features allow the die caster to monitor hundreds of parameters throughout the shot making cycle (Jackson, 2003).

The design of the back frame, or any other die cast part, should take consideration of the die cast process. Working with the die cast supplier and giving flexibility from a design standpoint will assist in preventing porosity. The die cast supplier can simulate material flow and predict areas of concern, before any tooling has started. If the design of the part cannot be modified, then it is up to the supplier to re-position gates or add extra venting.

During the prototype stage, fluoroscopic x-rays can be performed to ensure internal soundness (Jackson, 2003). Typically, there will always be some degree of internal porosity in a die cast component (Jackson, 2003). It is the Process Engineers job to reduce porosity from forming in critical areas and from expanding to its gaseous volumetric size (Jackson, 2003).

A control window is established prior to start of production. Data is collected using first off examinations. The cast cell's electronic surveillance system monitors each shot, and if the parameters exceed the window, an alarm is activated. The casting is then set aside for additional quality control inspection (Jackson, 2003).

CONCLUSION

The design and construction of the back frame can greatly affect the performance and craftsmanship of the seat back and seat complete. A cast magnesium back frame can save mass, provide reduced operating efforts of mating components and exceed customer appearance requirements over a traditionally welded steel back frame. This paper also shows that special lightweight material fasteners are not necessarily needed for attachment of mechanisms or other mating parts.

A magnesium back frame can be designed to do more than support the occupant; it can be an integral part of the seat back. The frame can define the STO of the seat back. The back frame can be designed to the hold the foam pad/trim cover in place for improved placement relative the cushion.

Part handling at a JIT plant is a factor not directly discussed in the main section of this paper, but the lightweight back frame is greatly appreciated from the plant personnel. If a cast magnesium back frame is within the limits of a program (mass, appearance, cost and effort targets) the benefits outweigh the negatives.

The magnesium back frames were successfully launched on a high volume production vehicle.

ACKNOWLEDGMENTS

We would like to thank the following personnel that helped with the design of the back frames that led to the writing of this paper:

Charles Young - Meridian
Hussein Sanaknaki - Meridian
Jeremy Bos – Meridian
Glen Simonds - Meridian
Al Mackey - Intier
Steve Reed - Intier
John Oilar - Intier
Bill Hagen - Intier
Ron Kaip – DaimlerChrysler
Ravi Hosalli – DaimlerChrysler

REFERENCES

1. The Online Material Database. 1996 – 2003. Automation Creations, Inc. 8 Sept 2004. http://www.matweb.com
2. Jackson, Ron. (2003, July). Controlling Porosity. Paper presented to Intier Automotive at design review meeting. Novi, Mi.

CONTACT

Matthew R. Abate – Product Engineer
matt.abate@intier.com

Mike C. Willman – Account Manager
mwillman@meridiansales.com

ADDITIONAL SOURCES

An overview presentation from Meridian Technologies on how magnesium is used in seat structures. The presentation was received on July 28, 2004.

2005-01-0724

Numerical Modeling of the Structural Behavior of Thin-Walled Cast Magnesium Components Using a Through-Process Approach

C. Dørum, O. S. Hopperstad and M. Langseth
Structural Impact Laboratory (SIMLab), Department of Structural Engineering
Norwegian University of Science and Technology

O.-G. Lademo
SINTEF Materials and Chemistry

S. Sannes
Hydro Magnesium

ABSTRACT

A through-process methodology for numerical simulations of the structural behavior of thin-walled cast magnesium components is presented. The methodology consists of casting process simulations using MAGMAsoft, mapping of data from the process simulation onto a FE-mesh (shell elements) and numerical simulations using the explicit FE-code LS-DYNA. In this work, generic High Pressure Die Cast (HPDC) AM60 components have been studied using 3-point bending and 4-point bending tests. The experimental data are applied to obtain a validated methodology for finite element modeling of thin-walled cast components subjected to quasi-static loading. The cast magnesium alloy is modeled using a user-defined material model consisting of an elastic-plastic model based on a non-associated J2-flow theory and the Cockcroft-Latham fracture criterion. The fracture criterion is coupled with an element erosion algorithm available in LS-DYNA. The constitutive model and fracture criterion are calibrated both with data from material tests and data from the process simulation using MAGMAsoft.

INTRODUCTION

As the lightest structural engineering metal available, magnesium is very attractive for structural automotive applications where weight saving is a matter of serious concern. With High Pressure Die Casting of magnesium and aluminum alloys, components with very complex, thin-walled geometry, like instrument panels, A and B pillars and front end structures, can be cast with a high production rate. The challenge with this production method is to optimize the process parameters with respect to the part design and the solidification characteristics of the alloy in order to obtain a sound casting without casting defects. Unbalanced filling and lack of thermal control can cause porosity and surface defects due to turbulence and solidification shrinkage. These defects can give low ductility compared to extruded materials, for instance. The HPDC method also leads to a "skin-effect", where the microstructure of the castings near the free surfaces differs significantly from the interior as the skin region has much finer microstructure [1].

Design and production of thin-walled cast structural components for the automotive industry are challenging tasks that involve the development of alloys and manufacturing processes, structural design and crashworthiness analysis. In order to reduce the lead time to develop a new product it is necessary to use finite element analysis to ensure a structural design that exploits the material. Accurate description of the material behavior is essential to obtain reliable results from such analyses. In order to minimize the weight of the structural component while maintaining the safety in a crash situation, the ductility of the material has to be utilized without risking un-controlled failure. Hence, a reliable failure criterion is also required, giving limits for the plastic deformations under various loading combinations. Very precise and validated constitutive models and failure criteria are available for materials such as extruded aluminum and rolled steel [2]. However, much work is still to be done in this area for thin-walled cast materials.

The long-term objective of this work is to develop design and modeling tools that allow the structural behavior of thin-walled cast components to be predicted when

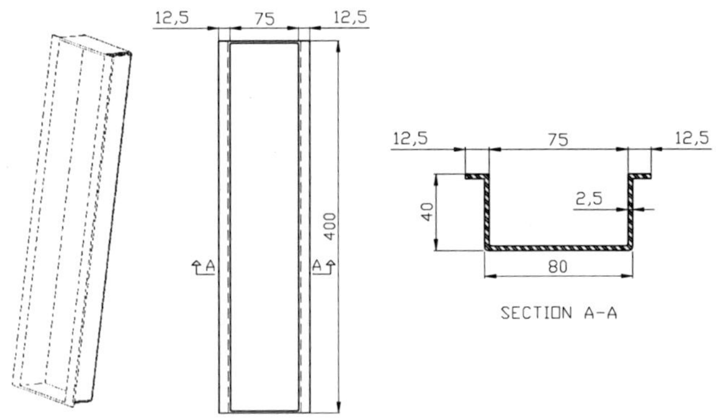

Figure 1: Generic profile geometry.

subjected to static and dynamic loads. In the current study, the structural behavior of generic structural HPDC components, shown in Figure 1, has been investigated using of 3- and 4-point bending experiments.

The components were cast of magnesium alloy AM60 at Hydro's Research Centre in Porsgrunn, Norway with a Bühler SC42D 420-ton cold chamber die casting machine.

MATERIAL CHARACTERIZATION

The first step in the development of the through process strategy for numerical modeling of thin-walled castings is material characterization. For the case of high pressure die-cast Mg alloys, the microstructure of the castings usually contains a fine-grained "skin" with a microstructure significantly different from the bulk material. In this work, the overall material hardening properties have been characterized using uniaxial tensile specimens cut from the 80 mm wide flange of the generic AM60 components illustrated in Figure 1. The specimens were aligned with the longitudinal direction of the components. Contrary to what is observed with extruded aluminum alloys, the test specimens were found to fail before the point of diffuse necking.

The AM60 material has also been characterized using uniaxial compression tests, since previous work [3, 4] revealed different tensile and compressive behavior for magnesium alloys. For details, it is referred to Dørum et al. [5]. The experimental true stress - true plastic strain curves for both uniaxial tension and uniaxial compression are shown in Figure 2. No significant differences were found between the specimens regarding the observed stress level. Thus, the stress-strain curves given in Figure 2, can be viewed as representative for the overall hardening properties.

Since the tensile tests [6] did reveal very small differences in stress level, it is assumed that the hardening properties are uniform throughout the component. Further, assuming that the fracture in the tensile tests starts in the bulk material, the tensile ductility of the bulk material is also characterized using these tests.

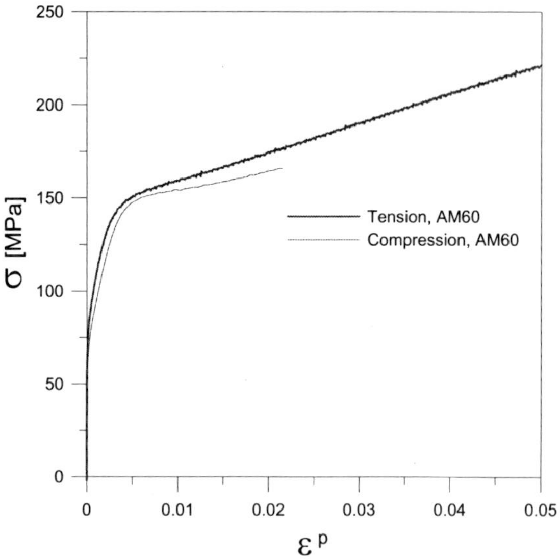

Figure 2: Typical experimental true stress – true plastic strain curves for AM60.

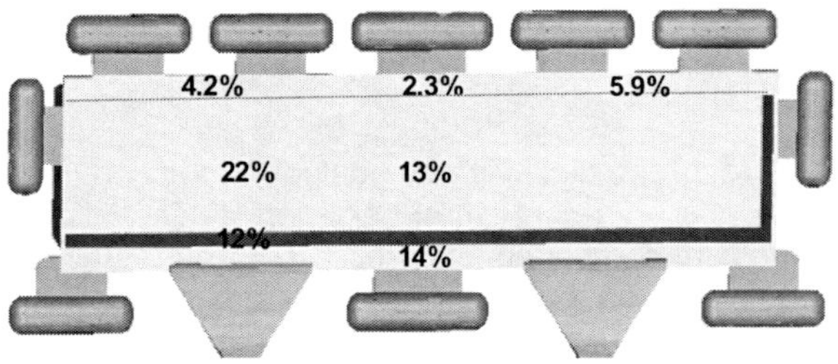

Figure 3: Experimental tensile elongation distribution.

The scatter in fracture elongation is quite large. The poorest area is the outer stiffener on the outlet side, where values as low as 2-3% (effective plastic strain) have been measured. On the other hand, the best areas were in the 80 mm flange in front of the gates, where fracture elongation values as high as 22% effective plastic strain have been measured [6]. An overview of the tensile elongation distribution is provided in Figure 3.

The ductility of the skin is characterized using experimental results from 3- and 4-point plate bending tests, illustrated in Figure 4, assuming that fracture in these tests starts in the skin material. The dimensions of the cast AM60 plates were as follows: 140 mm length, 100 mm width and 3 mm thickness. It should be noted that these are completely different castings, and not machined from the generic profile geometry shown in Figure 1.

As illustrated in Figure 5, the scatter in the experimental force-deformation-characteristics for the 3-point bending test is very small. Thus, the variations in tensile ductility are probably much smaller for the skin material than for the bulk material. For the case of 4-point bending, no fracture occurred in the cast AM60 plate.

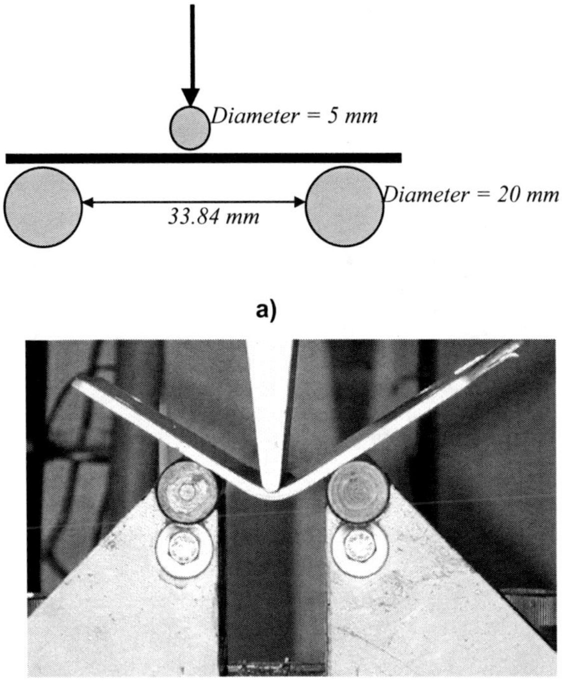

Figure 4: Test set-up for 3-point plate bending.

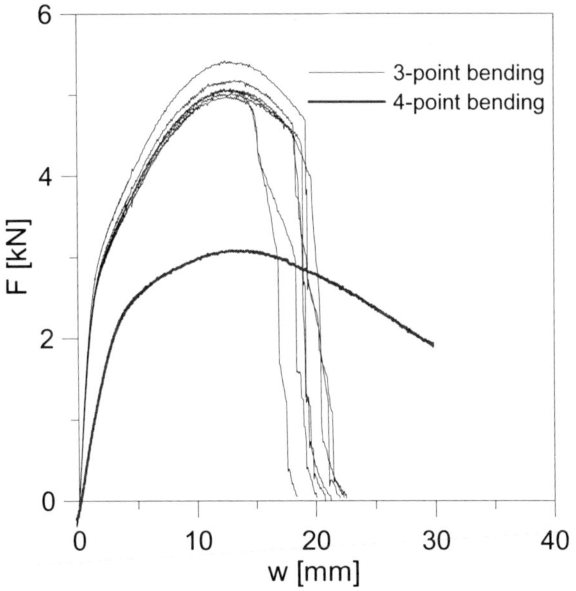

Figure 5: Force-deformation behavior for 3- and 4- point plate bending tests.

NUMERICAL SIMULATIONS

Numerical simulations of the generic structural High Pressure Die Cast components are performed using LS-DYNA. The components are modeled using shell elements. Both 3-point and 4-point bending tests are simulated, and the numerical results are then compared with experimental force-deformation characteristics.

CONSTITUTIVE MODEL

Due to the different types of behavior in compression and tension, it has been decided to model the structural behavior of the AM60 components using a non-associated extended J2-flow theory as proposed by Dørum et al. [7]. However, it should be noted that since the strength differential (SD) effect in the investigated AM60 alloy is rather small, approximately 10 % comparing the compressive and tensile yield stress, the classical J2-flow theory would give quite similar results. The non-associated extended J2-flow theory, which is adopted in this work, is designed to model the behavior of metals with strength differential effects. Therefore, the material model will now be referred to as the SD model. In the SD model, the von Mises yield surface is modified along the axis of normalized pressure such that different hardening properties can be given for two different user-defined values of triaxiality (e.g for uniaxial tension and uniaxial compression). Similar to the non-associative J2-flow theory discussed by Simo and Taylor [8], the yield surface for the SD model can be expressed as:

$$\Phi\left(\boldsymbol{\sigma},\varepsilon_e\right)=\sigma_e-\kappa\left(\mathrm{tr}\boldsymbol{\sigma}/\sigma_e,\varepsilon_e\right)=0$$

where σ_e is the effective von Mises stress, and ε_e is the corresponding effective plastic strain. The term $\mathrm{tr}\boldsymbol{\sigma}/\sigma_e$ represents a normalized pressure or a triaxiality parameter. κ represents the current yield stress, or the hardening properties, and is given as a function of both the effective plastic strain and the normalized pressure. The material model is calibrated based on stress-strain curves obtained from uniaxial tensile and compression tests with specimens machined from the mid-section of the 80 mm wide flange of the AM60 components. Consequently, the porosity and void growth in the cast material is directly taken into account in the plastic domain. For simplicity, the elastic properties are assumed to be unaffected by the porosity.

Fracture modeling

The constitutive model used for the numerical simulations, briefly discussed above, includes the Cockcroft-Latham fracture criterion [9]. The criterion can be expressed as:

$$W=\int\max\left(\sigma_1,0\right)d\varepsilon_e\leq W_c$$

where σ_1 is the maximum principal stress and ε_e is the true effective plastic strain. This criterion implies that fracture is a function of the tensile stresses and equivalent plastic strain and W_c has the dimensions of work per unit volume. If the maximum principal stress is compressive, fracture cannot occur. Furthermore, neither stresses nor strains alone are sufficient to cause fracture.

For the interior material, the critical value of "plastic work", W_c (or the corresponding value for the fracture elongation, ε_f), can then easily be calibrated against uniaxial tension tests. However, to take into account the variations in local tensile ductility throughout the cast AM60 components, a reliable distribution of W_c is required. Here, this distribution is based on a porosity distribution predicted by casting process simulations using MAGMAsoft [10]. An overview of the predicted porosity distribution is provided in Figure 6.

The predicted porosity distribution shows that the porosity mainly is below 1% throughout the whole component. Consequently, relating the numerical porosity predictions to the experimental results found from the uniaxial tensile tests, it is assumed that bulk material with 0% porosity has a fracture elongation of 22% while bulk material with 1% porosity has a fracture elongation of 1.5%. For porosity levels above 1%, no further decrease in tensile fracture elongation is assumed. For material with porosity between 0% and 1%, the fracture elongation is assumed to be linearly dependent on the porosity. This range of ductility can be regarded as a conservative assumption. It should be noted that the predicted porosity distribution is simply viewed as a defect map for the local ductility of the bulk material. The tensile ductility of the bulk material is probably not dependent only upon the porosity but on other variables such as grain size and oxide layers. However, it is assumed that these other "defects" are of second order importance.

The local tensile ductility of the skin material, and the corresponding value for the fracture parameter W_c is calibrated using the experimental results from the 3- and 4-point plate bending experiments. The calibration is performed using an inverse modeling approach; the experiments are simulated numerically using LS-DYNA and choosing the fracture parameter W_c such that the numerical force-deformation predictions fit the experimental results. The simulations are performed using plane strain elements and the implemented SD-model.

In the numerical force-deformation characteristics, shown in Figure 7, it is assumed that the skin layer has a higher strength than the bulk material. The plate has been modeled using 12 elements across the thickness, and it is assumed that the two outer elements represent the skin material, giving a skin thickness of 0.5 mm. The skin thickness is usually in the range of 0.05 mm – 0.5 mm. According to the results of Shan and Gokhale [1], a strength difference between skin and interior material of 20MPa has been chosen here. However, simulations of the 3- and 4-point plate bending without any skin effect give approximately identical results, and it can be concluded that the skin-effect is negligible with respect to the overall material hardening. The simulations show that assuming a fracture elongation of 22% gives good agreement with the most ductile 3-point plate bending experiments.

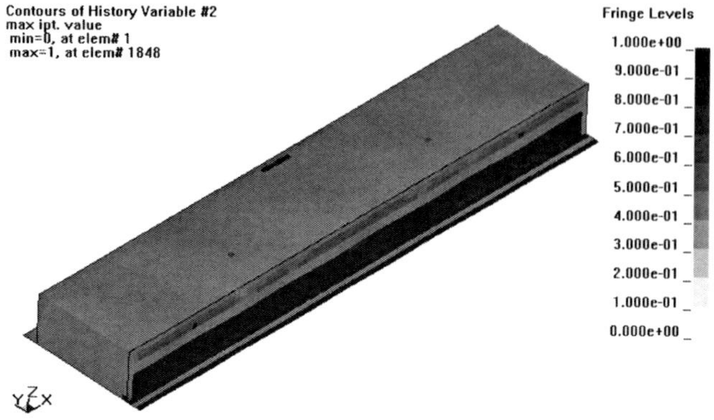

Figure 6: Predicted porosity distribution (component shown from outlet side).

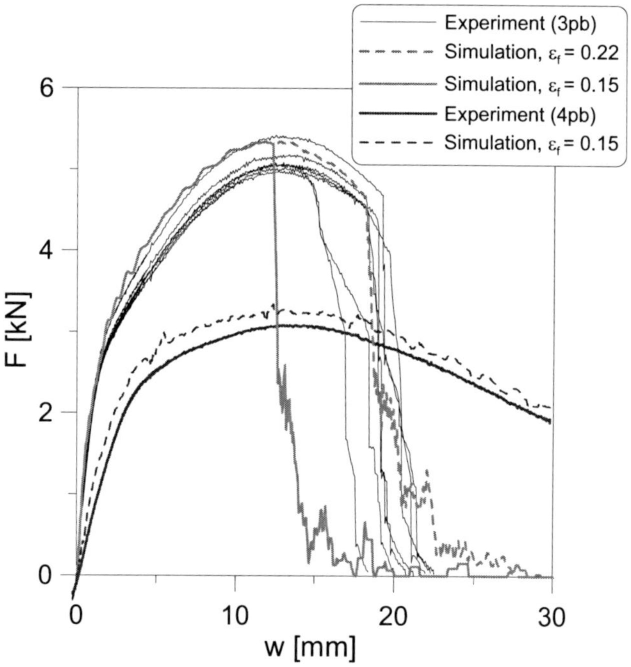

Figure 7: Comparison of numerical and experimental force-deformation behavior for 3- and 4-point plate bending.

Assuming a fracture elongation of 15% gives a somewhat too brittle prediction compared to the least ductile experiments. Further, it seen that the simulations predict no facture for the case of 4-point plate bending. It can therefore be concluded that the simulations fit well with the experimental results. For numerical simulations of the generic AM60 components, the ductility of the skin should have a reliable distribution. Here, it is assumed that the fracture elongation values (or corresponding W_c value) follow a surface quality map generated from numerical casting process simulations using MAGMAsoft [10]. An overview of this *quality-map* is provided in Figure 8, and upper and lower boundary for the fracture elongation of the skin material is assumed to be 22% and 15%, respectively. As for the ductility assumptions for the bulk material, this can be regarded as a conservative assumption.

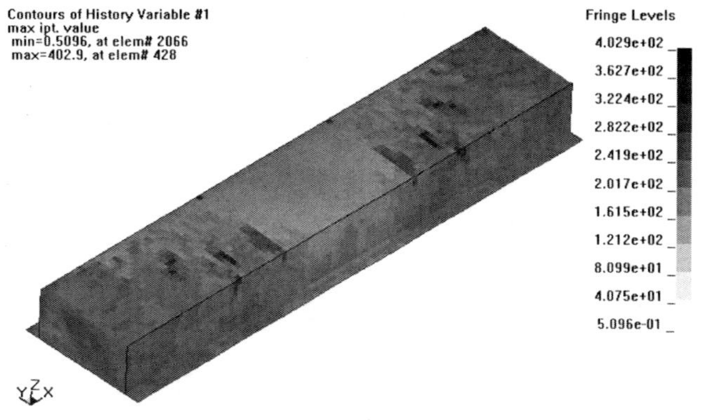

Figure 8: The surface quality-map distribution (component shown from outlet side).

GENERIC COMPONENT TESTS

The structural behavior of the cast generic AM60 components was studied using 3- and 4-point bending tests. For both the 3- and 4-point bending tests, two different load cases were defined. Depending on how the load was applied, on the 80 mm wide flange or on the small outer stiffeners, the load cases are referred to as the *n*-mode or the *u*-mode. The two different load cases are illustrated in Figure 9 and Figure 10, respectively. Numerical simulations were carried out using the explicit FE-code LS-DYNA [11]. The FE-model of the component consists of 5000 Belytschko-Tsay shell elements with one point integration in the plane and five through the thickness. This gives an average element size of approximately 4.0mm x 4.0mm. The relatively fine mesh was chosen to give a good description of buckling as well as being able to include a reasonable description of the material inhomogeneities.

In the simulations of the bending tests, the deformation was applied through rigid body motion of cylinders and plates, respectively, in which the material was assumed rigid. The deformation was applied smoothly but much faster than in the experiments. However, the analysis can be regarded as quasi-static, and therefore comparable with the experimental results, provided the kinetic energy and its variation are both negligible throughout the simulation. This was checked for all simulations.

The comparisons of force-deformation characteristics obtained from the numerical analyses and the experiments are shown in Figure 11 and Figure 12 for 3-point and in Figure 13 and Figure 14 for 4-point bending. The numerical data has been smoothened using a 9-point running average (total number of data points for each of the simulations is 1000) to give a better visualization of the force-deformation characteristics after fracture has occurred. It is seen from the experimental results that the components subjected to bending in the n-mode show a relatively large scatter in structural ductility. This is because the structural capacity of the components here is limited by tensile

Figure 9: Test set-up for 3-point bending, load case (n).

Figure 10: Test set-up for 3-point bending, load case (u).

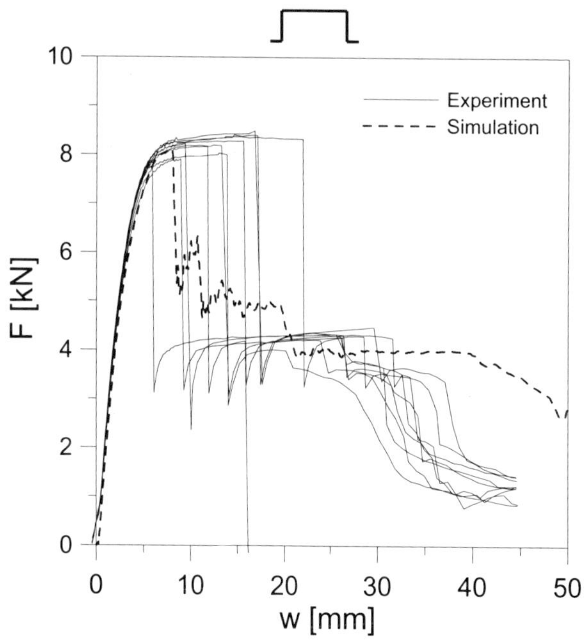

Figure 11: Force-deformation behavior for 3-point bending (n-mode).

failure of the small outer stiffeners. When loaded in the n-mode, the small outer stiffeners experience approximately uniaxial tension. The inhomogeneous micro-mechanical structure present in the HPDC AM60 components causes large differences in local tensile ductility. Consequently, as a large scatter was found in tensile elongation from uniaxial tensile tests [6], the scatter in structural response must be expected to be of the same order. Most significant is the scatter for the case of 3-point bending in the n-mode, where fracture always occurs at the small outer stiffeners at or close to the midpoint of component.

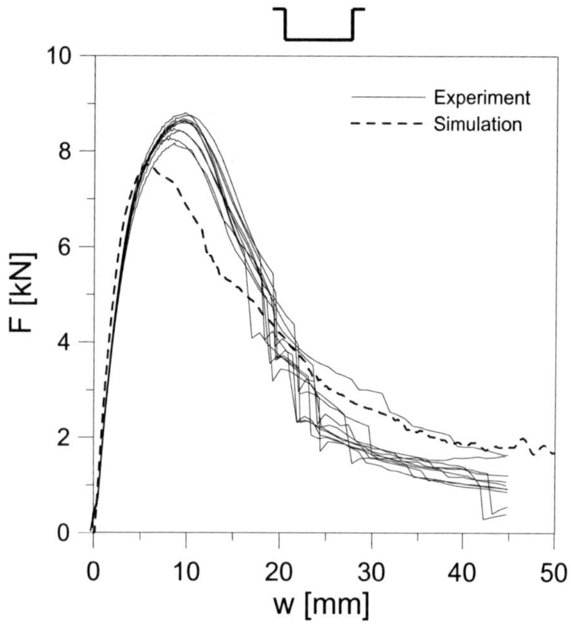

Figure 12: Force-deformation behavior for 3-point bending (u-mode).

For the case of 4-point bending, larger sections of the small stiffeners experience the same tensile loading. Owing to statistical effects, the probability for the critical section of the stiffeners to have a low tensile ductility is therefore larger for the case of 4-point bending than for 3-point bending. Thus, the reduced scatter in 4-point bending (n-mode) compared to the scatter in 3-point bending is reasonable.

For the components loaded in the u-mode, a very small scatter in force-deformation characteristics is found. Here, the first part of the force-deformation characteristics, i.e. before any fracture takes place, is limited by local buckling of the 40 mm side webs and the compression stiffeners. The buckling phenomenon is controlled by the material hardening, slenderness and initial imperfections of the profile. After the buckling load has been reached, the plastic deformation in the buckles increases rapidly as the buckles develop further and fracture occurs. Thus, relative to the global force-deformation behavior, the variations in local tensile ductility are of little importance, and the small scatter should be expected.

The figures further show that the structural response of the components subjected to bending is captured very well in the numerical simulations. Furthermore, it is seen that the failure criterion adopted in the SD model predicts failure quite accurately for both the cases of 3- and 4- point bending experiments in u-mode. For the 3-point bending loaded in n-mode, it is seen that the simulations using the SD model predict a structural response that is in the brittle range of the experimental results. This is in agreement with the choice of conservative assumptions for the fracture elongation distribution. It is seen that the fracture occurs at the small outer stiffeners that experience approximately uniaxial tension.

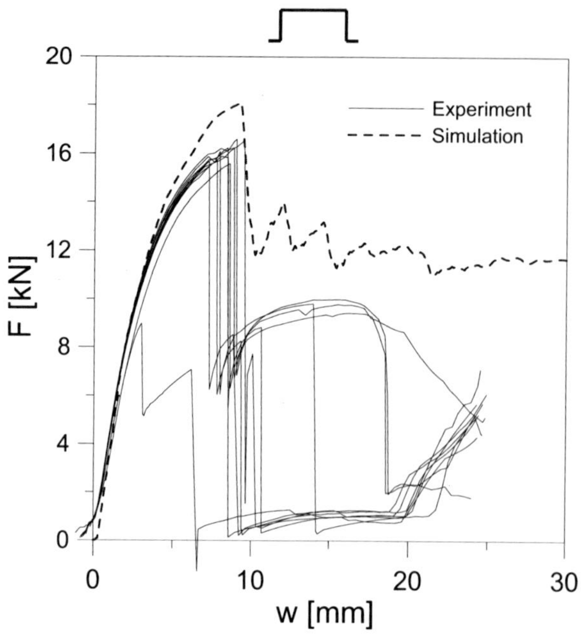

Figure 13: Force-deformation behavior for 4-point bending (n-mode).

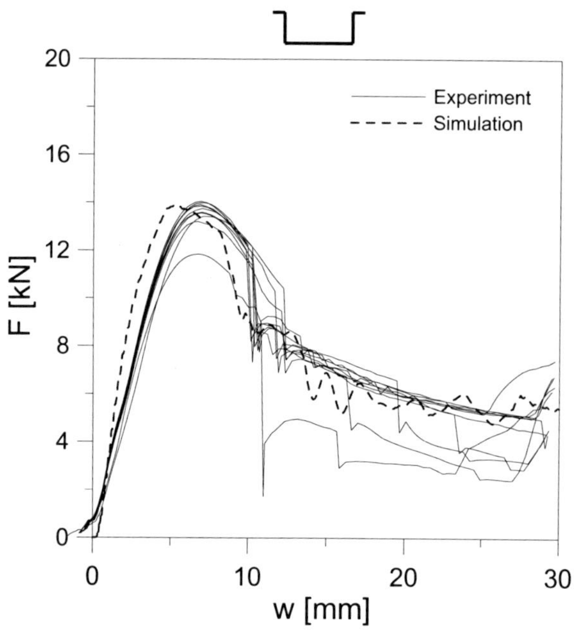

Figure 14: Force-deformation behavior for 4-point bending (u-mode).

Experimental work performed by Sannes et al. [6] show that the tensile ductility varies significantly throughout the cast component. The poorest area is the small stiffeners where elongations as low as 2-3% have been measured. The 80 mm wide flange, from where the specimens for calibration of the failure criterion were taken, showed much better properties with elongation in the range of 12-16%. Thus, in the bending experiments in n-mode, the largest strains in the components are found where the mechanical properties are poorest. For the case of 4-point bending in the n-mode, it is seen that the numerical predictions is somewhat too ductile.

Examination of the tested components in this case has shown that the fracture always occurred close to the gating channels, probably due to a notch effect. However, a proper modeling of the notch is not possible when using shell elements.

The fracture criterion proposed by Cockcroft and Latham seems to describe the structural ductility very well. More details concerning the experimental work is given by Dørum et al. [5] and Sannes et al. [6].

CONCLUSION

The long-term objective of this work is to develop design and modeling tools that allow the structural behavior of thin-walled cast components to be predicted when subjected to static and dynamic loads such as in crash situations. The approach consists of the following ingredients: casting of generic components relevant for the automotive industry, material and component testing, constitutive modeling and validation simulations using the finite element method.

In the present study, High Pressure Die Cast AM60 components were subjected to 3- and 4-point bending. Laboratory experiments and explicit finite element simulations with shell elements were performed. Only quasi-static loading conditions were considered. An elastic-plastic constitutive model based on a non-associated extended J2-flow theory, referred to as the SD model, was implemented in the FE-code LS-DYNA. The SD model is designed to model the behavior of metals with strength differential (SD) effects, and is calibrated against uniaxial tension and uniaxial compression tests. The fracture criterion proposed by Cockcroft and Latham [9] was adopted for the SD model, and was calibrated using both experimental results (uniaxial tensile testing and 3- and 4-point bending) and numerical predictions of the material quality resulting from casting process simulations. The experimental data from the uniaxial tensile tests was used together with porosity predictions to give a conservative map for the bulk material ductility. Similarly, using an inverse modeling approach for the plate bending tests together with surface quality predictions, also resulting from casting process simulation, a conservative map for the skin material ductility was created.

In general, experimental results presented both in this work and by Dørum et al. [5, 7] show that there exists a relatively large scatter in the force-deformation behavior when the HPDC components are subjected to deformation modes where the failure depends on the local tensile ductility of the material. For deformation modes where the force-deformation behavior is controlled by local buckling much less scatter is found.

The fracture criterion proposed by Cockcroft and Latham [9] linked to results from casting process simulation gave very promising results. Especially the structural ductility of components subject to bending in the u-mode was quite accurately predicted using a constant value of W_c. The reason for the good agreement in these cases is related to the evolution of local plastic strains with increasing global deformation after the buckling load has been reached. Here, a small increase in global deformation results in a relatively large increase in local plastic strains. Thus, the accuracy of the prediction of failure in terms of global force-deformation behavior is not very dependent on the accuracy of the Wc parameter. For the case of 3-point bending in n-mode, the simulations gave a prediction in the brittle range of the experiments. This is in good agreement with the conservative assumptions made for the material ductility. However, a somewhat too ductile response was predicted for the case of 4-point bending in n-mode. This was explained with the existence of a small notch remaining from the gating system.

In order to further improve the simulations with respect to the description of structural ductility, the quality of the material ductility map should be further examined.

ACKNOWLEDGMENTS

The authors would like to acknowledge Hydro Aluminium and the Research Council of Norway for their support through the NorLight project.

REFERENCES

1. Shan, Z., Gokhale, A. M., 2003. Utility of micro-indentation technique for characterization of the constitutive behavior of skin and interior microstructures of die-cast magnesium alloys, Materials Science and Engineering A361, 267-274.
2. Lademo, O.-G., 1999. Engineering Models of Elastoplasticity and Fracture for Aluminium Alloys, Dr. Ing. Thesis. Department of Structural Engineering, Norwegian University of Science and Technology.
3. Hydro Magnesium, 1998. Data Sheet, Die Cast Magnesium Alloys. Norsk Hydro.
4. Kelley, E.W., Hosford, H.F., 1968. The Deformation Characteristics of Textured Magnesium, Transactions of the Metallurgical Society of AIME 242, 654-660.
5. Dørum, C., Hopperstad, O.S., Lademo, O.-G., Langseth, M., 2003. Aluminium and magnesium castings - experimental work and numerical analyses, International Journal of Crashworthiness 8, 455-470.
6. Sannes, S., Svalestuen, J.M., Dørum, C., Lademo, O.-G., 2003. Die Casting of Magnesium Alloys - Crash Requirements and Testing of Part Performance. Magnesium Automotive and End User Seminar, Aalen.
7. Dørum, C., Hopperstad, O.S., Lademo, O.-G., Langseth, M., 2004. Numerical modelling of the structural behaviour of thin-walled cast magnesium components, accepted for publication in International Journal of Solids and Structures.

8. Simo, J.C., Taylor, R.L., 1985. Consistent Tangent Operators for Rate-Independent Elastoplasticity, Computer Methods in Applied Mechanics and Engineering 48, 101-108.
9. Cockcroft, M.G., Latham, D.J., 1968. Ductility and the Workability of Metals, Journal of the Institute of Metals 96, 33-39.
10. Sannes S, Hepp E., Dørum C., Lademo O-G., Lohne O, Gjestland H., 2005. Linking the HPDC Process Simulation and the FE Analysis through Quality Mapping, 2005 SAE World Congress, Detroit.
11. LS-DYNA, 1998. Theoretical Manual. Livermore Software Technology Corporation, Livermore.

CONTACT

Cato Dørum, Structural Impact Laboratory (SIMLab), Department of Structural Engineering, Norwegian University of Science and Technology, No-7491 Trondheim, Norway.

Norway, cato.dorum@ntnu.no

2005-01-0725

Mechanical Properties of AM60B Die Castings: A Review of the AUTO21 Program on Magnesium Die-Casting

J. T. Wood and R. J. Klassen
Dept. of Mechanical & Materials Engineering, The University of Western Ontario

M. A. Gharghouri
Dept. of Mining and Metallurgical Engineering, Dalhousie University

E. Maire
INSA de Lyon

G. Wang and R. Berkmortel
Meridian Technologies

ABSTRACT

This paper provides an overview of the research efforts within the AUTO21 program on magnesium die-casting. The objective of the program is to better understand the mechanical properties of magnesium high pressure die-castings (HPDC) and to develop the capability to predict the local mechanical properties over a given component based on its shape, the design of the mold and the casting parameters used in its production. The paper highlights the research group's findings related to the microstructure and mechanical properties of a full-scale instrument panel beam casting and outlines the goals of the continuing research program.

INTRODUCTION

Due to their low density, high stiffness/weight ratio, and high damping capacity, magnesium alloys have the potential to make a significant contribution to the weight reduction of vehicles. The current average North American vehicle contains approximately 1.5 kg of magnesium. Estimates suggest that this figure has the potential to reach 50 – 60 kg[1].

One of the problems currently limiting the use of magnesium components in automobiles is the variation in the local mechanical properties of magnesium die-castings. These local variations are generally attributed to two factors – the formation of casting defects and the removal of the as-cast skin when preparing tensile bars from a large casting. Although there is not yet a consensus as to the situations in which each of these factors dominates, a robust engineering design approach has been developed to deal with the variations [2].

The ultimate goal of the research program is to develop the capability to predict the local microstructure and, from this, the local mechanical properties in a complex magnesium die-casting based on the numerical simulation of the casting process. The work to date has involved a large-scale mechanical testing program to study the quasi-static [3] and fatigue [4] characteristics of the casting and an equally large microstructural characterization study. These two parts of the program have provided our group with a statistically significant map of both mechanical properties and microstructure over the entire casting. Current research is focused on modeling the non-equilibrium solidification in the die in order to predict the as-cast microstructure and the development of structure-property relationships to link the predicted microstructure to the local mechanical properties. This paper will describe our research group's findings to date and discuss the path for our future work.

EXPERIMENTAL METHODS AND RESULTS

This study has been conducted on fifty, sequentially cast instrument panel beams. The AM60B castings were produced using a 3000-ton cold-chamber die casting machine. Forty-seven individual locations were identified on the instrument panel beams that were suitable for producing standard tensile specimens. Based on their original location in the casting and the results of a numerical casting simulation, samples were categorized as being from either close-to-ingate, last-to-fill, last-to-solidify, knit line, or general areas of the casting. The samples from twenty-nine of the fifty IP beams were tested in tension (>1300 tensile and fatigue tests). The remaining specimens were subjected to other testing procedures not reported here.

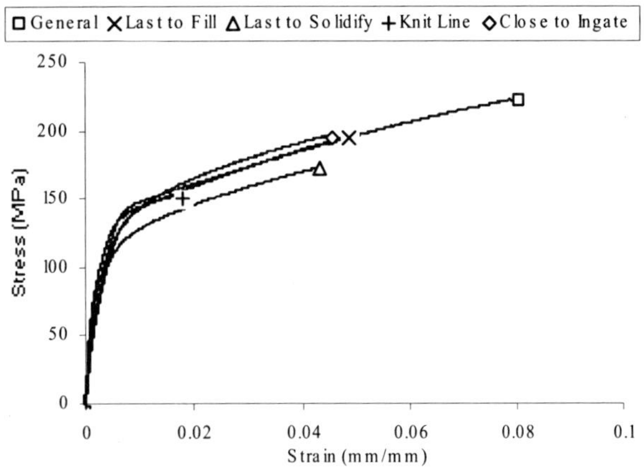

Figure 1 Typical stress-strain curves from characteristic locations on the casting. The yield strengths and work hardening rates are very similar throughout the casting.

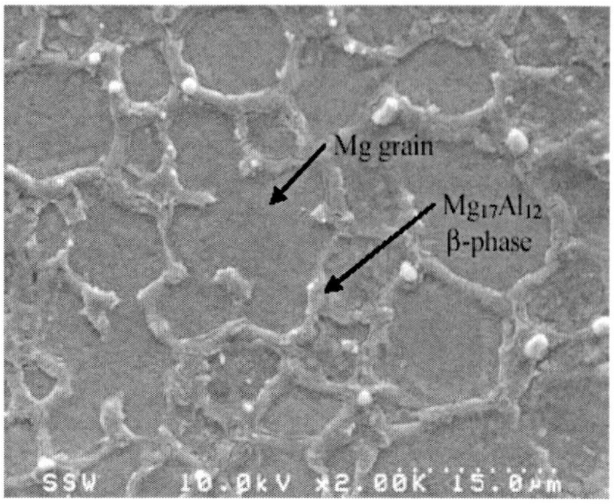

Figure 2 An SEM micrograph of the AM60B grain structure.

a)
b)
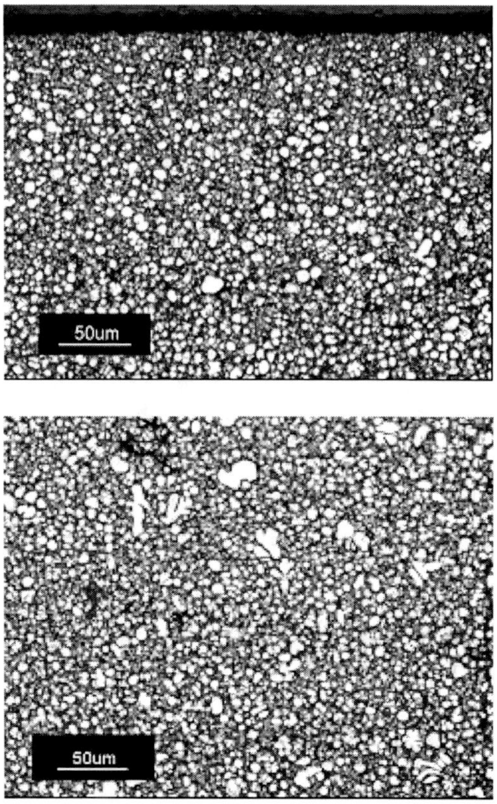

Figure 3 Optical micrographs of typical a) skin and b) core regions of the casting

PLASTIC DEFORMATION

The tensile flow curves were found to be remarkably consistent in many respects. Figure 1 shows tensile stress strain curves from a variety of characteristic locations on the casting. Both the yield strengths and the subsequent work hardening rates obtained from the tensile tests were consistent from location to location. The tensile flow curves were found to be suitably fit by a power law equation [3]:

$$\sigma = 400\varepsilon^{0.23}$$

This consistency in plastic behaviour suggests that any local variations in the microstructural features that control plastic flow (e.g. grain size, alloy and second phase content) are small enough to be ignored.

The microstructure of the castings is typical of a large die-cast component. In general, it consists of primary α-Mg grains surrounded by the β-phase (nominally $Mg_{17}Al_{12}$) and eutectic. An SEM micrograph of the typical microstructure is shown in Figure 2. The as-cast structure typically consists of dense, fine-grained skins surrounding a coarser grained core. Consistent with hypoeutectic solidification, finely dispersed microporosity is found throughout the casting.

The effect of grain size, however, can not be ignored if the surface of the casting is machined (such as for the production of uniform tensile coupons). Due to the higher rate of solidification adjacent to the mold wall, the as-cast structure typically consists of dense, fine-grained skins surrounding a coarser grained core. Typical skin and core microstructures are shown in Figure 3. The average grain size in the skin of these castings was found to be 6.5μm. The average core grain size was 14μm. Magnesium alloys have been shown to obey the Hall-Petch relationship and, thus, it is reasonable to expect that the yield strength of the casting will be governed by the fine-grained skins. Indeed, samples in which the skins had been machined to achieve a uniform cross-sectional area were found to have a reduction in tensile yield strength of as much as 20% [5].

FRACTURE PROPERTIES

The primary difference in the mechanical response of coupons cut from different locations of the casting is found in the elongation to fracture. In the absence of internal damage, the power law fit shown above suggests that AM60B should be capable of 23% uniform elongation. This is in contrast to handbook data that lists the fracture elongation of AM60 as 6% [6]. While seventy-five percent of the coupons had fracture strains in excess of five percent, six specific locations were found to have poor elongation. In considering these figures, it is important to recognize that all castings, including die-castings, are subject to defects that result from filling and solidification. It is the task of the mould designer to ensure that these casting defects are minimized by proper runner design and by incorporating features such as vents and overflows which force the defects into areas that are later trimmed and recycled. Further, good mould design practice should ensure that casting defects occur in areas of the casting that are either inconsequential (e.g. non-load bearing) or sections that are later removed. Each of the locations with poor elongation in this casting was in an area whose service life would not be affected by the diminished properties [2].

While the microstructural features controlling plastic flow are essentially invariant over the casting, the fracture results indicate that the features that control fracture (porosity, inclusions, etc.) [7] vary from location to location. Of particular interest to this study are these "non-standard" microstructural features that contribute to diminished mechanical properties. Thus far, the study has identified three such features:

Macroscopic shrinkage porosity, as in all types of metal casting, forms in areas of the casting that solidify without a continued source of liquid metal. In most cases, areas of the casting that are subject to this type of defect can be determined by visual inspection of a solidification simulation using commercially available finite element codes.

Knit lines are areas of the casting at which two "cold" liquid metal fronts meet. Knit line areas are normally found to contain the remnants of the oxide films that form on the advancing liquid metal fronts. The use of multiple in-gates to fill large die-castings makes this type of feature unavoidable. Proper mold and gating design and casting practice will minimize and shift the location of knit lines to inconsequential areas of the casting.

Gas porosity differs from shrinkage porosity in both structure and location in the casting. Typically spherical in shape, gas porosity can be found most commonly in areas of the casting that are last-to-fill or that coincide with the meeting of liquid metal fronts. Gas porosity is also found in locations of the casting in which the advancing liquid metal flow is disturbed, such as in the case of filling around sharp corners. This type of defect is minimized by the proper use of venting and overflows.

The last major step of the current research project is the development of numerical tools to predict the occurrence of these microstructural features based on a simulation of the mold filling and solidification processes.

Other "non-standard" microstructural observations include the presence of large, dendritic grains (presolidified grains) in the core of the cast wall in locations close to the ingates. Mechanical testing data,

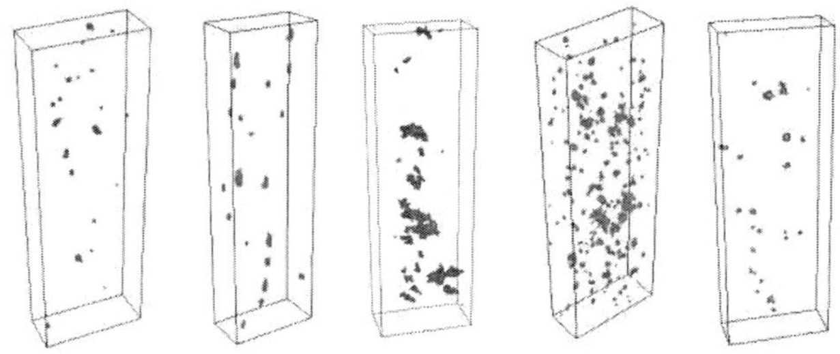

Figure 4 X-ray tomographic images of the gauge length of 5 different tensile samples.

however, suggest that these large grains have little or no effect on the strength or ductility of the casting. The "defect band" structure, sometimes found in die-castings between the skin and core, was not found in this casting.

Structure-Property Relationships

While the first phase of this project dealt with the characterization of the casting, a parallel research program has been involved in the development of relationships to predict mechanical properties based on the as-cast microstructure. Standard mechanical testing resulted in the determination of the power law hardening coefficients shown above. The more difficult task, however, is the prediction of the fracture strain.

Several studies have investigated the influence of porosity on the mechanical properties of die-cast magnesium alloys. Gutman et al. [8] find that the porosity and other casting conditions play an important role on the creep of AZ91. Mayer et al. [9] find that, during fatigue testing of several die-cast magnesium alloys including AZ91 and AM60, crack initiation occurs almost exclusively at porosity.

A major experimental difficulty in predicting the fracture of a component is the lack of knowledge of the size and shape of internal defects prior to failure. In a series of experiments employing x-ray tomography [10], the gauge lengths of five tensile samples were imaged prior to tensile testing at a spatial resolution of approximately 20 μm. Typical three-dimensional images from the tomographic study are shown in Figure 4. The imaged volume of each specimen is 3×6×25 mm.

This experimental technique yields the dimensions and location for each pore in the sample, enabling the calculation of the area fraction of porosity on any plane in the sample. This is the critical piece of information for the calculation of the fracture strain by the model of Caceres [12]. This model has been successfully adapted to the tensile results obtained in this study.

Figure 5 plots the fracture strain of the five samples as a function of the area fraction of porosity. The solid line is the predicted fracture strain. The good correlation between theory and experiment shown in Figure 5 indicates that, if the level of porosity can be predicted

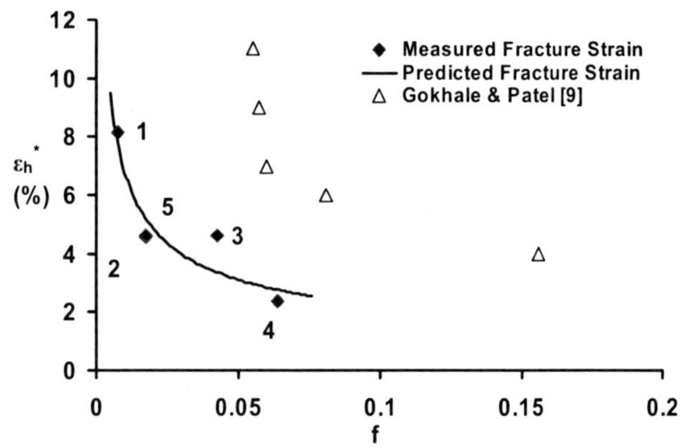

Figure 5 Predicted fracture strain ε_h, and the measure fracture strain plotted versus area fraction of porosity, f.

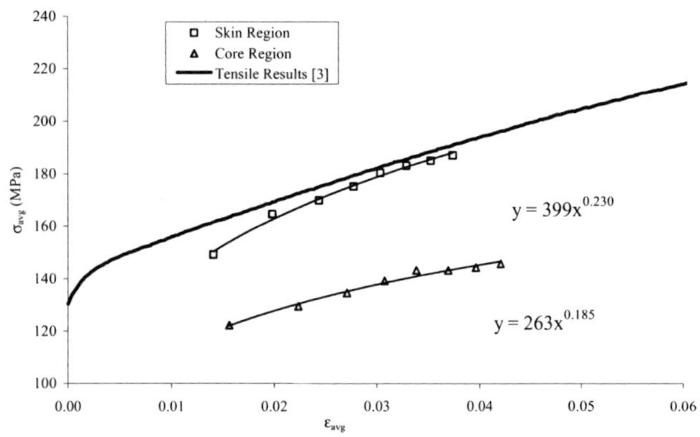

Figure 6 Comparison of a tensile stress-strain curve with flow curves determined from spherical micro-indentation.

with reasonable accuracy, the fracture strain is calculable. Also included in Figure 5 is the data of Gokhale and Patel [10] who correlated the fracture strain with the area fraction of porosity determined from the fracture surface. The data show similar trends but differ significantly in the fracture strain at a given fraction of porosity. This variation is likely due to a variety of factors including the inability of the tomographic measurements to resolve pores smaller than 20 μm, and the experimental difficulty in determining the area fraction of porosity on a fracture surface.

INDENTATION

A parallel study has been undertaken which employs microhardness measurements to characterize the local mechanical properties of the casting [13]. One advantage of microindentation is the ability to sample very small volumes of material. In this study, indentation was employed in order to investigate the difference in strength between the skin and core of the casting. By using a spherical indenter and multiple load-unload cycles to successively higher loads, we are able to reconstruct the tensile stress-strain curve. This also suggests that spherical microindentation may be a suitable, non-destructive method for evaluating the local material properties of Mg HPDC.

Table 1 Fatigue and quasi-static mechanical properties measured at one location on the casting.

Fatigue Limit	Yield Strength	Tensile Strength	Elongation
101 MPa	120 MPa	230 MPa	8%

Figure 6 plots the results of a tensile test (solid line) together with the flow curves constructed from indentation data. The indentation data for the skin compares well with the tensile curve, while that for the core is significantly lower. Although it is not yet clear, this difference could be attributed to the change in grain size from the skin to the core, to a gradient in alloy composition, or to a combination of the two factors.

Regardless of the reason, the measured change in flow strength between the skin and core is an important observation. When coupons are machined from castings, the skin is removed from at least the edges of the specimen. For this reason, it would be expected to see at least a small decrease in the measured mechanical properties.

FATIGUE PROPERTIES

Fatigue behavior is of great concern, due to the cyclic nature of the loads experienced by the components in service. Unfortunately, fatigue data on thin-walled die-cast magnesium alloys are sparse, as are studies of the microstructural damage mechanisms accompanying fatigue failure. Studies have shown that cracks initiate in grains favorably oriented for slip and at pores (in cast alloys) [14,15]. In cast alloys, it is generally found that surface and near-surface pores are the most common crack initiation sites [15,16].

To date, the fatigue limit (10^7 cycles) has been determined for a single location on the casting. The fatigue limit was determined using the staircase method on a total of 25, nominally identical fatigue specimens [17]. Constant amplitude, load controlled fatigue tests were performed using a stress ratio R=0.1, and a cycling rate of 25 Hz. Care was taken to preserve the as-cast skins on the broad faces of the fatigue coupons. In addition, the machined edges of the coupons were polished longitudinally to a 1200 grit finish. The quasi-static mechanical properties measured at this location are shown together with the fatigue limit in Table 1.

CONCLUSION

Thus far, the research program has developed a map of both microstructure and mechanical properties for a die-cast magnesium instrument panel beam. A useful indentation technique has been adapted for use in magnesium alloys for the prediction of the tensile stress-strain curve. The technique is also suitable for probing mechanical properties on a fine scale. In particular, it is useful for investigating differences on the scale of the skin and core of thin-walled die-castings. The structure-property relationship developed by Caceres for aluminum has been shown to quite accurately relate area porosity to fracture strain.

The continuing research is now focused on three areas:

1. Refining the necessary structure-property models, including an accounting for effects of residual strains in the as-cast structure.
2. The adaptation and development of models for non-equilibrium solidification that will be required for the numerical simulation of the die-casting.
3. The creation of the post-processing technique to predict local microstructure from the output of computational fluid dynamics models of the die-casting process.

ACKNOWLEDGMENTS

This research is funded by Meridian Technologies and the AUTO21 Network of Centres of Excellence, an automotive research and development program focusing on issues relating to the automobile in the 21st century. AUTO21, a member of the Networks of Centres of Excellence of Canada program is funded by the Natural Sciences and Engineering Research Council (NSERC), the Social Science and Humanities Research Council (SSHRC) and multiple industry and government partners.

REFERENCES

1. K. Buchholz, Automotive Engineering Intl., June 2000, pp. 44 – 45.
2. R. Alain, T. Lawson, et al. (2004) Proc. of SAE 2004 World Congress. Book #SP-1845, Doc. #2004-01-0131.
3. B.J. Coultes, J.T. Wood, G. Wang, and R. Berkmortel, in H.I Kaplan, ed. "Magnesium Technology 2003", pp.45–50 TMS, Warrendale, PA.
4. T. Koch, M.A. Gharghouri, in H.I Kaplan, ed. "Magnesium Technology 2003", pp.71–76 TMS, Warrendale, PA.
5. D. Yin (2004) M.E.Sc. Dissertation. The University of Western Ontario, London, Ontario, Canada.
6. ASM Metals Reference Book, 2nd Ed. (1983).
7. W. Sequeira and G. Dunlop (2004) Die Casting Engr., 48(2), 62-74.
8. E.M. Gutman, Y. Unigovski, M. Levkovitch, and Z. Koren (1998) J. Mat. Sci. Let. 17, 1787-1789.
9. H. Mayer, M. Papakyriacou, B. Zettl, and S.E. Stanzl-Tschegg (2003) Intl. J. Fatigue 25, 245-256.
10. A.M. Gokhale, and G.R. Patel (2001) in J. Hryn, ed. "Magnesium Technology 2001", pp.195-199, TMS, Warrendale, PA.
11. J. P. Weiler, J.T. Wood, R.J. Klassen, E. Maire, R. Berkmortel and G. Wang (2005) accepted for publication in Mat. Sci. Eng. A.
12. C.H. Caceres, B.I. Selling (1996) Mat. Sci. Eng. A, A220, 109-116.
13. J.P. Weiler, J.T. Wood, R.J. Klassen, R. Berkmortel and G. Wang. (2004) submitted to Scripta Mat.
14. V. V. Ogarevic and R. I. Sephens, (1990) *Annu. Rev. Mater. Sci.*, 20, 141-177.
15. H. Mayer et al., (1999) in X-R. Wu and Z-G. Wang, eds. *Fatigue '99: Proceedings of the Seventh International Fatigue Congress*, (Beijing, China:Higher Education Press, 2059 – 2064.
16. G. Eisenmeier et al., (2001) Mat. Sci. Eng. A, v 319-321, 578-582.
17. T.A. Koch (2004) M.A.Sc. Dissertation. Dalhousie University, Halifax, Nova Scotia, Canada.

2005-01-0726

Physical and Virtual Prototyping of Magnesium Instrument Panel Structures

John Z. Lin and Suresh Lanka
Delphi Corporation

Thomas Ruden
Lunt Manufacturing Co., Inc.

ABSTRACT

This paper reviews the current strategies for physical prototyping of Magnesium instrument panel (I/P) structures. Bottlenecks in the traditional physical prototype based product development process are discussed. As demand for fast-to-market and cost-reduction mounts, virtual prototyping becomes increasingly important in meeting the timing and performance goals. A virtual prototyping methodology is presented in this paper to enable high performance Magnesium I/P structures in Safety, NVH, and initial part quality aspects. Examples of Finite Element Analysis (FEA) results and correlations are included.

INTRODUCTION

Ever-increasing market competition has driven automakers to produce stylish, safe and durable vehicles at a faster-than-ever pace and reduced costs. While customers are attracted to niche styling design and more convenience provided by telecommunication and navigation devices as well as entertainment electronic packages, they have also become increasingly educated in safety and comfort related performance. In a sense, what safety means to automakers and customers is beyond Federal Motor Vehicle Safety Standards (FMVSS), which determine the salability of all vehicles marketed in U.S. It is the safety ratings such as NCAP (New Car Assessment Program) issued by NHTSA and J.D. Power Initial Quality Surveys that set the bar for high-performance cockpit modules.

As the backbone of an instrument panel assembly or cockpit – the cross-car I/P structures - are under pressure to support increasingly sophisticated electronics for telecommunication and entertainment while meeting styling requirement for roominess, from packaging point of view. On the other hand, federal motor vehicle safety requirements and star ratings dictate the viability and competitiveness. Also, the lead-time associated with the traditional prototyping of instrument panel structures has been proven to be unsatisfactory in providing lessons-learned for production design. Facing these challenges, it is very important for modular cockpit suppliers to develop a non-traditional methodology and use the right tools to design and validate a high performance product within the given program timing.

Die-cast Magnesium cross-car beams have been in production by all major carmakers due to its weight-saving nature and superior stiffness and strength. Recent reduction in raw material price is making Magnesium even more attractive for automotive applications. The AM60B grade has balanced stiffness and ductility and is commonly used for cross-car beam applications. This paper reviews the current strategies for physical prototyping of Magnesium instrument panel (I/P) structures. Bottlenecks in the traditional physical prototype-based product development process are discussed. As demand for fast-to-market and cost-reduction mounts, virtual prototyping becomes increasingly important in meeting the timing and performance goals. A virtual prototyping methodology is presented in this paper that would enable high performance Magnesium I/P structures in Safety, NVH, and initial part quality aspects. Examples of Finite Element Analysis (FEA) results and correlations are included.

PHYSICAL PROTOTYPING

Magnesium die-castings have commonly been prototyped using a variety of well-understood processes. The two most popular ones are sand and investment casting. The benefits and issues of these processes will be reviewed in this section.

PROTOTYPING CONCERNS

Ideally, the most direct and lowest cost path to produce a prototype magnesium cross-car beam would be to use the resulting math data from the structural analysis to create the die-cast tooling. However, the lead-time needed to produce a die-cast beam is typically 20 weeks. Due to project timing constraints this option may not be viable. As an alternative, low-pressure casting has been used. Currently, lead times required to produce a large sand cast prototype approach 8-9 weeks.

Tooling Lead Times

Perhaps more a strategic decision compared to budgetary constraints is the timing capacity of a process to meet a given deadline. This is an important factor to consider when planning for the prototyping phase of a program.

The time needed to develop a design for a specific process (die-casting, sand or investment) would be identical. Thus, the next consideration would be one of delivery of the prototype castings. In general, die-cast prototypes for complex cross-car beams have averaged 20-30 weeks in recent years – although 13-15 can be obtained at a premium. In comparison, low-pressure castings have averaged 11-15 weeks (see Table 1) and tooling costs are significantly lower since the tool material is not made from P-20 or H-13 steel.

Item	Phase	Average Lead Time	Comments
1	Pattern & Check Fixture	6 weeks	
2	Dimensional Check	1 week	Full dimensional check of pattern.
3	First Article	1 week	
4	Prototype Castings	1 to 4 weeks	Based on order number of prototypes required (up to 15 castings).
5	Secondary Finishing and Machining	2 to 3 weeks	
	Average Lead Time	**11 – 15 weeks**	

Table 1-Average Lead Times of Low Pressure Castings

Feature Sizes

A primary interest is the ability of alternate processes to replicate feature sizes desired in the production-intent die-casting. While magnesium die-castings are routinely cast to a section thickness of 1.0 – 2.5mm, alternate low-pressure casting processes such as sand and investment casting are usually limited to thicker section thickness of 2.5 – 3.0mm, or greater. This is due to the lower pressures used in the casting process and generally, lower fluidity and castability of the process. Additionally, features and section thickness of magnesium die-castings may be varied to a greater extent than low-pressure castings that, because of their lower fluidity, require a more uniform section thickness.

Feature size can influence the mechanical performance of a component. Thin wall die-castings are attractive to product engineers as they promote design efficiency (lightweight, small package size). However, processes that exhibit thicker section thickness not only promote a less efficient part (heavier, larger package size) but would contribute to a higher section modulus. This higher modulus promotes a part design having a higher stiffness and offering less deflection (especially during impact) than a part having a thinner, more efficient, section thickness.

Solidification Rates

There are distinct mechanical property differences between low-pressure castings and high-pressure die-castings. A primary reason for these variations in properties is due to the differences in the solidification rates between selected processes.

The high-pressure die-casting process uses high process pressures, short process cycles and rapid solidification rates of the molten magnesium. The result is a very thin (0.25-0.50mm), dense surface layer ("skin") on the surface of the die-casting. This dense layer exhibits a minimal amount of shrinkage and porosity. As a result of these influences, the surface layer of a magnesium die-casting is very strong

Low-pressure casting processes such as sand and investment casting use low pressure ("gravity") when pouring molten metal into the mold. Generally, these processes do not promote fast processing times and the molten metal needs time to cool and solidify. As a result, low-pressure castings generally require a larger, more uniform section thickness throughout the body of the casting [1-2]. However, in recognition of these uniform features, the mechanical properties of low-pressure castings may be modified by employing heat treatment. Depending upon the magnesium alloy used, various heat treatments are available to change the strength, elongation and toughness of a casting. A summary of the available heat treatment processes are shown in Table 2, along with a brief table (Table 3) of the properties available for the most widely used magnesium alloy for low-pressure casting, AZ91E.

Mechanical Properties

Magnesium cross-car beams are commonly produced in either AM50A or AM60B alloy, with the majority of beams die-cast in AM60B. Since a major objective of the prototyping process is to evaluate the load carrying capacity of the beam, it will required that the prototyping material match the mechanical properties of the production material. Varying results will occur if this is not taken into consideration. As shown in Table 3 (ref. [3]), properties of die-cast components do not replicate low-pressure castings. Key factors in alloy selection are combinations of strength, ductility ("elongation"), impact-strength and energy absorption.

Temper Designation	Sub-Temper	Description
F	-	As-fabricated, as-cast.
T	-	Heat treated to produce more stable tempers than F, O or H.
-	T4	Solution heat-treated, then aged at room temperature to a stable condition.
-	T5	Cooled from an elevated temperature shaping process (casting or extrusion), then artificially aged at an elevated temperature to improve properties or stability.
-	T6	Solution heat-treated, then stabilized at elevated temperature.
-	T7	Solution heat-treated, then stabilized at elevated temperature.
-	T8	Solution heat-treated, then cold-worked and artificially-aged at elevated temperature.

Table 2 – Summary of Heat Treatment Processes

Energy absorption capacity is the ability of a material to absorb total applied energy before it fails. This can be quantified as the area under the stress-strain curve of a material [4]. The greater the area under the curve, the more total energy a material is able to absorb before failure. This is also an indication of a material's "toughness." This aspect is especially helpful in the application of magnesium alloys for cross-car beams, or components subjected to significant external loads. Even though the true area under the stress-strain curve can be difficult to calculate, it can be approximated by using trapezoids (See Figure 1 below), expressed mathematically by Equation (1).

$$(TYS \times El.) + [(UTS-TYS) \times El. / 2] \quad (1)$$

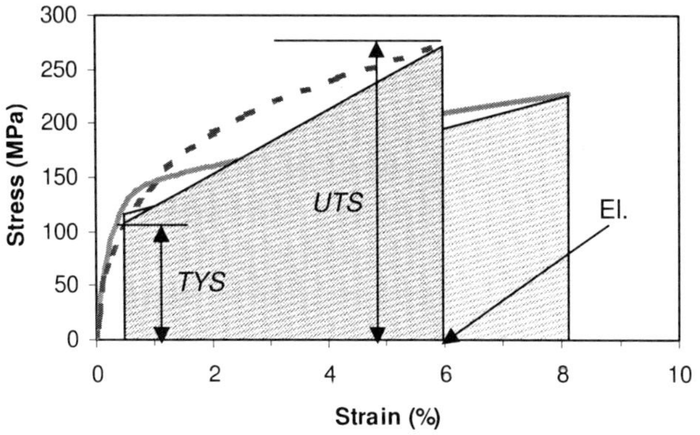

Figure 1. Effect of Mechanical Properties via Heat Treatment. Goal-Total Energy Absorption Capacity for Alloy "A" Approximates Alloy "B"

Where *TYS*=Tensile Yield Stress, *El.*= Elongation, and *UTS*= Ultimate Tensile Strength

It is interesting to note that Young's Modulus is similar for all magnesium alloys and processes in Table, and this is typical for the majority of Mg alloys and processes in general. Based on material property comparison, not all properties will equal each other. However, selection of the proper alloy and heat treatment of low-pressure castings can be made to approximate magnesium die-casting properties. Figure 2 shows the comparison among various as-cast and heat-treated alloys.

Property	Units	High Pressure Die-Casting			Low Pressure Casting		
		AZ91D F	AM60B F	AM50A F	AZ91E F	AZ91E T4	AZ91E T6
TYS	MPa	150	131	124	97	90	145
UTS	MPa	230	241	228	165	275	275
Modulus	GPa	45	45	45	45	45	45
Elongation	%	3	8	10	3	15	6
Impact Strength	J	2.7	2.8	3.0	0.8	4.1	1.4
Energy Absorption	10^4 J/m^3	570	1,488	1,760	393	2,737	1,260

Table 3 - Effect of Heat Treatment

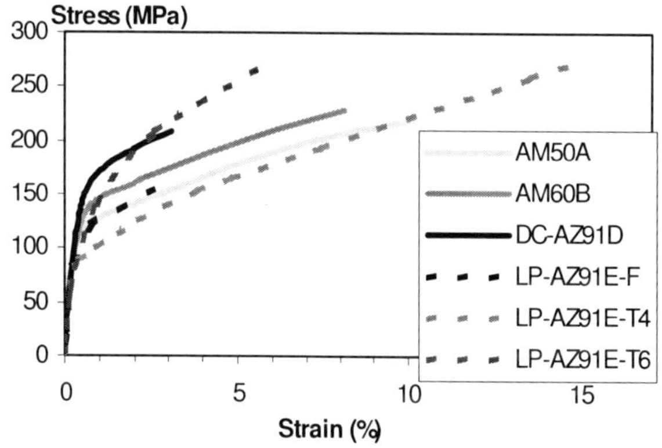

Figure 2. Comparison of Various Magnesium Stress-Strain Curves

Surface Finish

In addition to considering the differences in mechanical properties, surface finish also plays a role in the prototype decision-making process. It is well documented that surface finish and surface quality are important factors as fracture mechanics have an effect on mechanical performance, especially on highly and rapidly-loaded parts like a cross-car beam. In a practical sense, the smoother surface and tighter the grain size a part has, the greater resistance to premature surface cracking and failure it will offer when subjected to applied loads higher than the yield strength ("plastic region") of the material. Consequently, the surface quality and finish offered by magnesium die-castings is a key attribute of its use in energy-absorbing applications – such as cross-car beams.

Therefore, when prototyping magnesium die-castings it is more advantageous to select a process that approximates the overall quality and surface condition of a magnesium die-casting.

Process	Typical,Nominal Surface Roughness Height (μm)
Sand Casting	25.0-12.5 μm
Investment Casting	3.2-1.8 μm
Die-Casting	1.8-0.8 μm

Table 4 - Nominal Surface Roughness (height) Comparison of Processes

SUMMARY

The selection process for prototype Magnesium beams (parts subject to high strain rates) is as follows:

1. Design for a Mg die-casting. If timing allows, try to prototype using P20 or H13 tool steel.
2. If cannot prototype in tool steel, then select a low-pressure casting alloy and heat treatment process that would approximate the die-casting.
3. Refine the low-pressure casting design to replicate the features of a die-cast design. This would minimize any issues resulting in increased strength or stiffness from a prototype casting.
4. Process the low-pressure prototypes using smooth, fine materials for the casting material. This would provide the highest quality, smoothest surfaces possible to help create high integrity castings.

VIRTUAL PROTOTYPING

Finite Element Analysis (FEA) based math tools have been successful in evaluating design proposals and assessing various structural performances of all kinds of automotive components and systems [5-9]. The further success and significance of FEA-based digital product development process lies in its predictive ability and dramatic reduction of physical prototyping. The industry trend towards the goal of one hundred percent virtual prototyping is clear, and the push is not only felt at OEM's but also at suppliers because of the mounting demand for faster-to-market new products and cost competitiveness.

In this section, a general methodology for the digital product development of Magnesium I/P structures in outlined. Critical structural performance requirements – Safety, NVH and Static Stiffness – are summarized. Examples are provided to demonstrate the abilities of virtual prototyping.

GENERAL METHODOLOGY

It is always important to understand every aspect of performance requirements for any product design and development. Table 5 summarizes the engineering concerns need to be considered for I/P structures.

Category	Engineering Concern	Performance Measure
Static Stiffness	Column Mounting Stiffness	NVH
	Radio, HVAC Mounting Stiffness	NVH
	Sag, I/P Attachment Stiffness	Fit and Finish
Safety	Occupant Protection for Frontal Impact	FMVSS208, NCAP, IIHS
	Occupant Protection for Side Impact	FMVSS214, Side-NCAP, IIHS
	Occupant Protection for Head Impact	FMVSS201

Table 5 - Summary of Engineering Concerns For Magnesium I/P Structure Development

Given the current material cost situation, the general trend of Magnesium I/P structure applications remains to be for mid- to high-end passenger cars, where NVH and Safety performance can play a vital role in competition. In the safety area, side impact resistance is the most important performance since it is directly linked to FMVSS214 compliance and side NCAP ratings. The methodology outlined herein was developed with side impact resistance as top priority, and it should suffice for other stiffness and safety performance.

In virtual prototyping it is important to identify load transfer requirement for the cockpit cross-car beam. This has to be done on the full vehicle level, by either numerical simulations and/or physical barrier tests. It is possible to do so because at the stage of conceptualizing cockpit cross-car structure, the rest of vehicle body structures are usually determined digitally or hardware-wise (carry-over/common platform)[10]. After the load requirement is identified, one can proceed with FEA based simulations, examples of which are highlighted in the following sections.

SECTION ANALYSIS

This step is especially useful and effective to guide the initial concept design of the Magnesium beam. The goal is to have a continuous load carrying sections/paths defined.

Figure 3 shows an example of the outputs of such a section analysis, where moments of inertia of critical sections are charted It is important that a Magnesium beam provides local stiffening features to support the steering column for NVH performance and any local weakness that will potentially result in local buckling or cracking should be eliminated. It is also required that the beam provides sufficient mounting features for HVAC module attachment.

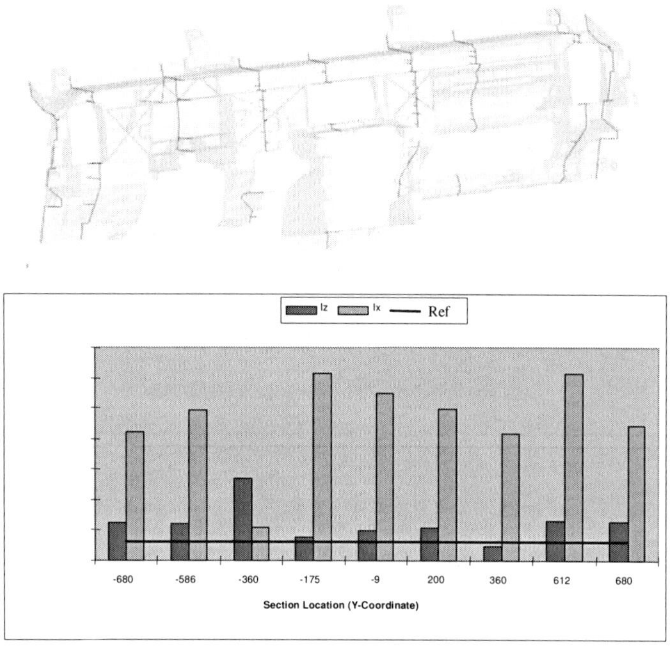

Figure 3. Typical sections and section properties of a prototype magnesium cross-car beam design

COMPONENT-LEVEL DROP-SILO SIMULATION

The purpose of drop-silo test and simulation is two-fold: (1) Determine the ultimate load carrying capacity of the beam. (2) Identify global and local failure modes if any.

Because of these reasons drop-silo tests are usually destructive. But it is often not the case in an actual side impact barrier test where loading of the beam is much less severe. Nevertheless, drop-silo simulations provide quick turn-around results and recommendations to the initial design and have been proven very effective in achieving the desirable full vehicle results.

Drop-silo simulations are done with *LS-DYNA3D™*. The loading and boundary conditions are setup to duplicate the exact physical test. Figure 4 shows a typical drop-silo set-up.

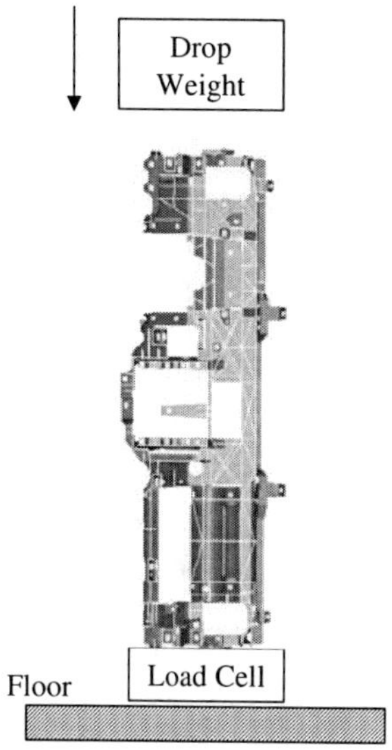

Figure 4. A typical drop-silo model

Figure 5 shows the correlation between the FEA result and physical drop-silo test result for axial load. It is evident that the data correlated well with both the peak load and timing of the peak. With the confidence in FEA models, drop-silo simulations play a central role in the DPD process of cockpit cross-car beam.

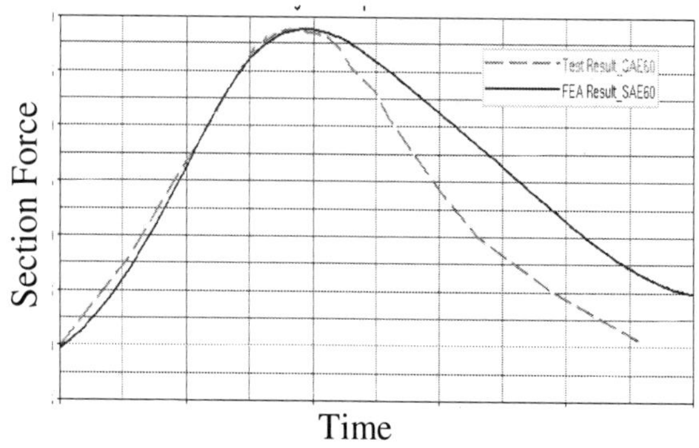

Figure 5. Correlation between the FEA results and physical drop-silo test results (filter: SAE60).

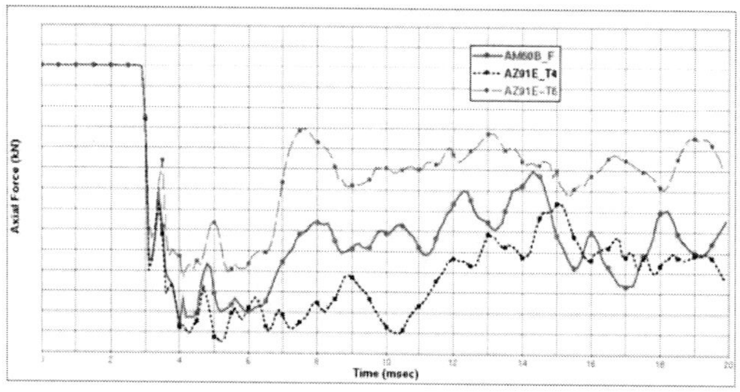

Figure 6. Unfiltered hit-end section force comparison among AM60B-F, AZ91E-T4 and AZ91E-T6

The key factors in achieving good correlation were: (1) Initial and boundary conditions defined accurately, and (2) Material model for the Magnesium has to include failure strain

Given equal part geometry and feature size, Figure 6 is intended to compare the load carrying capacities of different Magnesium alloys – AM60B-F (high pressure casting material with no secondary heat treatment), AZ91E-T4 and AZ91E-T6 (both sand-cast prototyping material with different heat treatment (see Table 2). The Magnesium beam shown has a nominal thickness of 3mm.

It is interesting to see that the commonly used AZ91E-T4 for sand-cast prototypes matched well with the production-intent AM60B-F in terms of peak loads. The AZ91E-T4 over-estimates the sustained load level due to its higher strain to failure. Nevertheless, the use of sand-cast AZ91E-T4 prototype seems to be justified even for critical tests such as drop-silo's, given the requirement that prototype parts have the same thickness as the production parts.

FULL VEHICLE SIDE IMPACT SIMULATION

Full vehicle barrier simulations involve test barrier, test dummies, full vehicle body, chassis, non-structural masses, tires, seats and cockpit. They are usually performed by OEM to access the performance of cockpit structure in term of load carrying ability and deformation. The main output of a full vehicle simulation is the axial section force time history in comparison with a targeted cross-car I/P structure. At the present, correlation for test dummy TTI numbers is difficult to achieve.

Figure 7 is a top view of a full vehicle side impact test configuration. Table 6 summarizes the criteria for side-NCAP ratings. Recently there have been on-going researches by NHTSA to promote head protection and possibly to include 5th% Female dummy in the testing, and major changes to FMVSS214 are expected soon, which have great potential to reduce death and injury for passenger cars and light trucks.

Figure 8 shows the section force history of the Magnesium beam (at y=223) in comparison with the result for a reference structure. Clearly, the load transferred through the Magnesium beam is higher. Given the same amount of energy input from the barrier, this in turn results in less intrusion towards the occupants and hence lower risk of severe injuries. The criterion is met according to our virtual prototyping methodology. Figure 9 (a) and (b) compares the deformed shapes of the Magnesium beam in the simulation and after the physical test, respectively. Good correlation is found.

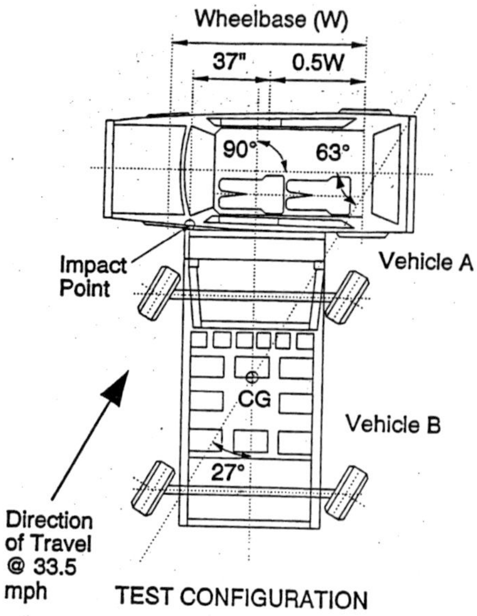

Figure 7. Overview of a full vehicle side impact test configuration

Side NCAP Star Rating	TTI
1 Star	>95
2 Star	85-95
3 Star	72 - 85
4 Star	59 - 72
5 Star	< 59

Table 6. Side NCAP star rating criteria (TTI –The Thoracic Trauma Index $=1/2\ (G_R + G_{LS})$ where G_R is the greater of the peak acceleration of either the upper or lower rib and G_{LS} is the lower spine peak acceleration, both expressed in g's)

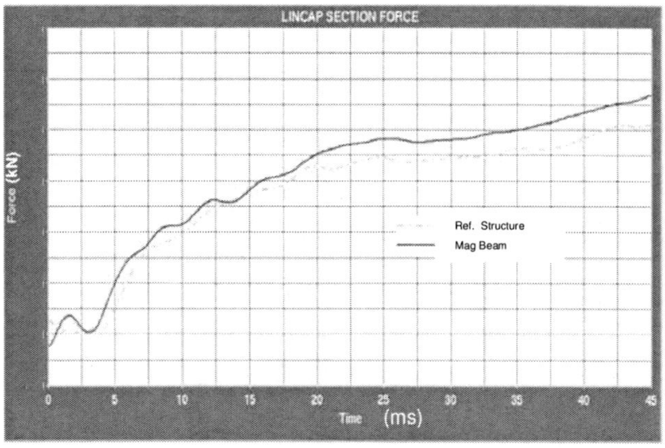

Figure 8. Section force history comparison: Magnesium beam vs reference structure

(a)

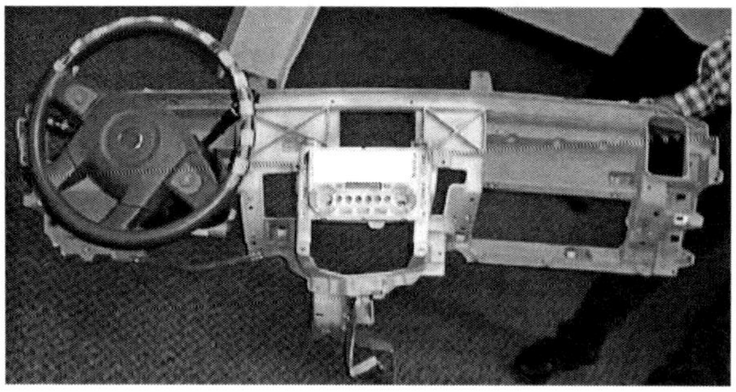

(b)

Figure 9 (a) Simulated deformation of the Magnesium beam in full vehicle barrier environment (b) Prototype part after a physical barrier test.

CASTING SIMULATION

In order to ensure part quality, casting simulations were carried out to provide a better understanding of mold filling, solidification, mechanical properties, thermal stresses and distortions. It is crucial in side impact applications that a cast Magnesium beam is free of defects (porosity, knit lines, etc.) that could give rise to premature local or global failure. Commercial softwares suitable for these tasks include MAGMASOFT®

CONCLUSIONS

A digital product development methodology is established for a high performance Magnesium cockpit structure. It has been proven to be successful in achieving the following results and was time- and cost-effective:

1. FMVSS214 compliance and 4-star side crash NCAP rating without side airbags
2. IIHS Offset Deformable Barrier (ODB) rating: Good.
3. 35+ Hz in-vehicle column frequency, world-class due to rigidity provided by the Magnesium structure

In the development of this actual mid-sized passenger car application, because of late-stage change and accelerated timing, the production design was completed in six months and all sand-cast, P20 and H13 tool releases occurred at almost the same time. All the structural performance targets were met with the first prototype sand-cast beam. In retrospect, the sand-cast and P20 events could have been eliminated to realize significant savings.

With regard to traditional physical prototyping, low-pressure casting, either sand casting or investment casting, offers shortest tool lead-time that are currently acceptable for most applications. It also has merit in terms of validation critical structural performances such as NVH and side impact resistance as long as the following guidelines are followed:

1. Select a low-pressure casting alloy and heat treatment process that would approximate the die-casting. Based on correlation between theoretical models and component testing, AZ91E-T4 would be a good choice for an AM60B production part.
2. Refine the low-pressure casting design to replicate the features of a die-cast design, especially at critical locations ("weakest links"). This would minimize any issues resulting in increased strength or stiffness of prototype casting. Nominal thickness of 3 mm or higher would make it easier to avoid overestimating performance.
3. Process the low-pressure prototypes using smooth, fine materials for the casting material. This would provide the highest quality, smoothest surfaces possible to help create high integrity castings.

The adoption of the above guidelines in conjunction of the virtual prototyping strategy results in a cost-effective and fast-to-market way to make high performance Magnesium I/P structures.

REFERENCES

1. T. Ruden, *"Prototype Alternatives for Magnesium Die Castings,"* Die Casting Management, pp.26-29, October 1991.
2. T. Ruden, D. Finkhousen and C. Engel, *"Prototyping Strategies for Magnesium Die-Castings,"* 52nd Annual International Magnesium Association Conference, New Orleans, LA. May 15, 2004.
3. "Magnesium and Magnesium Alloys," ASM International, 1999.
4. F. Crane and J. Charles, "Selection and Use of Engineering Materials," 1st Ed. Butterworths, 1987.

5. John Z. Lin and Stephen M. Pitrof, " Analytical Design of Cockpit Modules for Safety and Comfort", SAE paper 2004-01-1481.

6. Kumar B. Kulkarni and Raj S. Thyagarajan, "Effects of Body Attachment compliance on steering column in-vehicle modes", SAE paper 2000-01-2692.

7. Naganarayana B P, Shankar S, Bhattachar V S, Brines R S and Rao S, "N-hance: Software for identification of critical BSR locations in automotive assemblies using finite element models", #03NVC-283, Noise & Vibration Conference, SAE-2003, May 5-8, Grand Traverse Resort, Traverse City, MI, USA, 2003.

8. Mohamed El-Essawi, John Z. Lin, Gary Sobek, B. P. Naganarayana and S. Shanka, " Analytical Predictions and Correlation with Physical Tests for Potential Buzz, Squeak, and Rattle Regions in a Cockpit Assembly", SAE Paper 04AC-102.

9. Todd Kovacic, Suresh Lanka and Matthew Marks, " Development of an Integrated Structural HVAC Instrument Panel Cockpit System", SAE Technical Paper 2002-01-0309.

10. Mangala M. Jayasuriya. "2-Door Vehicle Body Local Force Evaluationwith the IIHS, EuroNCAP, and LINCAPSide Impact Barriers", SAE Technical Paper 2004-01-0333.

CONTACT

For further information, please contact any of the authors: John Z. Lin at john.z.lin@delphi.com or Suresh Lanka at suresh.s.lanka@delphi.com or Thomas Ruden at truden@Lunt.com

2005-01-0727

2004 Nissan 3.5L Cam Cover Material Study: Aluminum, Magnesium and Composite

Eric Herrera
Director Advanced Acoustics & Vibration for Freudenberg-NOK (FNGP)

ABSTRACT

The present study compares the NVH performance of three different materials used on cam covers in automobiles, Aluminum (Al), Magnesium (Mg) and Thermoplastic (TP). The cam cover design used for this comparison was the 2004 Nissan Maxima 3.5L production cam cover which is made of a thermoplastic (TP). The Al and Mg covers for this study were created by sandcast, due to time constraints, via laser scanning techniques using the 2004 Nissan Maxima 3.5L production thermoplastic cover design. Note that sand-cast covers generally provide a less quiet sound field than the standard casting method. The Nissan production cover comes with a production baffle made of a similar material as the cover. Testing was conducted with and without the production baffle for all covers. The study was conducted for the production boundary condition of a non-isolated cover and a Freudenberg-NOK (FNGP) partially isolated cover. Isolated bolt assemblies using elastomeric grommets were used to isolate the cover. Dynamic Mechanical Thermal Analysis (DMTA) of the thermoplastic part was performed and compared to thermo-set bulk moulding compound (BMC). A general correlation between damping was established revealing that for this application the material which allows the least damping provides the best NVH, data and analysis will be presented to support this. The results show that the isolation frequency and NVH levels (dB) for the metal covers are lower than for the thermoplastic for the semi isolated case and noticeably lower in the isolated case. Numerical and experimental methods were used in this study.

INTRODUCTION

It is well known that cam covers are one of the major contributors to the sound field of most automobiles (1, 2, and 3). With that attention brought to cam covers an important question becomes; which material leads to the optimum sound field, metals or plastics? Acoustics being one of the most interactive phenomena in nature does not generally resolve in a binary answer, metal or plastic. The cam cover is part of an integrated system, i.e. engine etc, with inherently unique geometries and unique boundary conditions. To investigate the question above further we need to normalize to these unique system parameters. To do this a geometry and boundary condition were picked to compare between metal and plastic cam covers. The 2004 Nissan Maxima 3.5L was picked because of premium Acoustic performance.

Figure 1: 2004 Nissan Maxima 3.5L V6 in FNGP Plymouth Mi Technology Chassis Dynamometer room

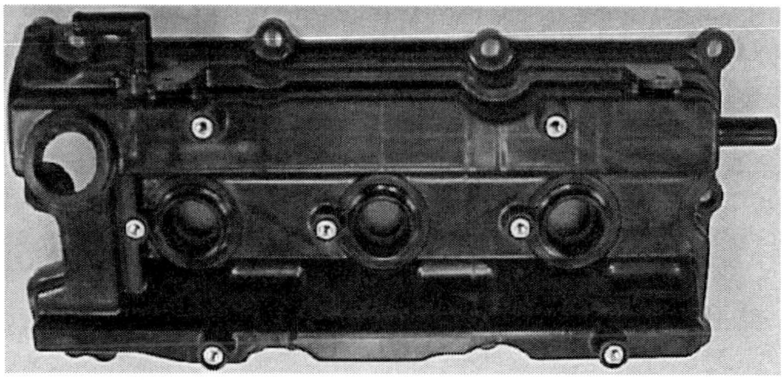

Figure 2: Production Nissan 3.5L cover

The Nissan cam cover and baffle were then used to make identical Aluminum and Magnesium cam covers. Identical bolt assemblies and gaskets were used. Each cam cover had its own respective bolt assemblies and gaskets to prevent cross data contamination. Of course there is an inherent mass difference between the covers. To make equivalent mass covers the wall thicknesses would need to vary and thus the geometry would vary. Since the mass effect is well understood already with respect to acoustics it was determined to investigate the other effects mentioned above.

The covers were tested on vehicle in the Freudenberg-NOK (FNGP) Plymouth Mi Technology Center chassis dynamometer room and on in the FNGP Plymouth Mi hemi-anechoic room. The test methodology used is unique to FNGP and uses the concept of the Log Transmissibility Ratio (LTR). The reason for using the LTR method was to reduce test time and to test configurations that may not be appropriate for vehicle test such as testing a cam cover with no baffle etc.

TEST ARTICLES: CAM COVERS

The Nissan 3.5L production cam cover was laser scanned to produce a point cloud to capture the exact geometry.

Figure 3: CAE Geometry capture for the Cam Cover

The baffle was also scanned.

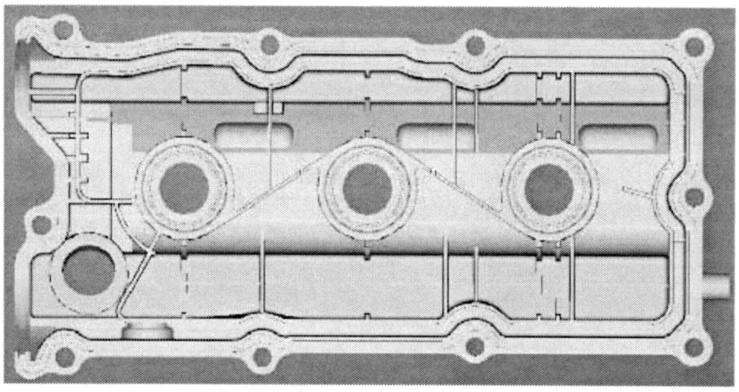

Figure 4: CAE Geometry capture for the Baffle

Cam cover wall thicknesses were taken into account by cutting sections and making measurements.

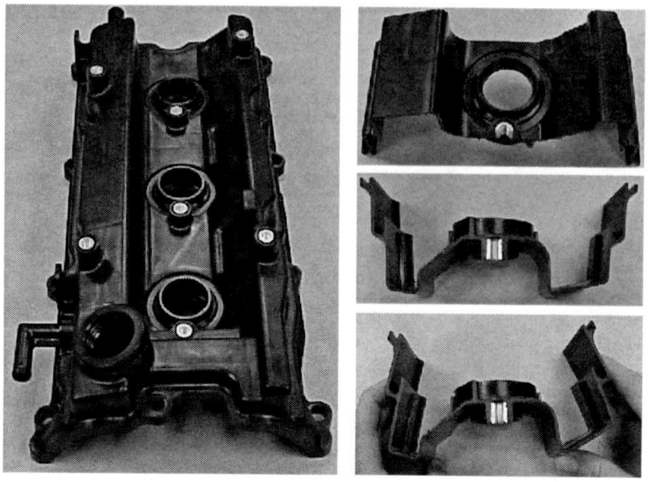

Figure 5: Cam Cover wall thickness

Once the geometry and wall thicknesses were determined the covers were sand-cast.

Figure 6: Al Nissan 3.5L cover

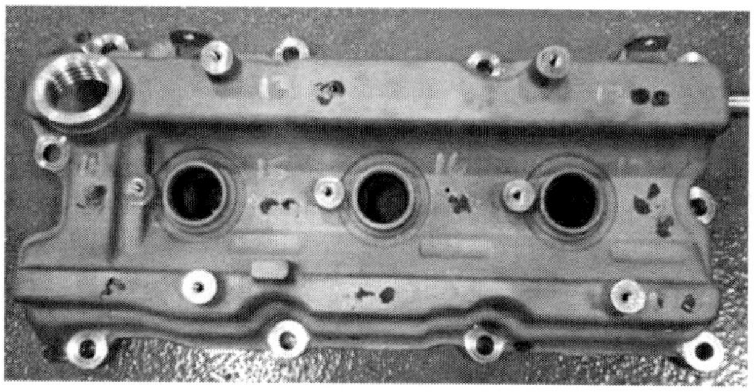

Figure 7: Mg Nissan 3.5L cover

CAM COVER BAFFLE

The production baseline baffle was made of composite material similar to the production cover. To characterize the effect of the baffle, tests were conducted with/without the baffle. To avoid damage to the engine etc the covers without baffle were tested using the FNGP LTR methodology in the hemi-anechoic room, which will be discussed in the next section.

TEST METHODOLOGY

Testing was conducted in vehicle and out of vehicle using the FNGP LTR methodology.

FNGP LTR METHODOLOGY FOR OUT OF VEHICLE INTEGRATED SYSTEM TEST

The FNGP LTR methodology has been in use for several years now, is continuously evolving and has been validated with OEM supplied on engine data. Describing the generic LTR concept in detail (4) is not the purpose of this paper, suffice it to say there is great confidence in the FNGP LTR methodology among the OEM's and they have adopted aspects of this testing in their facilities. Simplistically the methodology uncovers the transfer function between input (cylinder head etc) and output (cam cover etc). This was done here by a computer controlled shaker for excitation and numerous accelerometers strategically placed on the input and output for recovery. The excitation placement must be intelligently placed and should have the center of mass of "the integrated system" taken into account. Microphones are used for near (10cm) and far (100cm) field measurements. System checks for torque, gap height and etc were performed. During and after the data acquisition data quality checks were made, using coherence and etc. Data was recorded and LTR values calculated using the equation:

$$LTR = 10 \bullet Log\left(\frac{cover\ acceleration\ response}{head\ acceleration\ response}\right)$$

The acceleration response values in the equation above are spatial averages over each respective component. Documentation of this procedure with extensive details can be made available upon request to the author, see contact section.

Figure 8: FNGP LTR Setup out of vehicle in Hemi-Anechoic room

FNGP LTR METHODOLOGY FOR VEHICLE LTR TEST

The testing of the Nissan 3.5L in the chassis dynamometer room was setup so that accurate comparison could be made between it and the FNGP LTR testing in the Hemi-Anechoic room. Accelerometers were placed at the same locations for both tests. Microphone data for near and far field were also taken although microphone placements between tests were not the same.

Figure 9: Vehicle LTR Test in Chassis Dynamometer room

OUT OF VEHICLE TEST ARTICLE SETUP

It is important to recognize that the test article, cam cover here, is part of an integrated system. Of the utmost importance is to know where the influential components of the test article begin and end because this defines the "integrated system" to be tested. Defining the integrated system to test is important for both the out of vehicle and in vehicle tests.

BASELINE SETUP FOR OUT OF VEHICLE INTEGRATED SYSTEM TEST

For the Nissan 3.5L cam cover it was determined to include the cylinder head which incorporates what is often a separate front cover.

Figure 10: LTR Test setup; Cylinder Head Nissan 3.5L

Figure 11: LTR Test setup; Production Cam Cover and Cylinder Head Nissan 3.5L

The baseline setup also incorporated the production cover and bolt assembly. The production cover gasket was used for all tests. Each cover had its own gasket so that all gaskets would experience the same number and type of test.

<u>Al and Mg Setup</u>

The Al and Mg cover tests were the same setup as the baseline, only the cover material differed.

NO-BAFFLE TEST SETUP

Covers were made specifically for the no-baffle test with no baffles in them in order to eliminate possible introduction of spurious data upon removal of the baffle. The purpose of the no-baffle test was to characterize the effect of only the cover material.

ISOLATION SYSTEM SETUP

As noted the baseline setup used the baseline bolt assembly. As shown in the picture the baseline bolt assembly consisted of bolt and delimiter.

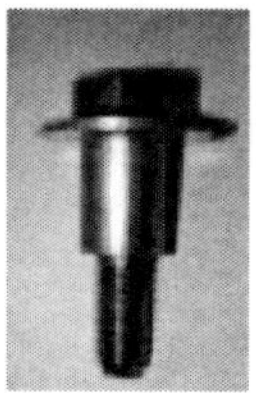

Figure 12: Bolt Assembly Baseline Nissan 3.5L

The FNGP isolation system here used a bolt with delimiter and 70 durometer HNBR grommet.

Figure 13: Bolt Assembly FNGP Isolation System

Typically the FNGP Isolation system consists of a specially designed gasket tuned to work with the grommet. These tests did not use a specially tuned FNGP gasket but used the production Nissan 3.5L gasket for all tests.

TEST ARTICLE MATERIAL PROPERTIES

Finite and Boundary element (FEM/BEM) modeling was performed in conjunction with experimental data acquisition. To accurately model the FEM/BEM several tests were conducted to characterize the materials. The most relative to the discussion of this paper is the Dynamic Mechanical Thermal Analysis (DMTA) which was performed to measure the elastic and plastic modulus of the Nissan 3.5L cam cover. The damping characteristics are revealed from the tangent delta which is shown in the figure. The yellow curve represents the BMC thermoset (TS) material and the pink curve is of the Minlon™ thermoplastic (TP) material from the Nissan 3.5L cam cover. The data shows the BMC (TS) as having a much larger tan delta than the Minlon™ (TP) which indicates the TS as more damped than the Minlon™ (TP). This is true for temperatures greater than 100 C (212 F) which is reached on a typical cam cover after ~ 5 minutes of engine running time on a standard day (~ 21 C or 70 F).

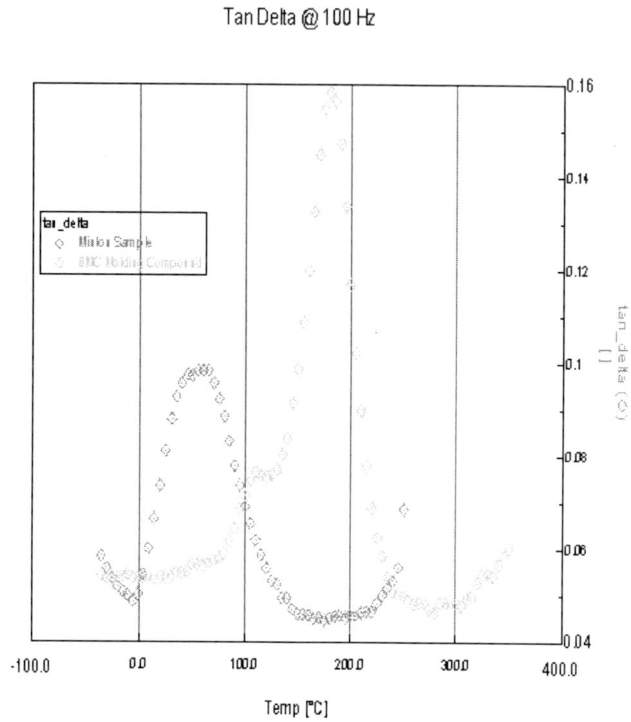

Figure 14: DMTA Thermoplastic VS Thermoset

Thermosets typically provide more damping than thermoplastics, as shown in the figure, because of the crosslinking character in the TS. Analogously the metallic bonds in Magnesium allow for more damping than Aluminum. Everything else equal (isolation system etc), Al provides improved noise reduction over Mg in cam cover applications. This is also shown in the graph below.

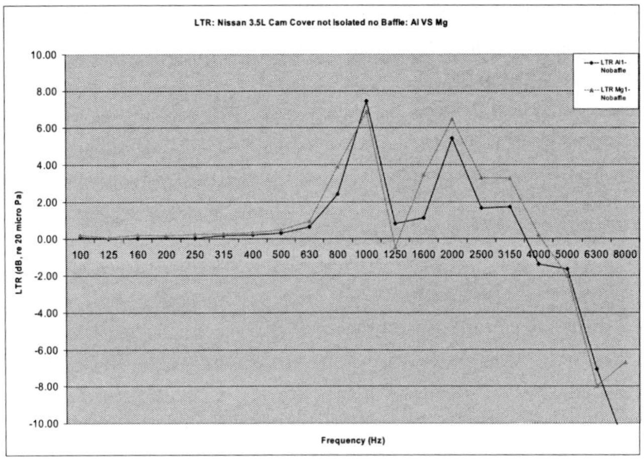

Figure 15: LTR; Nissan 3.5L can cover Al VS Mg Damping effect

The LTR graph shows the Al as having slightly lower dB levels than the Mg which can mostly be attributed to the damping effect. The trend to note here is that for this application the Al cover which allows less damping provides improved NVH over the Mg cover which allows more damping. Generally speaking and following the trend just discussed plastics allow more damping than metals and thus metals will provide better NVH than plastic covers. As shown in the DMTA graph the Minlon™ TP material from the Nissan 3.5L cam cover will provide better NVH than BMC TS material. The realization here is that a more flexural cover, less damped, will behave more like a speaker than a more stiff, less damped, cover.

RESULTS

PRODUCTION SYSTEM TEST WITH NO BAFFLE

Testing was conducted with no-baffle to determine the effect of the cover material with no influence due to a baffle. The no-baffle test could not be conducted on engine because of possible damage to the engine so the FNGP out of vehicle Integrated System test was used. The results of testing the Nissan 3.5L Al, Mg and production composite cover all without baffles are shown below.

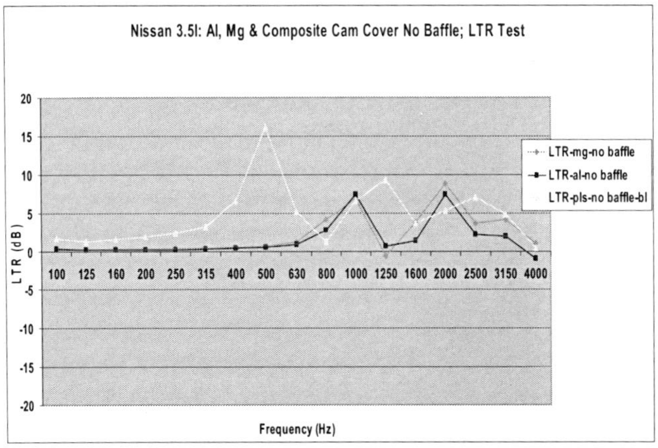

Figure 16: LTR; no Baffle Al, Mg and production covers with production bolt assembly and gasket

The LTR curves above show the ratio between the energy coming in through the cylinder head and associated components and emanating out through the cover and its associated components. The units are dB VS Hz. The production cover, composite plastic, yellow curve is shown to have upwards of 15 dB greater in energy than the Mg (purple curve) and the Al (blue). The curves do not cross the LTR axes until approximately 4,000 Hz. This denotes that the cover and head are in phase and thus an efficient radiator of noise and vibration, they are not isolated.

PRODUCTION SYSTEM TEST WITH BAFFLE

The cover material effect is influenced by the presence of a baffle. Once the baffle is added we no longer have equivalent test articles for comparing cover material differences. As mentioned above each cover was made with no baffle including the production cover, although the production cover did come with a baffle. The baffle effect for the Al and Mg for this study is somewhat misleading because the Minlon (TS) design of the Nissan production baffle was used in the Al and Mg covers. The following figure will make it obvious that the TS baffle's tested in the Al and Mg covers were not designed for those materials. This is further verified in the figure by noticing how the Production Nissan (yellow) curves follow each other for baffle/no-baffle test.

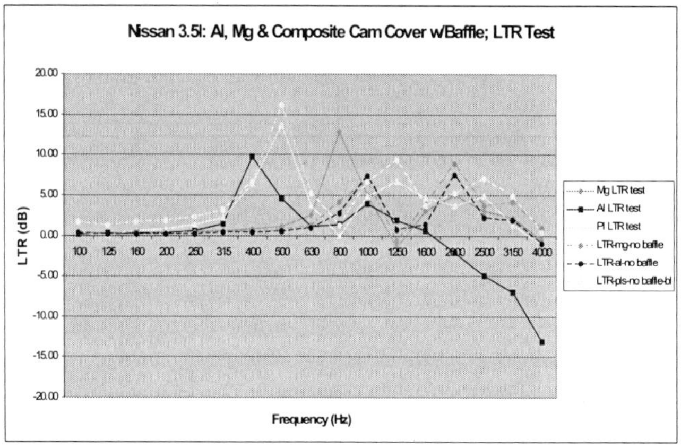

Figure 17: LTR; Production Baffle in Al, Mg and production cover with production bolt assembly and gasket (dashed line= no baffle, solid line = w/baffle)

The yellow curves, solid and dashed, follow one another in frequency while the purple and blue curves do not. This can be attributed to the fact that the yellow curves represent the Nissan production baffle/cover "tuned system" while the curves for Al and Mg do not represent a tuned system. The no baffle result for the production cover is greater in amplitude than the baffle result; this is expected that the baffle should help to reduce the NVH levels. But for the Al and Mg results it is seen that the baffle actually increases the NVH levels by at least 5 dB. The curves for Al (blue) and Mg (purple) were tested with the composite Nissan baffle which results in a "non-tuned system". The resulting dashed lines for Al and Mg show that the Nissan baffle is not compatible with these covers. The resulting cover/baffle system is not tuned and thus the baffle significantly degrades the Al and Mg NVH. Although the Al and Mg NVH is degraded due to the non-tuned baffle in them the NVH levels are still generally lower than the Nissan production cover as can be seen in the figure above.

FNGP ISOLATED BOLT ASSEMBLY SYSTEM TEST WITH BAFFLE

The results above address NVH which result from the production setup which is not an isolated system. This is obvious in the above figures since the curves do not isolate, cross the x axes, until upwards of 4,000 Hz except for the Al cover with no baffle. The picture below shows the isolated bolt assembly used for this test.

Figure 18: FNGP Bolt Assembly Isolation System on Nissan 3.5L

The figure below shows the resulting curve when using this bolt assembly with the production Nissan 3.5L gasket, as mentioned above the production Nissan gasket was used for all tests and each cover had its respective production gasket.

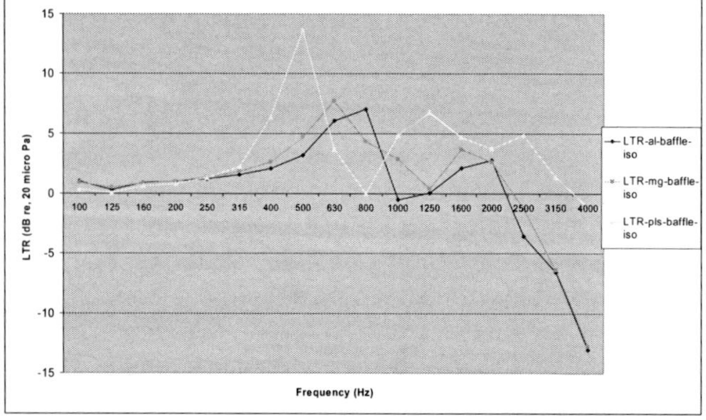

Figure 19: LTR Production Baffle in Al, Mg and production cover with production bolt assembly and gasket

The previous section showed the negative impact of using the production baffle in the Al and Mg covers. As shown in the above figure using the isolated bolt assembly allows the Al and Mg with the production baffle to isolate at a lower frequency and with lower amplitude than the Nissan production composite cover. Isolation is seen to occur at ~ 2,500 Hz for Al and Mg and still upwards of 4,000 Hz for the composite cover. Typically the isolation system incorporates a gasket tuned with the grommet. The effect of not incorporating a tuned gasket can be seen ~ 1,200 Hz where the Al and Mg covers closely isolate and then increase. This effect is due to feedback from the de-tuned gasket. With a tuned gasket typically we can lower the curves at the first isolation point (~ 1,200 Hz) by ~3dB; this would allow the Al and Mg to isolate there. Although the composite cover appears to start isolation at 800 Hz the tuned gasket effect of reducing by 3 dB would not be great enough to isolate the composite cover until upwards of 2,000 Hz.

ON VEHICLE PRODUCTION SYSTEM TEST WITH BAFFLE

The on vehicle test incorporated the baffle for all covers. As discussed above that has a negative impact on the Al and Mg covers. As discussed in the section "Production System Test with Baffle" the baffle was shown to actually increase the Al and Mg levels by ~ 5 dB. If we apply a delta of that magnitude (~ -5 dB) to the following curve it becomes obvious that the Al and Mg covers have much lower amplitude than the production composite cover.

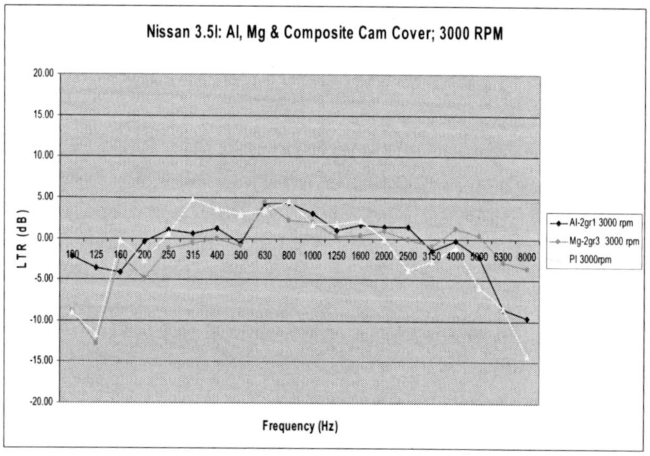

Figure 20: LTR on Vehicle Test

As discussed above applying the ~ -5 dB delta to the Al and Mg curves, due to the negative impact of the de-tuned baffle in the Al and Mg covers, it is seen in the above figure that the Al and Mg will isolate much earlier in frequency than the production Nissan 3.5L composite cover.

CONCLUSION

The 2004 Nissan Maxima 3.5L V6 has one of the quieter sound fields. This paper shows that this is not due to the composite (TP) cam cover alone. The beauty cover, which acts like a sound blanket, is likely the main reason for the resulting sound field.

It is shown here that for an equivalent plastic and metal cam cover "integrated system" the metal cam cover will provide the quieter sound field, metals are inherently better attenuators of noise and vibration than plastics. This can mostly be attributed to the mass differences between metals and plastics; in acoustics the mass effect is the strongest for the frequency range of interest here (5 and 6).

The cross-linking in plastics, which is an attempt to simulate the strong metallic bonds in the metals, has an important effect on damping. This effect was seen between the TS and TP materials tested, analogous to the Mg and Al covers respectively. For this cam cover application it was shown that the cover which allows more damping degrades the acoustic performance. This conclusion was shown to be carried over to plastic versus metals. A plastic cam cover which allows more damping will behave better as a speaker due to more flexural modes than a metal cover which is stiffer and allows less damping. A plastic cover, being less stiff than metal, will tend to remain in phase with the rest of the integrated system, i.e. cylinder head etc., longer than a less stiff metal cover and thus the isolation frequency for the plastic cover will be too high. Once again of utmost importance is to realize the cam cover is part of an integrated system. The damping due to the integrated system and not the cover alone is what is important.

Taking into account the concept of integrated system it was shown that when the bolt assembly is tuned to provide more damping the metals respond in a better sound field than the TP.

Which material leads to the optimum sound field for cam cover applications, metals or plastics? For all else equal, boundary conditions and etc, metal will provide a more optimum sound field than plastic. This is not to say that a plastic cover cannot be designed to be quieter than a metal one. But this argument can go on ad-infinitum because the metal cover can then be re-designed to the equivalent plastic cover and thus will be quieter as shown in this paper. The question needs to involve cost, aesthetic and mass considerations to be answered correctly.

ACKNOWLEDGMENTS

Mohamed Atali, Senior Staff Engineer FNGP, Plymouth Mi.

Kenric Smith, Engineering Technician FNGP, Plymouth Mi.

REFERENCES

1. Nashif, A. D., "Noise Control of Engine Structures",in "Noise Control in Internal Combustion Engines", Edited by Daxa, D.E., John Wiley and Sons, New York (1982)
2. Hong, K. L., Tathavadekar, P. and Raveendra, S. T., "On the Evaluation of Radiated Sound Field of Engine Covers", SAE Paper: 2001-0101412 (2001)
3. Challen, B.J., "Technique for Engine Noise Control", in "Engine Noise – Excitation, Vibration and Radiation", Edited by Hickling, R. and Kamal, M.M., Plenum Press, New York (1982).
4. Lu, Y.C. and Anderson, M.E., "On the Use of Spatial Transmissibility to Evaluate the NVH Performance of Engine Covers", SAE Paper, No. 2002-01-0458.
5. Beranek, L and Ver I., "Noise and Vibration Control Engineering Principles and Applications", John Wiley and Sons, (1992).
6. Morse and Ingard, "Theoretical Acoustics" Princeton University Press, (1986)

CONTACT

ERIC HERRERA; 541-921-8760, EZH@FNGP.COM

This is where main author information is typed, if desired, such as background, education, e-mail address, and web address. This is an optional section.

2005-01-0729

AJ (Mg-Al-Sr) Alloy Mechanical Properties: From Fatigue to Crack Propagation

Pierre Labelle
Noranda Inc.

Andreas Fischersworring-Bunk
BMW Group

Éric Baril
NRC-IMI (Formerly from Noranda Inc.)

ABSTRACT

In addition to the creep properties, the fatigue properties are essential for the design of a power-train component in Mg which is operated at elevated temperatures. In case of the new BMW I6 composite Mg/Al crankcase using the AJ alloy system, material testing focused on both subjects. The basic mechanical properties were determined from separately die cast samples and also from samples machined out from high-pressure die cast components. Tensile, high cycle fatigue properties, low cycle fatigue and crack propagation properties were established and analyzed within the technical context for power-train applications reflected in the temperature and load levels. The aspects of mean stress influence, notch sensitivity and crack propagation are evaluated to estimate the performances of the AJ62A alloy system.

INTRODUCTION

Generally speaking the technology used for fuel reduction, like the full variable valve train or direct injection systems, adds weight to the power train, which conflicts with the vehicle weight/weight distribution targets. The demanding technical specification in terms of fuel economy, weight and specific power output for the new BMW I6 engine (Figure 1) could only be achieved by a complete new design. With a weight of just 161kg (-7%), this engine offers outstanding performance with a specific power output of 63kW/dm^3 (+12%) and a 12% reduction in fuel consumption compared to the predecessor engine. To achieve the project goals, the fundamentally new design is based on the latest technologies, such as a composite magnesium-aluminum crankcase, electric coolant pump and a further development of BMW's VALVETRONIC variable valve timing (Ref.1)

The crankcase is the single heaviest component in the engine and is hence attractive for an efficient weight reduction. The composite design concept has been presented in Ref.2. The magnesium alloy optimization approach used to balance the creep properties and the high pressure die castability requirements of such an intricate design are described in Ref. 3. Limited data concerning the performance of the AJ (Mg-Al-Sr) die cast alloys, were available when Noranda introduced this patented alloy system.

Figure 1 View of a complete inline six engine

Consequently, a joint extensive effort was done to provide valuable data to understand the alloy in terms of performance, probability of survival in low and high cycle fatigue, with and without degradation due to corrosion and, when crack initiated, determine the crack grow rate under monotonic fatigue conditions.

MATERIAL

The material tested in this study came from same batches of samples that were used in the previous investigation (Ref. 3). The alloy corresponds to AJ62A specification and the details concerning chemical description and metallurgical aspects like grain size, micrographs, microstructure characterizations are available in references 3 and 4. Samples were machined from the HPDC engine block or from separately die cast samples using multiple sample types. The chemistry of all batches of material were monitored and found to be in the proposed ASTM B-94 limit of 5.5 to 6.6% aluminum and 2 to 2.8% of Sr, for the AJ62A die cast component. The AJ62A alloy was processed using typical diecast guidelines practices and parameters described in reference 5.

EXPERIMENTAL PROCEDURES

FATIGUE

Two series of samples were used to determined the fatigue limit of AJ62A alloy,

1. Axial fatigue using various stress ratios to establish fatigue curves at room temperature. These samples were machined from the chain wall cover of the engine block.

2. Rotating beam test (RTB) performed using separately die cast specimens, tested as received, as well as after 200 hrs in salt spray corrosion.

High cycle axial fatigue tests were carried out on more than 120 samples at room temperature, 120°C, and 150°C. Stress ratios (R) were 0.1, -0.5 and –1.0. Samples were machined and tested as per ASTM E466-96 (Ref. 6). Fatigue tests were conducted in load control on servo-hydraulic test equipment employing 45 Hz sinusoidal waveform. Run-out was defined as 10 million cycles. Due to the limited number of samples available run-out samples were re-installed and uploaded at higher stress to provide additional data points to build S-N curves. Fatigue limits were determined using the S-N curves for 10, 50, and 90% probability of failure.

Low cycle axial fatigue test were carried out on 10 samples at 150°C, under a completely reversed cyclic strain condition (R= -1.0). Samples were machined (Fig. 2) and tested according to ASTM E606-92 (Ref. 7). Fatigue tests were conducted in strain control on servo-hydraulic test equipment employing 0,5 Hz sinusoidal waveform. The total strain imposed ranging from 0.2 to 0.6%.

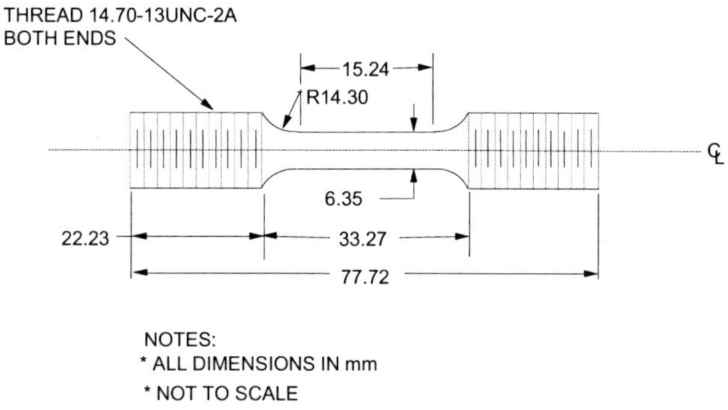

Figure 2 LCF axial fatigue samples

RTB fatigue was conducted using ‘Fatigue Dynamic’ standard sample geometry (Fig. 3) and fatigue test equipment. The tests were performed using the R.R. Moore procedure as described in Ref. 8. Only flashes were removed using a file, the die cast skin was not removed. Additionally from the same batches of material, a series of samples were pre-corroded for 200 hrs under ASTM B117 (Ref. 9) with 5% NaCl. RTB tests were performed at 3000 RPM .

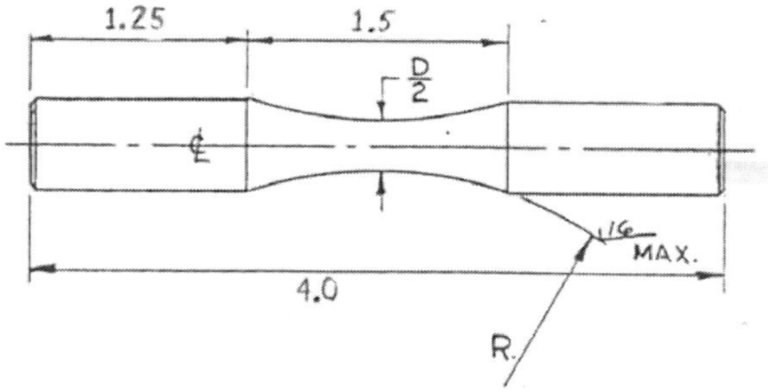

Figure 3 RTB Fatigue samples (Note all dimensions in Inches)

CREEP-FATIGUE

To assess a possible creep-fatigue interaction effect under typical operating conditions, axial fatigue tests were defined using a special test sample shape (Fig. 4) to avoid stress riser at the grip area. Fatigue tests were conducted in load control on servo-hydraulic test equipment employing 25 to 30Hz sinusoidal waveform.

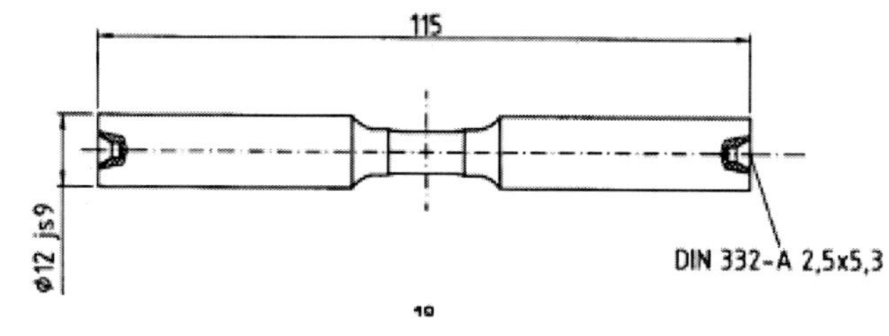

Figure 4 Creep-fatigue interaction sample

The test is defined by a cyclic hold time in seconds at both the maximum and minimum load level. The test cycle used has a trapezoidal shape with different loading/unloading rates (see figure 5). The detail test cycle is : +X, 30s(tensile hold), -X, -X, 30s(compressive hold), +X; where X, is the duration of the load variation and the +/- refer to the load direction: increase (+) or decrease (-) (Fig. 5),

- X=30 seconds for the standard test and
- X=3 seconds for the accelerated test

The equivalent load rate is therefore 1.67 MPa/s viz. 16.7 MPa/s for the 50 MPa load level and 2.33 MPa/s viz. 23.3 MPa/s for the 70 MPa level. In addition the influence of load ratio (R=0, -1) and load levels (50, 70 MPa) were investigated. Run-out was defined at 5000 cycles, resulting in a 41.5 hrs. hold time at both maximum tensile and minimum compressive stress conditions.

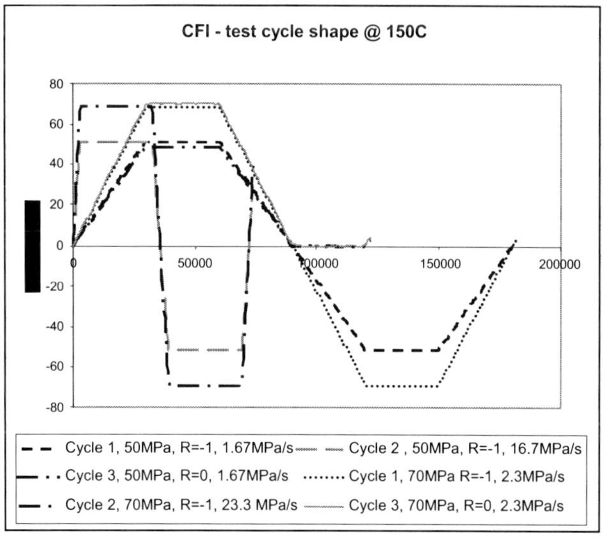

Figure 5 Load shape creep-fatigue interaction

NOTCH SENSITIVITY

Using the axial fatigue set-up and temperature previously described, with a stress ratio (R) of 0.1, samples with a notch machined with a stress concentration $K_t = 1.41$ in axial fatigue were tested. The notch was machined using a shape tool, which prevents surface defects in the machined area of the sample (Fig. 6).

According to Peterson Ref. 10, the notch sensitivity is defined by

$$K_f = 1 + (K_t - 1)\,q \qquad \text{Equation 1}$$

Where

K_t: Static stress concentration factor,

K_f: Fatigue stress concentration factor,

q: Notch Sensitivity, varies $0 < q < 1$, where 0 refers to a non-notch sensitive and 1 to a completely notch sensitive material.

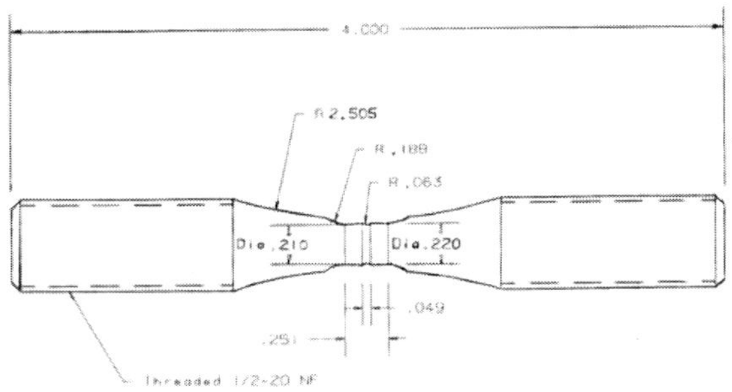

Figure 6 Notched Axial Fatigue samples (Note all dimension in inches)

FATIGUE CRACK GROWTH (FCG) RATE AND FRACTURE TOUGHNESS

Compact tension samples (Fig. 7) were machined from the component and crack growth tested per ASTM E647-00 (Ref. 11). Additionally, samples of typical AlSi17Cu4 hypereutectic aluminum alloy were machined. The AJ62A samples were tested at room temperature and 150°C under atmospheric conditions, the AlSi17Cu4 at 150°C only with a stress ratio (R) of 0.1. The objective of this test was to characterize the FCG-curve for region I (threshold), region II (Paris) and region III (crack instability) for simple FCG analysis (Fig. 8).

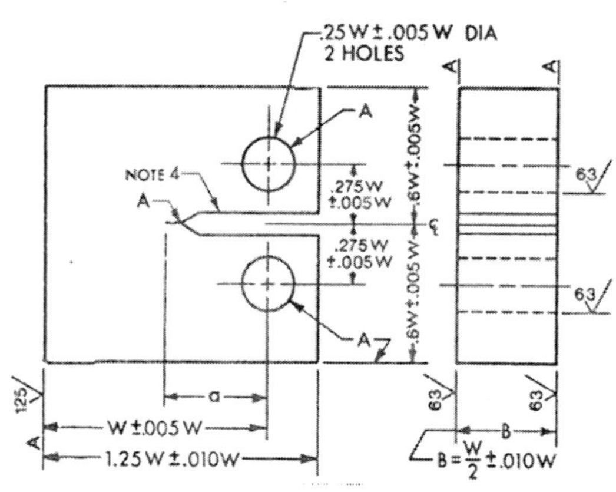

Figure 7 Crack Growth CT test sample

Samples were machined to the following dimensions: W = 18 mm, B= 8,95 mm, with surface finished as indicated on figure 7 (μ inch) and fatigue pre-cracked at 400 N to a final a/w of approximately 0.50, and then side grooved to a depth equal to 20% of nominal thickness (10% per side). Increasing and decreasing K technique was used, on a servo-hydraulic testing machine, to obtain:

Crack propagation threshold stress intensity factor range (ΔK_{th}). Below this threshold, cracks either remain dormant or grow at an undetectable rate. The threshold value is set by the testing limit of detection, which is in 10^{-10} m/cycle.

The da/dn vs ΔK curves, C and m of the Paris equation is

$$da/dn = C\,(\Delta K)^m \qquad \text{Equation 2}$$

for the region II.

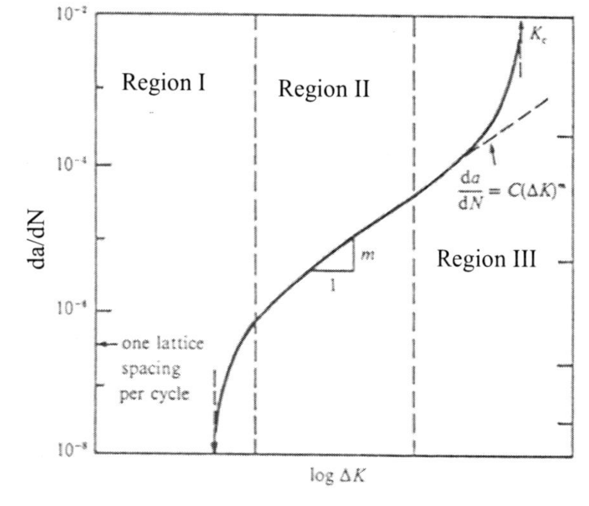

Figure 8 Schematic of FCG

- For fracture toughness was determined using J integral technique J_{Ic},

Due to the relatively thin sections of the crankcase geometry, and to the low Young's modulus of magnesium, it was not possible to machine a sample compliant with K_{Ic} requirements for plain strain condition. Consequently the goal was to obtain a valid value for K_{JIc} calculated from J_{Ic} which is calculated from the plain stress condition.

$$K_{JIc} = (J_{Ic} * E/(1-\nu^2))^{0.5} \qquad \text{Equation 3}$$

RESULTS AND DISCUSSION

HIGH CYCLE FATIGUE

The axial high cycle fatigue characteristics of AJ62A were presented previously (Ref. 3). Figure 9 compares the fatigue behavior of AJ62A with AZ91D, AM60B, AE42, AS21 and AlSi9Cu3 (Ref.12). All alloys show a pronounced fatigue limit and less than 1% of the failures occur beyond 2 x 10^7 cycles. The mean fatigue limits are 80 ± 10 MPa (AJ62A), 45 ± 10 MPa (AZ91D), 50 ± 11 MPa (AM60B), 42 ± 13 MPa AE42), 38 ± 10 MPa (AS21), 75 ± 18 MPa (AlSi9Cu3).

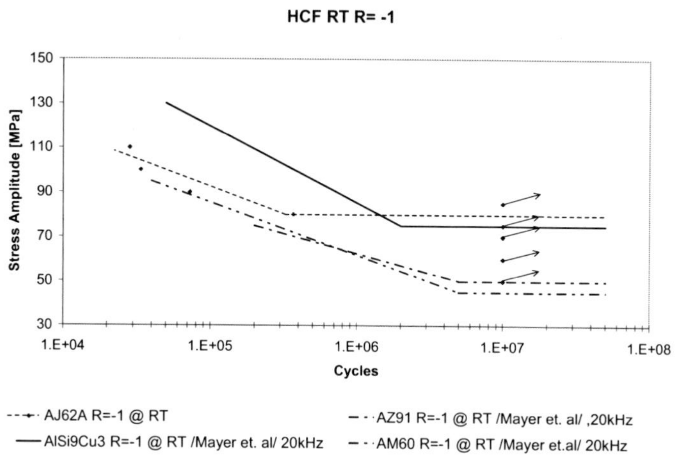

Figure 9 HCF at RT for AJ62A, AZ91, AM60 and AlSi9Cu3

However, in order to include statistical information and additional results, additional tests were performed at R > 0.5. Stress limits at 10% and 90% survival for 10^9 cycles were determined for various R values (Fig. 10) using the maximum-likelihood method applied to the full S-N line test results. The fatigue properties were evaluated using more than 300 samples. The additional testing includes component tests, which indicate that the Goodman model is appropriate for high R values, and that the Soderberg model is a conservative approach.

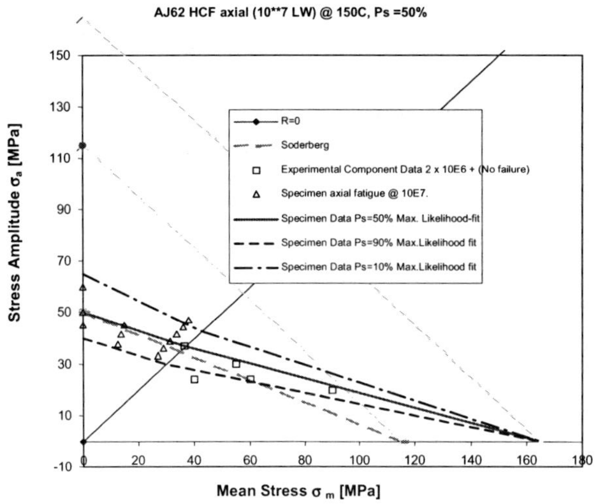

Figure 10 Axial Fatigue diagram of AJ62A with probability of survival at 150°C

Using fatigue properties tests methodology from MIL-HDBK-5F (Ref. 13), a comparison was also conducted between the rotating beam fatigue and axial fatigue at R=-1. At 50% survival probability the value of stress amplitude is 92 MPa for RTB (Fig. 11) and 80 MPa for axial fatigue. The ratio between the RTB and axial fatigue limits of 1.15 is less than the one previously published (Ref. 14,15), which vary from 1.5 to 1.9. It is suggested that the absence of brittle intermetallic such as $Mg_{17}Al_{12}$ and $Mg_{13}Al_3Sr$ improved the consistency of the fatigue properties to various loading conditions. Such behavior is reflected by the low value of notch sensitivity obtained as presented in the next sections.

The degradation of the as cast surface due to salt-spray exposition creates a reduction of fatigue life 34% at 61 MPa (Fig. 11) which is on the low end of the range observed, 35% to 50%, for samples subject to a sodium chloride solution attack prior to fatigue tests (Ref. 16). The observed HCF life reduction reflects the aspect of surface degradation with tentative pitting acting as crack initiation (Fig. 12).

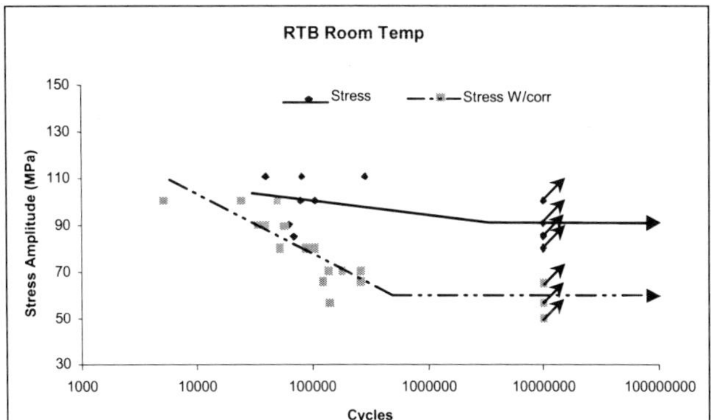

Figure 11 RTB Fatigue curve of AJ62A with 50% probability of survival at RT with and without salt spray exposure

However this degradation leads to a fatigue limit superior or equivalent to that of non-corroded magnesium alloys (Ref. 12); pre-corroded as cast AJ62A samples show better fatigue limit than other as cast die cast alloys like AZ91 or AM60 tested in normal conditions. Separate investigations have shown that the aspect of stress corrosion for the investigated stress levels and temperature is not an issue.

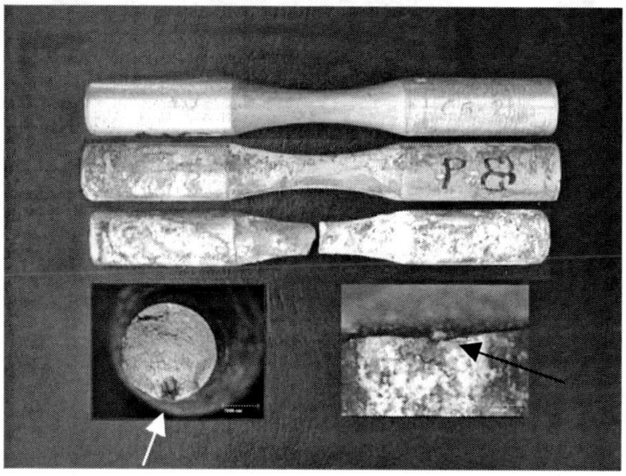

Figure 12 As cast & corroded samples prior & after test, surface fracture showing pitting action

LOW CYCLE FATIGUE

LCF tests at 150°C were made on two machined samples per condition. The results are summarized in table A1 (See Appendix). The variations between maximum and minimum stress in terms of number of cycles for different total strain ranges are summarized in figure 13. A cyclic softening behavior was observed during the initial short period for all strain levels; thereafter, the maximum stress dropped by about 10% to 15% and the minimum stress raised by the same increment, then they remain stable over the rest of fatigue life. The mean stress is always relatively small over the entire strain range.

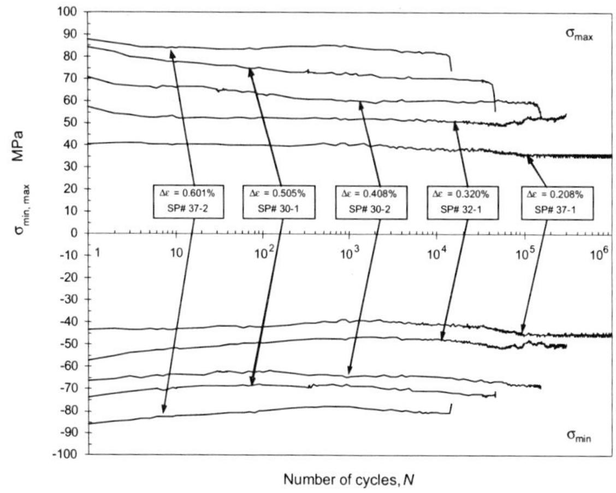

Figure 13 Maximum and minimum stress during the evolution of LCF tests at 150°C.

The relationship between the plastic and elastic components of the strain range, can be expressed using the following equations:

$$\Delta\varepsilon_{total}/2 = \Delta\varepsilon_{plastic}/2 + \Delta\varepsilon_{elastic}/2 \qquad \text{Equation 4}$$

The decomposed strain is illustrated in figure 14.

The decomposed strain can be detailed using Basquin's relationship:

$$\Delta\varepsilon_{elastic} = 0.0195\ N^{-0.1625} = 2\sigma'_f/E\ (2N_f)^b\ , \qquad \text{Equation 5}$$

where σ'_f = 436 MPa, E = 40 GPa, b = -0.1625

and using Coffin-Manson relationship:

$$\Delta\varepsilon_{plastic} = 0.1177\ N^{-0.4098} = 2\varepsilon'_f(2N_f)^c\ , \qquad \text{Equation 6}$$

where ε'_f= 0.059, c = -0.4098

ε'_f is the fatigue ductility coefficient (which is experimentally found to be approximately equal to the true fracture ductility ε_f in monotonic tension) and c is the fatigue ductility exponent (which ranges between –0.5 to –0.7 for most metals, Ref. 17).

σ'_f is the fatigue strength coefficient (which, to a good approximation, equals to the true fracture strength σ_f, corrected for necking, in a monotonic tension test for most metals) and b is known as the fatigue strength exponent (which ranges between –0.05 to –0.14).

For AZ91D, b and c parameters were -0.156 and –0.444 respectively at 130°C, (Ref. 18) which is in agreement with the value found for AJ62A at 150°C.

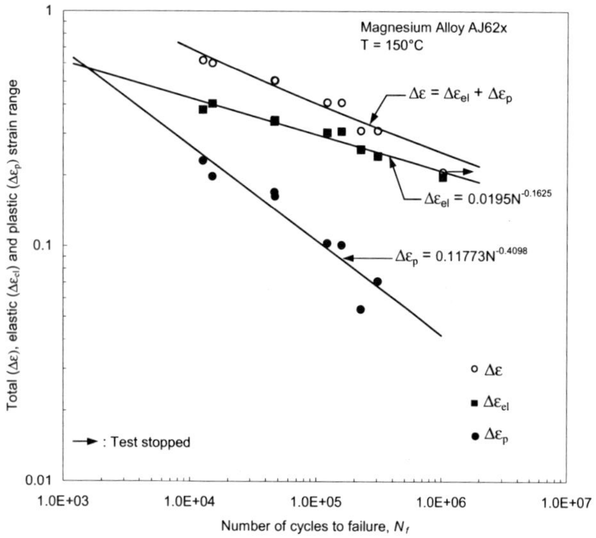

Figure 14 Total, Elastic and Plastic strain ranges as a function of the number of cycles at 150°C.

Finally, the stress amplitude can be adequately represented in terms of plastic strain amplitude in log-log coordinates by the equation 7 and illustrated in figure 15.

$$\sigma_a/2 = 160.62\ (\Delta\varepsilon_{plastic})/2)^{1/0.3185} \qquad \text{Equation 7}$$

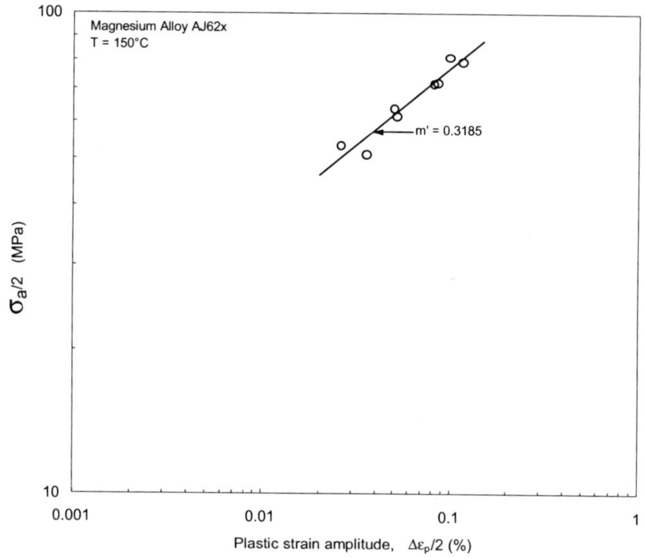

Figure 15 Total, Elastic and Plastic strain ranges vs. number of cycles.

CREEP FATIGUE INTERACTION (CFI)

As observed in figure 13 for the LCF test, for the first cycles at R=0 in the CFI test, a reduction in strain amplitude is also noticed. In general for all tests a minor cyclic increase in the strain amplitude, in this stress controlled test, is observed, which indicates cyclic softening. This is equivalent to the cyclic behavior observed in the strain controlled LCF-test.

A comparison of the stress-strain loops (cycles 2 and 1000) shows the influence of the load rate at R=-1 for the load levels of 50 MPa (Fig.16) and 70 MPa (Fig.17).

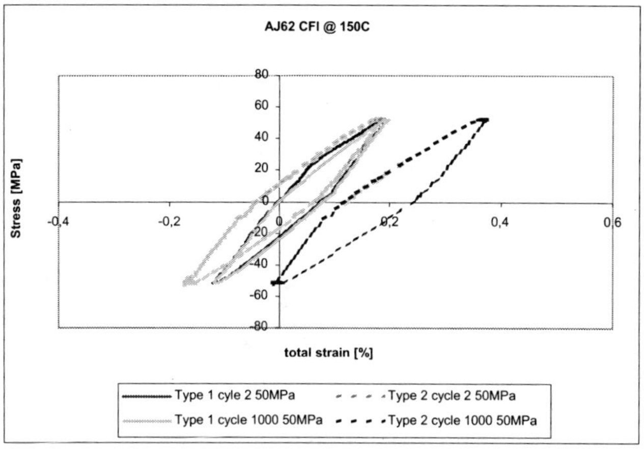

Figure 16 Stress-Strain loop for Type 1 (Standard, R=-1) and Type 2 (Accelerated R=-1) 50 MPa

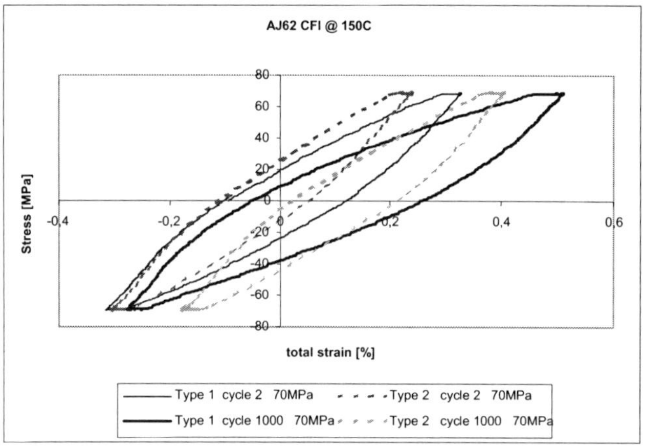

Figure 17 Stress-Strain loop for Type 1 (Standard, R=-1) and Type 2 (Accelerated R=-1) 70 Mpa

At a load level of 50 MPa type 1 (standard load rate) shows an almost constant stress-strain loop for the compared cycles with a reversed creep strain during loading and hold time in the tensile and compressive portions of the test. The plastic deformation for the accelerated test type 2 is less then expected due to both hardening effects and reduced exposure time to creep. Due to porosity a shift in the cyclic mean strain (ratcheting effect) with non-reversed creep strain and premature failure is noted. Also, at a load level of 70 MPa, a similar trend for the shift in mean strain is observed, suggesting that the creep strain from tension is only partly reversed under compression.

The effect of the load ratio on the stress-strain loop (cycles 2 and 1000) is shown in figure 18 by comparison of test type 1 and test type 3.

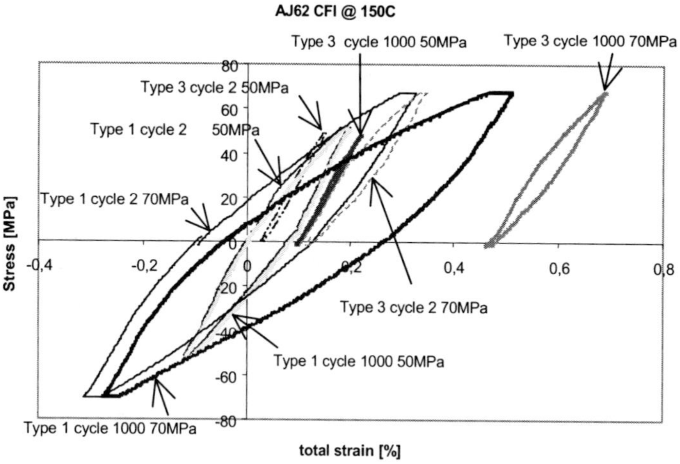

Figure 18 Stress-strain loop for Type 1 and Type 3 (Standard R=0) CFI test

The R=0 loading shows an increase in the mean strain due to the creep plasticity in the tensile hold time. However none of the specimens showed a premature failure under this condition. Generally speaking it can be concluded from these CFI tests, that creep has a significant contribution to the deformation for the load rates considered.

Figure 19 shows the results for the R=-1 and 150°C condition for the HCF, LCF and CFI tests. For the LCF and CFI test both stress and strain ranges at half-life are given for comparison. From the observed at the 140 MPs stress range (70 MPa R=-1), there seems to be only a limited effect of creep on the fatigue life, however this statement is only of limited validity due to the run-out limit of 5000 cycles, whereas for a stress range less than 100 MPa (50 MPa R=-1) tests need to be extended to the 10^4 to 10^5 cycle range to get a final confirmation of the creep fatigue interaction effect.

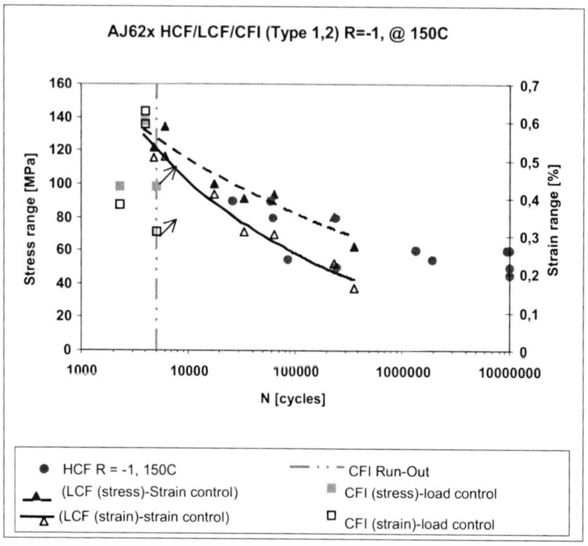

Figure 19 HCF, LCF and CFI results for 150°C

NOTCH SENSITIVITY

The sensitivity indices vary extensively for various magnesium alloys (Table 1); the measured notch sensitivity ranges from 0.15 to 0.55 (Ref. 16) depending on the casting process. Consequently, no specific value can be calculated from other properties. Using 28 samples from the same die cast batch of material a notch sensitivity q = 0.15 is computed. This value measured at a maximum stress of 90 MPa and 10 Millions cycles is the same than that previously observed at room temperature for die cast AZ91 and A8 (AZ80), (Ref. 16). Alloy AJ62A exhibits a limited effect of stress risers on fatigue behavior, consequently the same rules of design can be applied.

Table 1 Value of q notch sensitivity

Alloy	Casting Process	q
AJ62A	Die cast	0.15
AZ91	Die cast	0.15
A8(1) (AZ80)	Die Cast	0.15
A10v(1) (AZ91)	Sand Cast	0.15 – 0.4
AZ855(1) (AZ80)	Sand Cast	0.25 – 0.55

(1) Former British Standard Designation Ref. 16

CRACK GROWTH RATE AND FRACTURE TOUGHNESS

Fatigue crack growth was obtained for die cast AJ62A at room temperature and 150°C (Fig 20). However, at room temperature, for K decreasing tests, an important scatter in the data in transition from region I to II of crack growth in delta K varying from 1.5 to 3 is observed. This phenomenon can be attributed to various causes (Ref. 19) : (1) crack closure (induced by plasticity, environment effect or fracture surface roughness), (2) crack deflection from its normal growth plane which reduces the effective stress intensity factor at the crack tip, (3) resistance afforded by the grain boundary and the less favorably oriented neighboring grain which causes the crack to stop and become non-propagating, (4) recovery mechanism which reduces the effect of the plasticity left behind the advancing crack front and reduces the pressure on the crack, (5) twinning which favorably orients the neighboring grain and reduces the resistance to crack propagation.

A reduction of the growth rate in the mid-range of Stage II is observed (double S shape). These phenomena could be related to variations in grain size or heterogeneity in the macrostructure along the crack propagation path. It could also be related to the small size of the sample used for the test relative to the plastic zone and grain size.

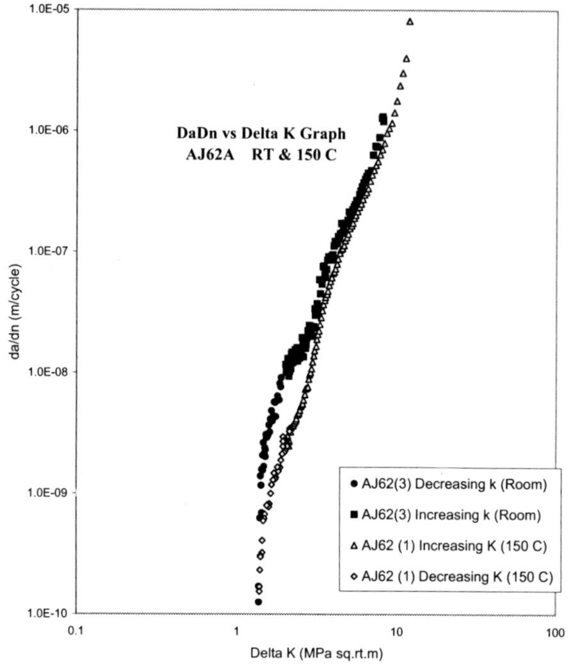

Figure 20 Da/dn VS Delta K at room temperature & 150°C for AJ62A

At 150°C, thermally induced mechanisms (recovery and twinning) help with the realignment of slip planes and reduce the plasticity-induced crack closure, which, even if the double S shape curve is maintained, reduces the scattering of data (Fig. 20). This reduced scatter in data has also been observed for the other fatigue test at 150°C compared to room temperature tests.

The crack growth rate was obtained under plain strain conditions according to ASTM E-647. To be able to compare AJ62A with AlSi17Cu4 hypereutectic aluminum alloy, ΔK was normalized by the Young's modulus (ΔK/E). The threshold value ΔK_{th} /E is lower for the AlSi17Cu4 alloy (Fig. 21). The slope of the AlSi17Cu4 is significantly steeper than for AJ62A, consequently the crack growth will be slower for AJ62A.

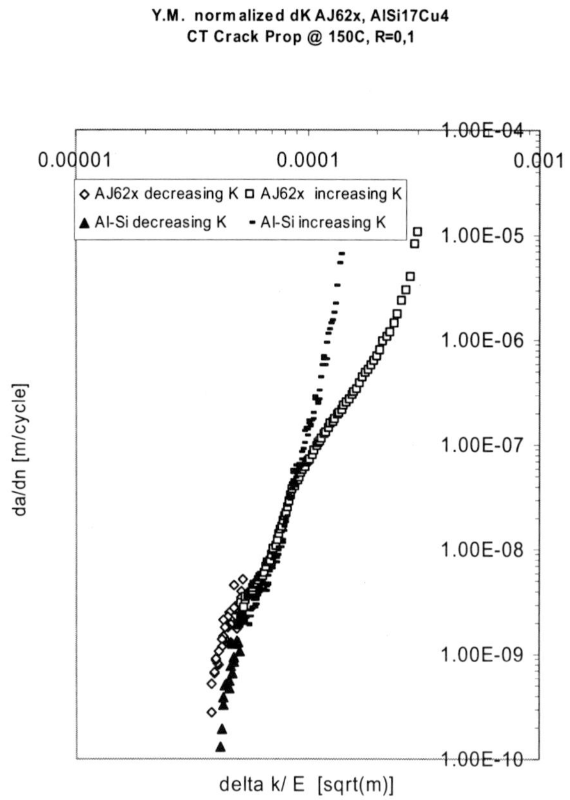

Figure 21 Da/dn for AJ62A and Al-Si Alloy at 150°C

However both materials have a different Young's modulus and different behavior for the same thickness. Due to the thin wall design, the magnesium component will tend to work in plain stress and the thick aluminum alloy components will tend to be subject to the plain strain condition. For comparison some additional data from the 'Nasgro' database (Ref. 20) has been included showing that the measured C and m are in alignment with the existing magnesium experimental data (Table 2). Using these value we are able to predict crack propagation. However, the R ratio dependency has not been determined, since all tests have been performed at R= 0.1.

Table 2 Value of C & m range of Da/Dn 10^{-10} to 10^{-9} m/Cycle

Alloy & Temp	**C**	**m**	**K_{th}** (Da/Dn at 1E-11 m/Cyc)
Magnesium alloys			
AJ62A RT	1.18 E-11	3.471	1.53
AJ62A 150°C	7.07 E-11	4.916	1.49
AZ91D RT [(1)]	2.97 E-11	3.6	1.3-1.55[(2)]
AM503[(3)] (Mg + 2% Mn) RT	9.62E-11	3.185	N/A
QE22A-T6 RT	8.45E-11	3.21	N/A
Aluminum Alloys			
A356-T60 RT	7.64E-17	4.98	N/A
AlSi17Cu4 150°C	1.97E-14	8.033	3.20
AlSi9Cu3[(2)] RT			2.45-2.70

Note: (1) Ref. 17, (2) Ref. 12, (3) British Standard Designation

FRACTURE TOUGHNESS

The fracture toughness values of AJ62A and AlSi17Cu4 alloys have been determined and compared with reference alloys (Table 3). It is observed that the fracture toughness increases with temperature, which is aligned with previous observations on crack growth and impact. It is well known that high silicon content aluminum alloys are brittle, and AJ62A shows superior fracture toughness resistance than Al380 and AlSi17Cu4. However, the values observed are consistent with other magnesium alloys and A356.

Table 3 Fracture toughness values

Alloy	**Temp** (°C)	**K_{Ic}** (MPa sqrt (m))
Magnesium Alloys		
AJ62A	RT	15,24*
AJ62A	150°C	24,53*
AM 503	RT	18,69
QE22A-T6	RT	19,76
Aluminum Alloys		
A356-T60	RT	17,58
AlSi17Cu4	150°C	7,29 **
A380	RT	8,71

Note: * Calculated from J_{Ic}
** K_q, some validity criteria not met

CONCLUSION

The focus within the engine crankcase development program was to have an Mg-alloy with a good balance between fatigue properties, creep properties, and castability; each of the properties having a minimum requirement to comply with the design and manufacturing concerns.

The fatigue properties are of paramount importance in the design of power train components. To establish the required fatigue material design database, a test program was defined to cover HCF, LCF, CFI and fracture mechanics properties. The load and temperature for the material testing conditions were based on expected engine operating conditions.

The properties measured in this study can be summarized as follows:

- An interesting result from the present investigation is the existence of a fatigue limit in the AJ62A, which was already observed for other Mg-alloys (Ref. 12). This property is not inherent to the investigated alloys but can be attributed to the casting process leading to a certain level of porosity. This is also supported by the reduction of the fatigue limit observed on corroded samples which can be related to the introduction of surface defects. This effect is reflected on the notch sensitivity measured on the AJ62A alloy.

- The low cycle fatigue total life is in good agreement with the laws of Manson-Coffin and Basquin. The parameters are in the range of reported parameters for magnesium alloys, which make the AJ62A alloy predictive in terms of fatigue behavior.

- It was found that creep and fatigue are not interacting for the load levels under investigation and therefore fatigue life is not reduced by creep and creep is not amplified by cyclic loading; consequently standard prediction approaches can be used to predict both creep and fatigue behavior.

- The crack propagation threshold is not influenced by temperature, which indicates that the environmental effects are equivalent at both room temperature and 150°C.

- The fracture toughness of AJ62A is superior to that of AlSi17Cu4 and A380 current casting and die casting alloys. In addition it is in the same range as A356 and magnesium alloy QE22A.

The results were used to evaluate the design using extensive finite element analysis of the component under in-service conditions in the early development stage to assess the viability of the chosen design concept. The design iterations have resulted in an optimized design satisfying the durability and NVH requirements proven later on by component and full engine testing within the engine development program.

With the introduction of the Mg-Al composite crankcase design in the new engine it is demonstrated that AJ62A has the capability to sustain such severe application.

ACKNOWLEDGMENTS

The authors would like to acknowledge BMW Group and Noranda Magnesium Inc., for sponsoring this project and for permission to publish this paper. The authors thank all the engineers and technologists from Noranda, BMW Group and external laboratories who were involved in die casting campaigns and performing the mechanical testing required by this program.

REFERENCES

1. C. Landerl, N. Klauer, M. Klüting, 'The Design Characteristics of the New BMW Six Cylinder Inline Petrol Engine', 13th Aachen Colloquium, October 4-6th , 2004
2. C. Landerl, R. Jooss, A. Fischersworring-Bunk, J. Wolf, A. Fent, S. Jagodzinski, 'Aluminum-Magnesium Composite Design – an Innovative Approach to Lightweight Crankcase Technology', 12th Aachen Colloquium, October 6-8th 2003
3. E. Baril, P. Labelle, A. Fischersworring-Bunk, 'AJ (Mg-Al-Sr) Alloy System Used for New Engine Block', SAE 2004-01-0659 (Detroit, MI:SAE2004 WorldCongress)
4. E. Baril, P. Labelle, M. O. Pekguleryuz "Elevated Temperature Mg-Al-Sr : Creep Resistance, Mechanical Properties, and Microstructure", *JOM*, Nov. (2003), pp. 34-39.
5. M. Pekguleryuz, P. Labelle, D. Argo "Magnesium Die Cast Alloy AJ62x with Superior Creep Resistance, Ductility and Die Castability" SAE 2003-01-0190, (Detroit, MI:SAE2003 WorldCongress)
6. ASTM E466-96 Std Practice for Conducting Force Controlled Constant Amplitude Axial Fatigue Test of Metallic Materials
7. ASTM E606-92 Std Practice for conducting Strain-Controlled Fatigue Testing
8. P. Labelle, D. Argo, M. Pekguleryuz, Y. Fasoyinu, R. Bouchard, M. Sahoo, 'Comparative Mechanical Properties of AE42 and AJ52x High-Temperature Diecast Magnesium Alloys for Elevated Temperature Applications' SAE 2003-01-0188 (Detroit, MI:SAE2003 WorldCongress)
9. ASTM B117-97 Std Practice for Operating Salt Spray (Fog) Apparatus
10. R.E. Peterson, 'Stress Concentration Design Factors' John Wiley & Sons Inc.
11. ASTM E647-00 Std Test Method For Measurement of Fatigue Crack growth Rates

12. H. Mayer et al., 'Influence of porosity on the fatigue limit of die cast magnesium and aluminium alloys', Int.J. of Fatigue 25 (2003), pp.245-256
13. Military Handbook Department of US Defense, 'Metallic Materials and Elements for Aerospace Vehicle Structure' MIL-HDBK-5F, 1992
14. P.D.D. Rodrigo, M. Murray, H. Mao, V. Chandrasekar, Y. Kisioglu, A. Deshpande, J. Brevick, C. Mobley, R. Esdaile 'Fatigue Properties of Die Cast Alloys' SAE 2000-01-1122 (Detroit, MI:SAE2000 WorldCongress)
15. E.F. Emley, 'Principles of Magnesium Technology' Pergamon Press 1966.
16. E.H.A. Beck 'The technology of Magnesium and its Alloys', F.A. Hughes & Co, 1943
17. S. Suresh, "Fatigue of materials", Cambridge University Press, Cambridge, 1991.
18. G. Eisenmeier et al. 'Cyclic deformation and fatigue behaviour of magnesium alloy AZ91D', Mat. Sc. & Engineering A, 319-321 (2001) pp. 578-582
19. P. Venkateswaran et al. "Fatigue crack growth behaviour of a die-cast magnesium alloy AZ91D", Materials Letters, 58 (2004), pp.2525-2529.
20. Nasgro Version 3.0.20, 'Fatigue crack growth computer program' NASA, Appendix G2, Materials constants SI (European Space Agency) Units, Lyndon B. Johnson Space Center, May 2002

DEFINITIONS, ACRONYMS, ABBREVIATIONS

CFI	Creep fatigue Interaction
FGC	Fatigue Crack Growth
HCF	High Cycle fatigue
HPDC	High Pressure Die Casting
LCF	Low Cycle fatigue
RTB	Rotating Beam
NVH	Noise Vibration Harness
σ_m	Mean Stress
σ_a	Stress Amplitude

Term: Definition for the term

AlSi17Cu4 is the European designation for an alloy similar to Aluminum A390

AlSi9Cu3 is the European designation for an alloy similar to Aluminum A380

A356-T60 is an alloy-temper designation similar to A356-T6. However; temper designations are not normalized by the Aluminum Association; thus differences with actual Aluminum Association's registered A356-T6 limits may exist.

APPENDIX

TABLE A.1

Fatigue test results on magnesium alloy AJ62A specimens at 150°C

$R_\varepsilon = \varepsilon_{min}/\varepsilon_{max} = -1.0$

SPECIMEN #	d (mm)	$\Delta\varepsilon$ (%)	E (GPa)	FIRST CYCLE						HALF-LIFE						N_f (cycles)	NOTES
				$\Delta\varepsilon_{el}$ (%)	$\Delta\varepsilon_p$ (%)	σ_{max} (MPa)	σ_{min} (MPa)	$\Delta\sigma$ (MPa)	σ_{mean} (MPa)	$\Delta\varepsilon_{el}$ (%)	$\Delta\varepsilon_p$ (%)	σ_{max} (MPa)	σ_{min} (MPa)	$\Delta\sigma$ (MPa)	σ_{mean} (MPa)		
2-3	6.34	0.618	41.9	0.410	0.208	87.7	-83.9	171.6	1.9	0.380	0.232	78.0	-81.2	159.3	-1.6	12 663	b
7-2	6.31	0.601	40.3	0.432	0.169	88.0	-85.8	173.8	1.1	0.404	0.198	82.3	-80.3	162.6	1.0	15 075	a
0-4	6.34	0.506	42.4	0.361	0.145	77.7	-75.4	153.1	1.2	0.339	0.171	68.9	-75.1	144.0	-3.1	46 708	a
0-1	6.33	0.505	41.8	0.379	0.126	84.6	-74.0	158.5	5.3	0.343	0.163	71.0	-72.3	143.3	-0.6	47 201	b
0-3	6.33	0.409	40.4	0.326	0.083	67.5	-64.0	131.5	1.7	0.304	0.103	60.5	-62.3	122.8	-0.9	121 862	a
0-2	6.34	0.408	41.4	0.332	0.076	71.0	-66.5	137.6	2.3	0.308	0.101	59.4	-68.1	127.5	-4.4	157 942	b
2-2	6.33	0.311	41.5	0.259	0.052	54.3	-53.1	107.4	0.6	0.257	0.052	48.7	-57.9	106.6	-4.1	224 946	a
2-1	6.34	0.320	42.1	0.273	0.047	57.4	-57.4	114.8	0.0	0.242	0.071	53.3	-48.7	102.0	2.3	306 169	a
2-4	6.35	0.221	41.7	0.219	0.002	44.4	-46.9	91.3	-1.3	0.204	0.008	40.7	-44.4	85.1	-1.9	221 110	c
7-1	6.33	0.208	41.0	0.205	0.003	40.4	-43.5	83.9	-1.5	0.197	0.009	36.2	-44.4	80.6	-4.1	1 024 231	e

a: Rupture in the gage length
b: Stop at 50% of σ_{max}
c: Rupture at the threads
e: Test stopped without failure

NOTATIONS

d: Specimen diameter at minimum section

E: Modulus of elasticity; determined from the elastic part of the stress-strain curve in the first tensile cycle (first cycle)

$\Delta\varepsilon$: Total strain range

$\Delta\varepsilon_{el}$: Elastic strain range

$\Delta\varepsilon_p$: Plastic strain range

σ_{max}: Maximum cyclic stress

σ_{min}: Minimum cyclic stress

$\Delta\sigma$: Stress range

σ_{mean}: Mean stress

2005-01-0730

Friction Stir Welding of Magnesium for Automotive Applications

K. Okamoto and F. Hunt
Automotive Products Research Lab., Hitachi America, Ltd.

S. Hirano
Hitachi Research Lab., Hitachi, Ltd.

ABSTRACT

Friction stir welding (FSW) shows significant advantages in the joining of magnesium alloys. In this paper, the feasibility of FSW for several kinds of the automotive magnesium alloys and the dissimilar material joint of magnesium to aluminum alloys were studied. Butt welded magnesium alloys of wrought and die-cast show advantages expected of the FSW process, i.e. good appearance and joint integrity. The 0.2% proof stress and ultimate tensile strength of friction stir welds in AZ31 are almost the same as the annealed parent metal. In any die-cast AZ91D, AM50A and AM60B alloys, the hardness in the stir zone is higher than the base metal. The proof stress and the tensile strength of the AM50A and AM60B joints are almost at the same level as the base metal. In the case of dissimilar material joint, the material mixing is limited and a more mechanical interlocking joint forms. Furthermore, intermetallic phases about 2μm thick are formed so that the ductility of the dissimilar material joint is inferior to those of the parent metal.

INTRODUCTION

Friction stir welding (FSW) is a novel joining process invented and patented by TWI in 1991 [1]. FSW is a solid phase process operated below the melting temperature of the material. A schematic illustration of FSW process is shown in Fig.1. A cylindrical rotating tool with a threaded pin and shoulder is plunged and feed to form a linear weld. The resultant weld joints show higher mechanical properties and are free of porosity or blowholes compared to conventional fusion techniques. Therefore, friction stir welding (FSW) has been remarkably applied as an innovative production method for several industrial sectors such as ships [2], railway cars [3], automobiles [4], airplane and rocket fuel tanks [5] because of its advantages to welding aluminum alloys.

Apart from aluminum alloys, application of FSW on magnesium alloy [6], copper alloy [7], ferrous materials [8][9] and titanium [10] have been reported. FSW of copper alloy has been already commercialized [7] but FSW of ferrous material is still in the research phase because of tool wear at high welding temperatures.

Magnesium is a light metal with density of 1.8g/cc that is 2/3s of aluminum while the specific strength of magnesium is the highest among any metal. In recent years, magnesium usage has increased significantly especially in automotive industry for weight reduction and improved fuel economy. However, magnesium is an active metal that oxidizes relatively easily at high temperature therefore inert gas shielding is required to prevent oxidation during fusion welding. Furthermore, thermal conduction and thermal expansion of magnesium alloy are higher than those of steel that causes higher welding distortion, also welds can crack with tight clamp.

Magnesium alloys have a relatively low melting point of about 650°C which make it an ideal material for FSW as well as aluminum alloy. In this report, several kinds of magnesium alloys of wrought AZ31 and die-cast AZ91D, AM50A and AM60B were selected to be FSWed. Joint properties such as: macrostructure, microstructure,

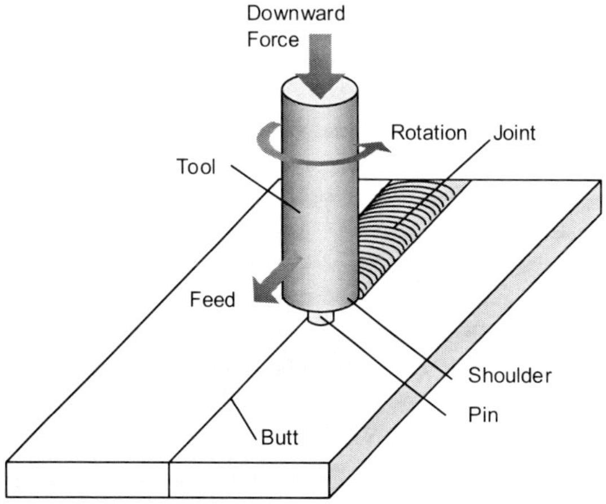

Fig.1 Schematic illustration of FSW for butt joint

hardness profiles and tensile strength were evaluated. Furthermore, FSW of magnesium and aluminum were investigated to study the possibility of joining dissimilar materials with this process.

EXPERIMENTAL

MATERIALS AND WELDING

Magnesium plates of AZ31 and die-cast AZ91D, AM50A and AM60B were utilized in this study. AZ31 sheet metal and die-cast AZ91D are Mg-Al-Zn alloy which were welded to themselves. Die-cast AM50A and AM60B are Mg-Al-Mn alloys that were also welded together. A1050 and A5182 aluminum alloys were also welded to AZ31. A1050 is a commercial pure aluminum that was welded to AZ31 to evaluate the microstructure of the joint. A5182 is a Al-Mg alloy which was also welded to AZ31 to evaluate the mechanical properties of these materials.

The weld test was conducted on sheet metal 1.2mm thick, welded at a tool feed rate of 200-800mm/min with tool rotating speed of 1500rpm and 3000rpm. On the other hand, all of the die-cast metals were 2.6mm thick, welded at 150-600mm/min and 2000-2500rpm. The plates were butt welded in the as-received condition. All of the welding trials were carried out on a CNC controlled 3 dimensional FSW machine (HITACHI GR-3DC5T) with 11kW servo drive motor shown in Figure 2. The tool, made of tool steel, was tilted 3deg. from vertical. The weld plates were tightly clamped during the welding process. The spindle speed was constant during the weld and the rotating tool pin was plunged into the plate until the tool shoulder came into contact with the surface of the material, which generates frictional heat to locally soften the material around the pin. Then the tool was traversed and retracted to complete the weld.

Fig.2 CNC controlled 3 dimensional FSW machine

JOINT PROPERTIES EVALUATION

Macrostructure and microstructure cross-sections of the weld joints were observed by optical microscopy. EPMA (Electron Probe Micro Analyzer) and TEM (Transmission Electron Microscope) observation were also carried out to see the detail microstructure at the dissimilar material interface. Furthermore, TEM-EDS (Energy Dispersive X-ray Spectrometer) was also utilized to measure the chemical composition at the weld interface.

Hardness tests, transverses 1mm below the surface, were measured by Vickers hardness tester using a 0.2 kgf load. Transverse tensile property was also evaluated at room temperature. The stain rate was set to $5x10^{-5}s^{-1}$.

RESULTS AND DISCUSSION

SHEET METAL: AZ31

Figure 3 shows the weld surface and cross-section of FSWed AZ31, as shown there are no volumetric defects at the surface of the weld. A smooth and glossy surface ripple appears on the weld beads and only a small amount of burr is observed on the retreating side of the weld. It is interesting to note that oxidation is prevented during the weld even without any shielding gas. Sound welds were available even at the higher feed rate of 800mm/min so that the weldability does not seem to depend on the feed rate. In the stir zone, equiaxial grains were observed. The average grain size in the stir zone was measured at about 15μm (same as the base metal).

Figure 4 shows the hardness profile of AZ31 joint. Hardness of the base material was 62-65Hv while that of the stir zone measured at 52-57Hv. The base metal of AZ31 is work hardened so that the dislocation density of the base metal seem to be higher than the stir zone since the grain size of the base metal and the stir zone are the same. The difference of the dislocation density may cause hardness drop in the stir zone. Furthermore, the hardness in the heat affected zone is dropped to 50Hv because of the thermal cycle during the process.

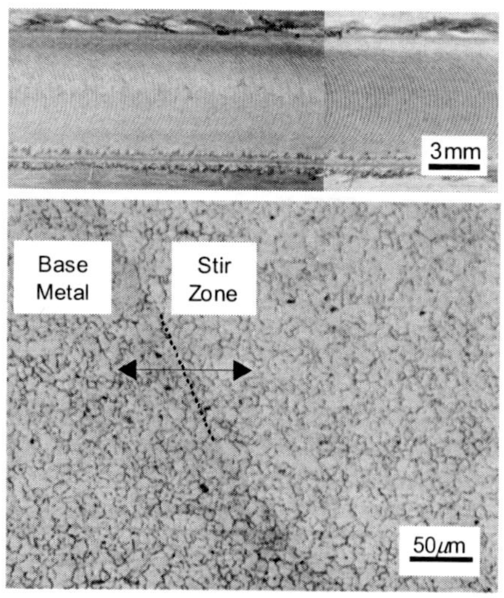

Fig.3 Weld surface and cross section of AZ31

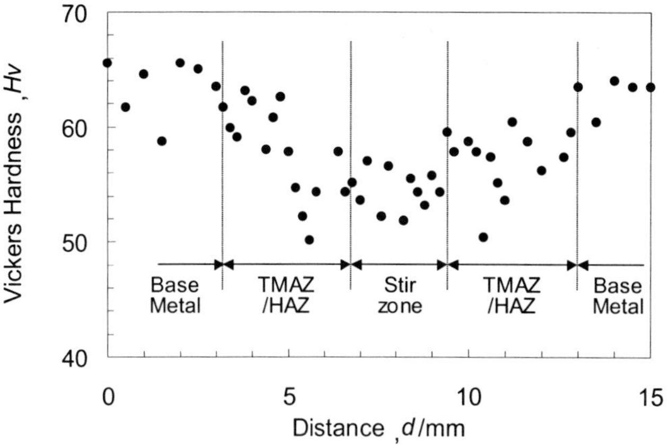

Fig.4 Hardness profile of friction stir welded AZ31

Tensile properties of AZ31 base metal and FSWed joints are summarized in Table 1. Proof stress and ultimate tensile strength (UTS) are relatively higher in the base metal since it is strengthened by work hardening. In the case of the joint, both uniform elongation and local elongation at the HAZ are observed before fracture. The HAZ has the minimum hardness of around 50Hv in the weld joint. It is considered that the HAZ has recovered by frictional heat during the FSW process. The transverse tensile failure occurred through necking at the HAZ that had the lowest hardness. Therefore, the tensile strength and the elongation of the joint were lower than the base metal.

Table 1 Tensile properties of AZ31 magnesium alloy

	0.2% Proof Stress / MPa	UTS / MPa	Elongation (%)
Base Metal	191	270	36.5
FSWed joint	135	211	18.5

DIE-CAST METAL: AZ91D, AM50A, AM60B

For die-cast magnesium alloys such as AZ91D, AM50A and AM60B, the effect of the welding parameters (feed rate and tool rpm) have been evaluated.

Figure 5 shows the effects of weld parameters on the weld surfaces of each alloy. No volumetric defects or oxidation were observed at the surfaces. However, it was sometimes observed that the material was adhered to the thread pin that removed some material out of the base material and caused surface defect (indicated by arrows) or root defect on the surface of the material. This phenomenon is very unique for magnesium alloy and further investigations are required.

Figure 6 shows the macrostructure of transverse cross section of the welds. Basically, sound welds were obtained for any of die-cast magnesium alloys. However, small defects (indicated by arrows) were observed in the stir zone if the heat input is relatively lower. As shown the stir zone is wider with low weld speed.

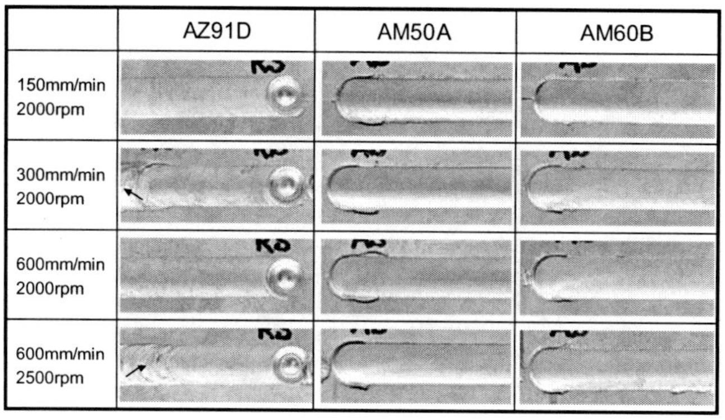

Fig.5 Weld surface of AZ91D, AM50A and AM60B

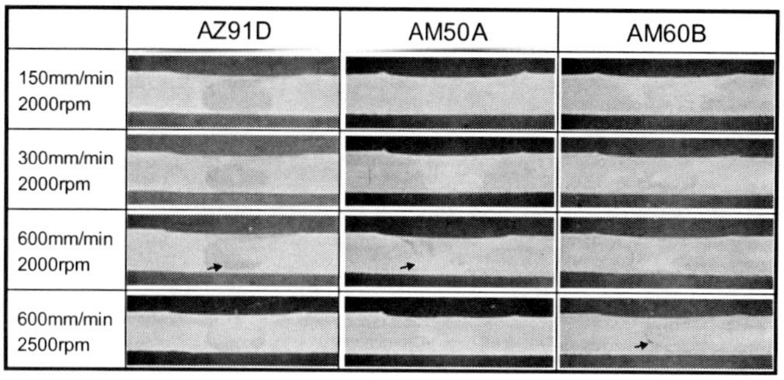

Fig.6 Cross Section of AZ91D, AM50A and AM60B

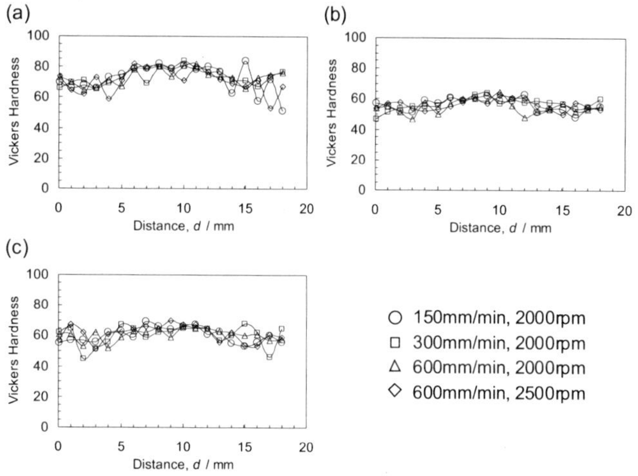

Fig.7 Hardness profile of friction stir welded (a) AZ91D, (b) AM50A and (c) AM60B

Figure 7 shows the hardness profile of the joints. In case of AZ91D, the weld joint demonstrates a 80-85Hv which is higher than the base metal (70Hv). Even though the graphs show some degree of scatter, the effects of the welding parameters on the hardness profiles were not significant. In case of AM50A, stir zone was hardened up to 60-65Hv while the base metal shows hardness around 55Hv. In case of AM60B, stir zone was hardened up to 65-70Hv while the base metal shows hardness around 60Hv. In any alloys, the hardness in the stir zone is increased and the effects of weld parameters on the hardness profiles are not obvious.

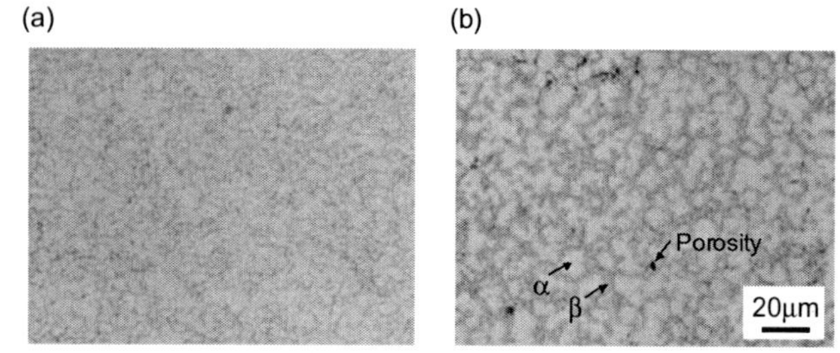

Fig.8 Microstructure of AZ91D in (a) stir zone and (b) the base metal

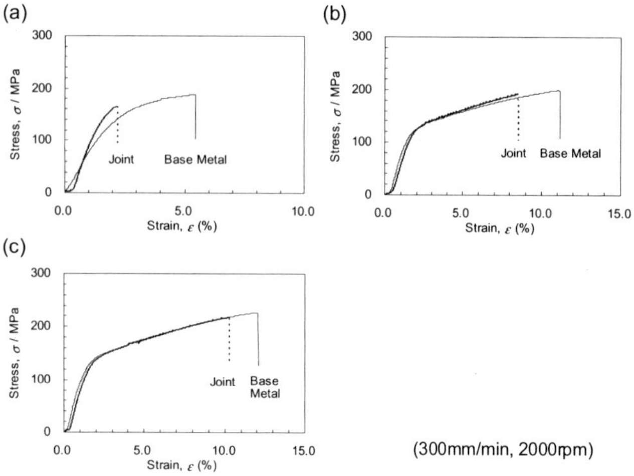

Fig.9 Stress-strain curves of base metals and FSWed joints of (a) AZ91D, (b) AM50A and (c) AM60B

Figure 8 shows the microstructure of the weld and base metal of AZ91D. T he average grain size in the stir zone is less than 5μm which is smaller that of the base metal. Furthermore, β phases observed in the base metal has been dispersed and partially solid solute into the α matrix in the stir zone. It is well known that current fusion welding techniques such as arc welding have some limitations for welding cast material that lead to high porosity or blowholes. However, porosities observed in the base metal were eliminated in the stir zone by the plastic deformation. The same phenomena were observed in any die-cast alloys. It seems that the hardness of the stir zone is increased since the stir zone has been grain refined and free of porosities or inter-dendritic β phases.

Figure 9 shows the stress-strain curves of the base metals and the FSWed joints of the die-cast alloys. The welding condition was selected to 300mm/min at 2000rpm. In the case of AM50A and AM60B, the proof stress and the tensile strength of the joints are almost at the same level as the base metals. The total elongation of the joint is lower than the base metal. The failure occurred out of the weld region since the hardness in the stir zone is higher than the base metal as shown in Fig. 7. On the other hand, the tensile strength and the elongation of the joint are relatively lower than the base metal in the case of AZ91D. Furthermore, the failure occurred out of the weld region. This evaluation is still ongoing.

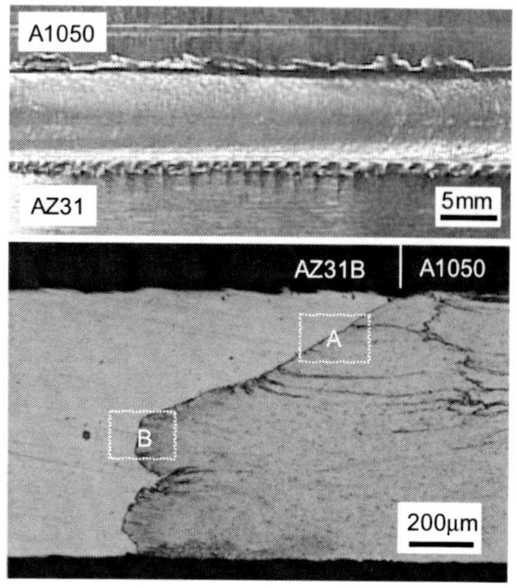

Fig.10 Weld surface and cross section of dissimilar material joint of A1050 and AZ31

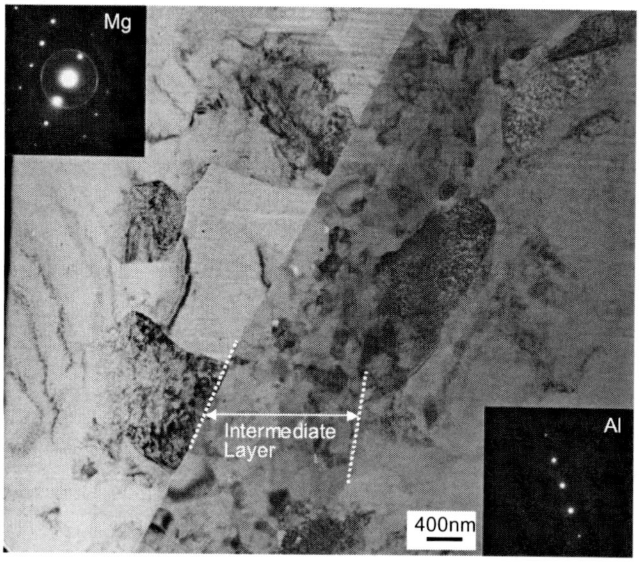

Fig.11 TEM image of joint interface between AZ31 and A1050.

DISSIMILAR MATERIAL WELDING

Figure 10 shows the weld surface and the cross-section of the dissimilar materials joining of A1050 and AZ31 welded at 200mm/min and 3000rpm. A relatively large amount of burr is observed on both the retreating and the advancing sides of the weld, however, no volumetric defects are observed in the cross section. The cross-section shows a different morphology from similar material joints, two phases of aluminum and magnesium are fully separated and the boundary looks like an S-shape. The interface on the magnesium side is quite flat while an indentation is observed on the aluminum side of the weld. According to EPMA, the overlap width of aluminum and magnesium is 3μm in Area A (See Fig.10). On the other hand, both aluminum and magnesium are detected to be overlapped within 20μm width in Area B (in Fig.10). Area B is in the stir zone so that the mixture of two materials has been enhanced compared to Area A.

Figure 11 shows TEM bright image of the interface between magnesium phase and aluminum phase. An intermediate phase 2μm thick was observed at the interface. The grain size of the intermediate layer is about 400nm that is finer than those of the magnesium or aluminum phases. The interface on the magnesium side is quite flat as observed by optical microscope shown in Fig.10. According to the diffraction patterns and EDS analysis, the intermediate layer contains 2 phases, i.e. $Al_{12}Mg_{17}$ on the magnesium side and unknown phase on the aluminum side. In the binary alloy phase diagram of Al-Mg, there are two eutectic reactions at 450°C and 437°C. However, no dendrite was formed so that these intermetallic compounds might be formed by a solid-state diffusion mechanism. Therefore, two phases are mechanical interlock jointed as reported by Smith [13].

Table 2 shows the tensile properties of dissimilar material joint weld of AZ31 and A5182 formed at 400mm/min and 1500rpm. The proof stress of the dissimilar material joint is higher than the base metal (BM) of A5182. During testing, A5182 yielded at first and necked but it did not fracture, finally fracture occurred at the interface where the intermetallic compounds were formed. The joint efficiency was 69% while the elongation was very poor at 3.3%.

Table 2 Tensile properties of the dissimilar joints of magnesium and aluminum alloys.

	0.2% Proof Stress / MPa	UTS / MPa	Elongation (%)
AZ31/A5182	159	205	3.3
AZ31 BM	220	290	35
A5182 BM	125	277	38

BM: Base Metal

CONCLUSION

FSW has been applied to automotive magnesium alloys of AZ31, die-cast AZ91D, AM50A, AM60B and dissimilar material joint of aluminum and magnesium alloys. The results obtained are summarized as follows:

1. FSWed joint of magnesium sheet metal AZ31 shows the ultimate tensile strength of 211MPa that is same as fully annealed AZ31.

2. In any die-cast alloys, the hardness in the stir zone is higher than the base metal since the stir zone has been grain refined and free of porosities or inter-dendritic β phases. The effects of weld parameters such as feed rate and rpm on the hardness profiles are not obvious. However, small defects were observed in the stir zone if the heat input is relatively lower.

3. In the case of AM50A and AM60B, the proof stress and the tensile the same level as the base metals. However, the tensile strength and the elongation of the AZ91D joint are relatively lower than the base metal even though the failure occurred out of the weld region.

4. Dissimilar material joining of magnesium and aluminum alloy is more likely mechanical interlocking with intermetallic compound layer that decreased the ductility of the joint.

ACKNOWLEDGMENTS

The authors acknowledge Mr. Thomas Ruden of Lunt Manufacturing Co., INC. (IL, USA) for the die-cast magnesium alloy supply.

REFERENCES

1. W.M.Thomas *et al.*: "Friction stir butt welding", International patent application PCT/GB92/02203.
2. O.T.Midling *et al.*: "Industrialisation of the friction stir welding technology in panels production for the maritime sector", Proc. of 1st Int. symp. on FSW, California, USA, 1999.
3. T.Kawasaki *et al.*: "Application of friction stir welding to the manufacturing of the next generation "A-train" type rolling stock", Proc of 2nd Int. symp. on FSW, Gothenburg, Sweden, 2000.
4. R.Sakano *et al.*: "Development of spot FSW robot system for automobile body members", Proc. of 3rd Int. symp. on FSW, Kobe, Japan, 2001.
5. M.R.Johnsen: "Friction stir welding takes off at Boeing", Welding Journal February, 1999, pp35-39.
6. R.Johnson *et al.*, "Friction stir welding of magnesium alloys", Proc. of 4th Int. Symp. on FSW, Utah, USA, 2003.
7. K.Okamoto *et al.*: "Fabrication of backing plate of copper alloy by friction stir welding", Proc. of 3rd Int. Symp. on FSW, Kobe, Japan, 2001.
8. C.Sorensen *et al.*: "Tool material testing for friction stir welding of high temperature alloys", Proc. of 3rd Int. Symp. on FSW, Kobe, Japan, 2001.
9. K.Okamoto *et al.*, "Metallurgical and mechanical properties of friction stir welded stainless steels", Proc. of 4th Int. Symp. on FSW, Utah, USA, 2003.
10. M.C.Juhas *et al.*: "Microstructural evolution in Ti alloy friction stir welds", Proc. of 2nd Int. Symp. on FSW, Gothenburg, Sweden, 2000.
11. L.S.Smith: Proc. of the First Osaka Univ. and TWI Joint Seminar, 22-23 March 2001, Osaka Japan, 63-72.

CONTACT

Dr. Kazutaka Okamoto, Automotive Products Research Lab., Research & Development Div., Hitachi America, Ltd., Farmington Hills, MI 48335,

Telephone: 248-442-6725, Fax: 248-473-8420, E-mail: kazutaka.okamoto@hap.com

2005-01-0732

Corrosion Protection of Structural Magnesium Alloys: Recent Development

Wenyue Zheng, C. DeRushie and J. Lo
CANMET Materials Technology Laboratory, Natural Resources Canada

R. Osborne
GMNA Vehicle Engineering Center, Materials Engineering

ABSTRACT

Magnesium is one of the viable materials for significantly reducing the weight of automotive component. However, wide-scale use of magnesium alloy is complicated by a number performance factors related to the intrinsic properties of these alloys such as corrosion resistance. The latest developments in coating technology and alloy chemistry design have removed much of the barrier and a number of chromate-free processes are commercially available for the parts that require a coating or coatings. As part of the Structural Cast Magnesium Development (SCMD) project, several types of newly selected coatings were evaluated for protection against general and galvanic corrosion. The latest results are presented in this paper.

Galvanic corrosion is a major concern for the parts that are joined with other metals, especially when there is salt-containing road poultice on the surface. While the use of compatible Al alloy isolators can significantly mitigate the problem, elimination of electrical contact or the electrochemical driving force between the dissimilar metals is essential for complete prevention of the problem. In the test in which a thin mylar film was inserted between a steel washer and the Mg surface, no visible galvanic corrosion was found after 40 cycles of GM9540P testing. Anodizing coating alone provides good resistance to general corrosion, but it does not seem be effective against galvanic corrosion.

The latest experimental work suggests that application of coatings can also beneficially affect the resistance of Mg alloy to stress corrosion cracking (SCC). When treated with Alodine 5200 and coated with a powder coat, samples of both AM50 and AE44 alloys showed drastic improvement in failure life, i.e., by at least 200%. Although the test conditions used are far more severe than that likely to be encountered during an automotive lifetime, these results have mechanistic implications for all applications where a high loading stress is present.

1. INTRODUCTION

Mg castings of high quality chemistries used in the automotive condition can provide adequate service life when the parts are shielded from aggressive corrosion-causing materials such as road salts. Many interior parts such as the instrument panel beams and the steering components, usually made by die casting of common AZ or AM alloys, require minimal surface treatment. Even for some exterior applications, magnesium parts can resist general corrosion when installed at particular locations. Figure 1 shows the appearance of an ABS modulator bracket (AZ91) after 7 years of service in the general area of northeast USA. Although the bracket showed signs of slight galvanic corrosion initiating around the washer, it exhibits little general corrosion on most of the surface.

In designing of new Mg alloys, rare-earth (RE) addition has been one of the effective approaches for improving creep and tensile properties [1]. These RE elements have also been known to have beneficial effects on corrosion resistance. For example, the corrosion rate in salt-spray environment of die cast AE42 alloy is less than 50% that of AM60B [2], despite the lower amount of Al in the AE42 alloy. Nevertheless, when Mg alloys are used as structural components [3], which could be under high level of stresses, corrosion prevention and protection against all possible forms of corrosion damage are still of paramount importance.

The work reported here is part of a large research effort currently taking place in North America. The general description of the two major ongoing projects organized by the United States Council for Automotive Research (USCAR) and the United States Automotive Materials Partnership (USAMP), i.e. the Structural Cast Magnesium Development (SCMD) [3] and the Magnesium Powertrain Cast Components (MPCC) [4], has been provided elsewhere. As the projects approach the completion point, many of the technical objectives have been achieved in assessing the technical feasibility and the cost to benefit ratio of lightweight magnesium chassis, body components, and power-train components. Prototype parts (engine blocks and engine cradles) have been made that can readily be used as lightweight replacements of their heavier Al counter parts [5,6].

The overall objective of corrosion and protection aspects of the Structural Cast Magnesium Development project [SCMD] has been described in a previous SAE technical paper [7]. In the initial part of the work, 17 different types of coatings for Mg alloys were screened by ASTM B117 and GM9540P test methods. In terms of general corrosion, it was found that anodizing coatings such as Tagnite and Anomag offered effective protection. On the other hand, lower cost conversion coatings such as Alodine 5200 with a powder coating can resist chloride attack just as well. A polyurea-based coating was also tested for its suitability as a protective topcoat that absorbs the impact action of the road sand and gravel [7].

The recent work in this area focused on assessment of galvanic corrosion between coated bolts/spacer assembles and coated or uncoated magnesium alloy plates. 'Novel' coating materials for magnesium were selected and tested, as these new coatings have shown potential for protecting other metals and they represent cost-competitive alternatives to those tested earlier. In addition, the effects of coating the alloy on the stress corrosion cracking (SCC) resistance of the base alloy were also evaluated using AM50 and AE44 samples and the GM9540P test solution.

2. GALVANIC CORROSION TESTS

2.1 Experimental Details

Corrosion testing was carried out using the GM9540P test method. Corrosion test plates of two magnesium alloys were used, namely, AM60B and AE44. Each test plate had a measured width of 100 mm, a length of 150 mm, and a thickness of 2 mm with five metric holes drilled prior to the coating application. Of the five holes, two M10 holes were located evenly spaced along the top half of the test plate while three M8 holes were located evenly spaced along the bottom half of the test plate.

The magnesium coatings evaluated included the following:

1. Magoxid with an epoxy powder coat
2. Magpass plus a powder coat
3. Magpass with a sealer (DC-80), plus a powder coat
4. Magoxid without a powder coat
5. Polyurea (200 and 400 microns thickness)
6. Hot wax (Nox Rust 1280HT by Daubert Co.)

Magpass is a pre-treatment process that produces a relatively thick (up to a couple of microns) chromate-free conversion layer [8]. Magoxid is a hard anodizing process that produces corrosion and wear-resistant surface on magnesium alloy [9].

Nox Rust 1280 is a type of anticorrosion paraffin that can resist both cold and hot climatic conditions. Widely used on steel body parts, specific formulations of this line of products are advertised to be able to survive 720 hours of salt spray testing [10]. The particular paraffin tested for this work was a Nox Rust 1280HT paint grade hot wax, and the coating process was adjusted, at Daubert Chemical Co., for the Mg test plates so that a three-mil layer was produced.

Polyurea is made of isocyanate prepolymer and a mix of primary and secondary amines [11]. The test samples were sprayed at room temperature and the curing took less than a minute. Two coating thickness were made for this work, namely, 200 microns and 400 microns.

Different types of washers were tested in combination with the above coatings. These are:

1. Al 5052 washers
2. Al 6061 washers
3. Steel washers plated with zinc
4. Steel washers plated with tin
5. Plain steel washer placed on top of a thin (25 microns thickness) mylar film

2.2 Test Results of Fastened Magnesium Plates

<u>2.2.1 The Magpass plates.</u> Digital photographs were taken of the test plates after each 10 cycles. Figure 2 shows the changes in the morphology of test plates (Magpass sealed with DC80 and then coated with epoxy and another coated with epoxy) as the testing proceeded.

Note that the white deposits on the plates after 40 cycles were essentially precipitated salts. To confirm this finding, cross-sections were made with the sectioning surface located right through the middle of each assembled bolt. None of the cross-sections demonstrated any depth-wise attack on the coated magnesium. As shown in Figure 3, all epoxy-magnesium interfaces stayed intact after 40 days of testing.

The other series of Magpass coated samples, which did not have a sealing treatment, exhibited very similar performances and the sealing treatment with DC80 did not seem to have any noticeable effects. Overall the Magpass samples coated with epoxy behaved very similar to Alodine 5200 samples that were coated with a top coat [7].

2.2.2 The Magoxid plates. The effect of epoxy coating in this series of samples is very dramatic. As shown in Figure 4, the Mg plate anodized by the Magoxid process all showed galvanic corrosion along the Al alloy (5052 or 6061) washers. The one without a washer (located in the top right corner, top side) showed most severe galvanic attack. Note that away from the washer and bolt, there is no sign of general corrosion, indicating that the coating is quite resistant to general corrosion when there is no anodic polarization of the Mg metal.

The application of an epoxy topcoat completely prevented corrosion attack, either in the form of general or galvanic corrosion. This finding is important in the sense that it demonstrates the insulating epoxy coating to be primarily responsible for preventing galvanic corrosion. Thus, it seems that the requirement for the anodizing layer would be mainly to form an intermediate pre-treatment with a good adhesion for a topcoat such as epoxy or polyester type polymers. In this case, a much less expensive anodizing version can be made for use in the automotive industry, as the coating cost is directly related to the coating thickness. A thinner coating anodizing means shorter time for this process.

2.2.3 The effect of tin, zinc and a thin mylar film

Figure 5 shows the appearance of various fastening units installed on a bare (AE44) test plate. A couple of results are shown in the Figures 5(a) and 5(b).

Firstly, the use of an insulating thin mylar film was very effective in stopping galvanic corrosion between a steel washer and magnesium. Although the uncoated washer exhibited significant rusting (Fig. 5(a)), the underlying Mg surface was intact when the mylar film was removed, Fig. 5(b).

This result would suggest that if an Al washer is coated with a thin layer of mylar material it should not cause any galvanic attack. Such a process would be much more affordable than a hard anodizing coating sometimes applied on aluminum alloy washers. As the film is so thin that its contribution to bolt-load relaxation should be minimal. In certain applications an insulating adhesive resin was applied between the joining faces of dissimilar metal. In addition to its mechanical functions, an insulating adhesive should also be beneficial for galvanic corrosion.

Secondly, both zinc plating and tin-plating on steel washers did not prevent galvanic corrosion of the magnesium alloy. Between the zinc-plated washer and the tin-plated washer, there is more visible attack around the zinc-plated washer.

2.3 Test Results for Hot Wax and Polyurea Coatings

As there might be conditions in an automobile where a magnesium part may be subject to certain degrees of wetness and humidity but not under the direct action of road salt, such as components under the hood or under the body, there is interest in knowing whether there exist low-cost alternative coating materials that could be sufficient for their intended purpose. Two alternative 'novel' coatings selected are a hot wax Nox Rust 1280HT product and a two-component polyurea coating.

Figure 6 shows the appearance of the polyurea coatings (200 micron and 400 micron) after 40 cycles GM9540P testing. No visible sign of corrosion can be seen on either test plate. There were no under-coating blisters either. Although earlier test [7] has shown that the coating does not have any self-healing properties, and corrosion can creep from the scribe line, the new results indicate that, for conditions where damages on the coating is not a concern, this product may well be a cost-effective alternative.

Figure 7 shows the appearance of the sample coated with Nox Rust 1280HT, before and after 20 cycles of salt-spray testing. Because the paraffin coating is soft, the as-received sample showed sign of rubbing and scraping during shipping and handling.

In the GM9540P test chamber, the sample surface stayed mostly free of water droplets as the surface is of a hydrophobic nature. After 20 cycles of testing, gray corrosion product can be seen to emerge in the areas of prior damage. Several scattered corrosion spots can also be seen on the test plate (Figure 7(b)), which are likely associated with pre-existing pinholes in the paraffin coating. While the hydrophobic properties of the paraffin film are definitively beneficial for corrosion protection, it seems the coating process needs to be better controlled to eliminate pinholes. The test plates were coated in as-received condition and the surface cleanliness may also have been a factor.

3. STRESS CORROSION CRACKING (SCC) TESTS

3.1 Experimental Details

SCC tests were carried out using LPDC AM50 and HPDC AE44 samples. Both series of the samples were pre-treated with an Alodine 5200 conversion process and then with an epoxy topcoat. In addition, bare tensile samples for each alloy are also tested to generate baseline data.

The maximum stress used was 130% of the yield stress and the R-value was 0.90. The load frequency used was

1.0 Hz. The solution used was of the GM9540P solution composition. In the SCC tests the whole gauge length was immersed in the test solution. All tests were conducted at room temperature.

The test conditions should be regarded as excessively severe in comparison with those likely to be encountered during the service life of an automobile. The main purpose at this stage was mainly to investigate the effects of the coating on the intrinsic resistance of the Mg alloy to this form of corrosion damage.

3.2 SCC Test Results

3.2.1 AM50 alloy

The uncoated samples showed significant cracking after about 0.4 million cycles. In fact, the sample ruptured when the cracks developed deep enough. Figure 8 shows a picture of the LPDC AM50 alloys containing stress-corrosion cracks. In contrast, the coated sample survived 2.4 million cycles without failure. However, during post-mortem metallographic examination, a small (less than 1.5 mm deep) crack was found that have initiated from an area of coating damage. Figure 9 shows a cross-section of the sample containing the crack.

3.2.2 AE44 alloy

Similarly the AE44 alloys also showed significant beneficial effects of the coating system on SCC resistance. Figure 10 is a comparison of the cycle-to-failure data for all samples tested. While the uncoated samples both failed under 40,000 cycles, the samples treated with Alodine 5200 and then coated with a powder coat both survived at least twice as long, under identical mechanical and chemical conditions.

It should be pointed out that there is a significant difference between the AM50 and AE44 alloys in the failure life under the test conditions used. This could be caused by a number of factors, including the shape of the tensile samples machined. For the AE44 sample, the gauge length, as shown in Figure 11, has a square cross-section in contrast to the round shape for the AM50 samples. Sharpe corners of the AE44 samples would likely facilitate crack initiation, and thus reducing the failure life. The focus of this work is on the effects of the coating on each respective alloy, and therefore the comparison is limited to within each alloy type.

Corrosion of magnesium is associated with liberation of hydrogen and hydrogen embrittlement is thought to be the responsible SCC mechanism in Mg alloys. The higher resistance to SCC of the samples coated with Alodine 5200 may be caused by the reduction in hydrogen generation rendered by the presence of the conversion product. Some of the conversion treatments are known to significantly reduce the formation of hydrogen gas on the Mg surface [12]. The relative contribution of the Alodine5200 conversion treatment versus the polymer topcoat is unknown at this point; this is an issue of interest for the on-going research.

4. SUMMARY AND CONCLUSIONS

Recent results from this on-going project on corrosion control of structural magnesium alloys are very positive and encouraging. More Mg coatings have been identified that can mitigate general and galvanic corrosion. The new SCC test results have uncovered another important benefit of magnesium coating, which could be useful for chemically and mechanically severe application conditions.

Specifically, the following summary can be made at this point:

1. There are competitive Mg coating systems for Mg alloys that can protect Mg alloys against aggressive testing conditions established by the industry. Two chromate-free conversion coatings, Alodine 5200 and Magpass, with a powder coating have both been found to be effective.

2. SCC tests conducted so far confirmed that an appropriate coating system could also improve the resistance of the Mg alloys to stress corrosion when high loading conditions are encountered in the presence of accumulated electrolyte. The beneficial effect of the coating on SCC resistance is likely a result of its effect in reducing hydrogen generation on the Mg surface.

3. A thin dielectric film, such as the mylar sheet used in this work, sandwiched between a steel washer and a magnesium plate, could effectively prevent the formation of a galvanic cell. A layer of powder coating on top of a Mg conversion film, which can be by either Alodine 5200, reported previously [7], or a Magpass process, functions by essentially the same principle. It is conceivable that a layer of appropriate adhesives can render similar performance if it has the necessary dielectric property.

4. Anodizing coating alone such as the Magoxid tested in this work can assist in preventing general corrosion. However, it by itself would not likely stop galvanic corrosion when joined with other noble metals.

5. There are also alternative lower cost coatings such as polyurea or paraffin products that can be considered for less demanding service conditions. A polyurea coating as thin as 200 microns was found to be able to protect Mg plates during 40 cycles of GM9540P test.

5. REFERENCES

1. A. A. Luo, "Recent magnesium alloy development for elevated temperature applications", International Materials Review, Vol. 49, No.1, 2004.
2. R. D. Carnahan, p94 in "Magnesium and Magnesium Alloys", ASM International, ASM 1999.
3. R. Osborne, G. Cole, B. Cox and P. Penrod, "USCAR Project on Magnesium Structural Castings", Proc. IMA 2000 Magnesium Conference, Vancouver, Canada, 21-23 May, 2000, pp 1-5,
4. Bob R. Powell, "The USAMP Magnesium Powertrain Cast Components Project", Proceedings of the 60th Annual World Magnesium Conference, May 11-12, 2003, sponsored by the International Magnesium Association, pp 44-51.
5. Joe carpenter, "The FreedomCAR and Hydrogen Fuel Initiative and Magnesium", proceeding of the 2004 International Confernce on Magnesium, Sept. 20-24, 2004, Beijing, China. (in Print)
6. N. Li, "Magnesium Advances and Applications in North America Automotive Industry", ibid. (in print)
7. W. Zheng, C. Derushie, R. Zhang and J. Lo, Protection of Mg Alloys for Structural Applications in Automobiles, SAE Paper # 2004-01-0133, Detroit, USA.
8. Michael Walter, "Magpass-CoatSr as a Chrome-Free Pre-Treatment for Paint Layers and An Adhesive primer for Subsequent Bonding", 2004-01-0134, SAE 2004, World Congress & Exhibition, March 2004, Detroit, MI, USA
9. Kurze, P; Banajee, D, "Magoxid-Coat: a novel anodic coating of magnesium based materials", Oberflachen Polysurfaces. Vol. 39, no. 2, pp. 7-11. Apr. 1998 (In French).
10. http://www.shpi-chem.com/products
11. J.A. Johnston and S. Smith, in Proc. Alliance for the Polyurethanes Industry (API) Conference, Salt Lake City, Utah, October 13-16, 2002.
12. En-Hou Han, Wanqiu Zhou, Dayong Shan and Wei Ke, Chemical Conversion Coating on AZ91D and its Corrosion Resistance, in Magnesium Technology 2004, Edited by Alan Luo, TMS 2004.

Acknowledgement. The authors acknowledge that this research was supported, in part, by department of energy cooperative agreement no. DE-FC05-02or22910, although such support does not constitute an endorsement by the department of energy of the views expressed herein. Support by the Canadian Lightweight Materials Research Initiative (CLIMRI) is also acknowledged. The authors wish to thank all SCMD project members who have provided materials, helpful assistance and ideas throughout the project.

6. Contacts

[a]wenyue@NRCan.gc.ca [b]richard.osborne@gm.com

Figure 1. Appearance of an ABS modulator bracket (AZ91) after 7 years of service

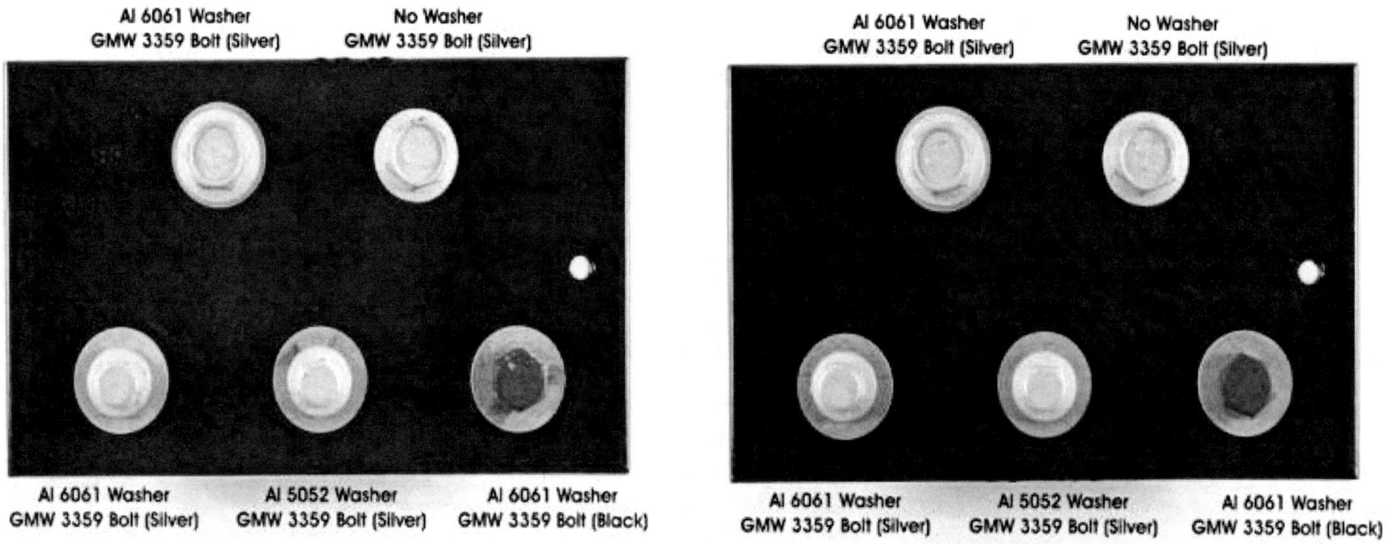

(I) MagPass with DC80&Epoxy

(II) MagPass with Epoxy

Figure 2(a) The appearance of test plates coated with Magpass sealed with DC80 then coated with epoxy in comparison with Magpass coated with an epoxy without a sealer: before GM9540P testing.

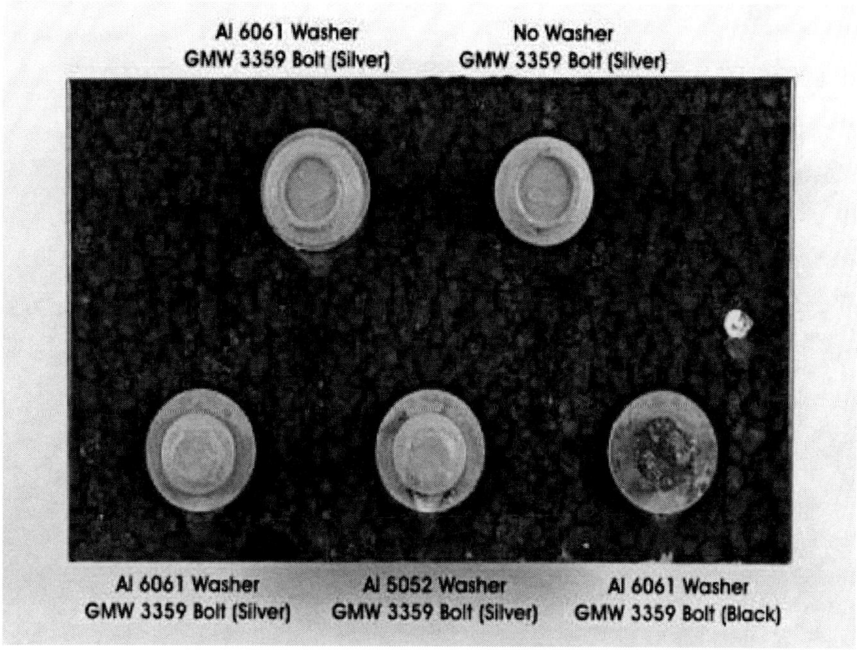

(I) MagPass with DC80&Epoxy

(II) MagPass with Epoxy

Figure 2(b) The appearance of test plates coated with Magpass sealed with DC80 then coated with epoxy in comparison with Magpass coated with an epoxy without a sealer: after 10 cycles of GM9540P testing.

Al 6061 Washer
GMW 3359 Bolt (Silver)

No Washer
GMW 3359 Bolt (Silver)

Al 6061 Washer
GMW 3359 Bolt (Silver)

Al 5052 Washer
GMW 3359 Bolt (Silver)

Al 6061 Washer
GMW 3359 Bolt (Black)

(I) MagPass with DC80&Epoxy

Al 6061 Washer
GMW 3359 Bolt (Silver)

No Washer
GMW 3359 Bolt (Silver)

Al 6061 Washer
GMW 3359 Bolt (Silver)

Al 5052 Washer
GMW 3359 Bolt (Silver)

Al 6061 Washer
GMW 3359 Bolt (Black)

(II) MagPass with Epoxy

Figure 2(c). The appearance of test plates coated with Magpass sealed with DC80 then coated with epoxy in comparison with Magpass coated with an epoxy without a sealer: after 40 cycles of GM9540P testing.

a) MagPass w/ DC80 & Epoxy

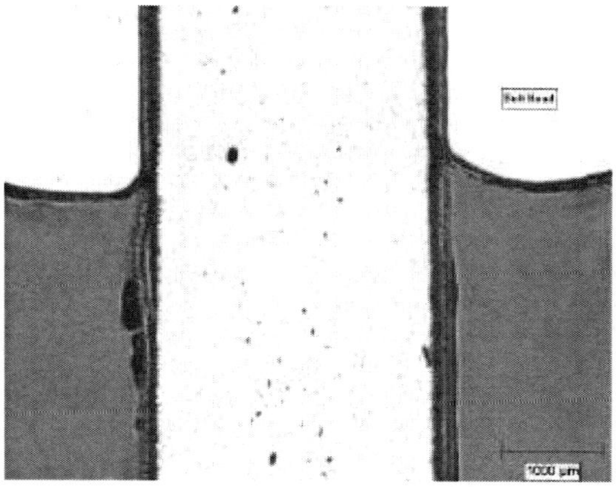

b) MagPass w/ DC80 & Epoxy

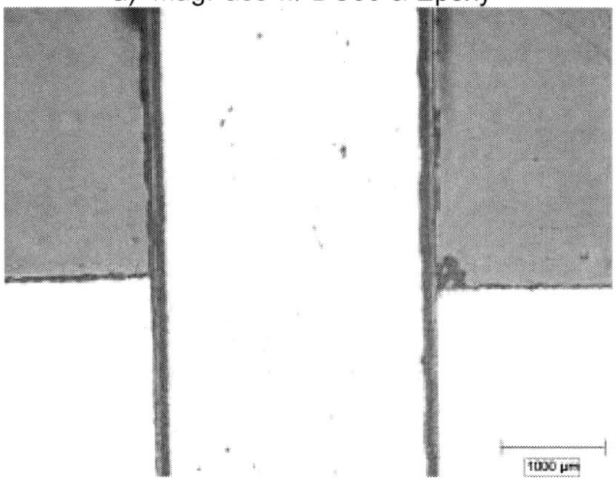

c) MagPass w/ DC80 & Epoxy

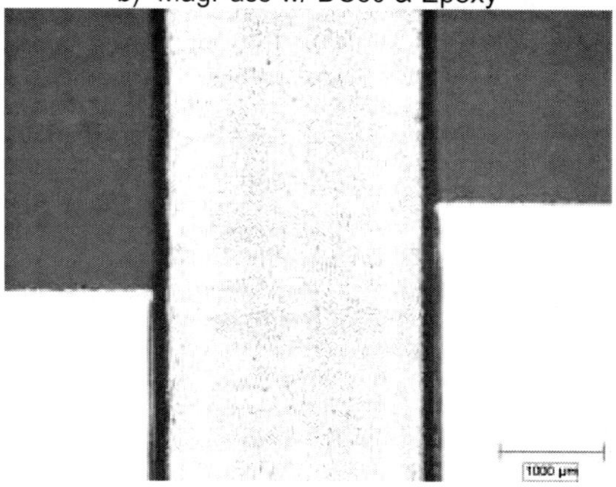

d) MagPass w/ DC80 & Epoxy

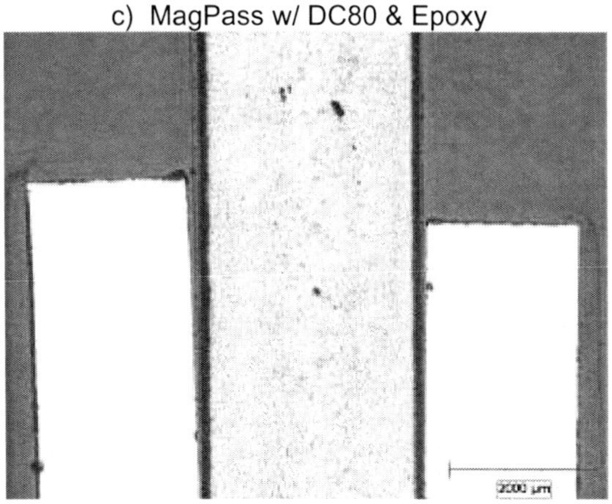

e) MagPass w/ DC80 & Epoxy

Figure 3. Cross-section images of each position of Magpass (w/ DC80 & Epoxy) coated magnesium test plates (40 Days of GM9540P testing).

a) MagOxid, Top side, 40 Cycles

b) MagOxid, Bottom side, 40 Cycles

c) MagOxid w/ Epoxy, Top side, 40 Cycles

d) MagOxid w/ Epoxy, Bottom side, 40 Cycles

Figure 4. Magoxid series plates after 40 cycles of GM9540P testing (all washers are of Al alloys, i.e., 5052 or 6061)

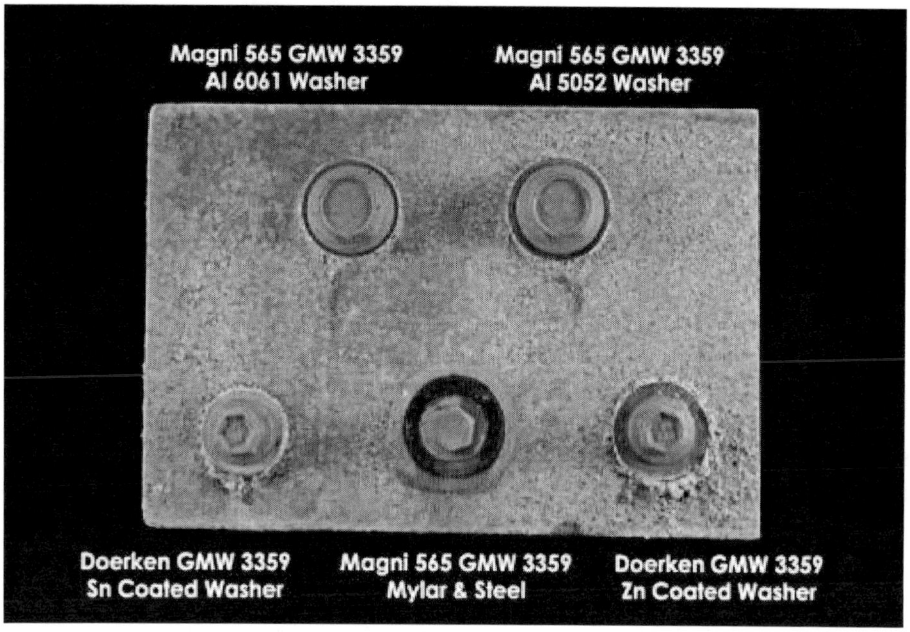

(a)

(b)

Figure 5. Effects of tin and zinc coating and of a mylar film on galvanic corrosion of AE44 (Coatings on the fasteners are either by Doerken or Magni).

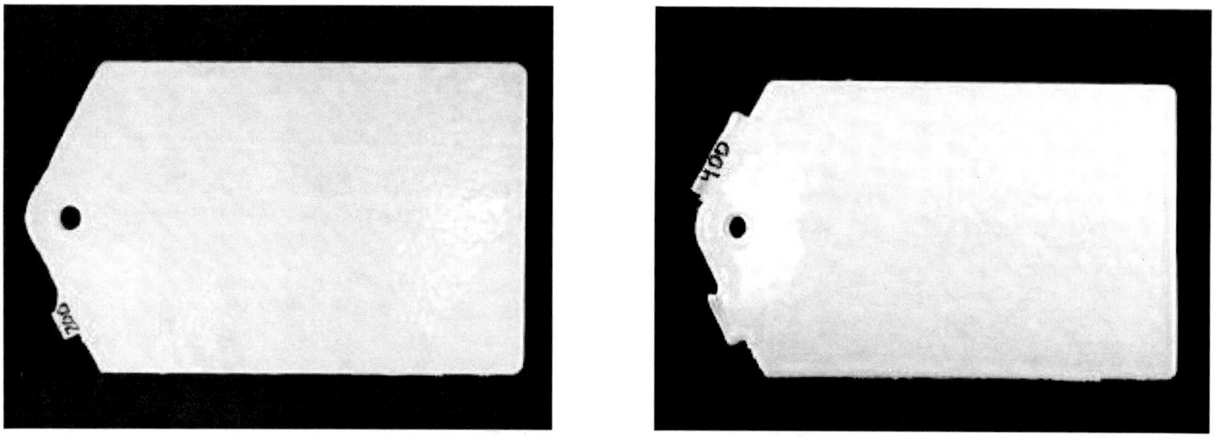

(a) (b)

Figure 6. Appearance of polyurea coating on Mg test plates after 40 cycle of GM9540P testing: (a) coating thickness 200 microns and (b) coating thickness 400 microns.

(a) (b)

Figure 7. Appearance of the Mg plate coated with Nox Rust 1280HT: (a) before testing and (b) after 20 cycles of GM9540P testing.

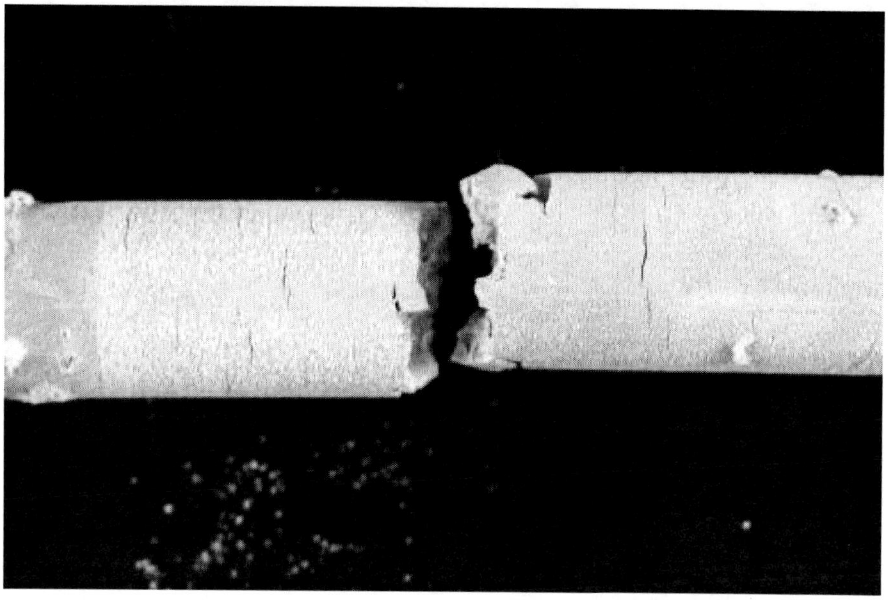

Figure 8. Uncoated SCC sample showing multiple secondary cracks near the main fracture surface

(400,000 cycles. LPDC AM50 alloy)

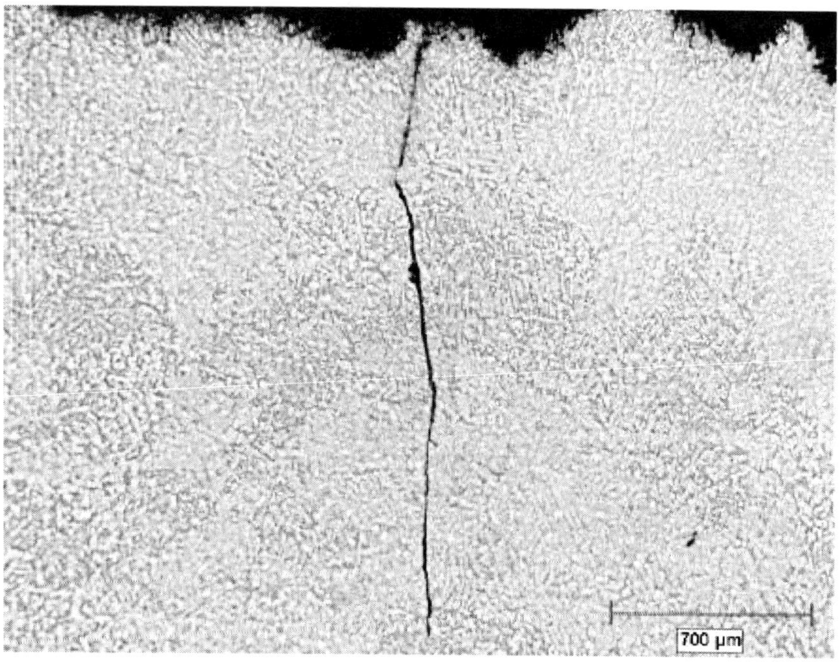

Figure 9 Cross-sectional view of the crack in Mg alloy after SCC testing (2,400,000 cycles. LPDC AM50 alloy treated with Alodine 5200 then coated with a powder coat.)

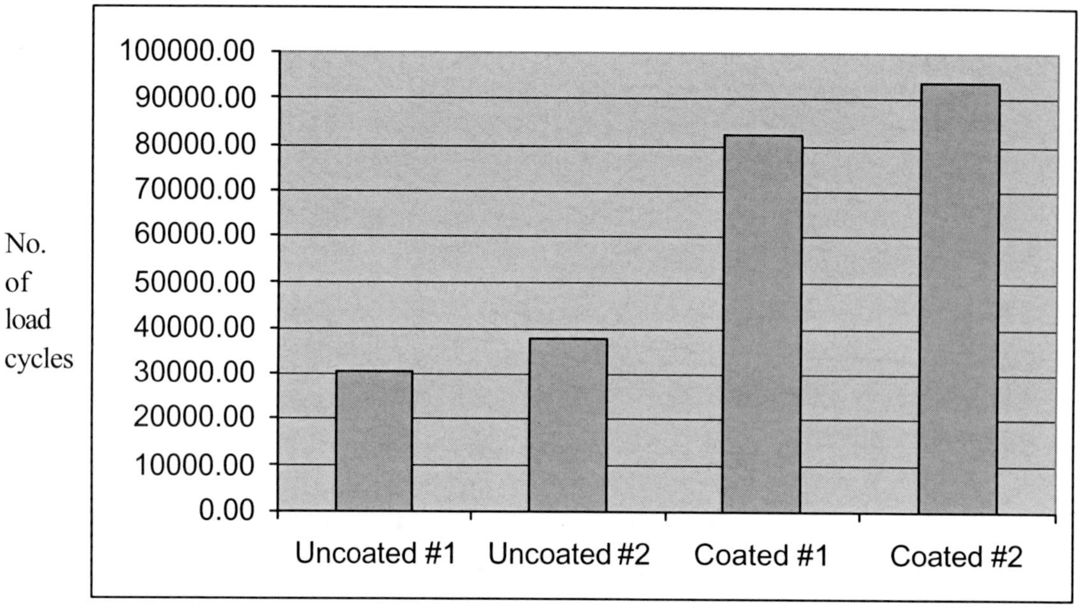

Figure 10 Comparison of the test failure life for coated and uncoated AE44 samples

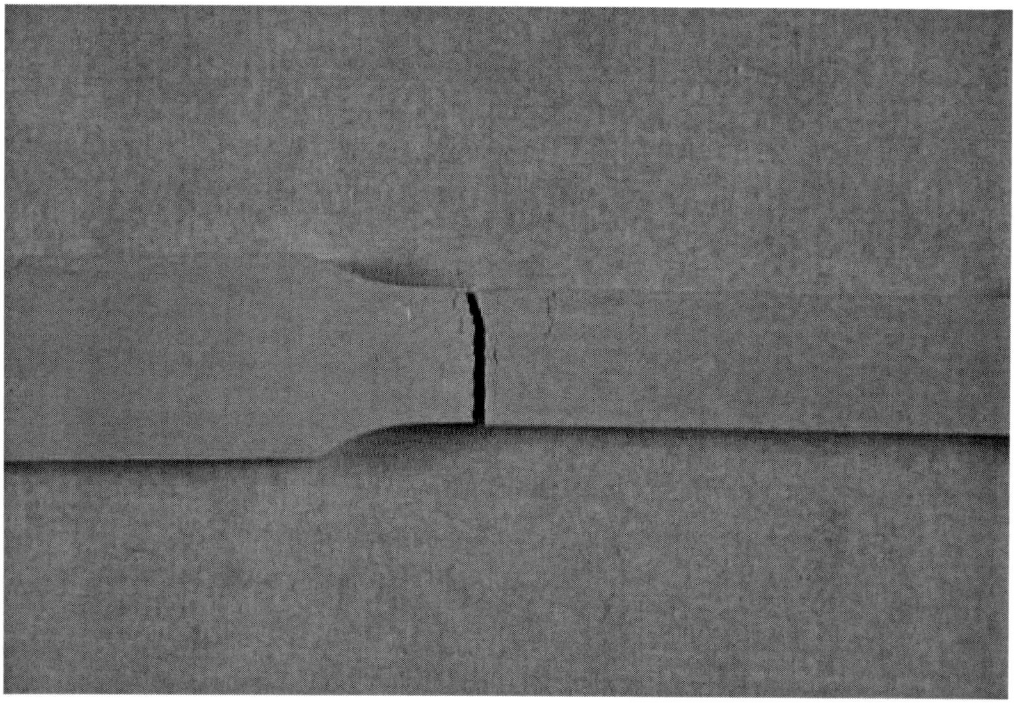

Figure 11. An AE44 ruptured sample showing the main crack and secondary cracks (Sample treated with Alodine 5200 and then coated with a powder coat)

2005-01-0734

Wrought Magnesium Alloys and Manufacturing Processes for Automotive Applications

Alan A. Luo
General Motors Research & Development Center

ABSTRACT

In this paper, the mechanical properties, structural performance and mass saving potential of wrought magnesium alloys are compared to several major automotive materials: mild steel, advanced high-strength steel, cast and wrought aluminum, cast magnesium, plastics and fiber-reinforced composites. Manufacturing processes including welding and joining of magnesium extrusions and sheet products are critically reviewed. The current and potential applications of wrought magnesium alloys in automotive interior, body and chassis areas are discussed. Technical challenges and research opportunities for these applications are identified.

INTRODUCTION

Magnesium components are being increasingly used by major automotive companies including General Motors (GM), Ford, Volkswagen and Toyota [1-7]. Current automotive magnesium applications are mostly die castings, such as instrument panel beam, transfer case, steering components and radiator support. The average magnesium content in a typical 2003 model year family sedan built in North America was only about 0.3% of the total vehicle weight, which compares to about 10% for aluminum and 8% for polymers and polymer-based composites [8].

While high-pressure die casting is the dominant process for current magnesium automotive applications, wrought magnesium alloys and their manufacturing processes are receiving increasing attention from academia and industries. As magnesium is expanding from interior components to more critical applications in chassis and body areas, there is a great need for developing wrought magnesium products and manufacturing processes to provide improved mechanical and physical properties, crash performance and corrosion resistance.

In this paper, the mechanical properties, structural performance and mass saving potential of cast and wrought magnesium components are compared with those of several major automotive materials: mild steel, advanced high strength steel (AHSS), aluminum, polymers and polymer composites. Manufacturing processes including welding and joining of magnesium castings and wrought products are critically reviewed. The current and potential applications of magnesium extrusions and sheet products in automotive interior, body and chassis areas are reviewed, and the technical challenges are also discussed.

MATERIAL PROPERTIES

Table 1 summarizes the mechanical and physical properties of typical cast and wrought magnesium alloys in comparison with other materials for automotive applications [9-12]. Being the lightest structural metal, magnesium has a density less than one fourth that of ferrous alloys (cast iron, mild steel and advanced high-strength steel). Magnesium offers similar mechanical and physical properties as aluminum alloys, but with about one-third the mass saving. For extrusion alloys, magnesium alloy AZ80 provides comparable tensile strength as aluminum alloy 6061, but is less ductile.

Magnesium sheet alloy, such as AZ31, offers similar strength but slightly less ductility compared to the commonly used 5xxx (e.g., 5182) or 6xxx (e.g., 6111) series aluminum sheet alloys in finished products. However, magnesium wrought alloys have much less formability than steel or aluminum at room temperature due to the hcp (hexagonal close-packed) crystal structure of magnesium.

Although slightly heavier than polymers or polymer composites, magnesium components are generally stiffer due to their higher elastic modulus. PC/ABS, a plastic blend of polycarbonate and acrylonitrile-butadiene-styrene, has a modulus of about 1/20 that of magnesium. This material has been used in a limited

Table 1. Comparison of mechanical and physical properties of various automotive materials [1, 7-12]

Material	Cast Mg		Wrought Mg		Cast Iron	Steel		Cast Al		Wrought Al		Polymers (PC/ABS)	GFRP[2] (glass/polyester)		CFRP[3] (carbon/epoxy)	
Alloy/Grade	AZ91	AM50	AZ80-T5	AZ31-H24	Class 40	Mild steel Grade 4	AHSS[1] DP340/600	380	A356-T6	6061-T6	5182-H24/ 6111-T8X[5]	Dow Pulse 2000	Structural (50% uniaxial)	Exterior (27%)	Structural (58% uniaxial)	Exterior (60%)
Process/ Product	die cast	die cast	extrusion	sheet	sand cast	sheet	sheet	die cast	P/M[4] cast	extrusion	sheet	injection molding	liquid molding	compression molding	liquid/compression molding	autoclave molding
Density (d, g/cm³)	1.81	1.77	1.80	1.77	7.15	7.80	7.80	2.68	2.76	2.70	2.70	1.13	2.0	1.6	1.5	1.5
Elastic Modulus (E, GPa)	45	45	45	45	100	210	210	71	72	69	70	2.3	48	9	189	56
Yield Strength (YS, MPa)	160	125	275	220	N/A	180	340	159	186	275	235/230	53				
Ultimate Tensile Strength (S_t, MPa)	240	210	380	290	293	320	600	324	262	310	310/320	55	1240	160	1050	712
Elongation (e_f, %)	3	10	7	15	0	45	23	3	5	12	8/20	5 at yield and 125 at break	< 1	2	< 1	< 1
Fatigue Strength (S_f, MPa)	85	85	180	120	128	125	228	138	90	95	120/186		N/A	N/A	N/A	N/A
Thermal Cond. (l, W/m.K)	51	65	78	77	41	46		96	159	167	123		0.6	0.3	0.5	0.5
Thermal Exp. Coeff. (d, μm/m.K)	26	26	26	26	10.5	11.70		22	21.5	23.6	24.1/23.4	74	12	14	2	4
Melting Temp. (T_m, °C)	598	620	610	630	1175	1515		595	615	652	638/585	143 (softening temp.)	130-160 (molding temp.)		175 (max. service temp.)	

1. AHSS: advanced high strength steel;
2. GFRP: glass fiber reinforced polymer
3. CFRP: carbon fiber reinforced polymer;
4. P/M: permanent mold;
5. T8X: simulated paint-bake (2% strain plus 30 min. at 177°C).

way in instrument panel beams where the design requirements are met through the geometry of the closed sections. Fiber reinforcements can be used to increase the modulus of polymers, as seen in many polymer-based composites such as the conventional glass fiber reinforced polymer (GFRP) and the advanced carbon fiber reinforced polymers (CFRP). Depending on applications, polymer composites can be classified into two groups:

- Structural composites (Structural-GFRP and Structural-CFRP) in which glass or carbon fibers are uniaxially orientated to provide unidirectional strength/stiffness for structural applications; and
- Exterior composites (Exterior-GFRP and Exterior-CFRP) in which glass or carbon fibers are properly aligned to provide "quasi-isotropic" strength/stiffness in planar directions but not in the thickness direction, for exterior panel applications.

Exterior-GFRP is used in many low/medium-volume body panel applications (e.g., Corvette and Saturn). While CFRP is much more expensive and is generally used in aerospace and super sports cars, a recent GM application of exterior-CFRP is the hood outer for the Commemorative Corvette Z06.

The properties of magnesium are isotropic compared to the directional properties of most polymer composites despite its lower modulus and strength. The much higher elongation compared to polymer composites makes magnesium alloys more attractive in applications requiring high ductility and crashworthiness. Also, polymers and composites are prone to creep, and thus not suitable for elevated temperature applications due to their low service temperatures (e.g., 143°C for PC/ABS and 175°C for CFRP). Forming and surface quality of these materials can also be a problem [9].

Other properties of magnesium, including the higher thermal expansion coefficients and lower thermal conductivity compared with aluminum, are also important in elevated temperature applications, and should be considered in designing magnesium components.

CORROSION AND CORROSION PROTECTION

Although the general corrosion resistance of high purity magnesium alloys is even better than A380 aluminum alloy in atmospheric or saline environments, galvanic corrosion remains a concern for design and applications engineers in developing magnesium components. The degree of galvanic corrosion attack on magnesium is strongly influenced by the compatibility of the second metal in contact with the magnesium parts, with aluminum alloy 5052 being the most compatible metal and steels the least compatible (Fig. 1 [13]). Proper joint design, combined with the use of non-conductive or compatible aluminum shims/washers, and Zn/Sn-coated bolts/nuts, is a very simple way to prevent galvanic corrosion when a magnesium part is attached to another metal. Such preventive measures have proved efficient in many magnesium applications including the 4WD (four wheel drive) transfer case in GM's and Ford's full-size trucks/SUVs which have been on the road for about 10 years.

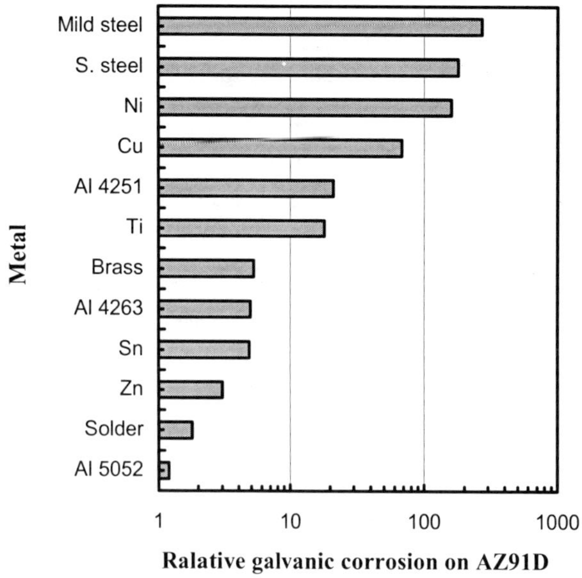

Fig. 1. Relative compatibility series of various metals coupled to magnesium. (Courtesy: Hydro Magnesium [13])

On the other hand, coatings including chemical conversation pre-treatment (ex. Alodine 5200), anodizing, E-coat and powder coat (polymer or epoxy) are also available for corrosion protection in magnesium applications, such as VW 3-Liter car's tailgate inner [14] and Ford F-150 truck's radiator support [15]. For coated magnesium parts, it is extremely important to avoid any coating damage during assembly and service, as this can cause locally accelerated galvanic corrosion, which makes the situation even worse than uncoated magnesium, due to the much larger area ratio of dissimilar metal to magnesium in the case of a damaged coating [16].

MANUFACTURING PROCESSES

EXTRUSION AND FORGING PROCESSES

Magnesium alloys can be warm or hot extruded in hydraulic presses to form bars, tubes, and a wide variety of profiles [17]. It is well known that a tubular section is significantly stiffer than a solid beam at the same mass. While hollow magnesium extrusions can be made with a

Table 2. Nominal compositions and typical room-temperature tensile properties of extruded magnesium alloy tubes [17-19]

Alloy	Temper	Composition, wt.%				Tensile Properties		
		Al	Zn	Mn	Zr	Yield Strength, MPa	Tensile Strength, MPa	Elongation, %
AZ31	F[1]	3.0	1.0	0.20	-	165	245	12
AZ61	F	6.5	1.0	0.15	-	165	280	14
AZ80	T5	8.0	0.6	0.30	-	275	380	7
ZK60	F	-	5.5	-	0.45	240	325	13
ZK60	T5[2]	-	5.5	-	0.45	268	330	12
AM50	*F*	*5.0*	-	*0.30*	-	*168*	*268*	*18*

[1] F signifies as extruded.
[2] T5 signifies artificially aged after extrusion.

mandrel and a drilled or pierced billet, it is generally preferable to use a bridge die where the metal stream is split into several branches and recombines before the die exit. Table 2 lists the nominal composition and typical room-temperature tensile properties of extruded magnesium alloy tubes [17-19]. Of the commercial extrusion alloys, AZ31 is most widely used in non-automotive applications. With higher aluminum contents, AZ61 and AZ80 offer higher strength than AZ31 alloy, but at much lower extrudability. The high-strength Zr-containing ZK60 was designed for applications in racing cars and bicycles, such as wheels and stems [18]. The extrusion speed for making ZK60A tubes is extremely low, rendering it uneconomical for automotive applications.

AM50 is a well-known die casting alloy with very high ductility (about 15% elongation) in the as-cast condition. Recent work at GM [19] succeeded in extruding this casting alloy into hollow tubes with excellent tensile properties as shown in Table 2. That work also examined the extrusion process parameters and compared the extrudability of several alloys. Fig. 2 shows that the maximum extrusion speeds for AZ61 and AM50 are significantly lower than that of AZ31 alloy, indicting higher costs in tube manufacturing. Table 2 and Fig. 2 indicate that AZ31 alloy offers the best combination of mechanical properties and extrudability among commercial magnesium alloys. However, the maximum extrusion speed of AZ31 alloy is only about a half of that of aluminum extrusion alloy 6063, which makes magnesium extrusions much more expensive due to the higher material and processing costs. Recently, a new magnesium wrought alloy, AM30 (Mg-3%Al-0.4%Mn), has been developed at GM [20]. Compared to AZ31 alloy, the new AM30 alloy has a 20% higher extrusion speed, 50% increase in room-temperature ductility with similar strength, and up to 30% ductility improvement at elevated temperatures. The alloy is currently being optimized for mechanical properties and formability.

Hydrostatic extrusion process, typically used for copper tubing fabrication, is a much faster extrusion process compared to the conventional direct extrusion. It was recently reported that seamless magnesium tubes were extruded using the hydrostatic process at speeds up to 100 m/min, due to the absence of friction between the billet and container since the billet is suspended in hydraulic oil [21]. Although the process is capable of extrusion ratios up to 700, the outer diameter of tubes produced by this process is limited to about 45 mm, even on a large 4,000-ton press [21].

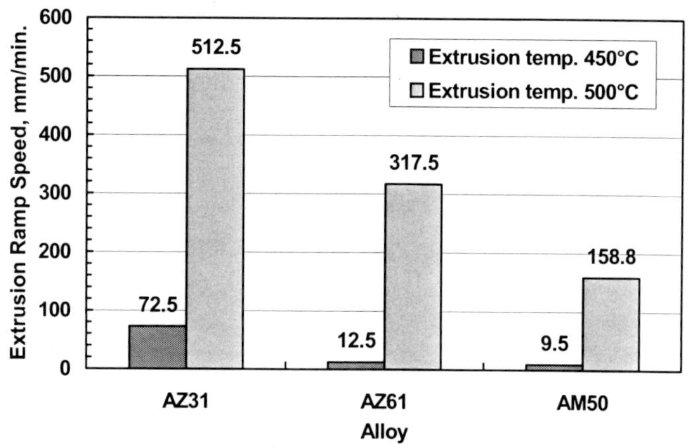

Fig. 2. Maximum extrusion speeds for various alloys: 25.4 x 25.4 mm square tube with 5 mm wall, extrusion ratio = 12.5.
(Courtesy: A.K. Sachdev [19]).

Commercial extrusion alloys listed in Table 2, AZ31, AZ61, AZ80 and ZK60, also can be forged into high-integrity components using hydraulic presses or slow-action mechanical presses. Forging is normally done within 55°C of the solidus temperature of the alloy. Corner radii of 1.6 mm, fillet radius of 4.8 mm, and panels or webs 3.2 mm thick can be achieved by forging. The draft angles required for extraction of the forgings from the dies can be held to 3° or less [17]. A recent successful forged part was a one-piece race car wheel developed Mazda using grain-refined AZ80 billets [54].

TUBE BENDING AND HYDROFORMING PROCESSES

Tube Hydroforming is a metal forming process that uses pressurized fluids such as water to make various perimeter shapes from tubes. Compared to stampings and castings, hydroformed tubular sections provide further mass savings for structural components. Hydroformed steel and aluminum tubes are currently used in many structural applications including frame rail, engine cradle, radiator support and instrument panel beam [22, 23]. Hydroforming of magnesium tubes is being studied at a number of universities in Europe and North America.

Tube Bending is generally needed as a pre-form step for hydroforming automotive parts, and bendability of magnesium extrusions at room temperature is very limited [24]. Recently, a moderate temperature (150-200°C) bending process has been developed at GM for magnesium alloy extrusions [25]. A bend radius twice of the tube outer diameter (2D) was achieved on magnesium alloy extrusions at 150°C, Fig. 3 [25]. The mechanical properties and microstructure of magnesium alloys at elevated temperatures indicate that moderate temperature hydroforming and other forming processes at this temperature range (150-200°C) should be explored [26].

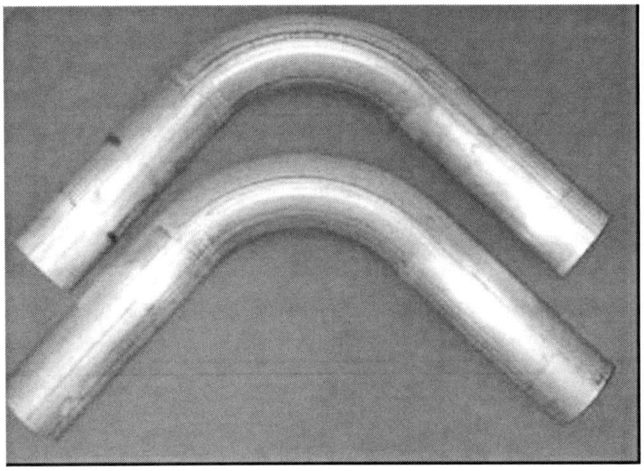

Fig. 3. Magnesium tubes bent (2D/90°) at 150°C.

SHEET PRODUCTION AND FORMING PROCESSES

Direct-chill (DC) casting process is generally used to cast magnesium slabs (about 50 mm thick) which are then **hot-rolled** at 315-370°C to produce magnesium sheet and plates [17]. Unlike aluminum for which cold-roll is usually the final step in sheet production, warm finish roll is applied to magnesium sheet products. Large grains of 200-300 μm in the slab can be reduced by warm rolling to fine recrystallized grains of 7-22 μm size at a sheet gage of 1.3-2.6 mm [27, 28]. Recently, **twin-roll continuous casting (CC)** process has been investigated for the production of low-cost magnesium sheet. Pilot plants have been established in Germany, Austria and Australia [30-32], and production plants are being planned in China and Korea. In this process, molten magnesium is poured into a gap between two rolls to produce a continuous strip about 2.5-10 mm thick, which is then warm-rolled to a final gage. Currently, magnesium sheet using this process is available from Salzgitter (Germany) with a maximum width of 1850 mm and a minimum gage of 1.0 mm [34].

Warm stamping (200-400°C) of magnesium sheet has been used to make complex products for aerospace and luggage components, and the optimum temperature for warm stamping was reported as approximately 350°C [28]. Fig. 4 shows a warm-formed AZ31 sheet pan drawn at 350°C to a depth of 83 mm [29]. Recently, it was found that stamping could be preformed at considerably lower temperatures of 150-175°C [31, 32], similar to the optimum tube bending temperature reported by GM [25]. **Sheet hydroforming** process differs from the conventional stamping process in that the solid punch (upper die) or female (bottom) die is replaced with a forming medium. When the female die is replaced with a fluid or a pad of flexible polyurethane, it is also called **hydro-mechanical drawing** [35, 36]. Another variant, **active sheet hydroforming**, is to use forming medium instead of a solid punch to directly press the blank against a die contour [37]. The optimum conditions for hydro-mechanical drawing magnesium sheet were reported as the temperature of 180-220°C and the pressure of 600-800 bar [35]. Another study [37] recommended active sheet hydroforming at a temperature of 300°C and a strain rate of 0.002 s-1. Obviously, forming at lower temperatures would be more economical and yield much better dimensional accuracy due to less thermal contraction and distortion during cooling to room temperature.

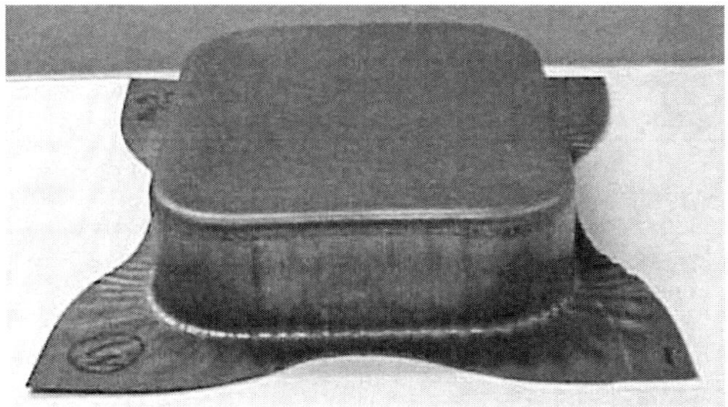

Fig. 4. Magnesium sheet stamped at 350°C.
(Courtesy: P.E. Krajewski [29])

Superplastic forming (SPF) is a forming process using one-sided tools (like sheet hydroforming) based on the superplasticity of many materials (aluminum, titanium, magnesium, etc.) at elevated temperatures and under controlled strain rates. SPF is often used to fabricate large and complex aluminum parts in the aerospace industry. Extensive R&D at GM on aluminum SPF has shortened the process cycle time to an acceptable level for automotive panels, with part consolidation offsetting costs of relative long forming cycles compared to the traditional stamping process. The high-volume production of this improved process, namely QPF (quick plastic forming), debuted on the 2004 Chevrolet Malibu Maxx aluminum liftgate [38]. Recently, it was reported that magnesium alloy AZ31B-H24 with an average grain size of 9 μm exhibited a total elongation of over 400% at temperatures above 450°C [28]. A door inner and a reinforcement panel were successfully formed by SPF at 475°C using this material, Fig. 6 [29].

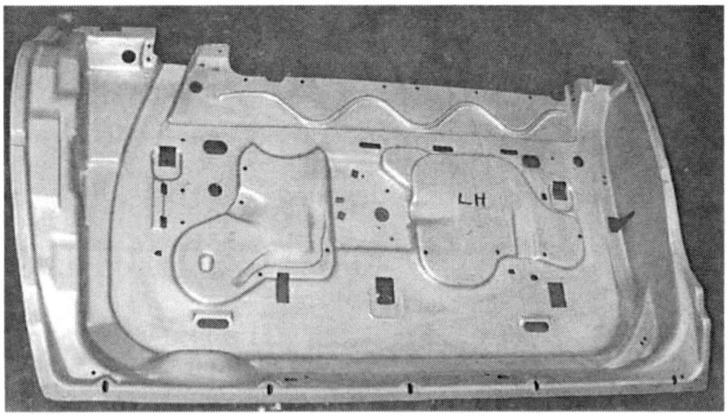

Fig. 5. Magnesium door inner superplastically formed at 475°C (Courtesy: P.E. Krajewski [29])

WELDING AND JOINING TECHNIQUES

Various welding processes can be used for joining magnesium to magnesium. Most magnesium alloys are readily fusion-welded with higher speeds than aluminum due to the lower thermal conductivity and latent heat of magnesium [17, 39-41]. While **gas metal arc welding** (**MIG** or **GMAW**) is used for joining magnesium sections ranging from 0.6 to 25 mm, **gas tungsten arc welding** (**TIG** or **GTAW**) is more suited for thin sections up to 12.7 mm [17]. It should be recognized that hot-shortness may produce cracks in welding magnesium alloys containing more than 1% zinc, which can often be overcome by using proper filler wires: ER AZ61A is preferred filler wire for welding wrought alloys containing aluminum, while ER AZ91A has been found to lower crack sensitivity in AZ and AM cast alloys [41]. Magnesium sheet and extrusions ranging from 0.5 to 3.3 mm can be joined by all types of **resistance welding**, including seam, projection, and flash, but the most common type is **spot welding** [17]. Higher speed welding of thin magnesium sheet can be achieved by **laser** or **electron beam** welding, due to its narrow heat affected zones and less weld distortion. Welding speeds of 2.5 to 9 m/min can be achieved with a 4 kW solid-state laser for welding AZ31 sheet 1.0 to 3.2 mm thick [32]. Non-vacuum electron beam welding of AZ31 alloy can reach as high as 12-15 m/min [32].

Fusion welding of magnesium components can be challenging due to the formation of brittle intermetallic phase ($Mg_{17}Al_{12}$) in the welds [39]. Solid-state welding techniques, such as **friction-stir welding** and **magnetic pulse welding** can be used to improve the weld quality of magnesium die castings [42-44]. While these solid-state welding techniques can potentially be used to join magnesium to dissimilar materials such as aluminum [42, 45], mechanical joining (**self-piercing rivets, clinching,** and **hemming**) and **adhesive bonding** are preferred for dissimilar material joining involving magnesium to aluminum or steel. Adhesive bonding of magnesium parts (with or without dissimilar materials) requires proper pre-treatment of the joint surfaces, which includes cleaning, etching and wet chemical passivation. Many adhesives including epoxy and polyurethane can be used as long as they are chemically stable and have aging stability [46].

STRUCTURAL PERFORMANCE AND MASS SAVING POTENTIAL

Materials selection for automotive structural applications is an extremely complex process in which component geometries, loading conditions, material properties, manufacturing processes and costs are all playing important roles [1, 10, 47]. As the bending mode is often the primary loading condition in many automotive structures such as instrument panel beams and frame rails, the following analyses on structural performance and mass saving potential of magnesium over mild steel (presently the dominant automotive material) are based on the bending stiffness and strength calculations.

SECTION THICKNESS AND MASS RATIOS

For a **panel** (plate) under bending loads, the minimum thickness (*t*) and mass (*m*) can be calculated using the "materials performance index" concept [47]. Designating steel and magnesium properties with subscripts *S* and *Mg*, the thickness ratios and mass ratios of components made of the two materials for an equal stiffness design may be expressed as:

$$t_{Mg}/t_S = (E_S/E_{Mg})^{1/3} \quad (1)$$

$$m_{Mg}/m_S = (d_{Mg}/d_S)(E_S/E_{Mg})^{1/3} \quad (2)$$

where E and d are the elastic modulus and density of the materials, respectively. Using the property data as shown in Table 1, the thickness and mass ratios of magnesium (AZ91 alloy) vs. a mild steel beam can be calculated:

$$t_{Mg}/t_S = 1.67 \quad (3)$$

$$m_{Mg}/m_S = 0.39 \quad (4)$$

Therefore, in order to achieve the same bending stiffness, a magnesium panel will be required to have 1.67 times the thickness of a steel one, with a mass saving of 61%. For bending strength-limited design (same bending strength at minimum mass), such ratios for AZ91 magnesium alloy vs. a mild steel become:

$$t_{Mg}/t_S = (YS_S/YS_{Mg})^{1/2} = 1.06 \quad (5)$$

$$m_{Mg}/m_S = (d_{Mg}/d_S)(YS_S/YS_{Mg})^{1/2} = 0.25 \quad (6)$$

where *YS* is the yield strength of the materials.

For a **beam** in an equal stiffness design, Eqs. (1) and (2) became:

Table 3. Thickness and mass ratios of various materials compared to mild steel for equal bending stiffness- and strength-limited design

Material		AHSS	Al (Cast)	Al (Wrought)	Mg (Cast)	Mg (Wrought)	PC/ABS	GFRP (Exterior)	CFRP (Exterior)
Equal Stiffness Panel	Thickness Ratio	1.00	1.44	1.45	1.67	1.67	4.50	2.86	1.55
	Mass Ratio	1.00	0.49	0.50	0.39	0.39	0.65	0.59	0.30
Equal Strength Panel	Thickness Ratio	0.73	1.06	0.81	1.06	0.81	1.84	1.06	0.50
	Mass Ratio	0.73	0.37	0.28	0.25	0.19	0.27	0.22	0.10
Equal Stiffness Beam	Thickness Ratio	1.00	1.72	1.74	2.16	2.16	9.56	4.83	1.94
	Mass Ratio	1.00	0.59	0.60	0.50	0.50	1.38	0.99	0.37
Equal Strength Beam	Thickness Ratio	0.65	1.09	0.75	1.08	0.75	2.26	1.08	0.40
	Mass Ratio	0.65	0.37	0.26	0.25	0.17	0.33	0.22	0.08

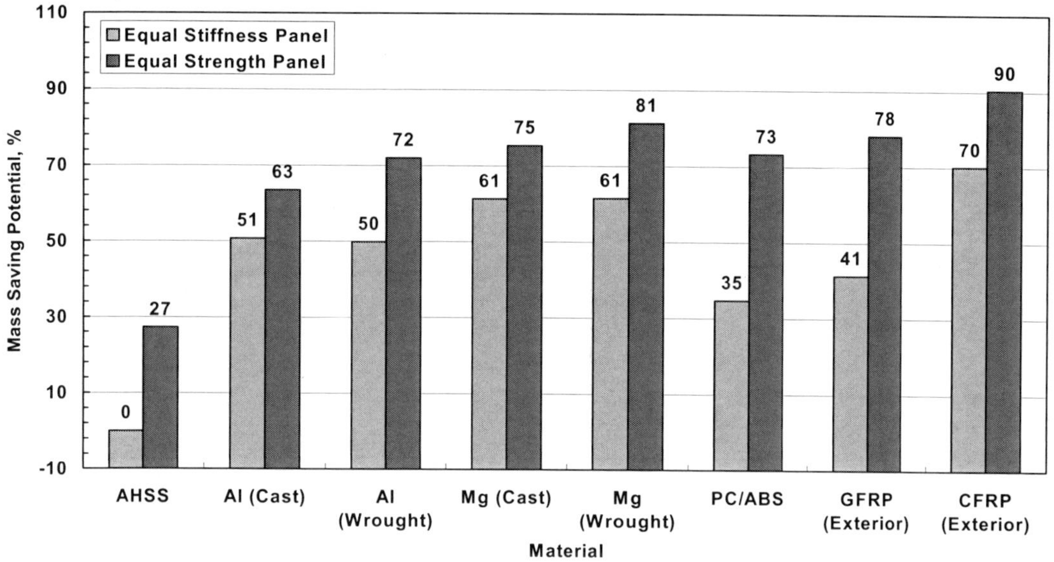

Fig. 6. Percentage mass savings of various materials vs. mild steel for designing a structural panel with equivalent bending stiffness or bending strength.

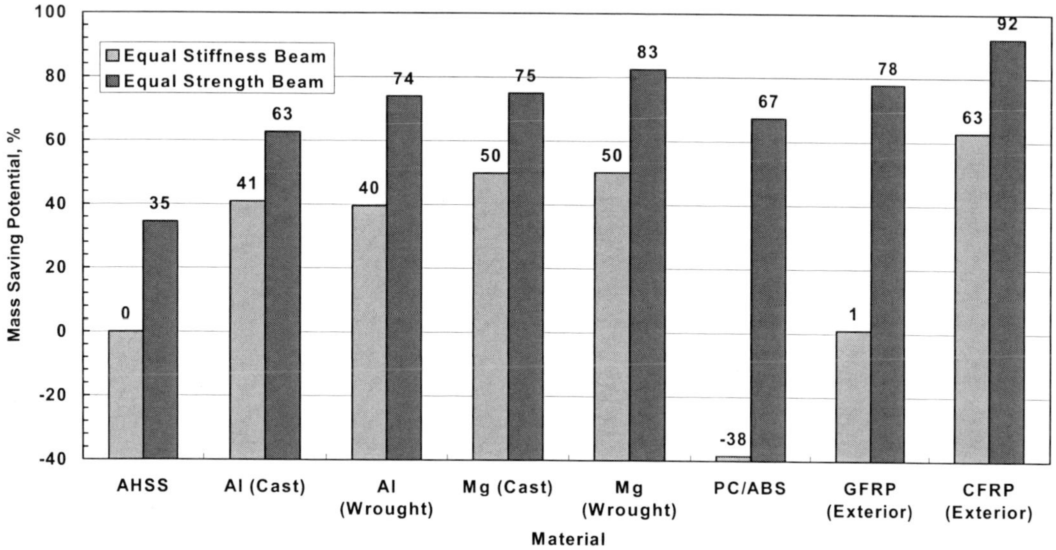

Fig. 7. Percentage mass savings of various materials vs. Mild steel for designing a structural beam with equivalent bending stiffness or bending strength.

$$t_{Mg}/t_S = (E_S/E_{Mg})^{1/2} \quad (7)$$

$$m_{Mg}/m_S = (d_{Mg}/d_S)(E_S/E_{Mg})^{1/2} \quad (8)$$

Similarly, for a solid beam in an equal strength design, Eqs. (5) and (6) became:

$$t_{Mg}/t_S = (YS_S/YS_{Mg})^{2/3} \quad (9)$$

$$m_{Mg}/m_S = (d_{Mg}/d_S)(YS_S/YS_{Mg})^{2/3} \quad (10)$$

Based on the above nomenclature and the property data in Table 1, Table 3 summarizes the thickness and mass ratios of various materials compared with mild steel, for solid panel and beam designs, respectively. For a generic comparison, the uniaxial properties of structural composites are not included in the analysis. Instead, the "quasi-isotropic" properties of exterior GFRP and CFRP composites are used to compare with the isotropic properties of other materials for mass saving analysis. Figs. 6 and 7 highlight the thickness ratios and the resultant percentage mass savings of the materials when replacing a mild steel component. These results show that magnesium alloys have higher mass saving potential when compared to AHSS, aluminum and polymers, in substituting mild steel structures for equal stiffness or strength. To overcome its much lower elastic modulus, a polymer part will have to be reinforced with fibers or a metal-polymer hybrid structure has to be used. Compared to GFRP composites, magnesium alloys have higher mass saving potential for equal stiffness and similar savings for equal strength. While CFRP has the highest mass saving potential, wrought magnesium alloys offer slightly less mass savings at a lower cost.

POTENTIAL AUTOMOTIVE APPLICATIONS

The current and potential automotive applications of magnesium castings have been reviewed by a number of authors [1-6]. While powertrain magnesium applications will remain mostly castings, Table 4 summarizes the potential application of wrought magnesium components in automotive interior, body and chassis areas.

INTERIOR

Since corrosion is of less concern in interior applications and many of those parts are not integrated in the vehicle structure, this area has seen the most magnesium applications, with the biggest growth in the instrument panels (IP) and steering structures. The first magnesium IP beam was die cast by GM in 1961 with a mass saving of 4 kg over the same part cast in zinc [6]. The design and die casting of magnesium IP beams have advanced dramatically in the last decade. For example, the current IPs normally have a thickness of 2-2.5 mm (compared to 4-5 mm for the earlier IP beams) with more part consolidation and mass savings. However, the use of cast magnesium IP beams is recently facing strong competition. IP beams made of aluminum extrusions are used by Mercedes in Europe. IP designs using bent steel tubes (with or without hydroforming) are slightly heavier than magnesium die casting, but significantly less expensive. While continuous improvement in magnesium casting design and thin-wall casting technology can bring incremental mass and cost savings, tubular IP designs using magnesium extrusions should also be explored.

The use of magnesium seat structures began in Germany, where Mercedes used magnesium die castings in its integrated seat structure with a three-point safety belt in the SL Roadster [48]. The latest example in North America is the "Stow-n-Go" seating and storage system for 2005 Chrysler and Dodge minivans, where the folding mechanisms require lightweight for easy operation, thus some aluminum is used in the second-row seats and the back frame of the third-row folding seats is magnesium casting [49]. Compared to magnesium castings, magnesium extrusions provide further mass saving opportunities for seat applications. Research is needed to reduce the cost of magnesium extrusions.

Table 4. Potential automotive application of wrought magnesium components

System	Component	Note
Interior	Instrument panel	Extrusion
	Seat components	Extrusion
	Trim plate	Sheet
Body	Door inner	Sheet
	Tailgate/liftgate inner	Sheet
	Door/roof frame	Extrusion
	Sunroof panel	Sheet
	Bumper beam	Extrusion
	Radiator support	Extrusion
	Shotgun	Sheet/extrusion
	A and B pillar	Sheet
	Decklid/hood inner	Sheet
	Hood outer/fender	Sheet
	Decklid/door outer	Sheet
	Dash panel	Sheet
	Frame rail	Extrusion
Chassis	Wheel	Forging
	Engine cradle	Extrusion
	Subframe	Extrusion
	Control arm	Forging

BODY

The use of magnesium in automotive body applications is limited but recently expanding. GM has been using a one-piece die cast roof frame since the C-5 Corvette introduction in 1997. Magnesium is also used in the 2004 Cadillac XLR roadster's retractable hard-top convertible roof and the roof top frame. The new Ford F-

150 has a coated magnesium casting for its radiator support. In Europe, Volkswagen and Mercedes have pioneered the use of thin-wall magnesium die castings in body panel applications. The one-piece die cast door inner for Mercedes S-Class Coupe is only 4.56 kg [6]. The key to manufacturing these thin-wall castings (approximately 1.5 mm) lies in casting design using proper radii and ribs for smooth die filling and to stiffen the parts. These thin-wall die castings, such as door inners, A and B pillars, can often offset the material cost penalty of magnesium over steel sheet metal construction due to part consolidation.

In body panel applications where bending stiffness is frequently the design limit, magnesium sheet metal can offer as much as 61% mass saving (see Fig. 6). However, elevated temperature forming of magnesium (warm stamping, warm sheet-hydroforming, or superplastic forming) and corrosion protection coatings further impose a cost penalty for magnesium sheet applications. In addition, the surface quality of the present magnesium sheet alloys needs improvement if magnesium is considered for outer panel applications. The development of new sheet alloys for near-room temperature forming is needed.

Prototype sheet magnesium hoods were made by GM as early as 1952 for the Buick LeSabre "Car of the Future" and later for a 1961 Chevrolet Corvette concept car [29]. Recently, a door inner was made from AZ31B-H24 sheet by GM using superplastic forming at 475°C [29]. More prototype parts including hoods and fenders were made in Germany using warm stamping. The prototype hoods (both inner and outer in magnesium sheet) made by Volkswagen showed similar crush test results to the aluminum design [3]. Despite these initial successes, current magnesium sheet and manufacturing processes remain too costly for volume production. Commercial applications of magnesium sheet panels will likely happen in 5 to 10 years.

Similarly, magnesium extrusions are used to make prototype parts such as bumper beams and most parts of a spaceframe for VW 1-Liter Car [3]. However, the use of magnesium tubes/extrusions in frame rails and other body applications would require at least 5 years to meet all structural and cost targets to develop a supply base for high volume automotive production.

CHASSIS

Cast or forged magnesium wheels have been used in high-priced racing cars or high-performance roadsters including GM's Corvette. However, the relatively high cost and potential corrosion problems of magnesium wheels prevented their use in high volume vehicle production. Cast magnesium engine cradles have been the development target of a USCAR (United States Council for Automotive Research) project titled "Structural Cast Magnesium Development", and a 10.2 kg magnesium casting (35% mass saving over aluminum cradle) is considered for production in 2005 for the Z06 Corvette.

Aluminum tubes are presently used in high-volume cradle production such as the welded extrusion design for GM's mid-size cars (Chevrolet Impala and Pontiac Grand Prix) and the hydroformed tubular subframe for BMW 5 and 7 series. Tubular designs are generally more mass-efficient than solid castings. Cradles using magnesium tubes would offer significant mass savings. The development of low-cost, corrosion-resistant coatings and new magnesium alloys with improved fatigue and impact strength will also accelerate the further penetration of magnesium in chassis applications.

CONCLUSION

Magnesium alloys have higher mass saving potential when compared to advanced high-strength steel (AHSS), aluminum, polymers, and glass fiber reinforced polymers (GFRP) for equal stiffness or strength. While carbon fiber reinforced polymers (CFRP) has the highest mass saving potential, wrought magnesium alloys offer slightly less mass saving but at a lower cost.

Despite the highest potential in mass savings, wrought magnesium alloys are currently too expensive for high-volume production. More research is needed to reduce the costs of these materials. Continuous casting for magnesium sheet production and hydrostatic extrusion process present great opportunities for low-cost magnesium wrought products. Forming processes at moderate temperatures (150-200°C) and new wrought alloys with improved formability and mechanical properties are being developed for automotive applications.

ACKNOWLEDGMENTS

The author wishes to acknowledge his colleagues at GM, Mark Verbrugge, Anil Sachdev, Paul Krajewski, Bob Powell, Hamid Kia, C.S. Wang, Howard Cox, Peter Foss, Ravi Verma, Jim Chen, Dave Barris, Larry Ouimet, Richard Osborne, Dennis Henderson, Don Seay, Bob Kubiak, John Roby, Mike Bride, Charlie Bruggemann, for many stimulating discussions. The author is grateful to Paul Krajewski, Anil Sachdev and Hydro Magnesium for providing graphs used in Figures 1, 2, 4 and 5. Lisabeth Riopelle of Hydro Magnesium, David Greer of Meridian Magnesium Products and Tom Ruden of Lunt Manufacturing are acknowledged for many helpful discussions.

REFERENCES

1. A.A. Luo, "Magnesium: Current and Potential Automotive Applications", *JOM Journal of Metals*, 2002, 54, (2), pp. 42-48.
2. G.S. Cole, "Issues that Influence Magnesium's Use in the Automotive Industry", Magnesium Alloys 2003,

Materials Science Forum, Trans Tech, Switzerland, 2003, vols. 419-422, pp. 43-50.

3. S. Schumann and H. Friedrich, "Current and Future Use of Magnesium in the Automobile Industry", *ibid*, pp. 51-56.
4. A.A. Luo, "Recent Magnesium Alloy Development for Automotive Powertrain Applications", *ibid*, pp. 67-66.
5. T. Kaneko and M. Suzuki, "Automotive Applications of Magnesium Alloys", *ibid*, pp. 67-72.
6. L. Riopelle, "Magnesium Applications", International Magnesium Association (IMA) Annual Magnesium in Automotive Seminar, Livonia, MI, April 20, 2004.
7. A.I. Taub, Presentation at SAE (Society of Automotive Engineers) World Congress, March 6, 2003, Detroit, MI.
8. American Metal Market, Metal Statistics, www.AMM.com, May 2004.
9. ASM, Metals Handbook, Desk Edition (ASM International, Materials Park, OH, 1998).
10. M.W. Verbrugge, Presentation at SAE World Congress, March 10, 2004, Detroit, MI.
11. Hydro Magnesium, "Data Sheet: Die Cast Magnesium Alloys", Hydro Magnesium, Porsgrunn, Norway, November 1996.
12. H. Kia, C.S. Wang and P.H. Foss, GM R&D Center, private communication, July 2004.
13. Hydro Magnesium, "Corrosion and Finishing of Magnesium Alloys", Hydro Magnesium, Porsgrunn, Norway, November 1997.
14. H. Schreckenberger, M. Papke and S. Eisenberg, "The Magnesium Hatchback of the 3-Liter Car: Processing and Corrosion Protection", SAE Technical Paper 2000-01-1123, SAE International, Warrendale, PA, 2001.
15. J.S. Balzer, P.K. Dellock, M.H. Maj, G.S. Cole, D. Reed, T. Davis, T. Lawson and G. Simonds, "Structural Magnesium Front End Support", SAE Technical Paper 2003-01-0185, SAE International, Warrendale, PA, 2003.
16. G.G. Wang, K. Stewart, R. Berkmortel and J.I. Skar, "Corrosion Protection for External Magnesium Automotive Components", SAE Technical Paper No. 2001-01-0421, SAE, Warrendale, PA, 2001.
17. ASM Specialty Handbook, *Magnesium and Magnesium Alloys*, eds., M.M. Avedesian and H. Baker, ASM International, Materials Park, OH, 1999.
18. *Timminco Magnesium Wrought Products*, Timminco Corporation Brochure, Aurora, CO, 1998.
19. B.L. Tiwari, A.K. Sachdev and J.L. Bommarito, "Magnesium Extrusions: Alloy and Process Development", GM R&D unpublished research.
20. A.A. Luo and A.K. Sachdev, "Magnesium Wrought Alloy Having Improved Extrudability and Formability", US Patent Application, March 4, 2004.
21. K. Savage and J.F. King, "Hydrostatic Extrusion of Magnesium", in *Magnesium Alloys and Their Applications*, ed. K.U. Kainer, Wiley-VCH, Weinheim, Germany, 2000, pp. 609-614.
22. C.J. Bruggemann, "Hydroforming of Structure Parts for Personal Cars", presented at International Conference on Hydroforming, Fellbach/Stuttgart, Germany, October 12-13, 1999, pp. 353-371.
23. A.A. Luo and A.K. Sachdev, "Strain Development and Microstructure Evolution in Aluminum Tube Bending and Hydroforming", in *Aluminum 2002*, eds., S. Das and M.H. Skillingbreg, TMS, Warrendale, PA, 2002, pp. 123-131.
24. J. Becker, G. Fischer and K. Schemme, "Ligh Weight Construction using Extruded and Forged Semi-Finished Products Made of Magnesium Alloys", in *Magnesium Alloys and Their Applications*, eds. B.L. Mordike and K.U. Kainer, Werkstoff-Informationsgesellschaft Frankfurt, Germany, 1998, pp. 15-29.
25. A.A. Luo and A.K. Sachdev, "Development of A Moderate Temperature Bending Process For Magnesium Alloy Extrusions", in Proc. of International Conference on Magnesium, Sept. 20-23, 2004, Beijing, China.
26. A.A. Luo and A.K. Sachdev, "Mechanical Properties and Microstructure of AZ31 Magnesium Alloy Tubes", in *Magnesium Technology 2004*, ed. A.A. Luo, TMS, Warrendale, PA, 2004, pp. 79-85.
27. J. Enss, T. Everetz, T.Reier and P. Juchmann, "Properties of Magnesium Rolled Products", in *Magnesium Alloys and Their Applications*, ed. K.U. Kainer, Wiley-VCH, Weinheim, Germany, 2000, pp. 591-595.
28. P.E. Krajewski, "Elevated Temperature Forming of Sheet Magnesium Alloys", in *Magnesium Technology 2002*, ed. H.I. Kaplan, TMS, Warrendale, PA, 2002, pp. 175-179.
29. P.E. Krajewski, "Elevated Temperature Behavior of Sheet Magnesium Alloys", SAE Technical Paper 2001-01-3104, SAE International, Warrendale, PA, 2001.
30. H. Palkowski and L. Wondraczek, "Thin Slab Casting as a New Possibility for Economic Magnesium Sheet Production", in *Magnesium – Proceedings of the 6th International Conference Magnesium Alloys and Their Applications*, ed. K.U. Kainer, Wiley-VCH, Weinheim, Germany, 2003, pp. 774-782.
31. A. Moll, M. Mekkaoui, S. Schumann and H. Friedrich, "Application of Mg Sheets in Car Body Structures", *ibid*, pp. 935-942.
32. S. Braunig, M. During, H. Hartmann and B. Viehweger, "Magnesium Sheets for Industrial Applications", *ibid*, pp. 955-961.

33. V. Kaese, L. Greve, S. Juttner, M. Goede, S. Schumann, H. Friedrich, W. Holl and W. Ritter, "Approaches to Use Magnesium as Structural Material in Car Body", *ibid*, pp. 949-954.
34. P. Juchmann and S. Wolff, "New Perspectives with Magnesium Sheet", *ibid*, pp. 1006-1012.
35. R. Neugebauer and M. Seifert, "Results of Tempered Hydroforming of Magnesium Sheets with Hydroforming Fluids", *ibid*, pp. 306-312.
36. G. Kurz, "Heated Hydro-Mechanical Deep Drawing of Magnesium Sheet Metal", in *Magnesium Technology 2004*, ed. A.A. Luo, TMS, Warrendale, PA, 2004, pp. 67-71.
37. K. Siegert and S. Jaeger, "Pneumatic Bulging of Magnesium AZ31 Sheet Metal at Elevated Temperatures", *ibid*, pp. 87-90.
38. General Motors Coropration Press Release, "Aluminum Forming Enables Innovative Manufacturing", January 27, 2004.
39. A. Stern, A. Munitz and G. Kohn, "Application of Welding Technologies for Joining of Mg Alloys: Microstructure and Mechanical Properties", in *Magnesium Technology 2003*, ed. H.I. Kaplan, TMS, Warrendale, PA, 2003, pp. 163-168.
40. S. Lathabai, K.J. Barton, D. Harris, P.G. Lloyd, D.M. Viano and A. McLean, "Welding and Weldability of AZ31B by Gas Tungsten Arc and Laser Beam Welding Processes", *ibid*, pp. 157-162.
41. K.G. Watkins, "Laser Welding of Magnesium Alloys", *ibid*, pp. 153-156.
42. R. Johnson, P. Threadgill, "Friction Stir Welding of Magnesium Alloys", *ibid*, pp. 147-152.
43. N. Li, T.-Y. Pan, R. Cooper and D.Q. Houston, "Friction Stir Welding of Magnesium AM60 Alloy", in *Magnesium Technology 2004*, ed. A.A. Luo, TMS, Warrendale, PA, 2004, pp. 19-23.
44. J.I. Skar, H. Gjestland, L.D. Oosterkamp, D.L. Albright, "Friction Stir Welding of Magnesium Die Castings", *ibid*, pp. 25-30.
45. A.C. Somasekharan and L.E. Murr, "Fundamental Studies of the Friction Stir Welding of Magnesium Alloys to 6061-T6 Aluminum", *ibid*, pp. 31-36.
46. L. Budde, J. Bischoff and T. Widder, "Low-Temperature Joining of Magnesium-Materials in Vehicle Constructions", ", in *Magnesium Alloys and Their Applications*, eds. B.L. Mordike and K.U. Kainer, Werkstoff-Informationsgesellschaft Frankfurt, Germany, 1998, pp. 613-618.
47. M.F. Ashby, "Performance Indices", ASM Handbook, Vol. 20, ASM International, Materials Park, OH, 1998, pp. 281-290.
48. A. Hector and W. Heiss, "Magnesium Die-Castings as Structural Members in the Integral Seat of the New Mercedes-Benz Roadster", SAE Technical Paper No. 900798, SAE, Warrendale, PA, 1990.
49. D. Alexander, "Intier Seats are Customer Driven", *Automotive Engineering International*, May 2004, pp. 22-24.
50. S. Schumann and F. Friedrich, "The Use of Magnesium in Cars - Today and in Future", in *Magnesium Alloys and Their Applications*, eds. B.L. Mordike and K.U. Kainer, Werkstoff-Informationsgesellschaft Frankfurt, Germany, 1998, pp. 3-13.
51. BMW, "BMW Pushes Lightweight Technology with Magnesium", www.GermanCarFans.com, July 21, 2003.
52. E. Baril, P. Labelle and A. Fischersworring-Bunk, "AJ(Mg-Al-Sr) Alloy System Used for New Engine Block", SAE Technical Paper No. 2004-01-0659, SAE, Warrendale, PA, 2004.
53. B.R. Powell, L.J. Ouimet, J.E. Allison, J.A. Hines, R.S. Beals, L. Kopka, and P.P. Ried, "The Magnesium Powertrain Cast Components Project: Part I - Accomplishments of Phase I and the Objectives and Plans for the Magnesium Engine in Phase II", in *Magnesium Technology 2004*, ed., A.A. Luo, TMS, Warrendale, PA, 2004, pp. 3-10.
54. M. Fujita, N. Sakate, S. Hirahara and Y. Yamamoto, "Development of Magnesium Forged Wheel", SAE Technical Paper No. 950422, SAE, Warrendale, PA, 1995.

CONTACT

The author can be contacted at alan.luo@gm.com

2004-01-0128

Energy Absorption Capacity for HPDC Components

Cato Dørum, Odd Sture Hopperstad and Magnus Langseth
Structural Impact Laboratory (SIMLab), Department of Structural Engineering, Norwegian University of Science and Technology

Odd-Geir Lademo
SINTEF Materials Technology

Stian Sannes
Norsk Hydro ASA

ABSTRACT

The long-term objective of this work is to develop design and modeling tools that allow the structural behavior of thin-walled cast components to be predicted when subjected to static and dynamic loads such as in crash situations. Here, the energy absorption potential of High Pressure Die Cast components made of magnesium alloys AM20, AM50, AM60, AZ91 and the aluminum alloy AlSi7Mg is investigated using a shear-bolt principle. For the AM60 alloy, single plates cast with different thickness have been tested in order to investigate the effect of plate thickness on the shear-bolt mechanism. It is found that this deformation principle gives an approximately constant mean force during the deformation process. The behavior seems to be very robust, especially for the magnesium alloys. A simple empirical model for prediction of the mean shearing force as a function of plate thickness and bolt diameter is proposed.

INTRODUCTION

Growing concerns for economy, environment and functionality have led to increased use of light-metals in the load carrying structure and safety components of cars. With High Pressure Die Casting (HPDC) of magnesium and aluminum alloys, components with very complex, thin-walled geometry, like instrument panels, A and B pillars and front end structures, can be cast with a high production rate. The challenge with HPDC is to optimize the process parameters with respect to the part design and the solidification characteristics of the alloy in order to obtain a sound casting without casting defects.

In order to achieve a robust design of HPDC load carrying structures, it is vital to control the defects present in the structure, and the influence the defects have on the structural behavior.

Figure 1: Generic profile geometry: width of flange = 80 mm, width of outer stiffeners = 10 mm, depth = 40 mm, thickness = 2.5 mm, length = 400 mm.

In the current study, generic u-shaped structural HPDC components, shown in Figure 1, made out of magnesium and aluminum alloys have been investigated. The components were cast at the Norsk Hydro Research Centre in Porsgrunn, Norway, with a Bühler SC42D 420-ton cold chamber die casting machine. Double components were manufactured by joining the single components with screws.

Experimental results from axial crushing and 3-point bending [1] show that the energy absorption capacity can be restricted by insufficient structural ductility. This can limit the use of HPDC components in safety applications that is required to absorb large amounts of energy. In addition, the ultimate load capacity and the maximum deformation are in certain deformation modes restricted by the local ductility of the component, e.g. a tensile failure when subjected to 3-point bending. With non-optimized process parameters, the casting process can lead to a large scatter in structural behavior. Consequently, there exists a need for a robust design that minimizes the effect of variations in process parameters.

To open up for the use of HPDC components in applications where extensive energy absorption is required, a new energy absorption mechanism for HPDC components has been studied, namely bolt shearing. In the automotive industry, this generic mechanism is already in use in extruded crash-boxes, but has not yet been utilized with respect to cast components.

SHEAR-BOLT PRINCIPLE

Purpose-built rigs are used to perform shear-bolt tests on components and single plates. In the experiments, holes are cut in the cast component before the component is inserted into the test rig, with circular bolts made of high strength steel then put through the holes. The bolts are then forced to shear through the component/plate and the force and displacement are measured.

COMPONENT TESTS

The component tests were performed using two bolts, giving four crushing/shearing zones. Components of magnesium alloy AM20, AM50, AM60, AZ91 and aluminum alloy AlSi7Mg were tested. The different magnesium alloys cover a range in uniaxial tensile mechanical properties from low strength/high ductility (AM20) to high strength/low ductility (AZ91) [2]. When loaded, a very complex stress situation arises in front of the bolts and it is clear that the sustainable force is a result of the interaction of several physical phenomena. In order to avoid local buckling of the profile, the maximum force in this shearing process must be below the local and global buckling force of the profile.

The components were tested in an Instron 1126 machine under displacement control, with a rate of displacement of approximately 50 mm/min. A rigid plate pushed the component downwards, forcing the bolts to shear through the cast material. Force and displacement were logged using INSTRON Labtronic 8800 with a frequency of 5Hz. For the AM60 alloy, both single and double components were tested with bolt diameters D equal to 5, 10 and 15 mm in order to assess the influence of varying boundary conditions. For the AM20, AM50 and AZ91 alloy, two parallels of single components were tested.

Single and double AM60 components subjected to shear-bolt deformation are shown in Figure 2. Figure 3 and Figure 4 show the force-deformation characteristics of single AM60 components and AlSi7Mg components, respectively. For the case of shear-bolt deformation with a 5 mm bolt, it is found that the force-deformation characteristics using single and double components are approximately equal. For the larger bolt diameters, the average shearing force obtained using double components is larger than the force obtained using single components.

Figure 2: Single and double AM60 components after testing.

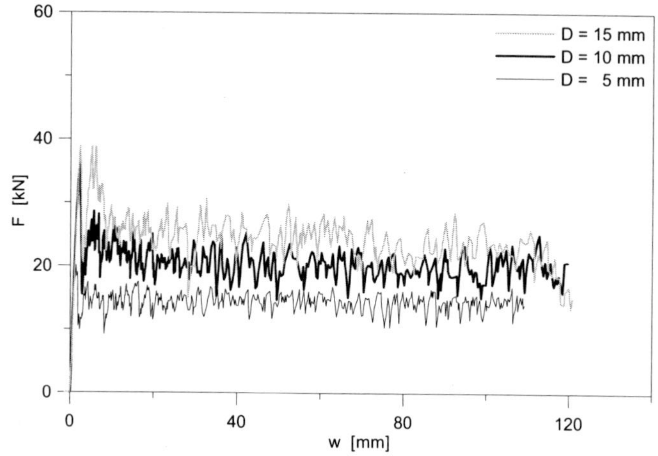

Figure 3: Force-deformation characteristics of single AM60 components subjected to shear-bolt deformation.

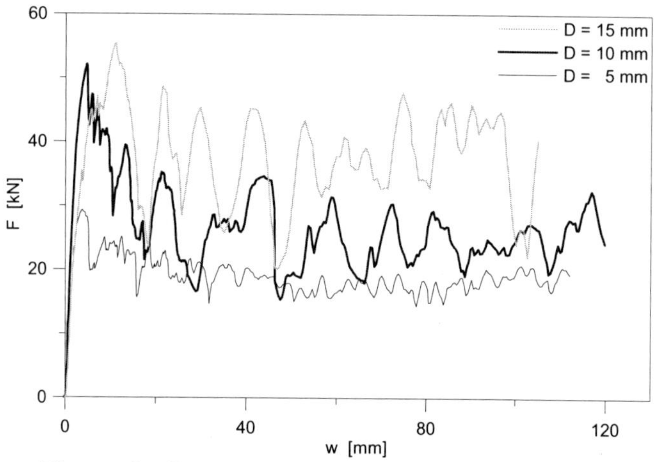

Figure 4: Force-deformation characteristics of single AlSi7Mg components subjected to shear-bolt deformation.

Comparing the force-deformation characteristics for the magnesium alloy AM60 and the aluminum alloy

AlSi7Mg, it is seen that AlSi7Mg give a higher load level. However, the variations in the shearing force are larger for the AlSi7Mg components, especially when using bolt diameters of 10 and 15 mm.

The mean shearing force for each of the magnesium alloys is presented in Figure 5. Higher values for F_m for bolt diameters of 10 and 15 mm are obtained for double components than for the single ones, probably due to the different boundary conditions. In the double components, the bolts are forced to shear through a plate with twice the width compared with the single components.

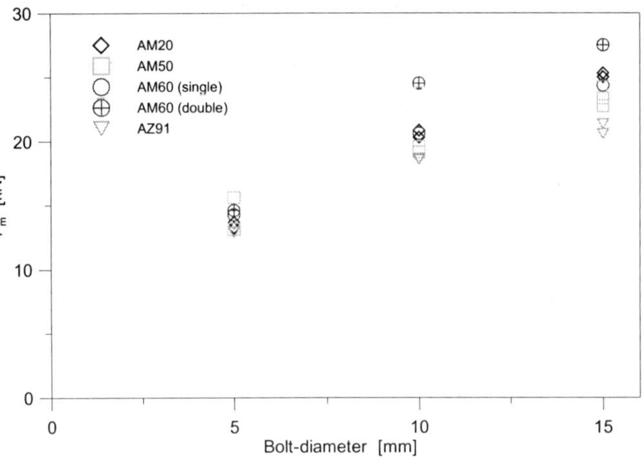

Figure 5: Mean shearing force for different HPDC magnesium alloys from component tests.

PLATE TESTS AND EMPIRICAL MODEL

Metallurgical investigations of the wide flange of the profiles [3] revealed that the micromechanical properties varied significantly along the length of the profile. In order to investigate the influence of this variation to the force-deformation characteristics found in the shear-bolt tests, single AM60 and AlSi7Mg plates were machined from the wide flange of the profiles and tested with bolt diameters of 5, 10 and 15 mm. The plates were of 240 mm length, 50 mm width and 2.5 mm thickness. In addition, single plates of 140 mm length, 50 mm width and 3.0 and 6.5 mm thickness were tested for the AM60 alloy to study the effect of thickness variations on the force-deformation characteristics.

The test procedure was the same as for the component tests. As seen from Figure 6, the force-deformation characteristics of the single AM60 plates machined from the components seem to be independent of the bolt's position along the length of the plate. Thus, the shear-bolt deformation is very robust with respect to micromechanical variations. From Figure 6 a), it is seen that the variations in the shearing force is of high frequency. This observed frequency is directly related to the deformation mechanism. As the bolt shear through the cast AM60 plate, small pieces of material are crushed under the moving bolt. Furthermore, by inspecting the deformed plate, it is seen that the number of material pieces crushed during the experiment equals the number of force peaks. Thus, each of the force peaks occurs when the plate material below the moving bolt reaches fracture.

Figure 7 shows the force-deformation characteristics for the single plates of the AlSi7Mg alloy. As for the components tests, much larger variations are observed in the shearing force compared to the shear bolt tests with AM60. The difference in force deformation characteristics is related to the difference in deformation and fracture mechanisms between the magnesium and the aluminum alloy. Compared to the component tests with two bolts and four shearing zones, the variations in shearing force seem to be larger and of higher frequency in the single plate tests using only one bolt and one shearing zone. Thus, the response seems to be smoothened as the number of shearing zones is increased.

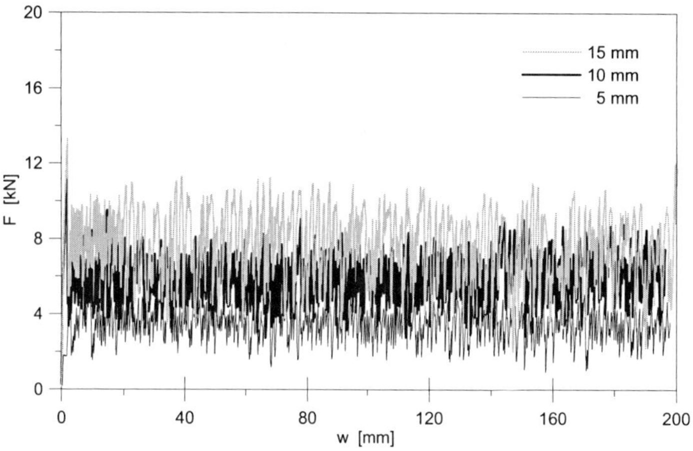

a)

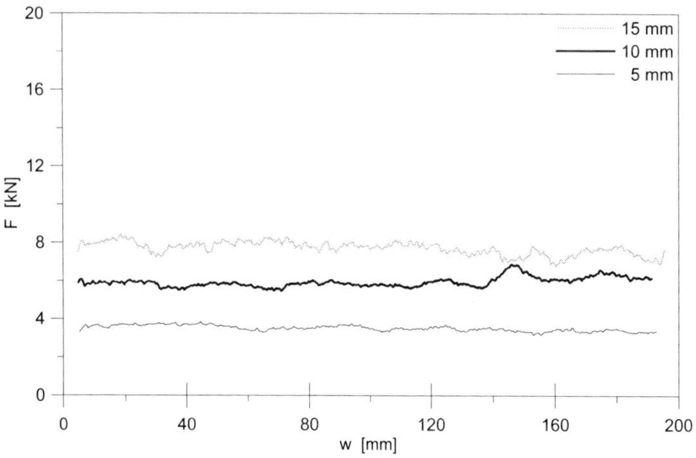

b)

Figure 6: Force-deformation characteristics for single AM60 plate of 2.5 mm thickness, a) raw data and b) smoothened (running average using 10mm averaging length).

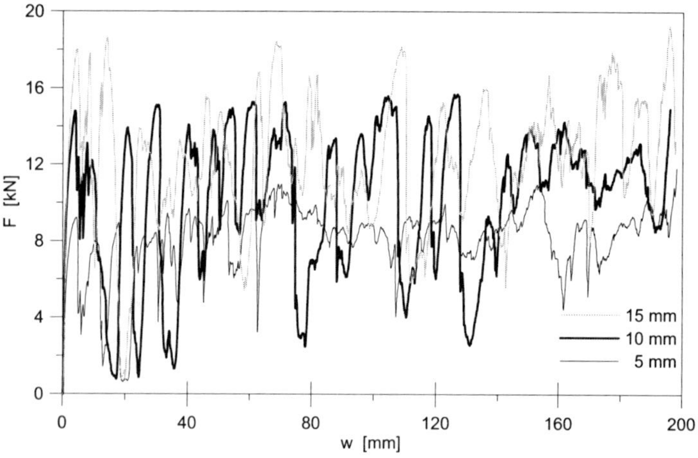

a)

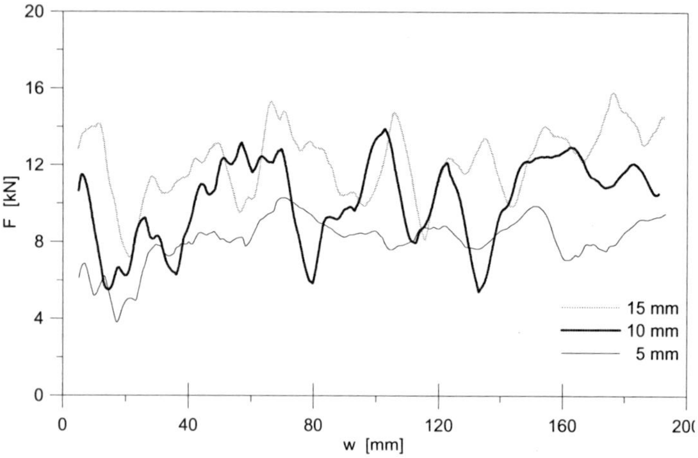

b)

Figure 7: Force-deformation characteristics for single AlSi7Mg plate of 2.5 mm thickness, a) raw data and b) smoothened (running average using 10mm averaging length).

As shown in Figure 8, the deformation/fracture path is much more regular and concentrated locally to the area close to the shear bolt for the AM60 alloy than for the AlSi7Mg alloy. It is assumed that this difference in deformation characteristics originates from the difference in crystallographic structure, that is hexagonal close packed (hcp) structure in the magnesium compared to the face centered cubic structure of the aluminum alloy.

By splitting the total absorbed energy by the shear-bolt deformation into plastic energy and fracture energy, a simple empirical model for the mean shearing force is proposed as a function of the bolt diameter.

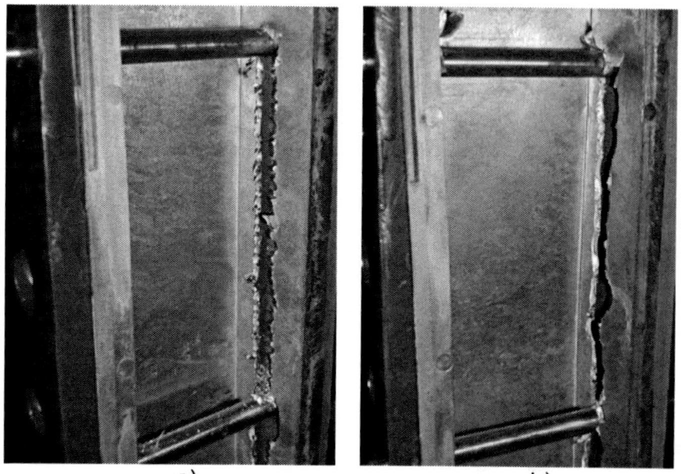

a) b)

Figure 8: Single components of a) AM60 and b) AlSi7Mg alloy subjected to shear bolt testing.

Accordingly, the total absorbed energy W reads

$$W = U_p + U_f = F_m \cdot w \tag{1}$$

where F_m is the mean force and w is the displacement of the bolt. The fracture energy U_f and the plastic energy U_p are then assumed on the form

$$U_p = C_1 \cdot V, \quad U_f = C_2 \cdot A \tag{2}$$

where C_1 and C_2 are fitting parameters, $A = 2 \cdot t \cdot w$ is the area of the shear zone, and $V = t \cdot w \cdot D$ is the volume of the material that is plastically deformed during the process. It follows that an empirical expression for the mean force per unit thickness is given by:

$$\frac{F_m}{t} = C_1 D + 2C_2 \tag{3}$$

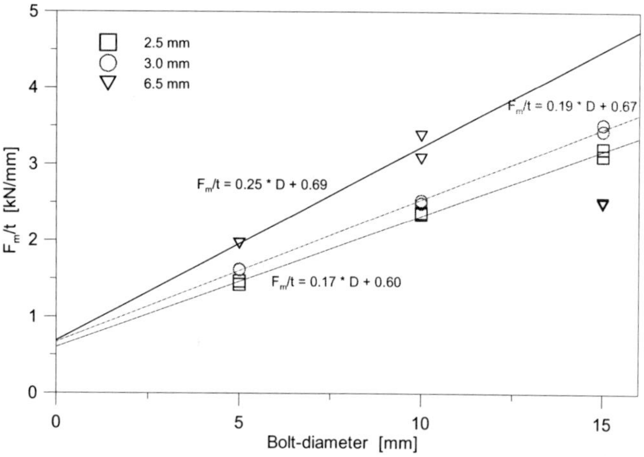

Figure 9: The mean shearing force per unit thickness for AM60 plates of 2.5, 3.0 and 6.5 mm thickness.

In Figure 9 and Figure 10, the proposed empirical model (3) is fitted to the experimental data using the method of least squares for the single plates of AM60 alloy and AlSi7Mg alloy, respectively. It is seen that the model fits well to the experimental results for the magnesium alloy AM60 with plate thickness equal to 2.5 and 3.0 mm. However, for the plate of 6.5 mm thickness, the model only fits for the smallest bolt diameters (5 and 10 mm). The constant C_2 related to the fracture energy as shown in (2) seems to be independent of the plate thickness, while the constant C_1 related to the plastic energy increases with increasing plate thickness. The simple empirical model also seems to fit to the experimental data for the AlSi7Mg. As for the AM60 alloy, C_1 increases with increasing plate thickness. However, the difference in the calculated C_2 values for the AlSi7Mg plates of thickness 2.5 and 3.0 mm is significantly larger than for the AM60 plates.

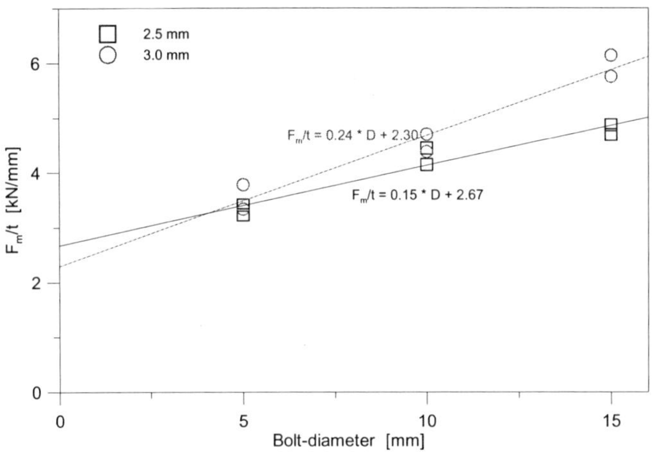

Figure 10: The mean shearing force per unit thickness for AlSi7Mg plates of 2.5 and 3.0 mm thickness.

SHEAR-BOLT PRINCIPLE IN AUTOMOTIVE DESIGN

For the different magnesium alloys that were tested in this work, the observed force-deformation behaviour is approximately constant throughout the deformation.

In an automotive application, utilisation of the shear-bolt principle gives the opportunity to quite exactly design thin-walled cast magnesium components with respect to force-deformation behaviour. Furthermore, referring to the IP-beam illustrated in Figure 11 and Figure 12, it seems to be quite easy to integrate the shear-bolt principle. With a proper design of the IP-beam, and the connection of the IP-beam to the A-pillar, the shear-bolt principle can be used as energy absorption mechanism both for side and front impact. A "programmable" combination of different force levels can be obtained by choosing the number of bolts, the bolt diameter and the thickness of the component.

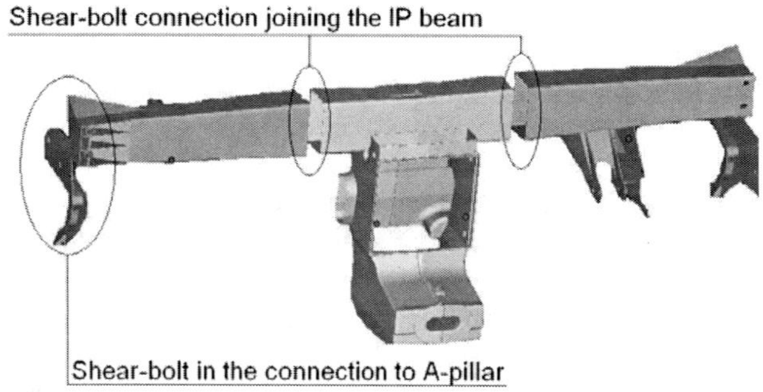

Figure 11: Example of IP-beam with possibility for shear-bolt connection for both side and front impact.

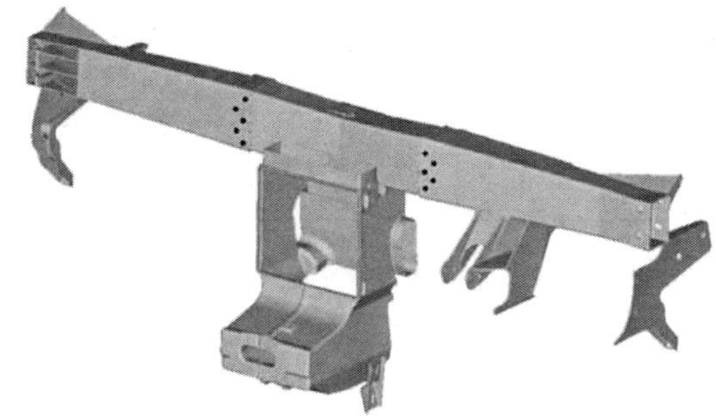

Figure 12: IP-beam joined with a shear-bolt connection for energy absorption in side impact.

CONCLUSION

The long-term objective of this work is to develop design and modeling tools that allow the structural behavior of thin-walled cast components to be predicted when subjected to static and dynamic loads such as in crash situations. The strategy of the work is to use an iterative procedure of casting of components with generic geometry relevant for automotive design combined with component and materials testing, constitutive modeling and validation simulations.

This work shows that use of the shear-bolt principle is a very robust deformation mechanism for magnesium castings. Components of magnesium alloys AM20, AM50, AM60 and AZ91 were tested with bolt diameters of 5, 10 and 15 mm. The average shearing force was approximately constant throughout the experiments for each of the tested magnesium alloys. Only small differences were found between the magnesium alloys with respect to force level.

Single plates of AM60 were then subjected to shear-bolt deformation in order to investigate the influence of micromechanical variations on the average shearing force. The experimental results from these tests showed that the average shearing force was approximately constant throughout the plate experiments, and it was concluded that the present micromechanical variations had no significant impact on the force level.

The shear-bolt principle also gave promising results for the aluminum alloy AlSi7Mg, especially when using bolt diameter of 5 mm. Other aluminum alloys should be tested in order to reduce the variations in force level during the deformation.

Based on the experimental observations found in this work, it is concluded that the shear-bolt principle is a very promising deformation mechanism for thin-walled cast components with respect to energy absorption. It should be noted that neither the geometry of the components nor the different alloys are optimized with respect to the investigated shear-bolt principle. Furthermore, the reported experiments are performed at quasi static conditions.

ACKNOWLEDGMENTS

The authors would like to acknowledge Norsk Hydro and the Norwegian Research Council for their support through the NorLight project.

REFERENCES

1. Dørum, C., Hopperstad O.S, Lademo O.-G., Langseth M., "Aluminium and magnesium castings - experimental work and numerical analyses", International Journal of Crashworthiness Vol. 8 No. 5 pp 001-016, 2003.
2. Hydro Magnesium, Data Sheet, 'Die Cast Magnesium Alloys', 2000.
3. Laukli, Lohne, Sannes, Gjestland, Arnberg, "Investigations of the Grain Size Distribution in a Complex AM60 Magnesium Alloy Die Casting", submitted for publication in the International Journal of cast Metals Research (Int. J. C. Met.R.) 2003.

CONTACT

Cato Dørum
Structural Impact Laboratory (SIMLab)
Department of Structural Engineering
Norwegian University of Science and Technology
7491 Trondheim, Norway.
cato.dorum@bygg.ntnu.no

2004-01-0130

A Case Study of a Die-Cast Magnesium Structure Supporting Transmission Shifter Mechanisms and Interfaced with other Structural Systems

Thomas Ruden
Lunt Manufacturing Company

Daniel Lenda
General Motors Corporation

ABSTRACT

During the last several years the use of magnesium die-castings for automotive applications has been on the rise. Magnesium's use in die-cast form has been expanding at an average growth rate of more than 15% a year. Reasons for the increase are both practical and economic. Magnesium die-castings offer components having the lowest mass when compared to almost any other structural material. Magnesium die-alloys exhibit properties that bridge the gap between engineered plastics and metals. The mechanical performance ratios (strength-to-weight and stiffness-to-weight) of magnesium also compete favorably with metals and plastics. Economically, magnesium alloys prices have fallen during the last several years making them extremely competitive with other materials. This combined with strategies that reduce total systems costs of a vehicle (such as part integration, net-shape part design, and lower vehicle assembly time) are making stronger business cases that justify magnesium's long-term use. In order to underscore the advantages for magnesium's use, this paper summarizes the recent decision to use a magnesium die-casting for a transmission shifter structure. The component is unique because it is a single magnesium die-casting that is used to tie together three distinct structural systems within the vehicle (cross-car beam, transmission shifter mechanism and body floor).

INTRODUCTION

This paper provides a general overview of the development of a magneisum die-casting for an automotive shift tower structure. The shift tower structure was first developed for the 2003 Saturn VUE, as shown in Fig. 1. Emphasis is placed on understanding the feasibility of magnesium die-casting for such an application. Included is an overview of the decision-making process that evaluates not only mechanical properties, but also takes into account other commercial and performance attributes. Within the paper several technical discussions are presented which offer an overview of the inherent performance features of magnesium for a shift tower structure.

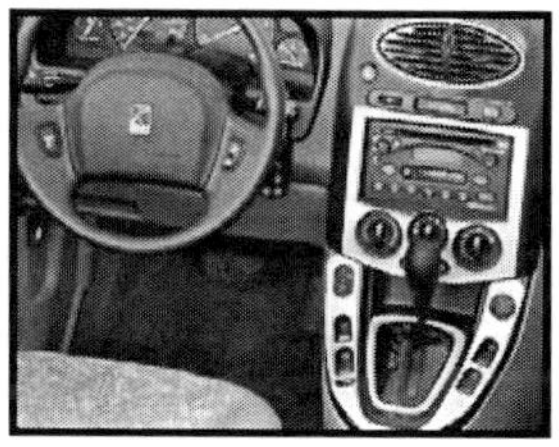

Figure 1. 2003 Saturn VUE & Center Console Area (shift tower located under center console)

SYSTEM REQUIREMENTS

DESCRIPTION

The shift tower was the first magnesium design to be accepted in this SUV. Although relatively small in comparison to other structural components (e.g. cross-car beam, seat frames and door panels), the magnesium die-casting is considered a keystone component as it ties several structural systems together within the vehicle interior. To keep program costs low, one shift tower design was desired to accommodate both manual and automatic transmission options. Homologation requirements were also required to accommodate the needs of European, Asian and North American markets.

In order to provide suitable stiffness for the body structure, the vehicle architecture design team required the shift tower be mechanically attached to several other systems. They included:

- High Mount Application: Tunnel – Body Floor
- Lower Center-stack of the Cross-car Beam
- Shifter Mechanism and Controls – Automatic or Manual Transmission Options
- Center Console Assembly
- Attachment of Body Electrical Wiring Harness.

FUNCTIONAL REQUIREMENTS

The performance requirement for a component that interfaces with multiple systems is considered to be very complex. A general overview of the specifications include the following elements within durability testing that met a minimum requirement of 100,000 miles of simulated customer usage;

1. Cross-Car Beam Center Stack
   - Support
   - Reaction Loads

2. Body / Floor Pan
   - Support for the Shifter Assembly
   - Shifter Reaction Loads
   - Body/Chassis Vibrations and Shock/Impact Loads

3. Powertrain / Transmission
   - Centering Torque of the Manual Shifter
   - Vibration
   - Reaction Force by the Transmission Case
   - Support for the Shift Cable Reaction Bracket

4. Electrical
   - Location and Support for Attached Body Harness
   - Vehicle Electromagnetic Compliance (EMI / RFI)

5. Center Console
   - Support the Mass of Attached Assembly
   - Reaction Loads

6. Crashworthiness Goals
   - LINCAP Performance
   - IIHS Performance

Shifter Loading Conditions

The shifter assembly places several overload conditions on the shift tower. In general, a design factor of 1.5 was used in the magnesium design and the structure did not experience failure during validation testing.

The various loading and test conditions included;

- Lateral Loading (-L / +R)
- Fore/Aft Loading (+F / -R)
- Vertical Loading (+V)
- Linkage Ultimate Strength (tension & bending)
- Overload Fastener/Stud Pull-Out (tension)

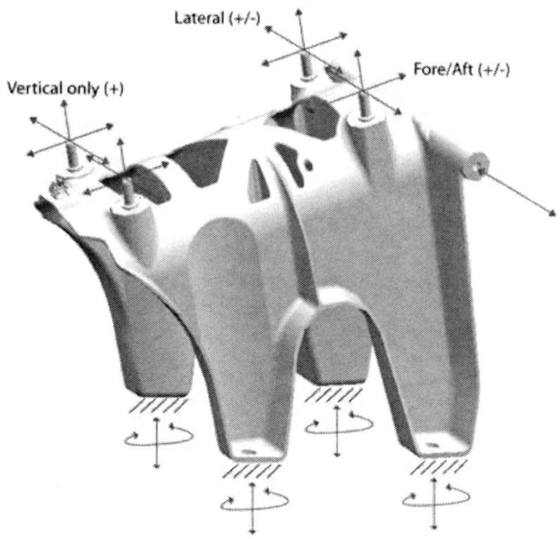

Figure 2. Force Diagram of Shift Tower Assembly

As evidenced by the multiple directions of the applied loads during testing and actual service, using a material having isotropic properties (uniform in every direction) is of benefit to such a multi-faceted system. The properties of aluminum and magneisum die-castings are considered isotropic.

In addition to the overload test requirements, the following performance specifications were also required of the shift tower within simulated environmental exposure of ten years.

Temperature and Humidity

Long term operating temperature range of the shift tower was specified between -40C to 80C, along with humidity from 0 to 100%. Minimum shipping temperature was specified at -50C.

Contamination Performance

The system specification required the shift tower to withstand contamination from dust, dirt, ash, coffee, cola and soft drinks, human skin oils, stomach acid, water, salt, fog, sand and ozone (O_3).

Component Drop Test

The shifter system was subjected to a 1.0 m (3.3 ft) drop test with the requirement of no structural loss or degradation after impact. The test was used to simulate the results a part being dropped in an assembly line.

Recyclability

General Motors required that the shift tower structure fully comply with its requirement for materials to be 100% recyclable and be labeled as such. Magnesium die-castings are fully recyclable and met this requirement.

Flammability

The magnesium shift tower was initially considered for flammability testing per GM specification (GM9070P) thus ensuring safety by design in the event of a vehicle fire. However, unlike plastics that have low ignition temperatures, magnesium has a much higher melt temperature and is not listed as a restricted or flammable material and therefore did not require testing.

Corrosion Resistance

Structural components designed for interior automotive applications require good general corrosion protection during the service life of all components. All potential materials considered for the design had to meet the program's minimum 10 year / 100,000 mile service life requirement.

Volume

Product packaging needs and overall size constraints required the first and second surfaces of the shift tower be designed as close together as possible. This general goal was achieved by using the relatively thin walls (2-3 mm) of a magnesium die-casting in place of aluminum (3-5 mm), steel (2-5 mm) or engineered thermoplastic (5-8 mm) designs.

Mass Targets

The target for the shift tower was to weigh no more than 1.4 kg (3 lbs). Of all the materials, magnesium die-casting offered the lowest weight at 1.1 kg (2.4 lbs). This was 20% less than the target weight. Steel, aluminum and nylon designs would not have met the weight target. Comparable aluminum and plastic designs were estimated to weigh 35% more, and steel 50% more than Mg. Lightening holes were used extensively throughout the Mg design to meet the weight target without compromising castability or structural integrity.

Noise, Vibration and Harshness

The shift tower experiences the greatest amount of vibration from being mounted between the vehicle body and transmission. The shift tower was not to induce or transmit vibration onto other components within the system. The first-mode frequency was targeted to be a minimum of 30 Hz. Validation test results confirmed the shift tower met this specification by not interfering with the natural frequencies of the vehicle or front-of-dash subsystem.

COMMERCIAL

Financial targets required the shift tower to have the lowest system cost while meeting product performance requirements. Although specific cost data is considered confidential, the total system cost included the following elements:

- Unit Price
- Investment (machines, tools & fixtures)
- Tool Life
- Design
- Prototyping
- Secondary / Ancillary Components
- Component Assembly
- Transportation / Logistics
- Manufacturing Assembly
    - Component Assembly
    - System Assembly
    - Ease of Handling
- Service & Warranty

THE DECISION PROCESS

CURRENT BENCHMARK ASSEMBLIES

Upon review of current shifter assemblies in the automotive industry, it was found the majority of shift towers are produced by steel fabrications. Further benchmarking showed shift tower structures were produced from steel stampings, injection-molded nylon, and some market penetration by die-cast aluminum. All of the designs were found to be stand-alone, non-integrated, assemblies and were attached to only one other system (usually to the body floor or transmission extension). No benchmark designs were found to integrate multiple systems like the intended design of the Saturn VUE shift tower.

Steel

Benchmarking results indicated that fabricated steel designs were used for the majority of shifter height distances between driver and vehicle floor. Steel designs were commonly used throughout General Motors for standard application – attach cable controls and limits of shifter movement. Such designs, like that shown in Figure 3, show that stampings are pervasive in the market. This is due to the low complexity and system cost associated with a simple design. However, it was found that when more complex designs and higher shifter heights were needed, steel designs were not used as often. Rather, at this transition point (increased complexity and higher system cost), there appeared to be a departure from the complex steel designs to reinforced plastics designs.

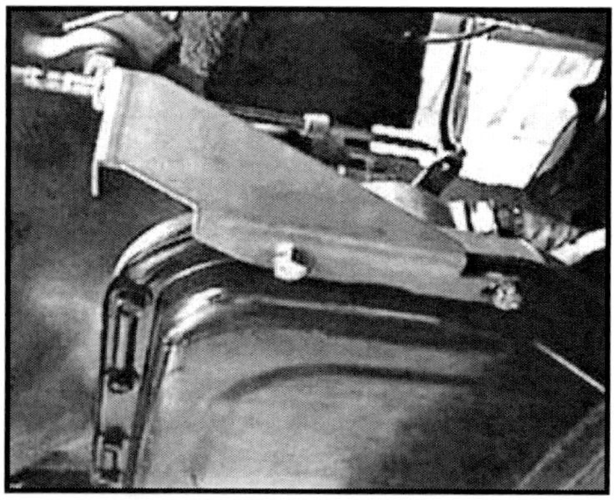

Figure 3. A Steel Shift Tower Assembly

Aluminum

Aluminum designs appeared to be limited to rear wheel drive and short-mount shifter applications. An example of this design is shown in Figure 4. There were two reasons for these limitations: (1) short distance from transmission to shifter, and (2) the majority of shift tower designs required toughness and ductility. In general, die-cast aluminum alloys are strong and stiff, but do not offer high amounts of elongation. Alternatively, steel and magnesium alloys generally offer higher amounts of energy-absorbing properties such as high elongation and ultimate tensile strength (see Table 1) that make them more suitable for tough, crash-worthy applications.

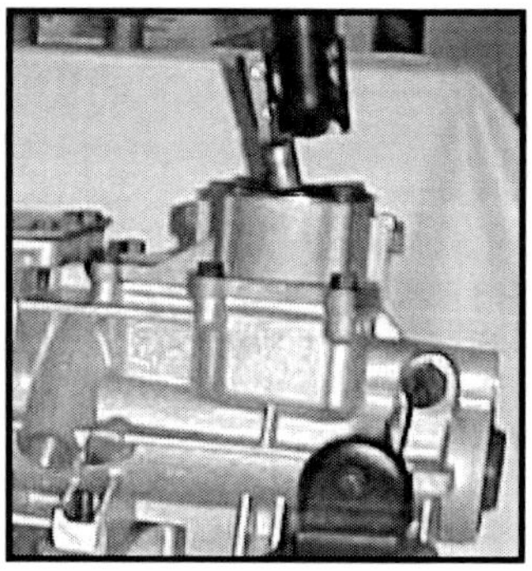

Figure 4. An Aluminum Shifter Bracket (top casting bolted to transmission extension)

Plastics

A reinforced-nylon shifter assembly is shown in Figure 5. Many shift tower assemblies were found made via injection-molded plastic using glass-filled nylon (PA66 + 33%GF). Although reinforced plastics had been considered to be low cost over the years, their low modulus limits their ability to provide the necessary stiffness to resist deflection under load. Further, a trade-off occurs with plastics in that as more fibers are added to the resin to increase strength and stiffness, the elongation is reduced. Another design approach to overcome plastics low stiffness is to increase the nominal wall thickness throughout the part. Steel inserts or bushings may be needed to improve stiffness and bearing strength of reinforced plastics. Nevertheless, these improvements add cost, weight and dimensional instability to the overall design.

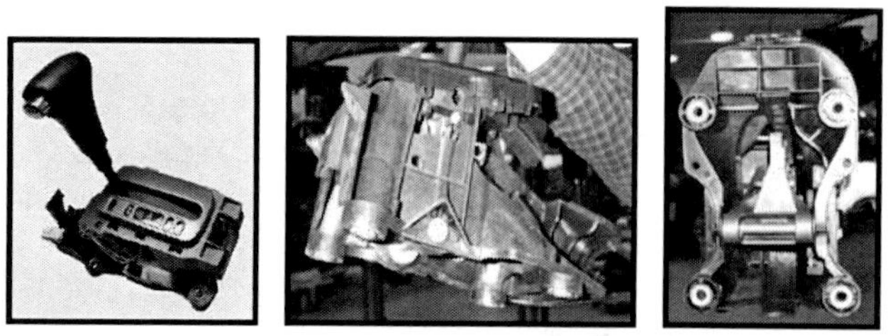

Figure 5. A Nylon Shift Tower Assembly

PROGRAM TARGETS

The benchmarking section found that all the designs were fastened to only one system and not integrated as a structural system. In addition to the need of being integrated with other systems, the shifter assembly for the Saturn VUE required the following elements:

- Low Cost (total system cost)
- Low Mass
- Structural Stiffness / Rigidity
- Dimensionally Stable
- Crashworthiness
- Enhanced Vehicle Stiffness
- Buzz, Squeak and Rattle (BSR)

Previous and current shifter designs did not show a clear, low cost way of accomplishing all these goals. The next phase then in the decision making process was to investigate alternate materials and designs.

FIRST DESIGNS

The initial shift tower design called for a one-piece steel stamping, as it was believed to offer the lowest system cost of all benchmarked materials. After several design iterations it became apparent the steel design would need multiple stampings and require additional joining operations to fabricate a single component that would interface with the other systems. The cost for multiple stamping dies and assembly fixtures were exceeding the program budget and the additional stampings were contributing to dimensional and tolerance issues. Additionally, the steel design was exceeding mass targets.

After preliminary vehicle crash analysis (using steel for the prototype shift tower) it was found the initial design was not producing optimum results. Furthermore, the steel cross-car beam was not producing optimum results with the initial steel shift-tower design. In order to compensate for the overweight designs, engineers investigated alternate materials to meet their goals.

The next step looked into reinforced plastic nylon as an alternate design. However, after preliminary designs and FEA analysis were conducted, it was found the nylon could not meet weight, stiffness and volume (packaging) targets. The plastic design further required a complex, 3-part tool in order to make the injection-molded part. The plastic would further need steel inserts to improve stiffness. The end result was viewed as a rapidly increasing system cost for plastic, without offering significant performance improvement over steel.

CONSIDERATION OF MAGNESIUM

There was no documentation of a magnesium die-casting having been used as a shift tower support structure. As a result, no reference design or historical performance data was available to base designs upon. Consequently, the decision to consider magnesium as a viable alternative came from benchmarking data from similar magnesium die-cast applications, such as:

- Ashtray Doors – Center Console
- Cross–Car Beams & Center Stacks
- Front Seat Armrest Support Bracket
- Seat Frames & Bench Seat Risers
- Shifter Consoles (Utility Tractor)
- Steering Column Support Brackets

COMPARING ALTERNATIVES

Based on current benchmark designs and similar magnesium applications, several alternate materials and designs were considered in the decision making process. They included:

- Die-Cast Aluminum (A380)
- Die-Cast Magnesium (AZ91D or AM60B)
- Injection-Molded Plastics (PA66 + 33%GF)
- Mild Steel Fabrication (SAE 1010)

Property Comparison of Materials

Table 1 compares key properties of alternate materials. Upon first review it appeared steel had a clear advantage over the other candidates. But after reviewing properties on a more specific (per unit weight) basis, magnesium alloys were seen to offer the greater combination of strength and stiffness. When evaluating designs having similar geometry and volume, specific ratios (such as strength-to-weight and stiffness-to-weight), provide better insight into the efficient use and selection of materials.

Table 1. Property Comparison of Materials
(key Mg crashworthiness properties in bold)

Property	*Magnesium*		*Aluminum*	*Plastic*	*Steel*
	AM60B	*AZ91D*	*A380*	*PA–33% GF*	*SAE 1010*
S.G.	1.79	1.81	2.76	1.34	7.87
TYS	**131**	150	150	145	305
UTS	241	230	325	206	365
% El.	**12**	3	3	4	20
Modulus	**45**	45	71	7	205
Specific Strength	**73**	83	63	154	39
Specific Stiffness	**25**	25	25	5	26

Key:
S.G. = Specific Gravity (g/cm^3)
TYS = Tensile Yield Strength (MPa)
UTS = Ultimate Tensile Strength (MPa)
% Elongation (%)
Elastic Modulus (GPa)
Specific Strength = Strength per Unit Weight
Specific Stiffness = Stiffness per Unit Weight

Product Performance Attributes

The next phase in the decision-making process reviewed the relative performance of attributes to one another. The result of this comparison is provided in Table 2. Each attribute is treated equally in contribution and is rated according to overall performance. The determining factor is then obtained by taking an average of the relative performance levels.

Table 2. Performance Matrix

No.	Attribute	A380	AM60B	PA-33%	SAE 1010
1	System Cost (less tooling)	2	**1**	3	4
2	Tooling & Machine Investment	3	**1**	2	4
3	Strength	3	4	1	2
4	Stiffness & Rigidity	1	2	4	3
5	Energy Absorption	4	2	1	3
6	NVH	2	**1**	4	3
7	Corrosion Resistant	3	2	1	4
8	Connection Integrity	2	3	4	1
9	Low Weight	2	**1**	3	4
	Average	2.4	**1.9**	2.6	3.1

Rank Key: 1 = Best

It is shown the magnesium design yields the best (lowest) average ranking. As well, magnesium ranked the lowest (best) in the four categories of system cost, investment, NVH and weight - along with offering the best overall performance.

Material Prices

List prices of raw materials are shown in Figure 6. On a per-pound basis, steel offers the lowest price compared to the alternates. However, on a unit volume basis magnesium offers the lowest price per cubic inch (see Fig. 7). This is important to emphasize, as competing designs typically occupy the same volume within a given environment. In this case, material use (specific-property-ratios) and price efficiency are key attributes in the decision-making process.

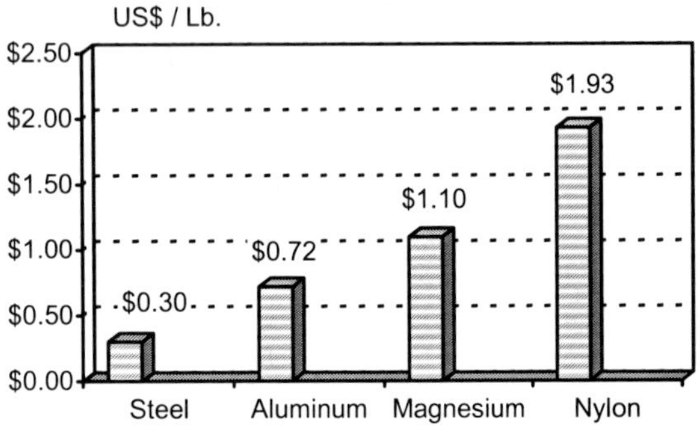

Figure 6. Price of Materials based on Weight

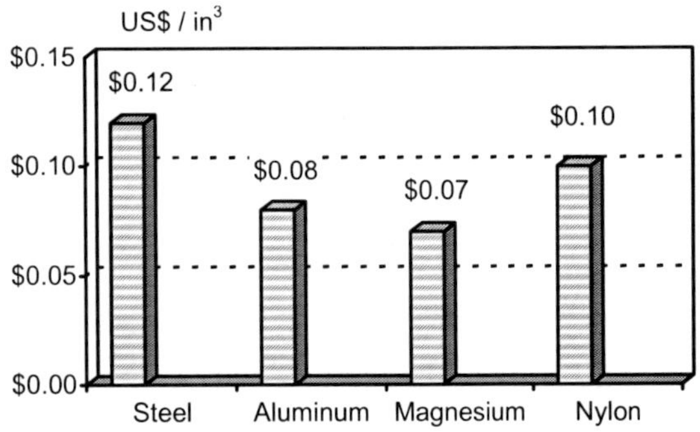

Figure 7. Price of Materials based on Volume

DEVELOPMENT

FINAL MATERIAL SELECTION

Based on the comparison of alternate material properties and system attribute performance, the die-cast magnesium alloy AM60B was selected as the production material for the shift tower. Selection was based on the alloy's ability to provide suitable strength and stiffness per unit cost, while offering a high degree of energy absorption. The AM60B alloy was also seen as having the greatest overlap between properties when compared to the other materials: aluminum, glass-reinforced nylon and steel.

DESIGN

The first and second surfaces of shift tower were initially designed to replicate features normally incorporated for an injection-molded plastic design. Uniform wall thickness was employed wherever possible, with smooth transitions at fillets and radii to minimize the influence of stress concentrations. After several design iterations, the Mg design began incorporating greater feature variation that is commonly associated with Mg de-casting. Overall, the magnesium design shows the ability to be die-cast with greater feature variation than aluminum, plastic or steel (ref. Figs. 3, 4, 5 and 8). This design flexibility allows for a one-piece magneisum die-casting, rather than the part being made from several components and requiring additional assembly.

Wall Thickness

Nominal wall thickness of the Mg die-casting was initially set at 2.5 mm. During the design phase, wall sections and component features changed considerably. Using magnesium's ability to cast complex sections, wall sections were reduced to as low as 1.5 mm, and increased to as much as 12.0mm (all within the same component). Figure 8 shows where larger sections were placed at the upper boss / stud locations to provide for additional stiffness and bearing strength for the shifter mechanism. Vertical ribs were added to the outer periphery of the structure to increase section modulus and lateral stiffness. Draft was set at 1.5 degrees, minimum. Analysis and physical tests confirmed the clearance holes and pads for the lower attachment points that did not require thicker walls. This was due to the orientation of the local walls. The design created, in effect, a stiff box section, surrounding the lower attachment planes.

Weight-reduction-holes were used throughout the design to minimize the mass of the die-casting and facilitate castability. A one-piece die-casting was made using an open architecture design that did not require the extensive use of slides, except for core pulls.

Attachment Points

Four, double-ended, self-tapping studs, were chosen to be automatically assembled to the upper boss areas. A 2.5 ratio of bolt-engagement-length to bolt-diameter (L/D ratio) was used for the studs assembled into the magnesium. These studs are used to attach either automatic or manual transmission shifter mechanisms. Each stud hole is cast to size. A rectangular clip was selected to be pressed-in to a net shape "square-hole" for attachment of the center console. Two tapped holes were incorporated at the front of the structure to attach the center steel H-bracket of the cross-car beam. Bearing diameters of the forward bosses were set at 12 mm to match up with the H-brackets.

Energy Absorption Requirement

The shift tower required a material that could absorb energy in the event of a vehicle crash. The area under the stress-strain curve offers a good approximation of a material's total energy-absorption capability. This factor is estimated by taking the products of total strain to stress of a material at failure, then divide in half to obtain the factor. Regardless of the amount and direction of applied load, in the event of failure the shift tower needed to fail in a ductile manner, as well as exhibit energy absorption. Since the direction of impact forces could come from any direction, isotropic properties of the material were desired over those that did not offer these characteristics. Magnesium die-castings offer isotropic properties.

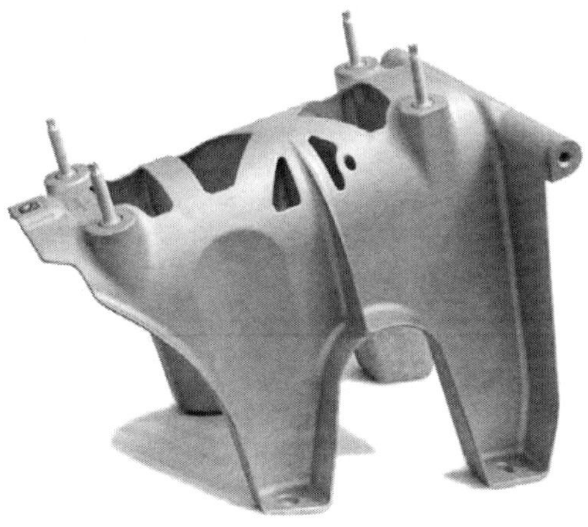

Figure 8. Die-Cast Magnesium Shift Tower

PROTOTYPING

Magnesium sand castings were selected to prototype the shift tower. Since AM60B was the chosen alloy for production, but was not used for sand-cast applications, a comparison was needed to select the proper alloy and temper that would best replicate key attributes of die-cast AM60B. Table 3 shows properties, processes and tempers of magnesium alloys. Based on comparative properties, the widely used sand cast alloy, AZ91E was chosen with a solution-annealed T4 temper for functional prototypes. First surface features of the sand-casting matched those of the die-cast surfaces. Nominal wall thickness of the sand-casting was 5 mm, compared to 2.5 mm for die-cast. Although the section modulus was greater for the sand-casting than the die-casting, vehicle crash tests at each key stages of development confirmed both sand-cast and die-cast designs performed to design targets.

Table 3. Property Comparison of Mg Alloys for Prototype Selection

Property	*Units*	*Die-Cast*	*Sand-Cast*		
		AM60B F	*AZ91E F*	*AZ91E T4*	*AZ91E T6*
TYS	MPa	131	97	90	145
UTS	MPa	**241**	165	**275**	275
Modulus	GPa	**45**	45	**45**	45
Elongation	%	**12**	3	**15**	6
Impact	J	**2.8**	0.8	**4.1**	1.4
Energy Absorption	[MPa • (mm/mm)]/2	**1,446**	248	**2,063**	825

DURABILITY

Vehicle testing simulating 10-year product life and 100,000 miles was performed at the General Motors proving ground in Milford, Michigan. Although not published, results show the magnesium casting met all mechanical, structural, corrosion, feature stability, NVH, and long-term durability requirements.

CRASH-WORTHINESS

The magnesium shift tower helped meet intrusion goals by maintaining the desired position of the cross-car beam, the body tunnel and body floor systems during crash analysis. This performance of the magnesium shift tower helped the vehicle meet crash-worthiness targets without requiring additional, heavy and costly reinforcements to the body structure.

BUZZ, SQUEEK and RATTLE (BSR)

In additional to the crashworthiness goals being met, the one-piece Mg die-casting met long-term durability and BSR goals.

CORROSION

General Corrosion Protection

It was well known that the general corrosion performance of magnesium alloys is superior to both aluminum and mild steel. Additionally, a bare magnesium casting met the requirement for the interior component after vehicle durability and corrosion testing,

Galvanic Corrosion Protection

Figure 9 shows the lower attachment areas of the shift tower as attached directly to the painted steel floor of the vehicle. After accelerated corrosion testing (GM-9540), minor products of galvanic corrosion are shown here as gray-white deposits near the washer/casting interfaces. This level of corrosion does not compromise the design. The carpeting is designed to a distance of approx. 10 mm away from the assembly so as not to wick moisture onto the magnesium/steel fastener interface.

Galvanic corrosion concerns at the magnesium / steel fastener interfaces are addressed by using a zinc-rich inorganic coating (GM–7114M) on the steel fasteners. This included the weld studs, double-ended studs and flange head nuts. This coating serves to insulate the steel fasteners from the magnesium and road salt (electrolytic solution). If left unprotected, the interface would see galvanic corrosion activity in a very short period of time. Test results confirmed this protection scheme met the performance requirements of the design.

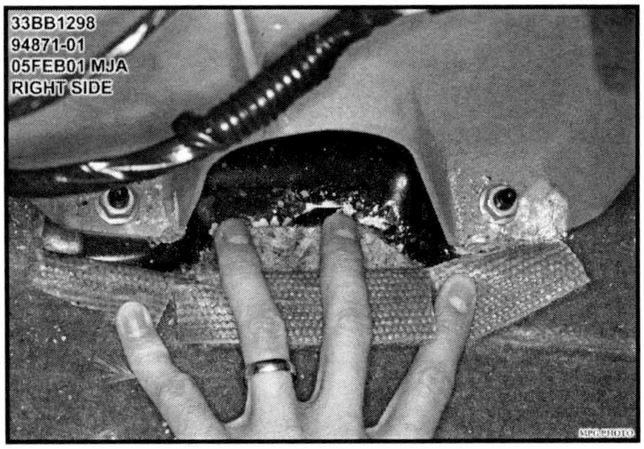

Figure 9. Galvanic Corrosion Test Results

Component Integrity

In addition to vehicle tests, non-destructive verification of casting performance was used to understand how the overall level of casting quality played to ensure a robust component. Non-destructive evaluation (such as x-ray analysis) was used and continued in the production quality assurance process to confirm the confidence level of producing high quality castings. Although not considered the cornerstone of non-destructive testing, Figures 10 and 11 indicate low levels of casting porosity and the absence of voids or inclusions. The casting design was proven to cast efficiently and cleanly with minimum porosity in key structural areas.

Figure 10. X-Ray of Lower Attachment Points

Figure 11. X-Ray of Upper Attachment Points and Square Hole

SUMMARY

1. A one-piece magnesium die-casting met component cost targets and also offered the lowest system cost.

2. Mechanical properties of magnesium were benchmarked and compared directly to A380, PA66-33%, and SAE-1010 using several functional and commercial attributes. AM60B was selected as the production material as it had the best overall rating and highest frequency of first-place ratings.

3. A magnesium die-casting met component-level performance requirements for the shift tower and also met system-level requirements for several interdependent structural systems within the vehicle. The result was increased safety and structural performance of the vehicle during a crash event.

4. The magnesium design was estimated to weigh 35% less than aluminum and reinforced nylon, and 50% lower than low carbon steel.

ACKNOWLEDGEMENTS

The authors would like to thank the management at General Motors and Lunt Manufacturing for their support in the development of this application. Special recognition is given to Tim Fillippi of GM and Dennis Debusschere for their creativity in designing the Mg shift tower, as well as Carlos Engel and Steve Gronlund of

Lunt Manufacturing for their expertise in magnesium die-casting and dedication to implementing this application.

SELECTED READINGS

Cameron, T., *"Use of Threaded Fasteners in Die Cast Magnesium,"* SAE Technical Paper No. 860289, Society of Automotive Engineers, Warrendale. PA (1986).

Crane, F. and Charles, J., *"Selection and Use of Engineering Materials,"* Butterworths & Co. (1987).

Hillis, J., *"The Effects of Heavy Metal Contamination on Magnesium Corrosion Performance,"* SAE Technical Paper No. 830523, Society of Automotive Engineers, Warrendale, PA (1983).

"Magnesium and Magnesium Alloys," ASM International (1999).

"NADCA Product Specification Standards for Die Castings," North American Die Casting Association, LaGrange, IL (1999)

Ruden, T., *"Magnesium vs. Plastics: A Comparison for Structural, Automotive Applications,"* International Magnesium Association, 14th Annual Magnesium In Automotive Seminar, Plymouth, MI (2003).

CONTACT INFORMATION

Thomas Ruden
Lunt Manufacturing Company, Inc.
816 E. Fourth Street
Royal Oak, MI 48067
Tel: (248) 399-2600
Fax: (248) 399-3797
eMail: truden@Lunt.com

Daniel Lenda
General Motors Corporation
3001 Van Dyke Ave.
Warren, MI 48090
Tel: (586) 492-3197
Fax: (586) 492-6014
eMail: daniel.lenda@GM.com

2004-01-0131

Robustness of Large Thin Wall Magnesium Die Castings for Crash Applications

Regis Alain, Tony Lawson, Pat Katool, Gerry Wang, John Jekl, Richard Berkmortel and Len Miller
Meridian Global Technology Center

Jorild Svalestuen and Hakon Westengen
Norsk Hydro

ABSTRACT

The robustness of large magnesium thin wall die castings has been proven through testing performed directly on components and supported by statistical analysis of the test data. In addition, an extensive investigation carried out on specimens cut from components tends to show that the material exhibits a composite failure mode. The skin shows high consistent elongations close to the intrinsic potential of the alloy, while the core properties vary with the micro-defects inherent to the process. Design data has therefore been developed by reverse engineering from component tests, considering local properties and loading mode.

INTRODUCTION

In the automotive industry, most of the structural crash components are manufactured from stamped steel with elongation properties that are high and relatively consistent. Die castings are often used for engine housings or brackets for which no elongation is required since these parts usually do not participate in the crash energy absorption requirements. The magnesium alloy used for these applications is AZ91 with handbook elongation values of about 3%.

Other die casting magnesium alloys offering higher elongations are extensively used for crash structural components like instrument panels, seats frames and steering wheel frames. These ductile magnesium alloys belong to the AM family with published elongation values ranging from 6-10%.

However, when testing properties on coupons (tests bars cut from the components), it can be seen that if average elongation properties are high, the associated scatter is also high, and varies from one location of the part to another. Even for a similar location, variation occurs from one casting to another. This issue raises the question of the robustness of magnesium die castings for crash applications.

Meridian's early experience in testing components showed significantly lower scatter, and acceptable performance for crash requirements. Therefore, to develop a better understanding of these issues, investigations have taken two routes. In the first route, different kinds of loading conditions have been simulated on components and multiple tests were performed to build reliable design data. In the second route, studies were carried out on coupons to explain the difference in elongation scatter, but also to grow a better understanding of the local properties and their correlation to product performance and microstructure.

ROBUSTNESS APPROACH & CORRELATION OF THE COMPONENT BEHAVIOR WITH FINITE ELEMENT ANALYSIS

Finite element analysis is extensively used during the development of structural high pressure die cast components, while component tests are designed to replicate major in-vehicle loading mechanisms. This is done to ensure that the failure mode under investigation is relevant to the development process. Defined here are the critical parameters used to understand the robustness approach and analysis.

STRESS STRAIN CURVE

It was found that the published tensile data associated with separately cast test specimens did not properly estimate the performance of large high pressure die cast components. Therefore, a reverse engineering approach was used to develop a virtual stress/strain curve that would allow good correlation. This curve is sketched in figure 1. The curve was first developed for simple tests and step by step optimised with real component tests. Note that the two elongation values given on the graph are different; one is the measured average tensile elongation on coupons, and the other is the equivalent effective strain criteria presented in this paper.

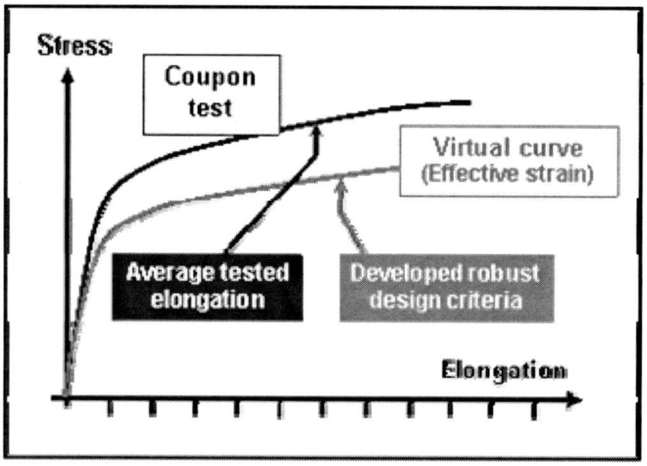

Figure 1: Sketch of the developed equivalent stress/strain curve reverse engineered from component test; comparison with coupon tensile curves.

The developed material property parameters are then used with LS Dyna3D to correlate the FE model with the actual tests. Correlation is considered achieved if the analysis predicts the correct deformation mode, failure location and force deflection curve [1].

Using this equivalent curve, good correlations are obtained between component tests and FEA as can be seen for example in figure 3.

FAILURE CRITERIA

In our analysis, 6% effective plastic strain failure criterion is used on this equivalent stress/strain curve. Initially, this value was an empirical value that had been defined by prior experience on component tests. To ensure the robustness of the product for crash applications, the reliability of the 6% equivalent effective strain was validated. In this paper "robust" is defined on the basis of the analysis of data through a Weibull distribution at 99% reliability with a confidence interval of 90% and compared with the required performance.

TESTING CONDITIONS

When dealing with scatter on component tests, great care needs to be taken when setting-up the tests; especially when it comes to large complex shapes like instrument panels. For this reason it is important to always use the same testing conditions, especially when products are tested at different times over the design and validation phase.

To get as close as possible to the real scatter in performance, it is crucial to mount parts in the exact same position. For instrument panels, a specific rig, mounted on a tensile machine, has been developed to accept different geometries as can be seen in figure 2. This rig is mainly used for loading the steering column area or center of the IP to reproduce frontal impact test loading mechanisms.

Both ends of the part are free to move in the Y direction and free to rotate along X, Y, Z axis. All movements can be locked or limited to meet specific conditions. All displacements are monitored for the FEA correlation work.

To apply the load, a specific bracket was mounted to the IP and linked to the load cell on the upper side and guided by a ram underneath. To insure correct positioning, all parts are drilled at both ends, located with pins and bolted to the movable end brackets of the rig.

Parts are randomly taken at the rate of one or two parts per shift at different times, to insure that the testing was representative of production. Testing was completed on 40 parts.

ANALYZED PARAMETERS

In the component testing, different parameters are statistically analyzed. Generally, the performance is a load level, a displacement at first crack or at full failure. The failure parameter has to be defined in agreement with the costumer.

RESULTS ON A CORRELATED COMPONENT TEST

An example of a correlated test setup is shown in figure 2, while the load / displacement data for this test is seen in figure 3.

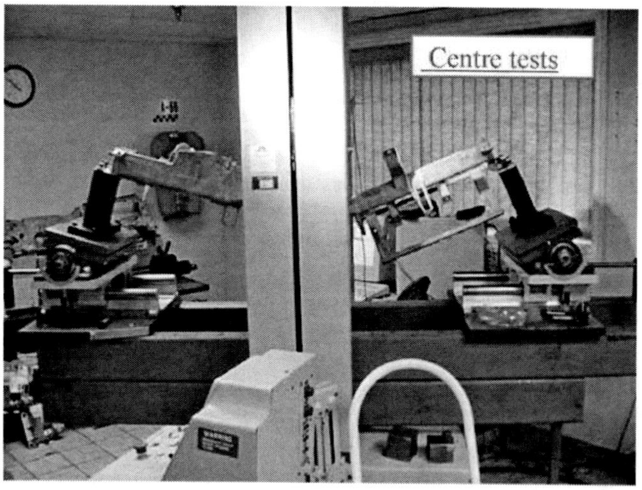

Figure 2: Center test on an instrument panel.

For the load levels not much difference is found when using a Normal or Weibull analysis to describe the variations. The load levels are at +/- 8% around the average at +/- 3 standard deviations. On the 40 tests presented on figure 3, the minimum and maximum measured load levels are ranging between +/- 2 standard deviations around the average.

Using the equivalent stress/strain curve, a good fit is achieved between the tested load curves and the FEA (see light blue curve on figure 3). The failure location was correctly identified (see figure 4), and the predicted failure was found below the minimum performance measured in tests. The most accurate prediction of failure has been found to be related to the presence of coincident hot spots of plastic strain and maximum principal stress. When this is identified, the level of plastic strain in this location is monitored and failure is then related to the failure strain design criteria.

Using the 6% failure criteria, the part is predicted to achieve 147 mm (see figure 3) of deflection. The Weibull analysis on the measured displacements at failure predicts a displacement of 160 mm at 99% reliability with 90% confidence interval (see figure 5). Therefore, the design criterion provides a conservative performance prediction since there is a security gap between the predicted failure and the statistically calculated minimum failure displacement.

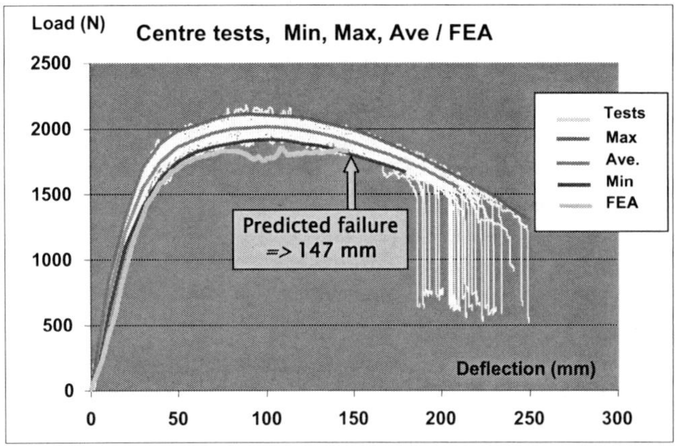

Figure 3: Centre tests, correlation between FEA and tested load/displacement curves.

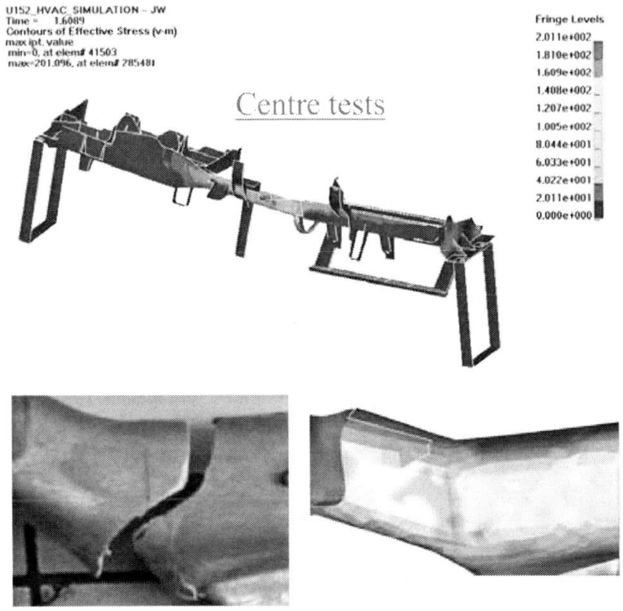

Figure 4: Failure location on centre tests [1].

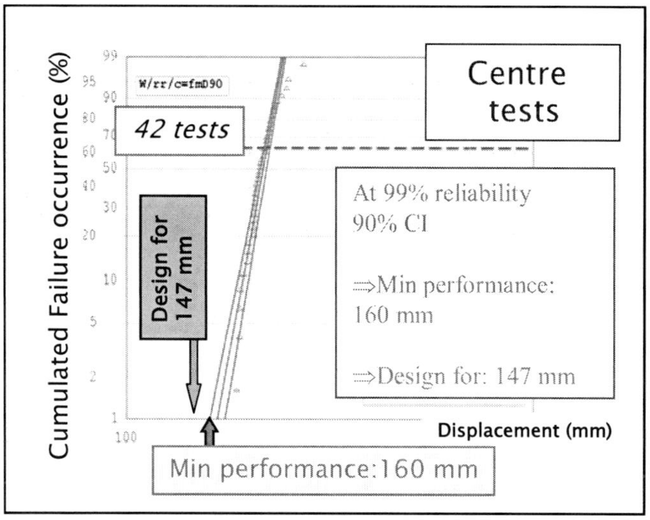

Figure 5: Weibull analysis on the displacement for failure for a centre test on an IP.

SECURITY GAP

Weibull is used to measure the performance at 99% reliability for 90% confidence interval. The extrapolation of this to the displacement at failure for reliability levels in the range of one failure per million can only be used as an indication. It is also worth noting that only a small percentage of the parts will actually be involved in a crash that generates this loading mechanism.

The gap, between the predicted deflection and the statistical minimum performance changes from one type of test to another. An example is presented on figure 6 which compares two load cases; steering column and center test, on the same part.

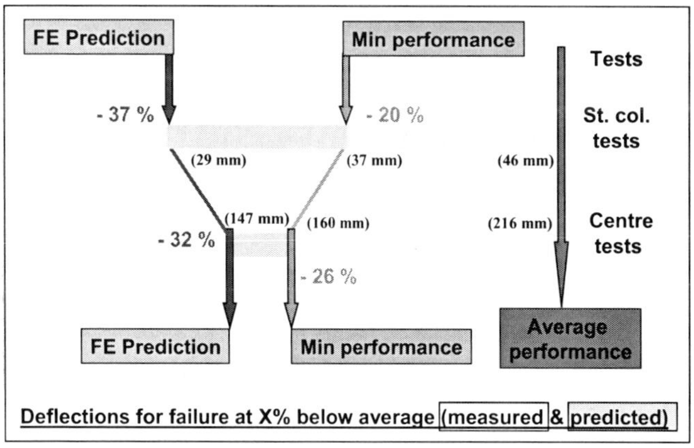

Figure 6: Comparison between the predicted deflection for failure and the minimum tested performance, for two different tests on the same component.

There are three main reasons why the security gap is not constant. The first one depends on the way the scatter is calculated. The scatter in performance, as defined by the difference between the statistical minimum and the average performance divided by the average performance, can also be seen in figure 6. As no failure is found in the elastic region, the scatter is triggered by the start of plasticity in the failure location. This localisation takes place at different displacements from one type of test to another. In a pure tensile load case, the failure location quickly moves into plasticity. In a bending test, most of the total deflection is created by local deformations all over the part prior to the final plasticity localisation and failure. Hence, the calculated scatter as a percentage of the total displacement is affected. The two other reasons why the security gap is not constant lie in the variation of the local properties over the part and the loading mode.

EFFECT OF THE PROCESS ON ROBUSTNESS

The most important factor that affects product reliability, in regards to the manufacturing process, is the design of the gating system. It drives local properties and major defects locations. Once the gating system is defined, the various process parameter variations within the defined control windows have little influence on the part performance. Some casting defects may occur but these have no influence on performance as long as they are not located in critical areas.

The robustness of the process parameters is often questioned by customers. The effect on performance or appearance of defects was assessed with a "design of experiment" study on the critical process parameters such as melt temperature, fast shot velocity, die lubricant ratio, and oil temperature. In this approach the parameters were adjusted within the limits of the process windows and the component performance was measured. The difference between the set up giving the lowest and the highest statistical minimum performance in failure deflection was only 15% (see example on figure 7).

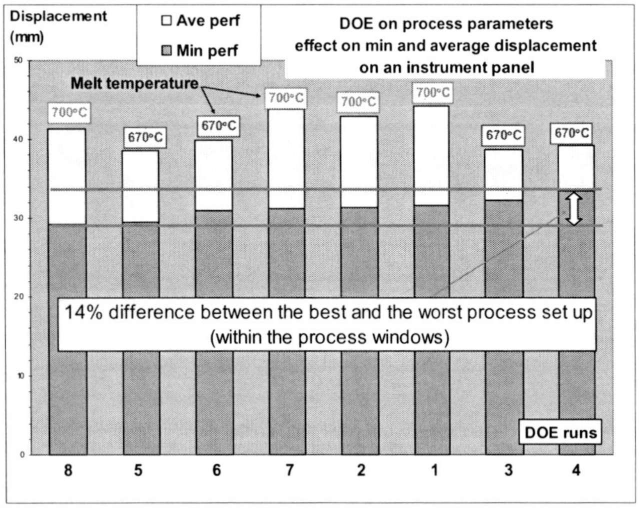

Figure 7: Example the effect of process parameters on average and statistical minimum performance.

In this experiment it can be seen in figure 7 that an increase of the melt temperature improved the average performance. This improvement was found to be related to the reduction of pre-solidified alpha phase, and is in agreement with the numerous process interactions [4] the minimum performance does not increase with this melt temperature.

EFFECT OF DIE CASTING DEFECTS ON PERFORMANCE

Even though the design is optimised, the prediction of local properties and defects from flow and thermal simulation is critical to a robust design. It is important to clarify the approach with regards to defects.

In the non-critical locations, defects have been shown to have no effect on crash performance through component tests and validation. The most detrimental defects that effect the performance are surface or through defects (shrinkage cracks, knit lines). By the utilisation of flow and thermal simulation defects can be minimised or predicted. This will also allow the design process to ensure these defects are not located in critical areas.

The core material will also contain other defects at a lower scale. These defects are micro-shrinkage of the pre-solidified alpha phase, and micro porosity. The effect of these micro-structural defects on properties is not yet fully understood because of the complex interactions. As these micro or macro scale defects are inherent to the die casting process, they are considered part of the "intrinsic" properties of the castings, and their effect on performance is taken into account in the design data and design approach.

MEASUREMENT OF LOCAL PROPERTIES WITH COUPON TESTS

CONSIDERATIONS ON THE RELEVANCE OF THE AMOUNT OF TESTS

The error that can be made when measuring the average elongation has been assessed on the basis of large amounts of tests, performed on different locations, through different production runs, and for different sizes of tensile coupons cut from parts. The evolution of the variation of the average versus the amount of tests was calculated and then normalised to one and presented in figure 8. It can be seen that it required about 30 tests to reach an accuracy of about +/-5% on the measurement of the average elongation. Measurements done on only 5 tests could lead to +/-50% variation on the average tensile elongation.

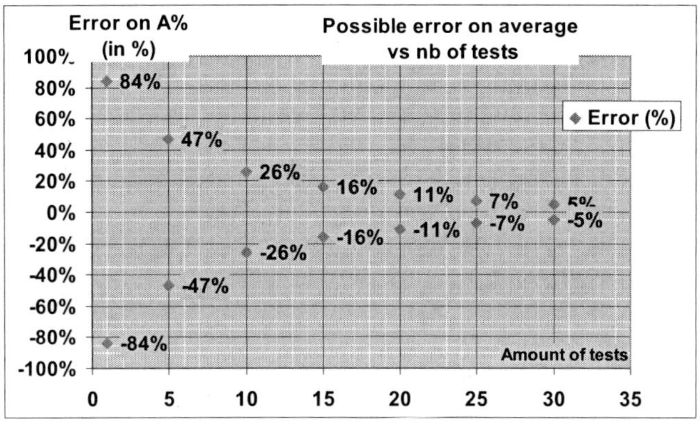

Figure 8: Assessment of the error on the average elongation vs. the number of coupon tests performed.

VARIATION OF THE ELONCATION OVER A LARGE CASTING

When a large amount of tests is done, it then becomes possible to better understand the variation of properties over a casting, and to initiate some correlation work between the local properties and flow and thermal simulation. Our experience is that property maps made with a small number of coupon tensile tests do not correlate.

It can be seen in figure 9 that the areas showing the lowest properties are the ones affected by defects and that these defects, like knit lines and air pockets, could potentially be estimated from flow and thermal simulation results.

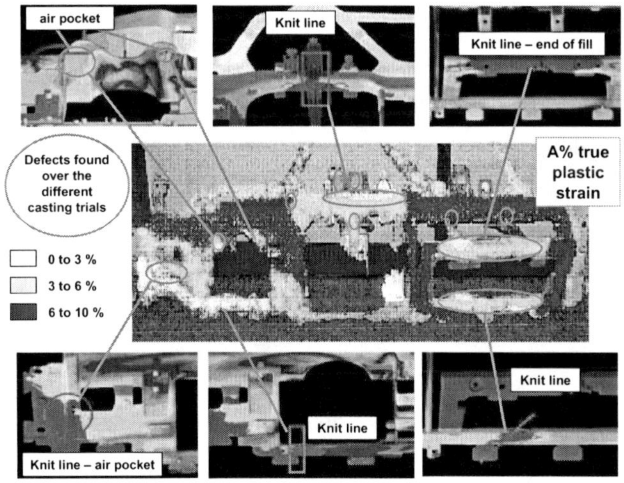

Figure 9: Simplified map of coupon tensile elongation, correlation with flow defects (air pocket, knit lines), [2].

Although a sufficient number of specimens will demonstrate a representative average elongation the overall scatter is relatively high. The high scatter results in a statistical minimum performance of up to -90% below the average (using 99% reliability with a Weibull distribution). This scatter is far above the scatter encountered on component tests. The reason for this difference in scatter between coupon tests and component tests is strongly influenced by the composite type characteristics of magnesium thin wall castings. With this identified composite structure, the measurement of the elongation is sensitive to the location, the loading mode, and to the effect of the edge machining when coupons are cut from components.

COMPOSITE FAILURE MODE

Average elongation and scatter are found to be different between four point bending and tensile coupon tests. In a preliminary study carried out in three locations of an IP, both tensile and bending tests have been performed. The results are presented in figure 10 and show that the average tensile elongation decreases from location A to C, while bending properties are similar for all locations.

This study was carried out on 1 inch (small) and 2 inch (large) gage length specimens to investigate the coupon size affect. No size effect was found on average properties or scatter.

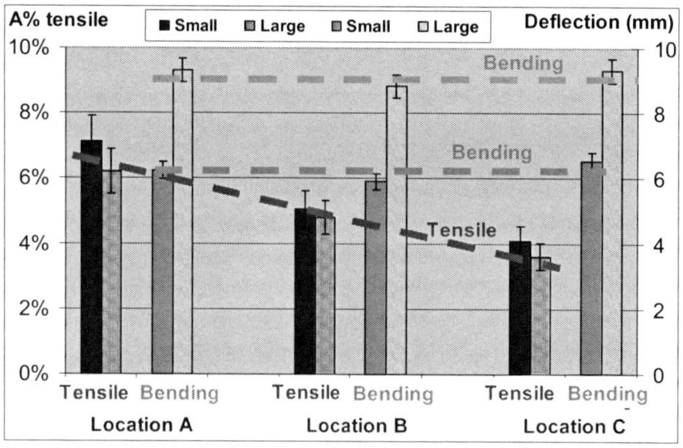

Figure 10: Comparison of tensile and bending properties at full failure, over 3 different locations of a part.

The tensile / bending tests comparison was extended to 37 locations [2]. The plastic strain measured at failure on the bending coupons was found to be more consistent than the tensile plastic strain. This trend is clearly visible on figure 11, where locations have been sorted from the lowest bending average elongation to the highest.

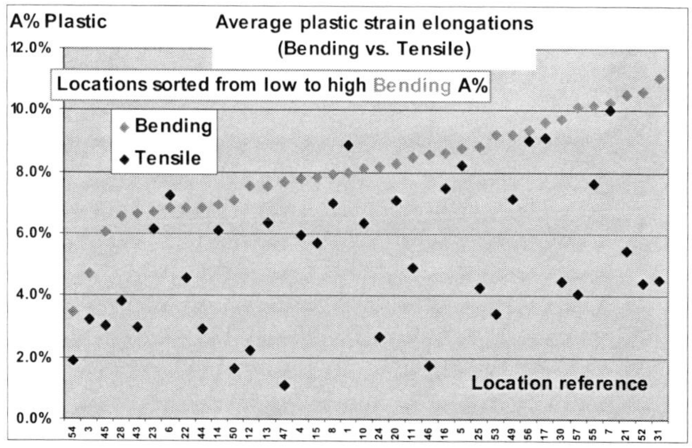

Figure 11: Comparison of tensile and bending properties at full failure, over of an instrument panel.

When comparing tensile and bending data, the scatter for the bending deflection at full failure was reduced compared to the tensile coupon elongation (see figure 12). The minimum performance at 99% reliability is approximately 80% to 90% below the average elongation in tension while, the minimum performance is approximately 20% to 40% below the average for the bending deflection at full failure. The bending scatter is similar to the scatter found in component tests.

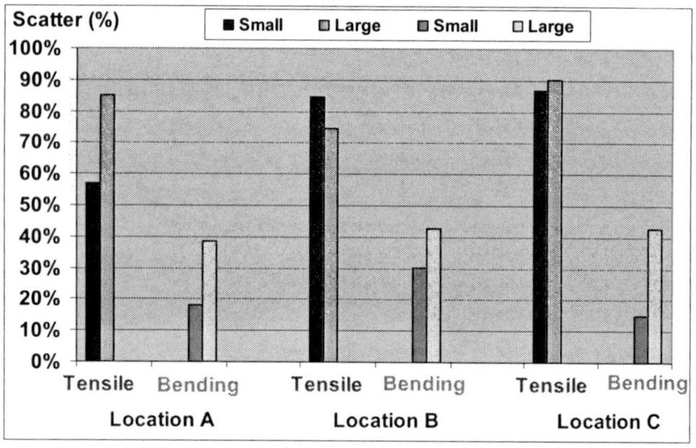

Figure 12: Comparison of the scatter for tensile elongation and bending deflection at full failure, over 3 different locations.

The material could be visualised as 2 skins with high elongation and low scatter, and a core showing lower properties and higher scatter (see figure 13). The component tests with bending mode failures will take advantage of the skin properties, while in the tensile load cases, stressing the core material will be more detrimental. Although bending represents 99% of the load cases in crash applications, local tensile hot spots have to be managed through design.

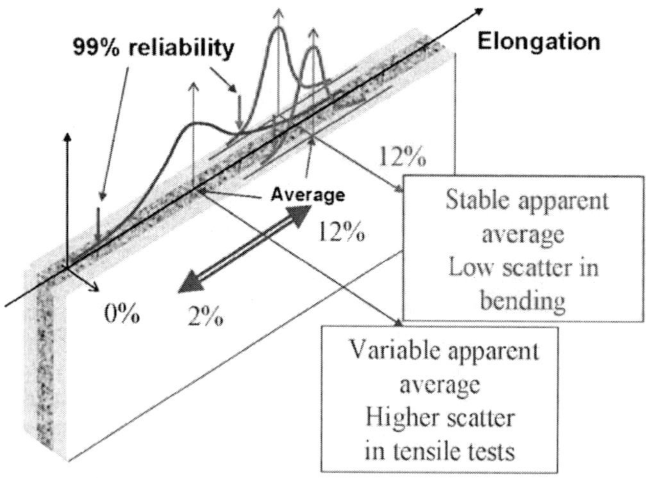

Figure 13: Simplified representation of the composite failure properties of thin wall magnesium die castings.

EDGE EFFECT IN TENSILE COUPON TESTS

When removing the complete skin in the range of 500 to 600 microns on large thin wall casting, an edge effect is found. This was found when comparing the tensile elongation values measured on a 2mm thick cast flat bar, and on similar bars of which the edges had been machined deep enough to open the core structure to the surface (see figure 14). It can be seen that both the average and minimum elongation can be affected. In the studies completed, the average elongation was lowered by 2% and the relative scatter is increased.

More work is required to further understand, but this could be due to the fact that, after machining, core defects are in a plane stress distribution while on a bulk component they are in a plane strain configuration. The size of the performance effect is dependent on the local core soundness.

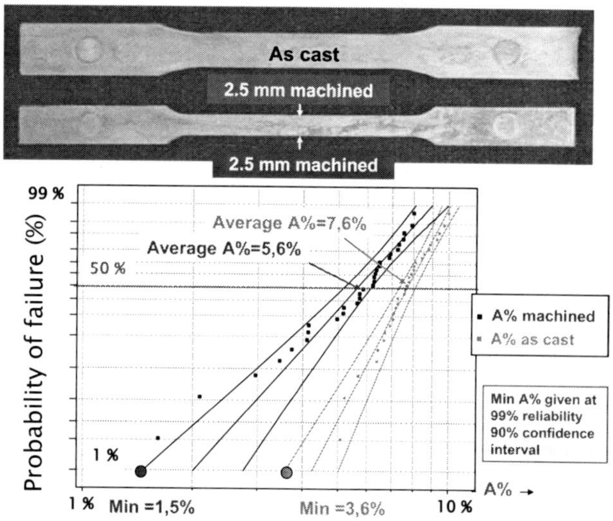

Figure 14: Edge effect on coupon tensile properties of thin wall Mg die castings; decrease of the average A%.

COMPONENT AND COUPON TESTS

VARIATION OF THE SECURITY GAP

As previously stated, the security gap depends on the local properties, the loading mode in the failure location, and the start of the local plasticity in the failure location.

In most cases the variation of the local properties is the most important parameter. Although the absolute values are affected by the edge effect, tensile tests performed locally can give an indication of different performance levels as shown in figure 15. The tensile coupon elongation is found to be 8% in the steering column area and 4% in the center tests failure location. The center elongations are lower due to the fact that two flows merge in this area. Despite this difference in tensile elongation, robustness was achieved in both failure locations using the equivalent 6% strain criterion.

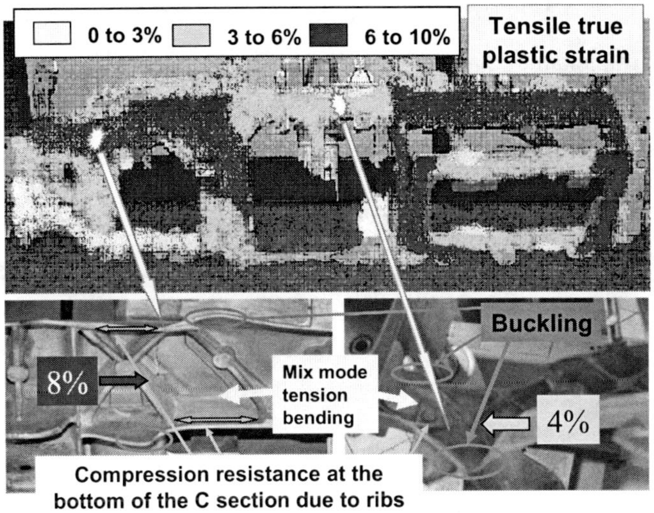

Figure 15: Local tensile coupon elongations.

Most failure modes on an IP are similar to the one shown in figure 15. The part starts buckling at the upper sides of the major "C section", initiating the bending of the bottom of the "C section". Due to the ribbing, the bottom side walls remain in compression, inducing some tension on the outside skin of the bottom "C section". The bottom "C section" is then subjected to a tensile/bending mode resulting in failure. The identification of the type of loading mode in the failure location is made by analyzing the distribution of the strain though the wall thickness.

EFFECT OF AN ALLOY CHANGE ON PERFORMANCE

In most cases, the skin will have a major effect on performance and scatter. The skin is believed to be close to the intrinsic properties of the alloy, while the core is more driven by micro-structural features and defects inherent to the process. Hence, an alloy change may have different effects depending on the loading mode and core soundness. In a bending failure mode the alloy change effect on performance will be significant as is illustrated in figure 16, where AZ91, AM60 and AM50 are compared. In these cases the improvement in deflection is very significant.

With a tensile failure mode, a change of alloy may not change the performance as core properties can be more driven by the process defects. If the core properties are low, little improvement on performance will be found with increased elongation alloys.

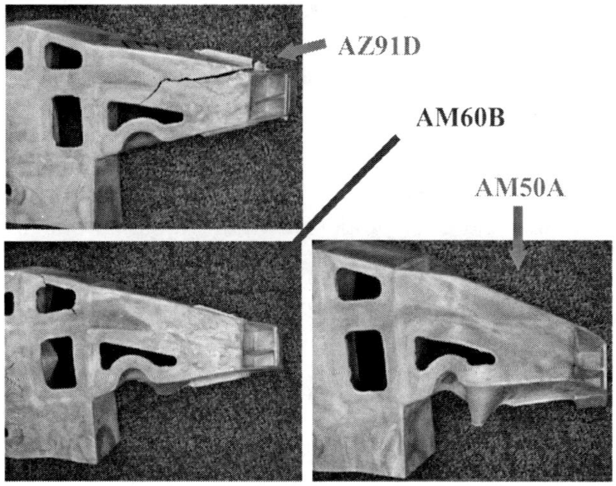

Figure 16: Effect of an alloy change for a bending driven failure mode.

CONCLUSION

Intensive investigations have been carried out to ensure the robustness of magnesium components for crash applications.

A reverse engineering approach was first used to develop an equivalent stress/strain curve. As a second step the same reverse engineering approach, supported with statistical analysis, has been used to define an equivalent strain that is conservative for common load cases.

To design for reduced scatter it is important to have the minimum number of plastic strain localizations and avoid stress balances across the component. If this is achieved then very little of the part plastically deforms during loading conditions and the process can be focused on achieving the best properties in these areas.

Majority of the load cases in crash situations are close to bending type modes which take advantage of skin properties in die castings. The scatter associated with component tests was found comparable to the scatter encountered in coupon bending tests.

The detailed investigation carried out on specimens cut from components tends to show that the material exhibits a composite failure mode. For magnesium thin wall castings, the material can be considered as two skins and one core sandwich structure. The skin properties appear to have high and consistent properties over the castings and are believed to be close to the intrinsic properties of the alloy. The core shows lower properties, higher scatter, and variable elongation over the casting. These properties are driven by micro defects inherent to the die casting process.

Component performance can not be assessed from coupon tests as the machining of the edges open the core structure to the surface and bias the measurement of the elongation.

The robustness of magnesium lies first in the robustness of the design, and then in insuring that the process generates the best properties where required and that there are no defects in the critical areas.

There is still a lot of work remaining to get a better understanding of the complex die cast magnesium properties and link to microstructures. Meridian will maintain its commitment to improve the knowledge in this area and to supply to its customers the tools and the data to develop robust magnesium die cast components.

ACKNOWLEDGMENTS

The authors would like to acknowledge the Advanced Engineering, Forward Program and Core Technology departments of the Meridian Global Technology Center organization for their support and contribution.

REFERENCES

1. Structural magnesium die castings, developing a robust solution, Glen Simonds, Joseph Wang, Meridian Technologies Inc, IMA Brussels, May 20th-22nd 2001.
2. Mechanical properties and microstructure of magnesium high-pressure die-castings, Brandon J. Coultes, Jeffrey T. Wood, Gerry Wang, Richard Berkmortel, Magnesium Technology 2003, San Diego, California, USA, March 2-6, 2003.
3. High pressure die casting – Getting the best out of magnesium; Haavard Gjestland, Stian Sannes and Hakon Westengen, Magnesium Materials Technology, Norsk Hydro, N3910 Posgrunn, Norway.
4. Development of magnesium die casting alloys, with emphasis on strength-ductility; Per Bakke, Ketil Pettersen, Hakon Westengen; Magnesium Materials Technology, Norsk Hydro, N3910 Posgrunn, Norway.
5. The influence of process conditions on microstructure and mechanical properties of magnesium die castings; S. Sannes, H. Westengen, Norsk Hydro Research Centre, Posgrunn, Norway; International conference on magnesium alloys and their applications, 28-29 April 1998, Wolfsburg.

CONTACTS

ralain@meridian-mag.com
tlawson@meridian-mag.com
pkatool@meridian-mag.com
gwang@meridian-mag.com
jjekl@meridian-mag.com
rberkmortel@meridian-mag.com
lmiller@meridian-mag.com
Jorild.M.Svalestuen@hydro.com
hakon.westengen@hydro.com

2004-01-0136

Reliable Light Weight Fastening of Magnesium Components in Automotive Applications

C. Friedrich
RIBE Verbindungstechnik GmbH

ABSTRACT

This paper deals with optimized fastening of magnesium components to overcome serious problems in practice. For this, fasteners made of aluminum have significant advantages compared to conventional steel fasteners. The reasons are shown to the reader; calculations and established examples of application from series production confirm the behavior in practice.

INTRODUCTION

Light weight design is of increasing importance for every development of a new vehicle, e.g. [13]. A lot of new materials, especially magnesium alloys have been developed and introduced in series production in the last years, e.g. [14, 15, 16]. With these materials the component weight is cut down, but in most cases, a reliable fastening of the components causes serious problems like significant preload relaxation or extensive galvanic corrosion [4]. Also the length of thread engagement normally has to be increased for magnesium parts which means large space requirements for fastening positions as well as more weight for screw and screwing boss. To overcome this, bolts and screws made of aluminum with a special profile of properties are very helpful. They are the ideal fastening element for magnesium – RIBE has made this technology available, **Fig. 1**.

This contribution deals with two main issues:

- proposing the calculated preload behavior for a fastening design with steel screws in a magnesium gearbox housing vs. aluminum screws. Main information is achieved with a new `preload history diagram´ for fastening systems. The preload history shows a clear advantage for aluminum screws which is also verified experimentally with a lot of preload measurements at various components. So, a real comparison of calculation and experiment is presented.

- demonstrating of recently launched examples of automotive applications with aluminum screws. The outlook to experience from practice is the best way to generate confidence in this light weight fastening technology.

Fig. 1: *range of RIBE© Aluform© Fasteners; covers bolts, nuts, studs, staking elements, rivets; covers all kinds of geometrical-, surface finishing-, locking- and sealing options; M5 to M18*

BENEFITS OF ALUMINUM FASTENERS

The basic background was presented at SAE congress 2000 [1], see also [2, 3]. Therefore, here only the main aspects are listed which are relevant for comparison of aluminum fasteners to steel fasteners:

1. weight saving of over 60% per each fastener, which leads to a range of `kilograms´ saved for typical engine attachments with flanges.
2. elimination of affecting galvanic corrosion in contact with magnesium alloys, so the danger of corrosive failures is overcome.
3. better fit of the mechanical properties to magnesium which leads to low contact pressures and small required length of thread engagement; this provides the chance for small fastening systems.
4. impressive reduction of extreme preload relaxation of steel screws after tightening during operation at elevated temperatures.

Fig. 2 shows in a simple chart the weight of typical bolts with metric thread profile and a flange head with hexalobular screw drive. It is confirmed that a flange of 20 fasteners M10 leads to a weight saving of app. 0.5 kg when using aluminum screws instead of steel screws. A secondary weight saving in the same range or even higher can be achieved, if the small length of thread engagement is taken into account in order to reduce the length of screwing bosses.

In the diagram of Fig. 2 also the usually used screw size of typical applications in the automotive sector is indicated like interior parts, manifolds, oilpans, housings (e.g. gearbox), attachments (e.g. fixing of alternator), mounts for engines or plug screws for closing openings of bores. With this diagram a design engineer can estimate the range of possible weight saving by aluminum screws for his application in some seconds.

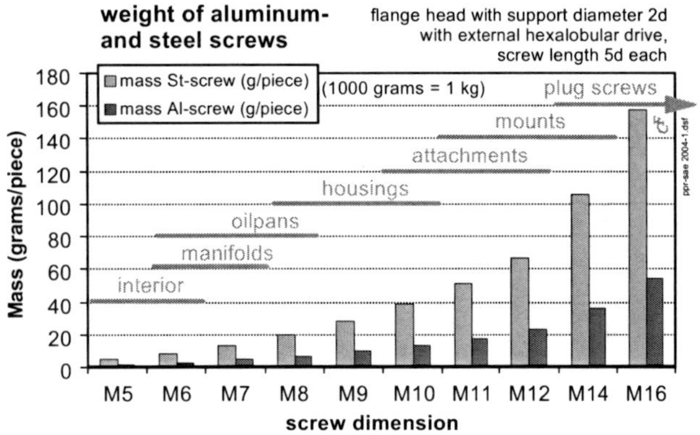

***Fig. 2:** weight of aluminum- and steel bolts dependent on their size*

Fig. 3 confirms the corrosive stability of aluminum screws with an image of a corroded component surface in contact with an aluminum screw made of the alloy AA 6056 (tensile strength of screw 400 MPa at minimum, EN 573-3 [10]). As it can be seen the aluminum screw appears almost unchanged after the applied salt spray test and there is no galvanic corrosion around the screw contact area exceeding the average amount of oxide products on the surface.

***Fig. 3:** image of corroded component after 150h of salt spray test with appearance of almost unchanged surface of the aluminum screw [2]*

A result of the much better matching of mechanical properties of aluminum than steel to magnesium is the possibility to create much smaller fastening systems, if an aluminum screw is used. **Fig. 4** emphasizes the details for this aspect.

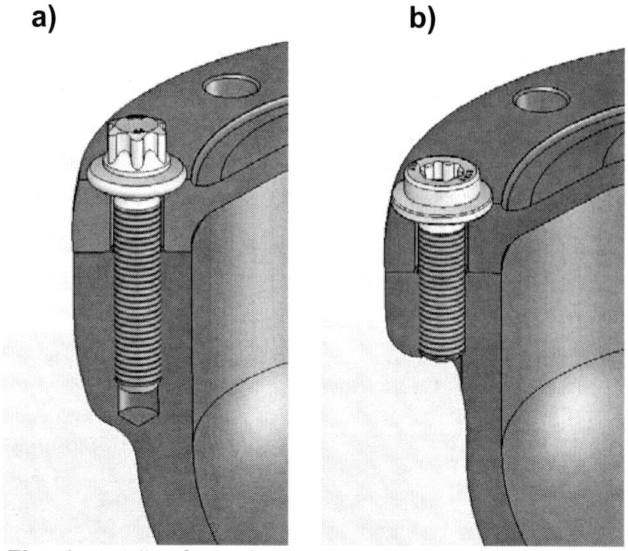

***Fig. 4:** smaller fastening systems with aluminum screw for components made of light metals like magnesium; a) steel screw 8.8, b) aluminum screw RIBE AL9, same or even better residual preload than a)*

DESIGN EXAMPLE

To give a direct impression of the design with aluminum screws M8 in a magnesium gearbox housing, we look to results of a design calculation with ʽpreload history´. This design is verified with preload measurement in experiment. It shows clearly the benefits of a strongly reduced preload relaxation. **Tab. 1** summarizes the main characteristics of the investigated fastening system including tolerances.

Figs. 5 and **6** give a compressed overview about the calculated and experimentally verified preload change in the fastening system for steel screw 8.8 (Fig. 5) and aluminum screw AL9 (Fig. 6), distinguished with 5 steps into

1. tightening (tightening preload at tightening temperature F_p)
2. seating after tightening (remaining preload F_{ps} after plastification of mechanical contacts at tightening temperature)
3. heating up to operating temperature of 140 °C (operating preload F_{pT} at 140 °C = 284 °F)
4. cooling down after step 3. to tightening temperature (stable preload at room temperature after short time thermal loading F_{pr} for some hours without creeping)
5. long time residual preload at tightening temperature after relaxation F_{pcr} (decisive for reliability)

These diagrams are obtained from the 'preload history' of the comprehensive design software 'Screw-Designer' [4]. This analytical software takes into account a possibly existing yielding of screw and/or clamped part, which can be generated by thermal and/or mechanical loading. Besides this, for this application it is necessary to consider exactly the different tightening methods used in the example (torque tightening 20 Nm for steel screw and angular controlled tightening specification of 10 Nm + 90° for aluminum screw). Screw-Designer allows directly the input of these tightening specifications - no other calculation tool can provide these features. Only using the linear equations from common used sources like VDI 2230 without yielding effects does not cover the real thermal behavior of such fastening systems.

Main characteristics of fastening system	value
Dimension of thread	M 8 x 1.25
material of screw	steel 8.8 and aluminum AL9
support diameter of screw head	16.7...17.5 mm
material of clamped part	magnesium AZ91
clamping length	23 mm
material of nut thread component	aluminum A 226 (Al Si 9 Cu 3)
length of thread engagement	20 mm
maximum temperature of operating	140 °C
tightening specification of steel screw 8.8	19...21 Nm
tightening specification of aluminum screw AL9	10...11 Nm + 80...100°
estimated frictional coefficients each	0.09...0.14
calculation with consideration of temperature dependent material properties incl. yielding	
situation with * is calculated with data from experience	

***Tab. 1:** main characteristics of exemplary fastening system M8 for calculation and preload measurement, results in Figs. 5 and 6*

Fig. 5 displays a calculated tightening preload F_p of 11...18 kN for the steel screw 8.8 (left bar). The verifycation in experiment from app. 500 assembly procedures shows a range of 10...20 kN in reality. This difference between calculation and measurement is due to the fact that for calculation standard frictional coefficients in the range of 0.09...0.14 have been taken, but these might have a wider spread in experiment. The calculated preload after seating (as expected) only shows a slight change compared to the tightening situation.

At 140 °C the preload is significantly increased to 18...23 kN which means a strong additional preload of app. 5 kN at the operating temperature (app. +50% of the tightening situation!). So, at the same time when magnesium has a reduction in material properties (by the temperature of 140 °C), it even gets an additional loading by the fastening system – no wonder that this leads to preload loss.

The quantification of residual preload is done by the two right situations in Fig. 5. The residual preload after short time thermal loading F_{pr} tells a minimum value of app. 10 kN, the long time stable residual preload F_{pcr} is determined with 0...1.2 kN which means a fastening system almost out of operation. This really unacceptable preload loss is confirmed by the measurements, where the results of residual preload after thermal exposure to 140 °C for 500h are in the range of 0.4...1 kN. This must be avoided – but how?

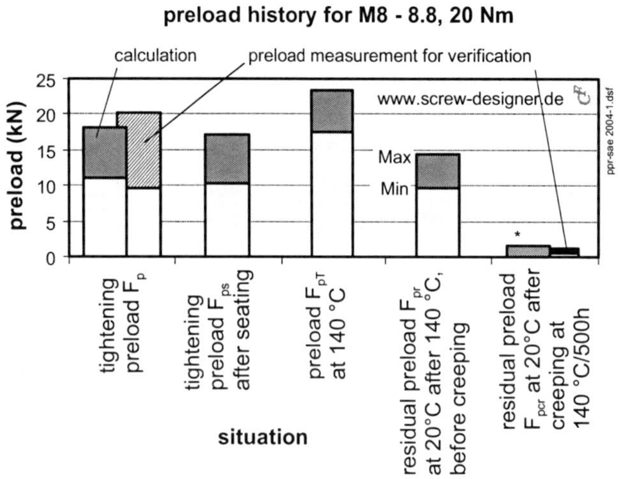

***Fig. 5:** preload history of the steel screw 8.8; detailed definition of fastening system in Tab. 1*

Fig. 6 shows a solution with the corresponding diagram to Fig. 5 for an aluminum screw made of AL9, which is tightened overelastically with 10 Nm + 90°. The tightening preload is calculated by 11...14 kN; the measurements result in 11...13 kN. Both, calculation and measurement, lead to a significantly reduced deviation of tightening preload at the same level of the minimum of steel screw 8.8 (it is 10 kN, comp. Fig. 5).

The difference in thermal behaviour compared to the steel screw 8.8 becomes obvious at 140 °C: the operating preload F_{pT} does not increase, but even goes slightly down to app. 9 kN at minimum by reason of the changing properties of the aluminum screw material. So, at 140 °C the magnesium material is prevented from additionally thermal loading, which leads to lower creeping effects. In consequence, the long time residual preload F_{pcr} is strongly increased from app. 1 kN (Fig. 5) to 4 kN (Fig. 6). Another aspect is the much better fit of mechanical properties like thermal expansion coefficient of steel screw and aluminum screw compared to the magnesium material [1] – the result can be seen from the difference between F_{pT} and F_{pr} which is very small (only app. 1 kN in Fig. 6, but even app. 8 kN in Fig. 5). Overall, Figs. 5 and 6 show that with aluminum screws the long time residual preload is increased by app. 4 times compared to steel screws.

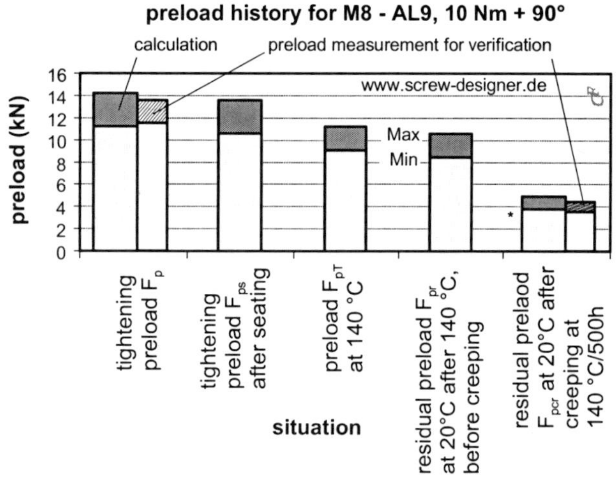

Fig. 6: *preload history of the aluminum screw AL9; detailed definition of fastening system in Tab. 1*

The preload measurement – here done with use of ultrasonics and a special evaluation-routine for overelastically tightened screws - is much more suitable to assess the quality and reliability of a fastening system than only determining torque values, because a high residual retightening torque can be generated with both, either by high residual preload and low residual friction (which is positive) or by low residual preload and high residual friction (which is negative). So, a high retightening torque says nothing about the really stable clamping force within the fastening system.

ESTABLISHED APPLICATIONS

Nowadays, numerous applications of aluminum screws in large series production of automotive components like gearboxes, manifolds or engine attachments confirm the benefits of RIBE Aluform coldformed parts [7] every day.

For the use of aluminum screws it is important to guarantee with high reliability the top level of mechanical properties, the right heat treatment for every part of the mass production as well as all sophisticated material forming and thread rolling processes without local overstressing during forming (see also [5]).

RIBE has done a lot of efforts to enhance all manufacturing activities due to optimized procedures for high strength aluminum: this looks quite easy at the beginning of dealing with aluminum screws, but leads to necessary investigations, reduced process deviations and improvements for nearly all manufacturing aspects; so, it is the best base to work together with a competent partner, if aluminum screws have to be introduced in new components.

Fig. 7 shows the new 7-speed automatic transmission from DaimlerChrysler which is launched in 2003. The housing is made of magnesium; it is using aluminum screws made of RIBE 63 (AL9) with hexalobular screw drive for fastening the different modules and gearbox mount as well as fastening of the exhaust system to the magnesium housing.

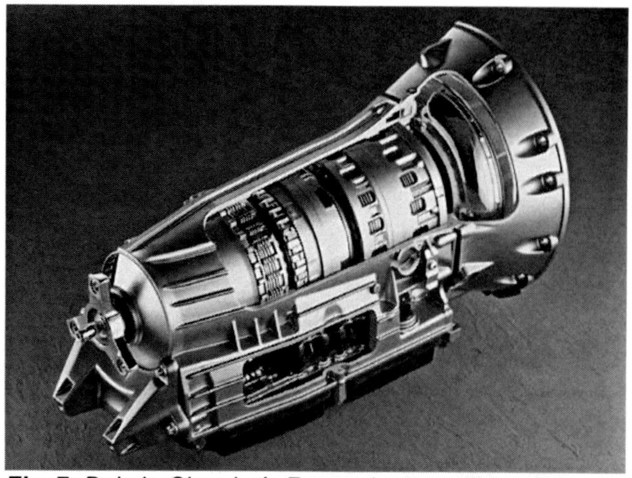

Fig. 7: *DaimlerChrysler´s 7-speed automatic transmission 7G-Tronic with Aluform screws, launched to market 2003 [8]*

Another example with aluminum screws in series production is ZF´s manual 6-speed gearbox ZF S 6-3-7 (**Fig. 8**). It is working with aluminum screws for fastening the different parts of the aluminum housing, positioning the devices for shifting and as bearing bolt for the shifting devices.

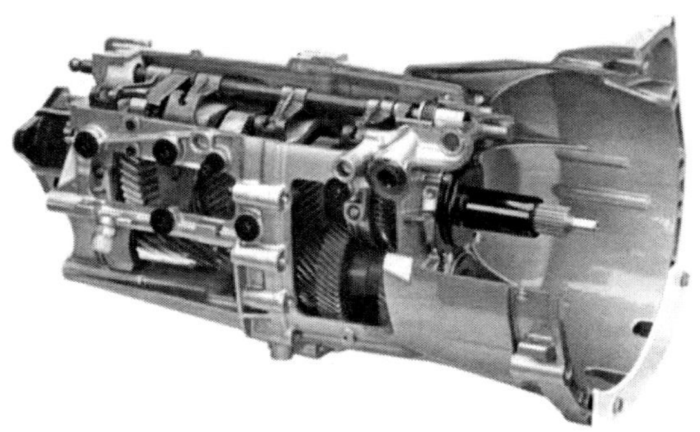

Fig. 8: *ZF´s manual 6-speed gearbox S-6-3-7 with Aluform screws for holding together the parts of the housing, internal fixing of shafts and devices, bearing bolts [9]*

Also the generation of continuously variable gearboxes like the ZF-CFT23 (**Fig. 9**) is working with aluminum screws made of RIBE63 (AL9) to reduce weight. It is manufactured at ZF´s Batavia plant in the U.S.

Fig. 9: ZF´s CFT 23 gearbox with Aluform screws; continuously variable gear ratio for passenger vehicles [9]

The innovative AUDI FSI technology with direct fuel injection to reduce fuel consumption and emissions while enhancing the available power of the engine is also using the benefits of aluminum screws in the field of the air intake system (**Fig. 10**).

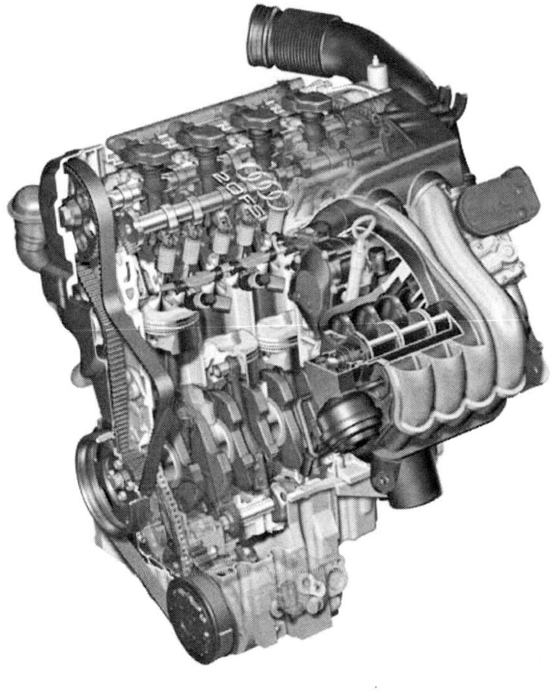

Fig. 10: AUDI´s engine FSI 2.0 with Aluform screws; direct fuel injection [11]

Besides the powertrain applications above, **Fig. 11** shows the intake manifold made of magnesium for the AUDI W12 engine used in their A8 model – aluminum screws are fastening this manifold to the cylinder head made of aluminum since several years.

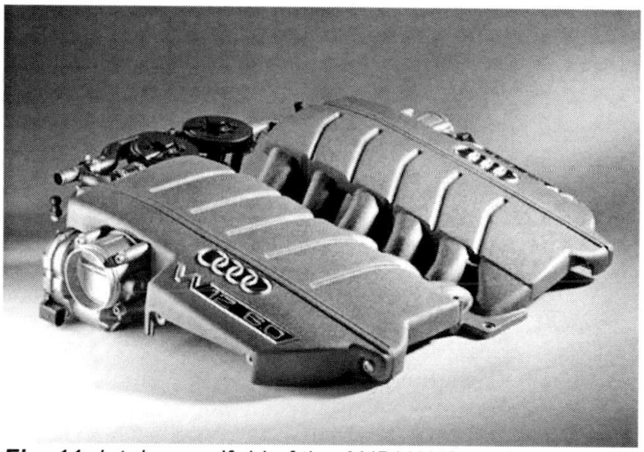

Fig. 11: Intake manifold of the AUDI W12 engine with Aluform screws [11]

Another example for application of aluminum screws will be the new BMW 6 cylinder water cooled engine with an engine block made of Magnesium with inserts of aluminum (**Fig. 12**). It can be read that it will be the lightest engine in this power range. Here a lot of different positions will use Aluform screws.

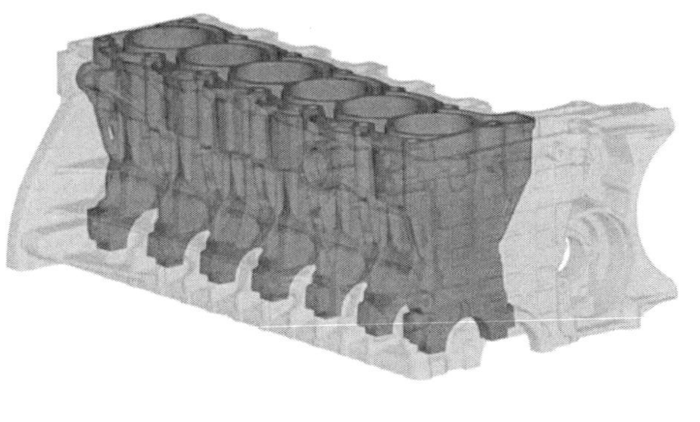

Fig. 12: public image of BMW´s new magnesium intensive 6 cylinder engine [12]

CONCLUSIONS AND OUTLOOK

As an overall result we can state that aluminum fasteners are the most suitable solution due to threaded fastening systems for contact with magnesium components. The stable corrosive behavior and the relatively high residual preload lead to a significant increased reliability without leaking or even breakdown of the lighter component. This was shown with an exemplary calculation which is experimentally verified. Several applications in the automotive market are using these benefits in series production

Meanwhile, all fastening experts have verified and accepted these aspects; therefore, inbetween a standardization of such fasteners made of corrosive stable high strength aluminum AL9 is done in numerous work standards of RIBE and OEM´s. It is discussed with the German secretariate of DIN Cologne to include the mechanical properties of the aluminum alloys AA 6013 resp AA 6056 as `AL9´ in DIN EN 28839 resp. ISO 8839 [6].

REFERENCES

1. C. Friedrich: Fastening of light weight components with elements made of aluminum. Paper 2000-01-1118, SAE Congress 2000, March 6-9.
2. C. Friedrich: Better fastening of light weight components with elements made of aluminum. Aluminum Technology Seminar, Troy, Michigan, 2001, May 9.
3. C. Friedrich: Designing Fastening Systems. In: Totten: Modelling and Simulation for Material Selection and Mechanical Design. Dekker 2004. ISBN 0-8247-4746-1.
4. C. Friedrich: Screw-Designer. Intelligent Design of Threaded Fastening Systems. 2004. www.screw-designer.de
5. K. Downs: Aluminum use in Fastener Manufacturing. Fastener Technology International, October 1996.
6. ISO 8839: Mechanical properties of fasteners – bolts, screws, studs and nuts made of non-ferrous metals.
7. RIBE Aluform brochure. VT/102/2/9811/2.0.
8. picture DaimlerChrysler, www.daimlerchrysler.com
9. picture ZF Friedrichshafen, www.zf.com
10. Standard EN-573-3: Aluminium and Aluminium alloys – chemical composition and form of wrought products.
11. picture AUDI Ingolstadt, www.audi.de
12. picture www.germancarfans.com
13. Towards the light – DaimlerChrysler´s research chief explains why steel has had its days and why finding lighter materials is a priority. Automotive Engineering, March 2001.
14. W. Diem: Magnesium in different applications. Auto Technology 1/2001.
15. R. Stauber and M. Baur: Advanced Material Technologies – a key factor for ultimate driving performance. Paper 2000-01-0418, SAE Congress 2000, March 6-9.
16. J.L. Broge: The Battle of the Metals. Automotive Engineering International, August 2000.

ACKNOWLEDGEMENT

Many thanks to the RIBE Aluform Team for valuable discussions and engaged support, especially to Dr W Thomala and M Leng.

CONTACT

Dr C Friedrich
Richard Bergner Verbindungstechnik GmbH
D 91124 Schwabach
Germany
Phone ++49-9122-87-1471
Fax ++49-9122-87-1504
christoph.friedrich@ribe.de

2004-01-0654

Progress Toward a Magnesium-Intensive Engine: The USAMP Magnesium Powertrain Cast Components Project

Bob R. Powell and Larry J. Ouimet
General Motors Corporation

John E. Allison and Joy A. Hines
Ford Motor Company

Randy S. Beals and Lawrence Kopka
DaimlerChrysler Corporation

Peter P. Ried
Ried & Associates

ABSTRACT

The US Automotive Materials Partnership (USAMP) and the US Department of Energy launched the Magnesium Powertrain Cast Components Project in 2001 to determine the feasibility and desirability of producing a magnesium-intensive engine; a V6 engine with a magnesium block, bedplate, oil pan, and front cover. In 2003 the Project reached mid-point and accomplished a successful Decision Gate Review for entry into the second half (Phase II) of the Project. Three tasks, comprising Phase I were completed: (1) evaluation of the most promising low-cost, creep-resistant magnesium alloys, (2) design of the engine components using the properties of the optimized alloys and creation of cost model to assess the cost/benefit of the magnesium-intensive engine, and (3) identification and prioritization of scientific research areas deemed by the project team to be critical for the use of magnesium in powertrain applications. Phase II of the Project also comprises three goals: (4) casting and dynamometer-testing of the magnesium components in assembled powertrains, (5) developing a powertrain magnesium alloy design database and common alloy specification for magnesium powertrain alloys, and (6) promoting the strengthening of the scientific infrastructure for magnesium in North America to enable even more advanced powertrain applications in the future.

INTRODUCTION

The earliest known application of magnesium alloys (Mg) in automotive transportation was in the 1930's when castings were used for the lower crankcase in city buses in London, England [1]. In that same time period over 500,000 produced [1]. Since then, the first and, to date, only major commercial use of magnesium in powertrain systems was its use by Volkswagen (VW). VW produced both crankcases and transmissions with high-pressure die cast AZ81 (Mg-8^{w}/oAl-1^{w}/oZn). Together these components weighed 17 kg, which was 50 kg less than the iron components they replaced. In the 1960's, 4000 vehicles per day were produced, requiring the equivalent of 30,000 tonnes of magnesium per year [1]. Production reached at 42,000 tonnes per year in 1971, but diminished thereafter [2].

Unfortunately, several factors were emerging that combined to cause the reduction and eventual elimination of magnesium as a structural powertrain component. These factors included: (1) the conversion of the engine from air to water cooling, (2) the increase in the power requirements of the engine, which increased both its operating temperature and load, and (3) the increasing cost of Mg relative to Al [2,3]. The use of water-based coolants put magnesium at a disadvantage compared to aluminum and iron because of its poor corrosion resistance. At that time the accelerating effect of iron, copper, and nickel impurities (in ppm amounts) on the corrosion of magnesium had not been recognized [4]. By the time it was, in the 1980's, the other factors of high temperature and cost were closing the door on magnesium usage. For example, as the necessary operating temperature of the engine increased it became apparent that AZ81 did not have adequate strength or creep resistance at these temperatures. This led to the development of lower-aluminum-content Mg alloys, first AS41 (Mg-4Al-1Si) and later AS21 (Mg-2Al-1Si) [3]. However, magnesium alloy cost gradually increased and by 1982 the VW no longer used magnesium in the powertrain.

Magnesium has yet to break back into the mainstream of structural powertrain materials. While there are some exceptions such as the VW Passat and Audi gearbox housings [4], the volume of these applications is small. The real growth in automotive magnesium use has been in non-powertrain applications. In the 1990's, magnesium usage in these applications grew significantly. These applications, which accounted for a better than ten percent annual growth in magnesium usage per year throughout the 1990's, included instrument panel structures and cross-car beams, seat frames, steering wheels, intake manifolds, and numerous brackets and covers [5].

In the non-powertrain applications, magnesium alloys have demonstrated their ability to significantly reduce weight at acceptable automotive costs. However, the potential weight-reduction benefit of magnesium has not yet been realized in powertrain components. There are four reasons.

- The higher cost of magnesium relative to aluminum and iron, the traditional powertrain materials, is further increased due to the requirement of using rare earth additions in order to achieve high-temperature creep resistance of the alloys [6]. The non-powertrain alloys are AZ91D (Mg-9Al-1Zn), a high-strength and low-ductility alloy, and AM50/60 (Mg-5Al-0.4Mn and Mg-6Al-0.4Mn), lower strength and higher ductility alloys, all of which are limited to applications for which operating temperatures are below 125°C.
- There is limited powertrain design experience with magnesium alloys. While the major auto companies have experimented with magnesium engine blocks in the past, the scope of the work has been narrow and short-lived; e.g., resulting in small gains in design knowledge.
- There is no long-term field validation or controlled-fleet testing data of magnesium structural powertrain components, with the exception of the previously mentioned VW experience.
- There is limited scientific infrastructure in North America that is directed to developing the necessary understanding of magnesium alloys and casting processes.

Recently, several developments have taken place that may change the potential for powertrain applications of magnesium alloys. First has been the development of several potentially low-cost, creep-resistant alloys [7-11]. Second, the extensive utilization of magnesium alloys in non-powertrain components has resulted in the development of design experience and the beginnings of a mechanical properties database. Furthermore, a program to formally develop a magnesium design database was started in 2000. This program, the USAMP Structural Cast Magnesium Development Project (SCMD), seeks to resolve the critical issues that limit large-scale applications of magnesium castings in automotive body and chassis components [12]. The third development has been a series of magnesium technical symposia held in North America at the 2000, 2001, 2002, and 2003 TMS annual meetings [13-16]. These meetings, as well as the symposia held at the SAE annual meetings have contributed to greater awareness of the issues and opportunities for automotive magnesium applications and research needs in North America.

The convergence of these developments, coming at a time of increasing fuel costs, provided a unique opportunity and the need for the automotive industry to address the question of a magnesium-intensive engine. To this end, DaimlerChrysler, Ford, General Motors, the Department of Energy, and a substantial portion of the magnesium community agreed to work together to develop an ultra-light-weight powertrain based on magnesium alloys. In 2001, the Magnesium Powertrain Cast Components Project was launched under the organizational leadership of The United States Council for Automotive Research (USCAR) and The United States Automotive Materials Partnership (USAMP). This paper provides an overview of the Project and its accomplishments at mid-point in the Project.

THE MAGNESIUM POWERTRAIN CAST COMPONENTS PROJECT

The Magnesium Powertrain Cast Components Project (MPCC) was organized to realize the vision of a magnesium-intensive engine that has demonstrable cost-effective performance benefits and meets the manufacturability and durability requirements of the automotive industry. To accomplish this vision, the Project must answer two questions.

1. Are magnesium powertrain components technically feasible?
2. Does the cost-to-benefit ratio of magnesium in powertrain components warrant its use?

If the answer to either question is "no," then a third question will have been answered.

- What is necessary to make magnesium powertrain components technically feasible or their cost-to-benefit ratio good enough, such that they are desirable to the automotive industry?

To answer the above questions the MPCC Project team redesigned a V6 engine from 319 aluminum alloy to magnesium. The engine chosen was the 3.0L Ford Duratec, which is shown in Figure 1. A necessary compromise was made in arriving at this selection. A production engine was chosen in order to allow production parts to be used in the assembled engine. Whereas a green-field engine would enable greater opportunity to design for the advantages of magnesium alloy, the economics of this level of effort were prohibitive. Furthermore, designing a V block in itself represented a significant technical reach for magnesium because a V block does not allow through bolting to be used. Though bolts are thought to be necessary to provide structural stiffness to the engine; but only inline engines are amenable to this approach. In addition to having the necessary structural stiffness, a successful engine design would have to overcome other anticipated technical challenges such as engine stability and durability at temperature, cooling the engine with a non-corrosive coolant, and developing fastening, gasketing, and bore/piston/ring strategies. The engine was to be designed to have four magnesium components: the engine block, the bedplate (or lower crankcase), a structural oil pan, and the front cover.

Figure 1. The Ford Duratec engine block and bedplate.

The MPCC Project enabled sharing the high costs and risks of this large, complex objective. It has already renewed interest in using Mg in automotive powertrains within the Mg supplier and manufacturing community, as well as in the automotive industry. The project is promoting greater competition and rapid development of Mg sources, manufacturing capability, and engineering expertise. The Project has already encouraged the development of scientific research projects into the properties, processing, and behavior of magnesium at various universities and national laboratories in North America. The USAMP Magnesium Powertrain Cast Components Project was launched in January 2001 and is being implemented in two phases. In Phase I the alloys were evaluated to find the most promising among them and the engine design was completed. In October 2003 the Project Team successfully passed the Decision Gate Review for undertaking Phase II of the Project: casting of the engine components, the testing of assembled engines, the completion of the alloy design database and the initiation of scientific research activities. The success at the Gate Review was based on the predicted feasibility and cost-effective performance benefits of the engine. A detailed description of the Project follows.

MPCC TASKS IN PHASE I

The objective of Phase I of the MPCC Project was to design the magnesium engine, predict its cost-effective performance benefit, and determine the properties magnesium alloys must have to make the engine technically feasible. It was also to evaluate the advanced Mg alloys and identify those areas where the scientific understanding of magnesium alloys, their behavior, engine design and casting technology was either marginal or inadequate to support the MPCC engine. These goals were embodied in three tasks of Phase I of the Project: (1) evaluation of the new magnesium alloys, (2) FEA design of the components and the creation of a cost model of the engine, and (3) identification of the critical scientific knowledge necessary for use of magnesium in powertrain components, including advanced versions of the engine designed in this project. These tasks are described in some detail below.

TASK 1 – EVALUATION OF THE NEW GENERATION OF CREEP-RESISTANT MAGNESIUM ALLOYS - The goals of Task 1 were to:

- Cast, test and select the alloys, which are most suitable for the magnesium-intensive engine, and
- Develop a cast-specimen material properties database containing the new generation of "low-cost" magnesium powertrain alloys to support the engine design effort (see Task 2).

In recent years, several high-temperature, creep-resistant magnesium die casting alloys have been developed and have been proposed as candidates for powertrain components. Magnesium sand casting alloys have also been developed. Unfortunately, comparing these alloys with each other was difficult for several reasons. The amount of data available for each alloy varied from alloy to alloy. The specimen geometries and test conditions varied considerably among the alloys, which made it challenging at best to interpolate within the available data and even harder to extrapolate to test conditions that are representative of engine operating requirements. There were many engineering properties necessary for the design process which had not been measured for the alloys. Finally, realistically, there were likely to be differences in the quality of the data that was available.

To facilitate comparison of the alloys, it was recognized during planning of the Project that it would be necessary to cast the alloys in common dies to generate specimens for testing and castability evaluation. Thus the database resulting from Phase I of the Project would be a cast-specimen database. A database comprising excised specimens will be created using the alloys that are selected for Phase II of the Project. The format for the database will be common with the database that is being developed for the body and chassis magnesium alloys in the Structural Cast Magnesium Project [12]. This database format evolved from one that was developed for aluminum during an earlier USAMP project, the Design and Production Optimization for Cast Light Metals Project (DPO), [17]. That all three databases are common is an important objective of this activity. Its accomplishment will facilitate the building and use of magnesium design experience. The title page of the

MPCC database is shown in Figure 2 and a working page containing a typical data set is shown in Figure 3.

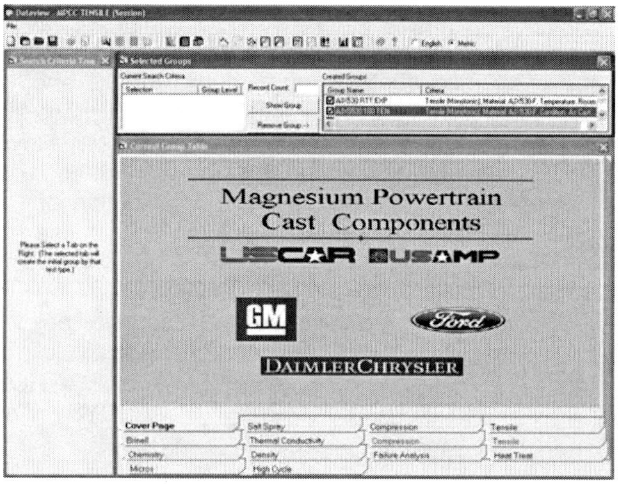

Figure 2. The title page of the MPCC Project cast specimen database.

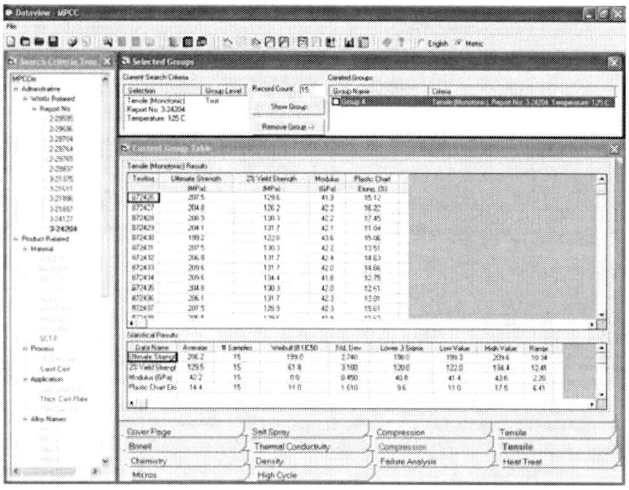

Figure 3. A working page from the MPCC database.

A key aspect of the architecture of the database is its versatility. The working page shows the range of properties and data that is contained the tree architecture that was developed. Each test result can be shown, specimen by specimen, as well as the statistical summary of the entire test set. All data and graphs are exportable to facilitate use by either the design engineer or the research scientist. The procedures, specimen geometries, and sample histories are also contained in the database.

To accomplish the goals of Task 1, the MPCC team conducted a literature search of the known magnesium powertrain alloys. Eighteen HPDC alloys and eight sand casting alloys were considered. The results comprised data from technical journals, conference proceedings, and alloy supplier public literature. Additionally, suppliers provided internal test results to the members of the task team. On the basis of these results, and other factors such as probable alloy cost and availability, the list of potential alloys was shortened to seven HPDC alloys and three sand casting alloys. One HPDC alloys did so poorly in the initial casting trials, however, that it was dropped from further consideration. The alloys that were evaluated in Task 1 of the Project are shown in Table 1.

Table 1. Alloys cast and tested in the MPCC Project.

Alloy Name	**Company**	**Casting Process**
AJ52X	Noranda	HPDC
AS21X	Norsk Hydro	HPDC
AS31	Avisma	HPDC
AXJ530	General Motors	HPDC
MRI153 M	Dead Sea Magnesium	HPDC
MRI230D	Dead Sea Magnesium	HPDC
ML10	Solikamsk	Sand
MRI202S	Dead Sea Magnesium	Sand
SC1	Australian Magnesium Corp.	Sand

At the same time as the literature search was being accomplished, a team comprising DaimlerChrysler, Ford, and GM (OEM) powertrain engineers developed a test matrix that would be used for evaluating the alloys for powertrain components. The matrix included tensile properties, hardness and density, creep and corrosion behavior, fatigue, bolt load retention, impact strength, and various thermo-physical properties. The specimen geometries, test conditions, and test protocols were also defined. The test conditions were chosen to be reflective of the powertrain operating environment.

Upon completion of the test matrix development, the Task 1 team began the alloy casting trials. There were two kinds of trials; those that generated specimens for the test matrix and those that enabled the MPCC team to evaluate the individual alloys on the basis of their castability in thin and thick wall parts. Thin wall parts would indicate the castability of the alloys in the structural oil pan and the front cover and thick wall parts would indicate the castability of the alloys for the bedplate and the bulkhead regions of the engine block.

A die was selected and modified to enable high-pressure die casting (HPDC) of all the necessary specimens in that single, multi-cavity die, see Figure 4. To evaluate castability, a thin wall part (a Ford transfer case) was selected. The transfer case is shown in Figure 5. For sand casting alloy evaluation, a plate casting was designed and built to evaluate the alloys at low solidification rates, see Figure 6. It was intended to be representative of the thicknesses in the redesigned engine block. A Chevrolet V6 block pattern, see Figure 7, was used to evaluate the castability of the sand casting alloys. Fill and solidification modeling was also performed using the specimen die to acquire analytical experience with the alloys. This in turn, will facilitate designing the actual component dies that will be built in Phase II of the Project.

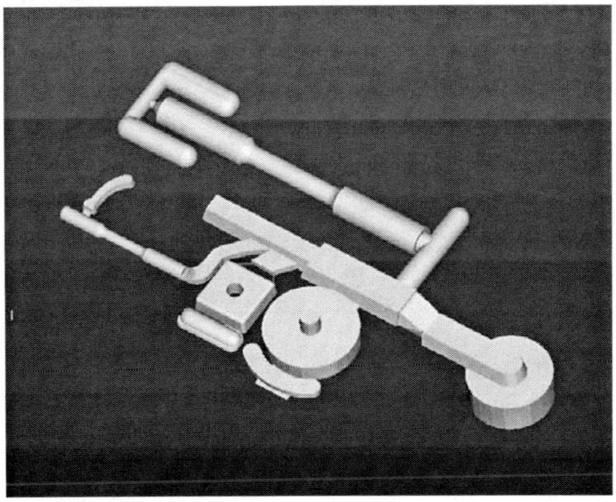

Figure 4. The shot from the four-cavity die used to cast the specimens for the alloy evaluation test program.

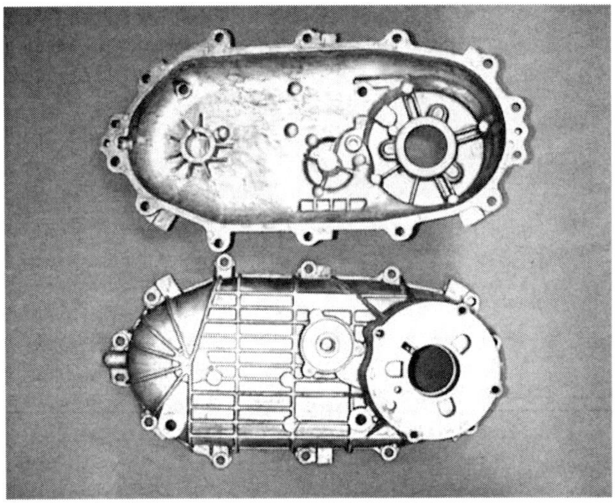

Figure 5. The Ford transfer case used for the castability trials.

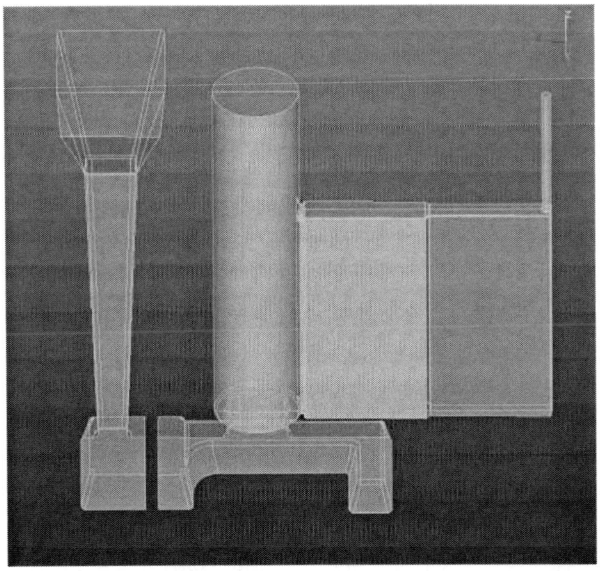

Figure 6. The plate pattern for the sand casting alloys.

Figure 7. The Chevy V6 block.

The casting trials were conducted in the following manner. Each casting trial was a team effort comprising the die caster, the alloy supplier, and one or more members of the OEM project team. Casting parameters and melt conditions were allowed to be adjusted to achieve the best casting results possible. The history of each casting trial was thoroughly documented to allow an assessment of the castability to be made. The development of the casting assessment protocol benefited from published, independent castability studies [18-22]. In brief, the MPCC castability assessment sought to distinguish those results for each alloy that would be related to the casting learning curve for the alloy in a new die and a new HPDC machine as separate from the "learning curve plateau performance" of the alloy. The factors that would have an impact on the cost of the manufacturing process, such as required die and furnace temperatures, and cover gas requirements were also documented for consideration in the cost model. The casting documentation will be made available to the fill and solidification modeling companies who are participating in the Project so that they will gain knowledge that will help their design efforts in Phase II.

Having completed Task 1, the Project Team accomplished a design database of magnesium alloys based on HPDC-cast specimens and specimens machined from sand castings. The database format is common with that of the SCMD and DPO projects and the content of the database and the castability assessments was used along with the cost model and engine design results to select two HPDC alloys and one sand casting alloy for Phase II of the Project.

TASK 2 – FEA DESIGN AND COST MODEL OF THE POWERTRAIN COMPONENTS – The Task 2 goals were:

- A complete FEA design of the Duratec engine with four magnesium components, and
- A cost model of the engine which contained the effects of magnesium cost and the technology requirements of the engine that distinguished it from the aluminum version.

The technical details of the engine design and many of the design decisions are proprietary to USAMP and the MPCC Project team so only an overview of the accomplishments of the design team is presented. Magna Automotive Testing (Magna Steyr) was chosen to do the FEA design. In support of their effort, Ford, General Motors, and DaimlerChrysler engineers participated and agreed on the design approach, the underlying materials assumptions, and the various technology options for the engine. Several non-OEM members of the Project team also made valuable technical contributions to these activities. Ford provided the initial CAD files for the Duratec engine (see Figure 8) and a considerable amount of technical background as the design effort progressed.

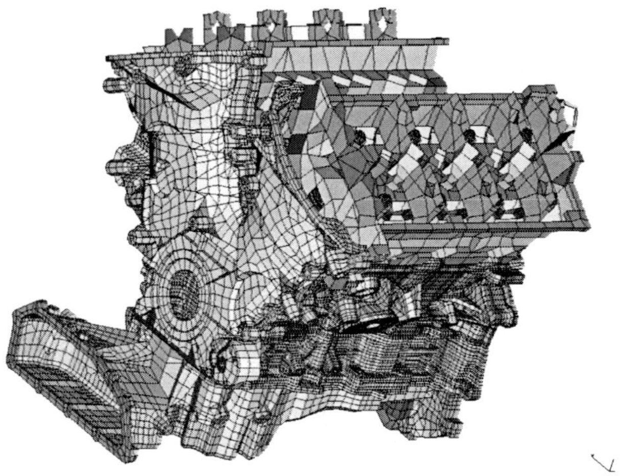

Figure 8. The working mesh for the Ford Duratec engine.

The scope of the Task 2 goals required the formation of several teams within the Task 2 organization. These included:

- Task 2.1 – Engine Block, Bore, and Journals – identify and prioritize the technical options for the FEA design and cost modeling teams,
- Task 2.2 – Fasteners, Gaskets, and Sealing – address the mechanical design aspects of fastener systems and gaskets for the block and ancillary structures,
- Task 2.3 – Coolants and Corrosion – assess the state of the technology for coolant strategies, identify the best coolant system, and address the corrosion issues for the magnesium alloys. These included general corrosion inside the engine, outside the engine in the under-hood area, and galvanic corrosion that might arise due to contact with internal materials such as gaskets, pumps, and other dissimilar materials.

Four major coolant manufacturers participated in the MPCC Project. Each of them had been developing advanced anti-corrosion coolant/additive formulations. As with the creep-resistant alloys, this project team sought to determine the state of the art with respect to coolants

coolants. The Coolants and Corrosion team worked with the coolant manufacturers and the magnesium alloy suppliers to define a coolant evalution protocal and to establish a common platform for testing the alloys and the coolant/additive packages. Three ASTM test procedures were adopted for the evaluation:

- ASTM B117-97 – Standard Practice for Operating Salt Spray (Fog) Apparatus,
- ASTM D1384-01 – Standard Test Method for Corrosion Test of Engine Coolants in Glassware,
- ASTM D4340-96 – Standard Test Method for Corrosion of Test Aluminum Alloys in Engine Coolants Under Heat-Rejecting Conditions.

An example of the D4340 test results is shown in Figure 9, in which both the alloys and coolants are coded.

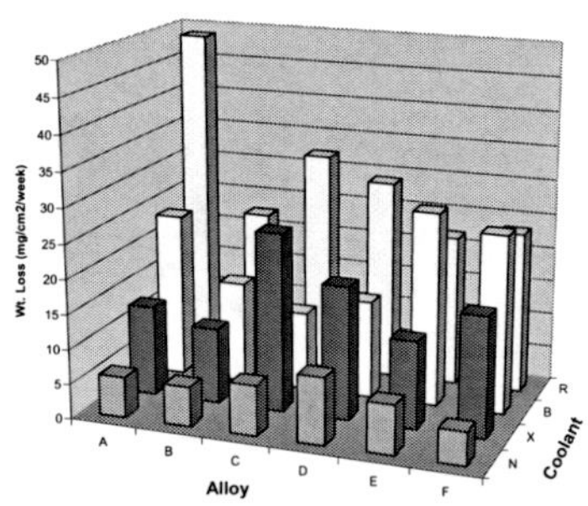

Figure 9. ASTM D4340-96 test results showing corrosion rates of alloy-coolant combinations.

The independent testing house (Amalgatech) used by the Task 2.3 Team evaluated each of the candidate alloys under the test conditions and accomplished this using a "double blind" approach to respect the intellectual property of each individual coolant supplier. The results showed that some alloy coolant combinations were quite good, whereas others were poor. In Phase II of the Project, the opportunity for the coolant suppliers to introduce better coolants based on their experiences with the Phase I testing should contribute significantly to the reliability of the engine.

Three other Task 2 teams were also formed:

- Task 2.4 – Design Integration and Analysis
- Task 2.5 – Component Casting and Casting Analysis
- Task 2.6 – Cost Modeling.

The Design Integration and Analysis team was responsible for coordinating the Task 2 input to Magna Steyr and for the development of the FEA model. Once the FORD CAD files for the engine were delivered to Magna Steyr, meshing of the components to be designed in magnesium was completed. Computational Fluid Dynamics and heat transfer analyses were accomplished and were used to develop a complete structural analysis of the engine, which considered assembly, bolt loads, primary and secondary creep behavior of the magnesium, and the cylinder gas forces as well as the inertial mass forces. Fatigue analysis of the block to determine safety factors was also accomplished. Powertrain bending analysis and acoustic analysis of the assemblies were also completed. The baseline for comparison in bending and acoustic analysis was the engine using the current aluminum components.

The Component Casting and Casting Analysis team is responsible for the design of the dies and mold tooling for casting the components in Phase II of the Project. During Phase I the team began learning about the casting behavior of the magnesium alloys by participating in the casting trials and modeling the fill and solidification behavior of the alloys in the specimen die. As soon as the FEA designs of the magnesium engine components were completed, the respective CAD files were delivered to the fill and solidification modeling teams so that the Phase II activities of designing the dies and patterns could be expedited.

The Cost Modeling team contracted IBIS Associates to develop a cost model of the engine. This sophisticated model enabled the Project Team to assess:

- The cost impact of the alloy based on both intrinsic alloy costs and manufacturing process costs such as furnace and die life predictions and environmental and recycling considerations
- The cost of engine technologies required for successful operation with magnesium components; e.g., coolant costs and unique fastening and gasketing requirements
- The benefits of reduced weight and design advantages that are enabled by the use of magnesium.

IBIS Associates produced two independent models, one for the die casting alloys and one for the sand casting alloys. The models take the form of interactive spreadsheets, which are fully accessible with respect to the algorithms and constants used. They are "portable" and were used by each of the OEM organizations to determine engine costs. These models will be further developed in the course of Phase II, as more is learned about all aspects of producing the magnesium-intensive engine.

TASK 3 – IDENTIFICATION OF THE FUNDAMENTAL SCIENTIFIC CHALLENGES FOR USING MAGNESIUM ALLOYS AND CASTING PROCESSES IN POWERTRAIN COMPONENTS - The Task 3 objective of the MPCC Project is the legacy of the Project. It seeks to promote the establishment of a scientific basis for future magnesium alloy and component developments by identifying the critical gaps in our understanding of magnesium alloys and casting processes, with respect to powertrain applications and other applications which require high-temperature alloys. Obvious needs included determining detailed mechanisms of creep and fatigue behavior of the magnesium alloys at temperature and establishing thermodynamic and phase equilibrium databases. One of the reasons that the Task 1 casting trials were necessary was due to the lack of knowledge of the creep mechanisms and kinetics, i.e., creep maps. This prevented extrapolation of creep rate behavior from one range of test conditions to those that would be representative of engine operating temperatures and loads. Thermodynamic and phase equilibrium data is essential for further alloy development and for predicting alloy creep and corrosion kinetics. As an example, recent work has shown that the creep rate of the rare earth-containing magnesium alloy AE42 increases at temperatures above 150°C because of a phase change in the microstructure which eliminates a creep-inhibiting phase and creates a creep-promoting phase [23]. An analogous increase in the creep rate of another high temperature magnesium alloy has also been reported [24]. Another area of understanding that is inadequate is the interpretation of bolt load retention tests, which are used to predict the performance of magnesium alloys when the alloy is used as either a threaded boss or an unthreaded flange [25, 26].

The implementation of Task 3 was initiated in Phase I and will proceed through Phase II. In Phase I, the Project Team documented the scientific needs as they were recognized. In October 2003 a workshop was held at USCAR in Michigan to which North American universities and national laboratories were invited. At this workshop, the Project Team described the critical research areas and the specific needs within those areas. These areas, which were then discussed with the workshop participants, are:

- Phase Equilibrium and Thermodynamics
- Corrosion Behavior and Protection
- Casting, Solidification Behavior, and Microstructure
    - Hot tear and melt quality issues
    - Protective gases
- Alloy Development and Structure-Property Relationships
    - Fatigue and Static Mechanical Properties
    - Creep and Bolt-Load Retention
- Recycling

The information presented at the workshop will be made public at technical meetings and through other means to call attention to the research needs. Workshop participants and others were invited to submit brief proposals to the Project Team for consideration for funding in Phase II. Ultimately, more formal requests for proposals (RFP) to address the research needs will be distributed to national laboratories and North American universities. The selected research proposals will be implemented in Phase II. The funding available with the MPCC Project budge will support only a few projects. However, it is anticipated that the promotion of the recognition these scientific needs will facilitate the funding of magnesium research from other organizations in North America. Furthermore, additional USAMP projects could be created to support more work.

THE DECISION GATE REVIEW

The MPCC Project passed its review in October 2003 and proceeded into Phase II. The success of the review was based on the predicted weight reduction benefit of the engine, the estimated cost-to-performance ratio and durability, and the determination that there exist alloys with the properties required by the design. The review board made several suggestions which will be incorporated in the Phase II plan. The design of the magnesium-intensive engine is shown in Figure 10.

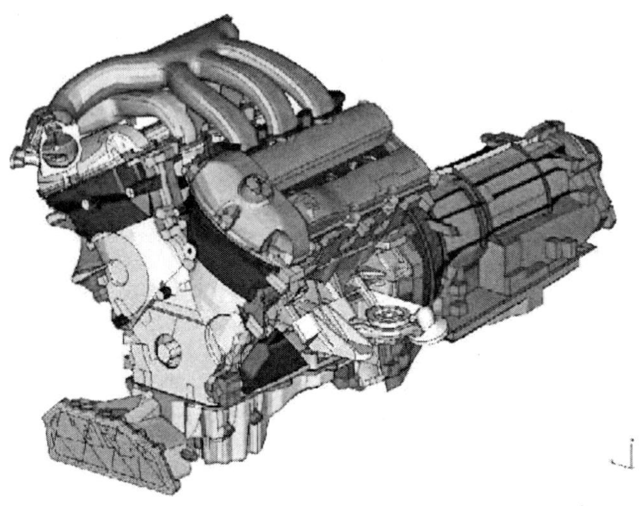

Figure 10. The magnesium-intensive Duratec powertrain.

MPCC TASKS IN PHASE II – VALIDATING THE FUTURE OF MAGNESIUM POWERTRAIN COMPONENTS

Entry into Phase II was contingent on the results of the Task 1 and Task 2 activities. A reconsideration of the specific objectives and plans for Phase II was part of the Decision Gate review and those plans were expanded as a result of suggestions by the review board. Phase II is also divided into three tasks: (4) die/pattern design, die build and casting of the components, (5) completion of the alloy property database using specimens excised from the cast components, and (6) assembly and testing of complete powertrains containing the magnesium components.

TASK 4 – CASTING THE MAGNESIUM POWERTRAIN COMPONENTS – In Task 4, dies and patterns for the magnesium engine components will be built using designs based on fill and solidification models. Much of the knowledge used to design the dies and the pattern will be the direct result of modeling of the specimen and castability dies that was done in Task 1 of Phase I of the Project. Additionally, some consideration is being given to expanding the number of parts to be converted to magnesium.

The magnesium components may require some design engine, the alloys, or the fill and solidification modeling results. Thus all aspects of this task will be carefully re-evaluated before proceeding with die and mold construction. The cast parts will be subjected to dimensional validation, X-ray inspection, and leak test validation (following machining). Provisions for up to three iterations of casting and die modification are included in the project timing. As in Task 1, casting results will be thoroughly documented to extract and preserve as much information as possible about the behavior of the alloys.

TASK 5 – COMPLETION OF THE MG ALLOY PROPERTY DATABASE – EXCISED SPECIMENS - It was noted that the cast specimen property database developed in Task 1, was by necessity, only a step toward the goal of a magnesium powertrain alloy design database. Task 5 will compile a comprehensive thermo-mechanical property and corrosion database for the magnesium alloys chosen for the casting the engine components in Phase I of the Project. In this case, the specimens tested will be excised from the cast components rather than being separately cast specimens. This difference will yield "actual" properties of the alloys and is expected to be more meaningful in the long run to the design community. It is also possible that the test matrix comprising the database will be changed or expanded to comprehend additional information that is found to be necessary. The results of Task 5 are likely to identify additional areas of technical challenge along with specific needs for scientific research into the properties and processing of the alloys, thereby furthering the objectives of Task 3.

TASK 6 – VALIDATION TESTING OF THE ASSEMBLED POWERTRAINS - The cast components from Task 5 will be assembled into complete powertrains for dynamometer testing to validate the design, performance, and durability of the components. All tested engines will be torn down for inspection. As with all other areas of the Project, OEM consensus on all aspects of testing and post-test analysis will be achieved. A final report containing these results will be produced.

MPCC Project Deliverables

The overall objective of the Project is to determine and enhance the feasibility and benefit of using magnesium in powertrain components. There are six specific deliverables of the Project:

- The design of the engine components using the best low-cost, creep-resistant magnesium alloys.
- The magnesium design database, comprising the results of the original literature search, the cast-specimen test results, and the excised-specimen test results. These, combined with the castability assessments will form a design and casting guide for magnesium powertrain components.

- The dynamometer test results for engine performance and magnesium component durability.
- The cost model of the engine to enable a determination of the cost-to-benefit ratio.
- An OEM-common material specification for Mg powertrain alloys, including the procedures to be followed by suppliers to certify their alloys.
- A start toward building the magnesium research infrastructure in North America.

CONCLUSION

The USAMP Magnesium Powertrain Cast Components Project was forecast to be a forty-eight month effort. It is now anticipated that Phase II of the Project will be completed in 2005. At that time key questions about the future prospects for a magnesium-intensive powertrain will have been answered and any scientific and economic implementation barriers identified. Programs to overcome the scientific barriers will have been proposed by interested universities and laboratories and may have already commenced.

A question that has been asked of the MPCC participants during the course of the Project is the incentive for participation in Phase II of the Project by suppliers of the alloys that are not selected for Phase II. The answer to this question is, that by continued participation in the Project, each supplier will be able to follow the development of dies and molds, the casting of the components, and have access to the test results. In doing so, the actual alloy property requirements and design and operating requirements will be revealed. This information will provide all participating alloy suppliers with the opportunity to assess and focus further development of their own individual alloys. Continued participation by all of the suppliers will also benefit the project by their valuable technical contributions to the Project.

Finally, if the Magnesium Powertrain Cast Components Project realizes its vision, a magnesium-intensive engine that is cost-effective and has demonstrable performance benefits, then it will be up to DaimlerChrysler, Ford, and General Motors to take the next step: production implementation of the magnesium-intensive engine. This will be a revolutionary change for both the automotive and magnesium communities.

ACKNOWLEDGEMENTS

The authors thank their colleagues at DaimlerChrysler, Ford, and General Motors for their support in the development and implementation of this project. These colleagues include Paul Bujalski from Daimler Chrysler and Jim Quinn from General Motors. The active participation of all of the non-OEM companies, Tables 2-5, who are part of this project both enables the Project and makes accomplishing the Project goals a stimulating and enjoyable activity. Their support and that of the US Department of Energy is gratefully acknowledged.

This research was supported in part, by Department of Energy Cooperative Agreement No. DE-FC05-95OR22363 (and/or for work conducted after 12-31-01, DE-FC05-02OR22910). Such support does not constitute an endorsement by the Department of Energy of the views expressed herein.

Table 2. Alloys, casting houses and recycler participants in the MPCC Project.

Mg Alloys	Casting	Recycling
Australian Mg Corp.	Eck Industries	Amacor
Dead Sea Magnesium	Gibbs Die Casting	
General Motors	Hayes-Lemmerz Intl.	
Hydro Magnesium	Intermet	
Noranda Magnesium	Lunt Manufacturing	
Solikamsk Mg Works	Meridian Technologies	
VSMPO-Avisma	Nemak, S. A.	
	Spartan Light Metals Products	
	Thixomat	

Table 3: Tooling, testing, and modeling participants in the MPCC Project.

Tooling	Testing	Modeling
Becker CAD CAM CAST	Amalgatech	EKK
Delaware Machinery and Tool	CANMET Materials Technology Lab	Flow Science
EXCO Engineering	Climax Research Services	Magma Foundry Technologies
H.E. Vannatter	Quasar Intl.	Technalysis
	Westmoreland Mechancial Testing and Research	ESI-Group

Table 4: Design, fasteners and gasket participants in the MPCC Project.

Engine Design	Fasteners	Gaskets
Magna Automotive Testing	Ribe	Victor Reinz

Table 5: Coolants, coatings, and association participants in the MPCC Project.

Coolants	Coatings	Associations
Ashland/Valvoline	Henkel Surface Technologies	IMA
Chevron/Texaco	Keronite	NADCA
Honeywell Intl.		
Intac Automotive Products		

CONTACT

Dr. Bob R. Powell, MPCC Project Leader
Senior Staff Research Scientist
Materials & Processes Laboratory
M/C 480-106-212
General Motors Research & Development Center
Warren, MI 48090-9055
Tel: 586-986-1293
Fax: 586-986-9204
Email: bob.r.powell@gm.com

REFERENCES

1. E.F. EMLEY, "*Principles of Magnesium Technology*," first edition, Pergamon Press Ltd., Oxford, England 1966 p 828.
2. S. SCHUMANN AND F. FRIEDRICH, "The Use of Magnesium in Cars – Today and in the Future," in "*Magnesium Alloys and their Applications*," edited by B.L. Mordike and K.U. Kainer, Werkstoff-Informationsgesellschaft mbH, Frankfort, Germany 1998, pp 3-13.
3. F. HOLLRIGL-ROSTA et al., "Magnesium in Volkswagen," Light Metal Age, August 1980, pp 22-9.
4. J.E. HILLIS, "The Effects of Heavy Metal Contamination on Magnesium Corrosion Performance," SAE Technical Paper No. 830523, SAE, Warrendale, PA, USA, February 1983.
5. D. ALBRIGHT, "Global Automotive Applications Review," presented at the 13th Automotive Magnesium in Automotive Seminar, International Magnesium Association, Plymouth, MI, April 10, 2002.
6. W.E. MERCER, "Magnesium Die Cast Alloys for Elevated Temperature Applications," SAE Technical Publication No. 900788, SAE, Warrendale, PA, USA, 1990.
7. B.R. POWELL et al., "Development of Creep-Resistant Magnesium Alloys for Powertrain Applications," in *Light Metal Applications for the Automotive Industry: Aluminum and Magnesium,* SAE Special Publication SP-1622 Paper No. 2001-01-0422, SAE, Warrendale, PA, USA, 2001.
8. B.R. POWELL et al., "Creep-Resistant Magnesium Alloy Die Castings," United States Patent No. 6,264,763, dated July 24, 2001.
9. M. PEKULERYUZ AND E. BARIL, "Development of Creep Resistant Mg-Al-Sr Alloys," in *Magnesium Technology 2001*, edited by J.N. Hryn, TMS, Warrendale, PA, USA, 2001, pp 119-126..
10. F. VON BUCH et al., "New Die-Casting Alloy MRI153 for Powertrain Applications," in *Magnesium Technology 2002*, edited by H.I Kaplan, TMS, Warrendale, PA, USA, 2002, pp 61-67.
11. K. PETTERSEN et al., "Creep-Resistant Magnesium Alloy Development," *Magnesium Alloys and Their Applications,* ed. K.U. Kainer (Frankfurt, Germany, 2000).
12. R.J. OSBORNE et al., "Structural Cast Magnesium Development," in "*FY 2001 Progress Report for Automotive Lightweighting Materials,*" edited by R. Kost, U.S. Department of Energy January 2002, pp 61-66.
13. "*Magnesium Technology 2000*," ed. by H.I. Kaplan, J.N. Hryn, and B.B. Clow, TMS, Warrendale, PA, USA 2000.
14. "*Magnesium Technology 2001*," ed. by J.N. Hryn, TMS, Warrendale, PA, USA 2001.
15. "*Magnesium Technology 2002*," ed. by H.I. Kaplan, TMS, Warrendale, PA, USA 2002.
16. "*Magnesium Technology 2003,*" ed. by H.I. Kaplan, TMS, Warrendale, PA, USA 2003.
17. R.J. OSBORNE et al., "Design and Product Optimization for Cast Light Metals, USAMP-LMD 110 Project," in "*FY 2000 Progress Report for Automotive Lightweighting Materials,*" edited by R. Kost, U.S. Department of Energy January 2001, pp 39-45.
18. J.J. BERKMORTEL et al., "Die Castability Assessment of Magnesium Alloys for High Temperature Applications: Part 1 of 2," SAE Technical Paper No. 2000-01-1119, SAE, Warrendale, PA, USA, 2000.
19. A.P. DRUSCHITZ et al., "Evaluation of Structural and High-temperature Magnesium Alloys," SAE Special Publication SP-1683, Paper No. 2002-01-0808, SAE, Warrendale, PA, USA, 2002.
20. D. ARGO et al., "Process Parameters and Die Casting of Noranda's AJ52 High Temperature Mg-Al-Sr Alloy," in *Magnesium Technology 2002*, edited by H.I Kaplan, TMS, Warrendale, PA, USA, 2002, pp 87-93.
21. B.R. POWELL et al., "The Die Castability of Calcium-Containing Magnesium Alloys: Thin-Wall Computer Case," in *Magnesium Technology 2002*, edited by H.I Kaplan, TMS, Warrendale, PA, USA, 2002, pp 123-129.
22. M.J. Dierks, "The Castability of Creep-Resistant Magnesium Alloys," to be published in the Proceedings of the 61st Annual Meeting of the International Magnesium Association, New Orleans, LA, May 2003.
23. B.R. POWELL et al., "The Relationship Between Microstructure and Creep Behavior in AE42 Magnesium Die Casting Alloy," in The Journal of Minerals, Metals, and Materials (JOM) 54 [8], 34-38, August 2002.
24. M.O. PEKGULERYUZ, personal communication, February 19, 2002.
25. X. SU et al., "Finite Element Modeling of Bolt Load Retention of Die-Cast Magnesium," SAE Technical Publication No. 2000-01-1121, SAE, Warrendale, PA, USA, 2000.
26. E.G. SIERACKI et al., "Compressive Stress Retention Characteristics of High Pressure Die Cast Magnesium Alloys," SAE Technical Publication No. 960421, SAE, Warrendale, PA, USA, 1996.

2004-01-0655

Powertrain Components - Opportunities for the Die Cast AE Family of Alloys

Per Bakke, Håkon Westengen and Stian Sannes
Magnesium Competence Centre, Norsk Hydro

Darryl Albright
Hydro Magnesium Marketing

ABSTRACT

By selecting the right combination of alloy and processing method, a wide range of temperature exposed drive train parts can be made out of die cast magnesium, including engine blocks and automatic transmissions as probably the most demanding components. Successful new alloys for these purposes must fulfill a multitude of requirements to offer a viable solution, including mechanical properties, corrosion properties, die castability and recyclability. Therefore, selection of alloys must be based on the customers' requirements, at the same time as other factors are optimised. In this paper, results from the ongoing alloy development work by Hydro Magnesium are presented, focusing mainly on creep resistant alloys within the Mg-Al-RE system. High temperature tensile data, tensile creep-, stress relaxation- and bolt load retention results from a selection of AE alloys and reference alloys are presented.

INTRODUCTION

By selecting the right combination of alloy and processing method, a wide range of temperature exposed drive train parts can be made out of die cast magnesium [1,2]. A schematic illustration of the creep/strength performance vs. temperature requirements is shown in Fig. 1, indicating that the engine blocks and the automatic transmissions may be the most demanding components. For high power engines and automatic transmissions, the operating temperatures may exceed 150°C and the moderate creep resistance of AZ91D becomes critical. This is in line with the experience with the VW Beetle engine that was initially produced in AZ81, and later changed to AS41 and further to AS21 as the engine power was increased [3]. Unfortunately, none of these alloys, which all show excellent castability performance, are suitable for service at temperatures of 150°C or above. For applications requiring <150°C and moderate loads, an inexpensive, castable and recyclable alloy like AS21X may be considered [4]. For more demanding applications AEX522 [5], AXJ530 [6] and AJ52 [7] may withstand temperatures up to 175°C, but poor castability and recycling problems are arguments against them. The creep properties of AE42 at 175°C probably represent the lowest values, which would be acceptable at this temperature.

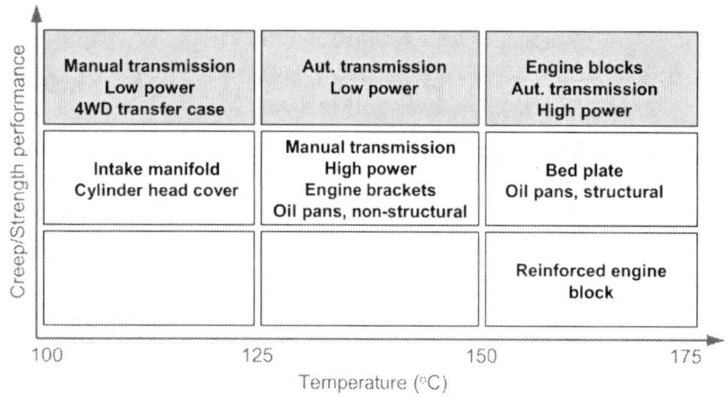

Figure 1. Creep strength/performace vs. temperature requirements for some automotive drivetrain components.

The elevated temperature market must therefore be approached from a pragmatic angle, realizing that successful new alloys for this purpose must fulfil a multitude of requirements to offer a viable solution [1,2,8,9,10]. This is illustrated in the Alloy Property Profile chain (Fig. 2) where all issues must be fulfilled for an alloy to penetrate the market. One weak link is enough to disqualify a new alloy. As an example, the section of Figure 2 entitled Molten Metal Handling includes consideration of whether or not the concentration of alloying elements changes during normal practices, such as holding at constant temperature and freezing/ remelting during/after weekends. It also implies that the molten alloy should not require special equipment or procedures that adds extra cost. Recyclability means that the alloy can be recycled back to its original composition with well-established recycling techniques (flux, flux-free). Die casting performance depends upon the alloy composition necessary to fulfil the requirements, either related to high temperature creep, or on strength/ductility. The key challenge is to identify alloys fulfilling the property requirements, and at the same time demonstrate a die castability, which is compatible with the geometry of the targeted components. For all the alloys the properties depend strongly on the processing parameters. Grain size, porosity distribution, segregation, oxides, casting defects, etc. all have significant impact on the properties of die cast specimens [11-15].

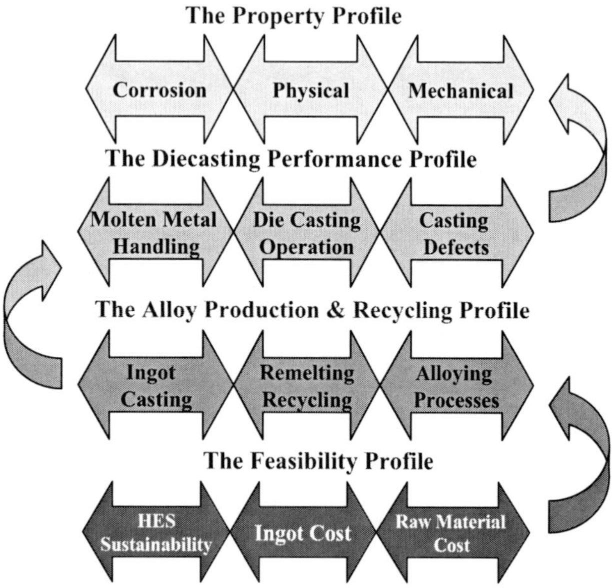

Figure 2. Elements in the alloy development process – The Alloy Property Profile chain.

Since so many factors need to be addressed, and each end user application has its own requirements, the focus is to select alloys based on the customers' requirements and at the same time optimise other factors. In this paper the AE family of alloys is highlighted, indicating some possible paths in the development of new alloys with good high temperature properties and castability. The letter E denotes rare earth elements or misch metal, in this case a standard grade Ce based alloy (>50% Ce) with La (20-35%), Nd (10-20%) and Pr (4-10%).

EXPERIMENTAL

Test bars and plates of different experimental AE alloys were die cast on a 420 ton Bühler SC42D Evolution machine, with a five cavity plate/bar die (version 0), Fig. 3. This die contains two 3 mm corrosion test plates, a 6 mm cylindrical tensile creep bar, a full size 6 mm cylindrical tensile bar, and a rectangular impact test bar. Heats of reference die casting alloys AZ91D, AS21X, AE42, AJ52, AJ62 and AXJ530 were prepared and cast in the same die. Permanent mould cast medallions were withdrawn from each alloy and chemically analysed by ICP-AES. The first 15-20 castings from each heat were always rejected. inspection. The mechanical testing was done within weeks after casting, hence any possible aging effects were considered insignificant.

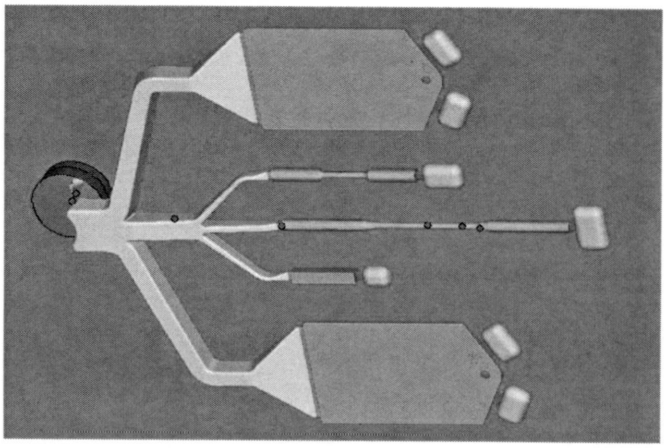

Figure 3. The five cavity plate/bar die for the Bühler SC42D Evolution die casting machine.

Standard 6 mm test bars from each composition were tensile tested in a 10 kN Instron machine at temperatures ranging from room temperature and up to 210°C in accordance with ISO 6892. In all tests a strain rate of 1.5 mm/min was used until 0.5% strain, whereafter the strain rate was increased to 10 mm/min.

Two tensile creep specimens from a large number of alloys were tensile creep tested for minimum 150 hrs in accordance with ASTM E139, at loads and temperatures relevant for powertrain components; 150°C/90 MPa and 175°C/75 MPa.

Stress relaxation testing was done at 175°C in accordance with ASTM E328-86. Initial loads were 12 kN and 6 kN corresponding to 106 and 53 MPa, respectively. The specimens (12 mm dia., 5 mm length) were cut from one arbitrary end of a tensile creep bar.

Solidification analysis was done in a thin walled crucible equipped with two 3 mm type K thermocouples, one in the centre and one near the wall. The crucible was pre-heated to the alloy temperature in an adjacent furnace before it was filled with metal. The solidification rate was controlled to within the order of 1 K/s by air cooling. The temperatures were recorded at a frequency of 10 Hz.

RESULTS AND DISCUSSIONS

TENSILE TESTING

Tensile yield strength (TYS) as a function of temperature for some experimental AE alloys is shown in Fig. 4. AE42 and aluminium alloy A380 are shown for comparison. It is seen that TYS for the experimental AE alloys is well above that of AE42. Specifically, at 200°C, TYS for four of the AE alloys is in the range of 105 -115 MPa. For AE42, TYS at 200°C is measured to 90 MPa. On the other hand, TYS for A380 at 200°C is as high as 175 MPa. Notice that throughout all figures the order of the legends follows the order of the curves at RT.

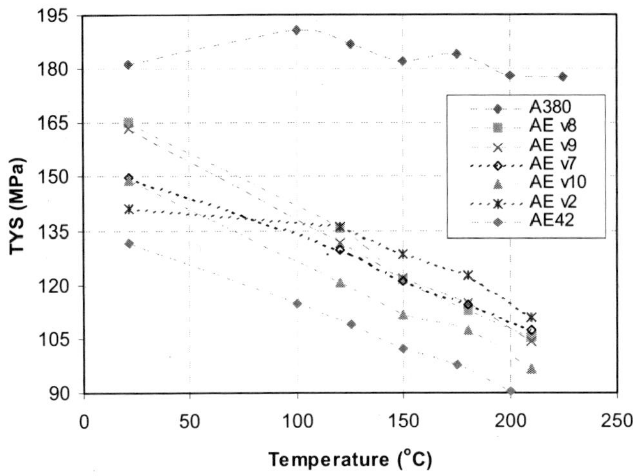

Figure 4. Tensile yield strength (TYS) as a function of temperature for some AE alloys, AE42 and aluminium alloy A380.

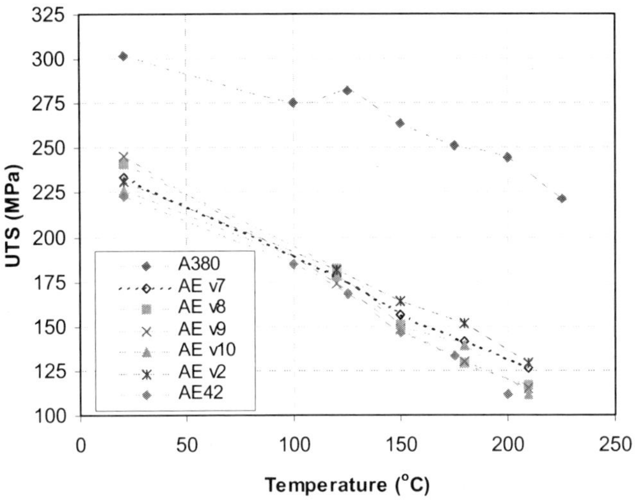

Figure 5. Ultimate tensile strength (UTS) as a function of temperature for some AE alloys, AE42 and aluminium alloy A380.

Ultimate tensile strength (UTS) for the same alloys is shown in Fig. 5. It is seen that at room temperature all AE alloys tested exhibit UTS in the 225-250 MPa range. The experimental AE alloys shown perform better than AE42 at all temperatures. At 200°C, UTS of the AE alloys is in the range of 120-135 MPa, while for A380 it is 240 MPa.

The range of alloying elements in AE type alloys is quite wide due to the extensive liquid solubility [12]. Compared to other dispersoid forming alloying elements like Sr and Ca, RE elements have a less harmful effect on ductility/fracture toughness. Elongation at fracture as function of temperature is shown in Fig. 6. It is seen that the elongation of the AE alloys is better than the Sr and Ca containing reference alloys (AJ52 and AXJ53) at room temperature as well as at the higher temperatures.

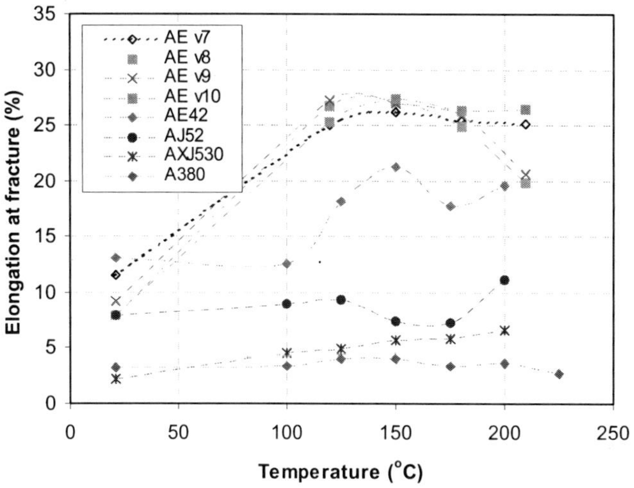

Figure 6. Elongation at fracture as a function of temperature for some new AE alloys, AE42, AJ52, AXJ530 and aluminium alloy A380.

It should also be noted that the value of the yield stress is very dependent on the strain rate, especially at elevated temperatures, and direct comparison of values from different sources may be misleading. In order to seek a more accurate comparison, the only data used for alloys reported in Figs. 4-6 are from internal Hydro Magnesium testing.

Generally accepted values for the modulus of elasticity at room temperature are 45 GPa for Mg alloys [16] and 72 GPa for aluminium alloys [17], slightly decreasing with increasing temperatures. The results obtained in this work indicate no significant deviations from these values.

TENSILE CREEP

In Figs. 7-8 creep strain as a function of time are shown for selected alloys at 150°C/90 MPa and 175°C/75 MPa, respectively. It is seen that at both these relatively demanding conditions the creep curves of some of the new AE alloys are significantly better than for example

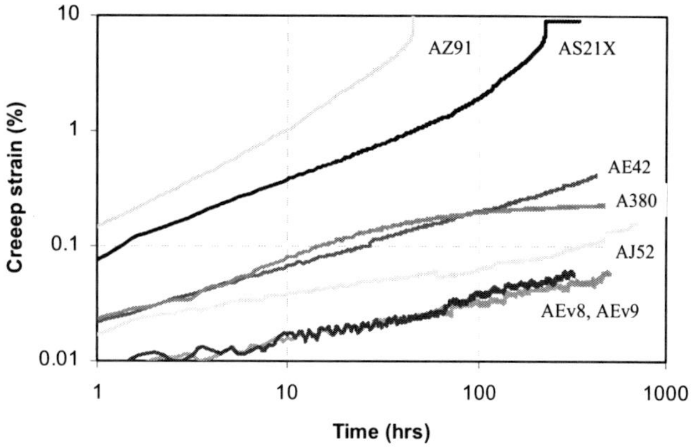

Figure 7. Creep strain as a function of time at 150°C/90 MPa for selected alloys. Notice that both axes are exponential.

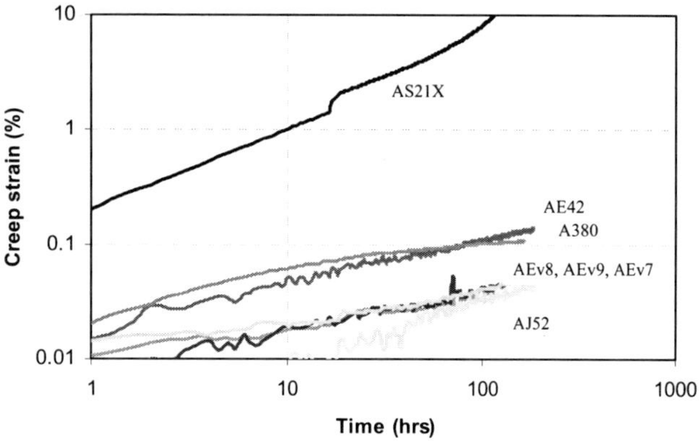

Figure 8. Creep strain as a function of time at 175°C/75 MPa for selected alloys. Notice that both axes are exponential.

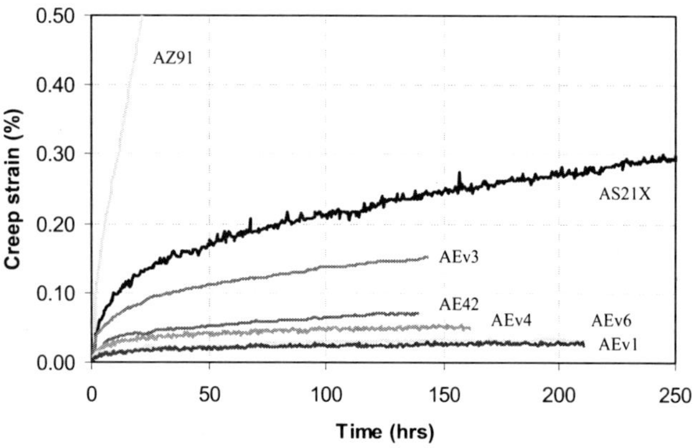

Figure 9. Creep strain as a function of time at 150°C/50 MPa for selected alloys.

AE42. The new AE alloys are matching AJ52, which is known for outstanding creep performance up to 175°C (but poor die castability) [7]. For applications exposed to such conditions AS21X is no longer an option, although it may be an exceptionally good solution for modest temperature/load range applications.

Total creep strain at 100 hrs seems to be better for the new AE alloys than for A380. This because the at these load and temperature conditions all these alloys, and A380 in particular, have not reached their steady state creep regimes. As can be seen from Fig. 7, the creep rate of A380 beyond 100 hrs is likely to be as low as the best Mg alloys shown in Figs. 7 and 8.

Total creep strain for selected alloys at 150°C/50 MPa is shown in Fig. 9. Also at this condition we se there are alloys outperforming for example AE42. Notice also AE v3, which experiences creep strain between AE42 and AS21X. As will be discussed later, this alloy exhibit excellent castability and this alloy could be a candidate for medium temperature applications.

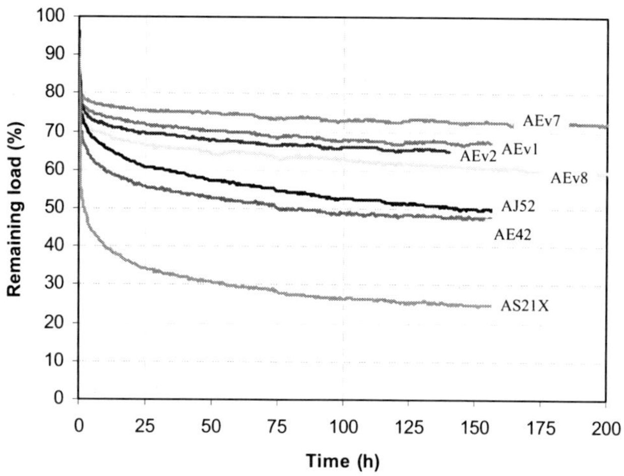

Figure 10. Remaining load as a function of time for some AE alloys at 175°C with initial load 12 kN (106 MPa). The alloys AS21X, AE42 and AJ52 are included for reference.

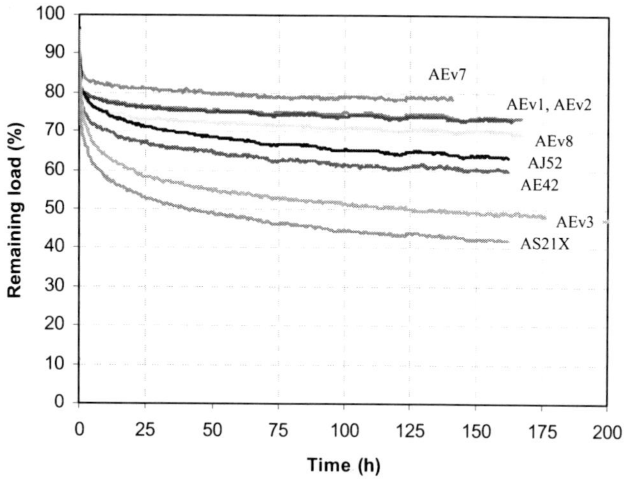

Figure 11. Remaining load as a function of time for some AE alloys at 150°C with initial load 12 kN (106

MPa). The alloys AS21X, AE42 and AJ52 are included for reference.

STRESS RELAXATION

In the stress relaxation test a compressive initial load is applied, and as the alloy deforms due to creep, the remaining load is measured. Bolted joints experience the same type of behaviour in components [18]. Results of the stress relaxation tests are shown in Figs 10-12. At 175°C, 12 kN initial load (Fig. 10) there are AE alloys outperforming the reference alloys AJ52 and AE42. Specifically, at 100 hrs, the alloy AE v7 has >70% initial load remaining, while for example AJ52 and AE42 has about 55% and 50% load remaining, respectively. The internal ranking of the alloys tested is the same at at 150°C, 12 kN (Fig. 11) and at 175°C, 6 kN (Fig. 12). In Fig. 11 it is noticed that AE v3 performs better than AS21X but not as good as AE42, in line with the tensile creep data at 150°C/50 MPa.

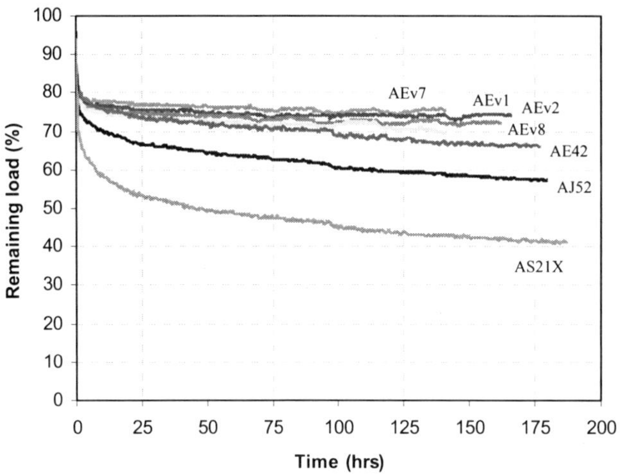

Figure 11. Remaining load as a function of time for some AE alloys at 175°C with initial load 6 kN (53 MPa). The alloys AS21X, AE42 and AJ52 are included for reference.

Notice that the tensile yield strength is an absolute upper limit for the load that should be applied in a creep test. The initial load of the stress relaxation test (106 MPa at 175°C) is well above the tensile yield strength of for example AS21X [9]. This means that the results of AS21X at 106 MPa/175°C are misleading as remaining load quickly drops to below the yield strength.

THERMAL ANALYSIS

Thermal analysis has been undertaken on a number of AE alloys and AJ52 to study the solidification characteristics. The recorded temperature vs. time data that were recorded were treated according Backerud and Tamminen [19] to obtain the a) liquidus temperature, b) solidus temperatures, and c) fraction solid as a function of temperature. Figure 12 illustrates these properties for AE42 using the method cited.

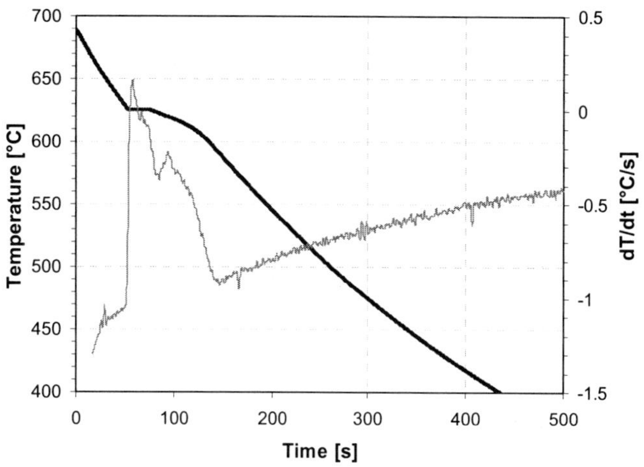

Figure 12. Thermal analysis data of AE42 showing temperature and its derivative as functions of time.

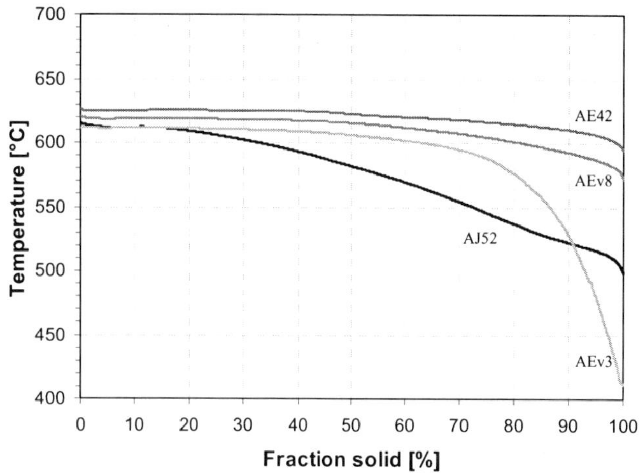

Figure 13. Fraction solid as a function of temperature for three AE alloys including AE42, and AJ52.

As mentioned earlier, fraction solid as function of temperature has been obtained from the time derivative of the temperature. Examples of some alloys are shown in Fig. 13. It is seen that AE42 has a narrow solidification range of about 30°C. The solidification range for the different versions of the AE alloys can be varied from the narrow interval of AE42 to the wide solidification interval of the AE v3 alloy (200°C). In comparison the AJ52 alloy shows a solidification interval of about 115°C.

The solidification range is a result of the chemical composition, in this case the RE and Al contents, which of course also determines the mechanical properties of the alloy. Unfortunately, it often turns out that good high temperature properties are achieved at the expense of castability, and vice versa. This is partly attributed to the use of elements like Ca and Sr which both have very strong affinity to oxygen, combined with a relatively low Al content. Due to the different nature of the elements involved, it is therefore difficult to obtain a one-to-one correlation of the solidification range and castability.

Increasing the Al content is usually detrimental for the creep properties due to an increasing volume fraction of Mg-Al eutectic. This eutectic constituent, which begins to melt at 437°C, contributes to both grain boundary migration and sliding. In some cases it is also embedding the dispersed particles, for example Mg_2Si in the AS alloys, reducing the pinning effect on grain boundary sliding these otherwise can have. As a result, a high Al content can be projected to contributing strongly to the creep deformation process.

For all alloys shown in Fig. 13, except AE v3, the solidification process has stopped before it reaches the Mg-Al eutectic. With the AE alloy family different solidification characteristics can be obtained. This feature can be helpful when developing alloys with a property profile tailormade for certain applications.

CONCLUSION

Propertywise, magnesium is not a lighter version of aluminium, and it will never be, regardless of alloy development efforts. Nevertheless, the differences can largely be accounted for in the design, and the present tensile-, creep- and stress relaxation results demonstrate that AE alloys show a remarkable potential among the Mg-based die casting alloys, especially for drivetrain applications. Based on a comprehensive set of property data, tailored alloys can be picked to meet the various requirements related to strength, creep resistance and castability. Hence the AE alloys should become a very attractive lightweight solution for the automotive industry.

ACKNOWLEDGMENTS

Sten Bjørneboe, Jan Ivar Klovland and Vidar Sjøberg are acknowledged for alloy preparation, die-casting and thermal analysis. Per R. Gjertsen is acknowledged for tensile testing, creep testing and stress relaxation testing.

REFERENCES

1. H. Westengen, P. Bakke, "Magnesium Die Casting Alloys for Elevated Temperature Applications: Recent Developments", *SINOMAG Conf.*, Beijing, China, Nov. 2002
2. H. Westengen, P. Bakke, "Magnesium die casting alloys for use in applications exposed to elevated temperatures: Can they compete with aluminium?", *Magnesium Alloys 2003, Proc. 2nd Intl. PSTAM Conf.*, (Eds: Y. Kojima et al.), Trans Tech Publications, Switzerland, 2003. 35-42
3. F. Hollrigl-Rosta, E.Just, J.Kohler, H.-J.Melzer, "Magnesium in Volkswagenwerk", *Metall*, 34 J, H12, Dec. 1980.1138-1141
4. K. Pettersen, H. Westengen, J.I. Skar, M. Videm, L.-Y. Wei, "Creep resistant alloy development", *Magnesium Alloys and their Applications*, (Ed.: K.U.Kainer) Wiley-VCH, Weinheim, Germany, 2000. 29-34
5. S. Koike, K. Washizu, S. Tanaka, T. Baba, K. Kikawa, "Development of lightweight oil pans made of a heat resistant magnesium alloy for hybrid engines", *SAE Technical Paper No. 2000-01-1117*, 2000
6. A.Luo, B. Powell, "Tensile and compressive creep of Magnesium-Aluminium-Calcium based alloys", *Magnesium Technology 2001*, (Ed.: J. Hryn), TMS-AIME, 2001. 137-144
7. M. Pekguleryuz, E. Baril, "Development of creep resistant Mg-Al-Sr alloys", *Magnesium Technology 2001*, (Ed.: J. Hryn), TMS-AIME, 2001, 119-125
8. P. Bakke, H. Westengen, "Die Casting for High Performance – Focus on Alloy Development", *Magnesium Alloys and their Applications*, (Ed.: K.U.Kainer) Wiley-VCH, Weinheim, Germany, 2003
9. P. Bakke, H. Westengen, K. Pettersen, D. Albright, "Die cast magnesium alloys for high temperature applications", *SAE Technical Paper 2003-01-0189*, 2003
10. P. Bakke, H. Westengen, K. Pettersen, "Enhanced ductility and strength through RE addition to Magnesium die casting alloys", *Magnesium Technology 2003*, (Ed.: H.I. Kaplan), TMS-AIME, 2003. 171-176.
11. H. Gjestland, S. Sannes, H. Westengen, "High pressure die casting – Getting the best out of magnesium", *10th Mg Automotive and End User Seminar,* Aalen, Germany, Sept. 2002.
12. K. Pettersen, P. Bakke, D. Albright, "Magnesium Die Casting Alloy Design", *Magnesium Technology 2002*, (Ed.: H.I. Kaplan), TMS-AIME, 2002. 241-246.
13. S. Sannes, H. Westengen, "The influence of process conditions on the microstructure and mechanical properties of magnesium die castings", *Magnesium Alloys and their Applications*, (Ed.: B.L. Mordike, K.U. Kainer), Wiley-VCH, Weinheim, Germany, 1998. 223-229
14. P. Bakke, K. Pettersen, S. Guldberg, S. Sannes, "The impact of metal cleanliness on mechanical properties of die cast magnesium alloy AM50", *Magnesium Alloys and their Applications*, (Ed.: K.U. Kainer), Wiley-VCH, Weinheim, Germany, 2000. 739-745
15. M.W. Dierks, C. Kuhn, J. Etling, "Enhanced mechanical properties of die cast AM series magnesium; through part design, die design and process control", *SAE Techn. Paper 940410*, 1994
16. R.S. Busk, *Magnesium Products Design*, Marcel Dekker, Inc., New York, NY, USA. 1987. pp.240-241
17. J.R. Newby et al., Editors, *Metals Handbook Ninth Ed. Vol. 8 Mechanical Testing*, American Society for Metals, Metals Park, OH, USA. 1985. pp.22-23
18. K. Pettersen, S. Fairchild, "Stess relaxation in bolted joints of die cast magnesium components", *SAE Technical Paper 970326*, 1997
19. J. Tamminen, "Thermal Analysis for Investigation of Solidification Mechanisms in Metals and Alloys", PhD. Thesis, Dept. of Structual Chemistry, University of Stockholm, Sweden, 1988.

2004-01-0656

High Temperature Mg Alloys for Sand and Permanent Mold Casting Applications

B. Bronfin, E. Aghion and N. Fantetti
Dead Sea Magnesium Ltd

F. Von Buch, S. Schumann and H. Friedrich
Volkswagen AG

ABSTRACT

The need to reduce weight of large and heavy components used by the automotive and aerospace industries such as engine block, cylinder head cover and helicopter gearbox housing has led to the development of new Mg gravity casting alloys that provide adequate properties and cost effective solution. The new Mg gravity casting alloys are designed for high stressed components that operate at a temperature up to 300°C. These new alloys exhibit excellent mechanical properties and creep resistance in T-6 conditions. The present paper aims at introducing three new Mg gravity casting alloys designated MRI 201S, MRI 202S and MRI 203S, which were recently developed by the Magnesium Research Institute of DSM and VW. Apart from the excellent high temperature performance of these alloys, they provide adequate castability and dimension stability along with good weldability and corrosion resistance. In addition, the paper will also introduce practical applications that demonstrate the capabilities of the newly developed alloys.

INTRODUCTION

Magnesium is one of the most promising lightweight metals and therefore can play the crucial role in the automotive and the aerospace industries enabling mass savings and consequently fuel economy and reduction of CO_2 emission. Powertrain components represent a significant portion of the weight of vehicles. Therefore, the development of new creep resistant and cost effective magnesium alloys will resolve several critical issues that limit the large-scale application of magnesium castings in automotive industry. Recently several creep resistant die casting alloys were developed [1,2]. However, for particularly large and heavy components such, for example, engine block, high-pressure die casting may not always be the best solution. In fact, most of such parts are being produced by gravity casting (sand and permanent mold) and low-pressure sand and permanent mold casting. All creep resistant, gravity casting magnesium alloys contain zirconium and consequently are aluminum free. Zirconium is a unique grain refiner of magnesium. It also benefits corrosion resistance of the alloy and helps prevent porosity in castings [1,3]. However, zirconium can serve as a powerful grain refiner of magnesium only in the case when the molten metal is saturated with soluble Zr at high temperature range of 770-800°C. This results in strong oxidation of the molten metal followed by loss of magnesium and expensive alloying elements such as Y, Gd, Dy, Yb, Nd etc. Furthermore, the existing technology of alloy preparation is based on single batch process with the use of small crucibles up to 300-400 kg. The process efficiency is relatively low due to the need to retain a significant amount of heel in the crucible in order to prevent the entrapment of settled inclusions and excessive Zr compounds in the castings. Thus, the preparation process and the use of expensive alloying elements lead to relatively high cost of existing creep resistant gravity casting alloys both in ingot form and in final castings. Therefore, those alloys are mainly dedicated to the production of high performance complicated castings for the aerospace industry and for premium and racing vehicles. In such applications, the performance is of prime importance, while cost is regarded as a secondary factor. The research activities in the field of gravity casting alloys were very limited since the 1980's when Magnesium Electron Ltd has developed WE43 and WE54 alloys [1,3,4] These alloys combine both excellent creep and corrosion resistance and are considered as the best performance high temperature Mg alloys. However, the extremely high cost of WE alloys restricts their use to specific applications that can afford such high cost.

Currently the largest portion of gravity casting applications is covered by ZE41 alloy [3], which has moderate strength and creep resistance, combined with good castability. Although this alloy exhibits poor corrosion resistance (especially in marine environment), it is still most widely used as a preferred material in various helicopter transmission applications. The significant gap, which exists between high performance and extremely high cost of WE43 alloy and the moderate cost but limited creep resistance and poor corrosion properties of ZE41 alloy, was recognized in the last 3 years. This situation stimulated the new activities in the development of high temperature magnesium alloys for sand and permanent mold casting applications [5-7].

ALLOYING PRINCIPLES UNDERLYING THE DEVELOPMENT OF CREEP RESISTANT GRAVITY CASTING ALLOYS

The development of magnesium alloys for high temperature applications is a very complicated task. It requires an understanding of the basic principles of creep behavior of alloys and in particular the factors and the mechanisms affecting the creep resistance of magnesium alloys. In general, alloying principles for designing creep resistant magnesium alloys for gravity casting applications after T6 heat treatment that includes solid solution treatment with subsequent aging for formation of fine precipitates can be reduced to the following:

- The main alloying elements should have significant solid solubility in magnesium, which decreases with decreasing the temperature. This factor is very important to provide a marked response to aging. Solubility limits for binary magnesium alloys can be found in [8]
- Alloying elements should have a low diffusion coefficient, provide strong interatomic bonds and enable solid solution, which has no response to aging over serves conditions.
- Alloying elements should provide the formation of thermally stable intermetallic compounds that have the coherency with the matrix, are effective obstacles to deformation and strengthen grain boundaries.
- Alloying elements should provide good castability (increased fluidity, cracking, reduced porosity and greater casting integrity) combined with improved corrosion resistance and affordable cost.

It is well known [1,3,4,8,9] that Mg-Zr binary system should be selected as the basic system for high temperature gravity casting alloys. Zirconium has a grain refining effect when added to magnesium leading to the greater casting integrity, and improved mechanical properties. Mg-Zr alloys have more consistent properties through thin and thick sections, and are not prone to through-wall porosity, which can cause lubricant leakage [1,3,4].

In comparison to Mg-Al alloys, all Mg-Zr alloys have the tendency to be intrinsically high purity material. This is due to the highly reactive nature of zirconium, which ensures that virtually all common impurities found in magnesium alloy melts are precipitated as separate compounds. Those compounds are settled down before any zirconium dissolves in molten magnesium (the maximum solubility of zirconium in molten magnesium is 0.6%). This leads to the fact that content of impurities such as Si, Cu, Fe and Ni is invariably below 50 ppm and typically in the range 10-20 ppm. The most practical elements compatible with Zr are Zn, Rare earth (La, Ce, Pr and Nd), Y and Ag. The soluble Zr fraction mostly provides the fine grain size obtained in Zr contained alloys [9,10].

Since the development of practical alloys is usually a compromise between the required performance and the affordable cost, the main challenge is to find the optimum combination of the above factors. This was addressed by DSM and VW in the attempt to develop new creep-resistant magnesium alloys for sand and permanent mold casting applications.

PROPERTIES AND PERFORMANCE OF NEW MRI'S ALLOYS

The systematic study initiated by DSM and VW has resulted in the development of 3 new creep resistant alloys designated MRI 201S, MRI 202S and MRI 203S. All of these alloys are well suitable for sand and permanent mold casting applications and for associated low-pressure variations. The alloys have to be used only in the T6 condition in order to generate the optimum combination of ambient strength, ductility and creep properties. Typical mechanical properties of new MRI alloys at room and elevated temperatures are listed in Table 1 in comparison with commercial alloys WE43 and ZE41. It is evident that alloy MRI 201S-T6 exhibits a combination of properties similar to that of WE43-T6 alloy. Therefore, based on the chemical composition of both alloys, one can consider MRI 201S as a low cost substitution for WE43 alloy. The alloy MRI 202S is a creep resistant alloy with moderate strength and good corrosion resistance. It outperforms all common sand casting alloys, except for WE43 and WE54 alloys in creep and corrosion resistance.

The alloy MRI 203S is a low strength gravity-casting alloy, which exhibits creep and corrosion resistance superior to those of ZE41A alloy. Since the production cost of MRI 203S is considered similar to that of ZE41A alloy, this new alloy can be used as a high performance substitution for ZE41A alloy.

Figs. 1 and 2 illustrate tensile and compressive elevated temperature properties of MRI 201S and MRI 202S alloys. Both alloys exhibit excellent strength retention at temperatures up to 250°C (maximum temperature at which the properties were tested). In contrast to this, the aluminum alloy A319.0-T6 (AlSi6Cu3.5), which is commonly used as sand casting alloy for crankcase manufacturing, showed significant loss in strength at temperatures higher than 150°C.

As can be seen from Fig. 3, all new MRI alloys also outperform the alloy A319.0 in creep strength at the temperature range of 150-200°C.

It is well known from WE54 experience [1,3,4] that one of the important requirements to high temperature alloys for crank-case and gearbox applications is their long-term stability at temperatures up to 200°C, particularly around 150°C.

Table 1. Typical Mechanical Properties of New MRI's Alloys

Properties	MRI 201S T6	MRI 202S T6	MRI 203S T6	WE43 T6	ZE41 T5
TYS (MPa)					
at 20 °C	170	150	125	180	140
at150 °C	170	145	120	178	120
at 175 °C	165	140	115	175	110
UTS (MPa)					
at 20 °C	260	250	210	260	220
at 150 °C	245	220	190	210	170
at 175 °C	240	215	180	205	150
E (%)					
at 20 °C	6	7	4	6	5
at 150 °C	11	15	8	7	22
at 175 °C	12	16	12	11	25
CYS (MPa)					
at 20 °C	190	145	125	190	140
at 150 °C	190	140	120	185	115
at 175 °C	185	136	113	185	110
Fatigue strength (MPa) (1) at 20 °C	110	85	-	100	95
Corrosion Rate (2) (mg/cm2/day)	0.10	0.12	0.14	0.10	3.1
Stress [MPa] to produce0.2% creep strain					
at 175 °C	185	155	110	190	70
at 200 °C	160	100	70	160	50
at 250 °C	75	40	30	60	20

(1) Rotating bending 5X10 ^{7}Cycles

(2) 100hr. Salt Spray Test (ASTM Standard B-117)

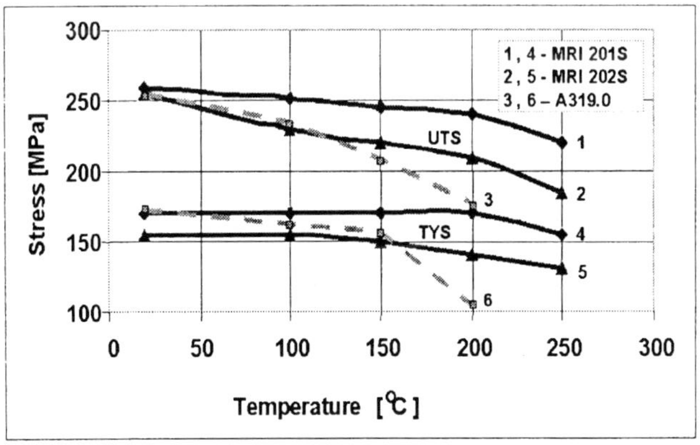

Figure 1. Tensile Strength of MRI 201S, MRI 202S & A319.0 vs. Temperature

It was found, for example, that alloy WE54 showed significant degradation in room temperature ductility (less than 2%) after long-term (beyond 1000h) exposure at 150°C.

Tables 2 and 3 illustrate the effect of long-term exposure at 150°C and 200°C for 1000 hours on the tensile properties of MRI 201S and MRI 202S at room temperature and 175°C. It is evident that both alloys are not prone to embrittlement after long-term exposure. In addition, it should be noted that both alloys exhibit excellent strength retention at 175°C also after long-term exposure at 150-200°C

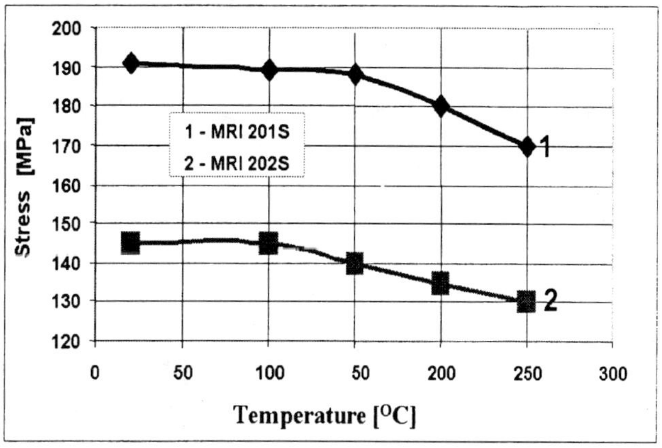

Figure 2. Compressive Yield Strength of MRI 201S & MRI 202S vs. Temperature

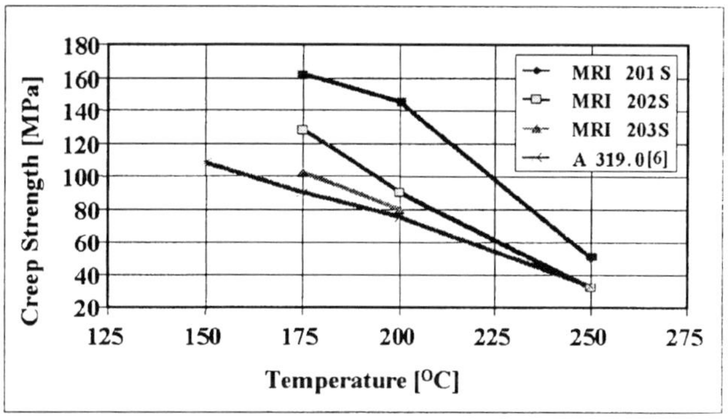

Figure 3. Creep Strength of new MRI Alloys in Comparison with A319.0*

** The Strength to produce 0.1% creep strain for 100h*

Table 2. Effect of long-term exposure for 1000 h at elevated temperatures on tensile properties of MRI 201S-T6 and MRI 202S-T6 tested at room temperature.

Exposure temperature, °C	MRI 201S TYS MPa	MRI 201S UTS MPa	MRI 201S E %	MRI 202S TYS MPa	MRI 202S UTS MPa	MRI 202S E %
20	170±7	260±10	6±1	150±5	250±9	7±1
150	170±9	265±8	6±1	146±8	250±7	7±1
200	165±8	262±10	6±1	139±7	242±8	7±1

Table 3. Effect of long-term exposure for 1000h at elevated temperatures on tensile properties of MRI 201S-T6 and MRI 202S tested at 175°C.

Exposure temperature, °C	MRI 201S TYS MPa	MRI 201S UTS MPa	MRI 201S E %	MRI 202S TYS MPa	MRI 202S UTS MPa	MRI 202S E %
20	164±3	240±8	12±1	140±3	230±9	16±1
150	163±4	242±7	11±2	138±6	228±8	17±1
200	162±5	240±9	11±2	136±5	232±6	16±2

All MRI's newly developed alloys also exhibit excellent castability, pressure tightness and weldability. This was demonstrated by several successful casting trials of various engine blocks and cylinder head covers.

For example, Fig.4 illustrates the crankcase for the 3 cylinders TDI Lupo 3l Volkswagen car. This 14.25 kg demonstrative crankcase was produced by low-pressure sand casting from the alloys MRI 202Sand MRI 203S.

Figure 4. A demonstrative crankcase for the 3 cylinders TDI Lupo 3l Volkswagen car (Low – Pressure Sand Cast of MRI 202S and MRI 203S)

It should be also noted that automotive producers always require from new materials the highest performance combined with the affordable cost. As can be seen from Table 4, the alloys MRI 201S and WE43 exhibit practically similar performance in terms of strength, creep behavior and corrosion resistance.

Table 4. Combined rating of MRI's alloys compared to commercial alloys

		Creep Resistance	Tensile Yield Strength	Corrosion Resistance	Performance Score	Cost	Total Score
Alloy	Weight	10	7	9		10	
MRI 201S	Score	10	9	10		4	
	W x S	100	63	90	253	40	293
MRI 202S	Score	9	8	10		7	
	W x S	90	56	90	236	70	306
MRI 203S	Score	8	5	9		9	
	W x S	80	35	81	196	90	286
WE43	Score	10	10	10		2	
	W x S	100	70	90	260	20	280
ZE41	Score	7	8	3		10	
	W x S	70	56	27	143	100	243

The other two new alloys exhibit lower performance compared to WE43 alloy but outperform ZE41 alloy. However, if the cost is also taken into consideration, it is clearly evident that all the newly developed alloys provide the best combination of properties in terms of value to cost (Fig.5)

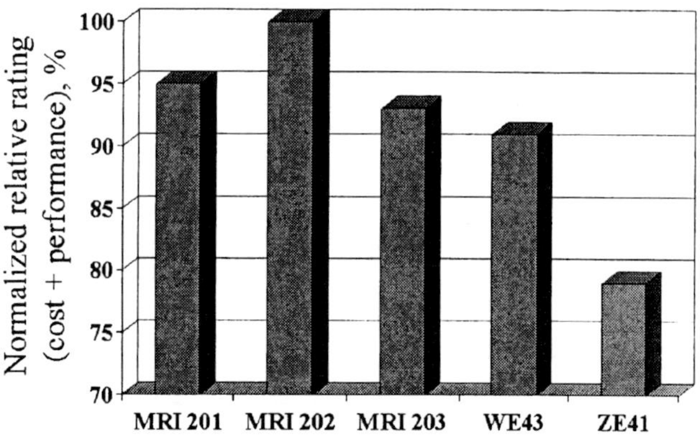

Figure -5 Normalized relative rating (cost + performance) of MRI's alloys compared to commercial alloys

CONCLUSIONS

Three newly developed magnesium alloys MRI 201S, MRI 202S and MRI 203S were introduced as a higher-performing, more economical solution to current low-pressure-cast magnesium alloys. These new alloys are designed for sand and permanent-mold casting ,low pressure sand and permanent-mold casting as well as squeeze casting The results obtained with cast specimens and practical applications have clearly shown that these new alloys exhibit excellent high-temperature properties and corrosion resistance combined with affordable cost. The outstanding castability, pressure tightness and dimensional stability of new alloys were manifested by the production of demonstrative components such as crankcases and cylinder heads for various vehicles.

REFERENCES

1. J.F. King, "Development of Practical High Temperature Magnesium Casting Alloys", Magnesium alloys and their applications, ed. K.U. Kainer (Weinheim, Germany: Wiley-VCH Verlag GmbH, 2000), pp.14-22.
2. A.Luo "Recent magnesium alloy development for automotive powertrain applications" Materials science Forum. Magnesium alloys 2003,Osaka. January 2003,p.p.57-66.
3. I.J.Polmer, Metallurgy of Light Alloys, Halsted Press, New York, 1996, pp.196-247.
4. J.F.King, P.Lyon, K.Savage,"Influence of rare earth elements and minor additions on properties and performance of magnesium-yttrium alloys in critical aerospace applications", Proceedings of 59th World

Magnesium Conference, Montreal, May 19-21, 2002, pp.15-22.
5. E.Aghion et al, "The art of developing New Magnesium Alloys for high temperature applications", Materials Science Forum. Magnesium alloys 2003,Osaka.January 2003,pp.407-417.
6. C.J.Beetles et al," AMC-SC1: A new magnesium alloy suitable for powertrain applications", SAE technical paper 2003-01-1365.
7. B.Bronfin et al," High temperature resistant magnesium alloys", Patent application EP 1329530.
8. I.J.Polmear," Grades and Alloys", Magnesium and Magnesium Alloys, ASM Specialty Handbook, ASM International, 1999,pp.12-25
9. E.F. Emley, Principles of Magnesium Technology, Pergamon, Oxford, .
10. Ma Qian et al "Effect of soluble and insoluble zirconium on the grain refinement of magnesium alloys Material Science Forum. Magnesium Alloys 2003,Osaka January 2003, pp.593-598.

2004-01-0657

Evaluation of High Temperature, Cast Magnesium Alloy, Front Engine Covers

Alan P. Druschitz and Eric Showalter
Intermet Corporation

ABSTRACT

A variety of new magnesium casting alloys specifically designed for high temperature applications are currently available. However, there is little published data from component tests or from test specimens sectioned from component castings. In this study, the mechanical properties (tensile, bracket thread integrity, bracket distortion and fastener/attachment point acceptability) of front engine covers made from three magnesium alloys (AZ91D, AJ62x and MRI-153M) and from aluminum alloy 380 are presented.

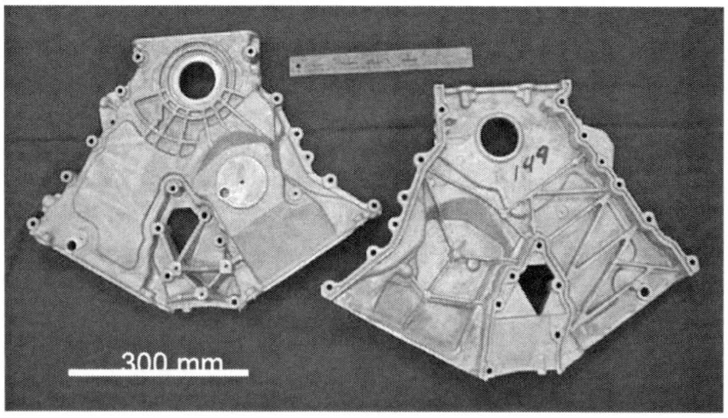

Figure 1. Front engine cover (obsolete Ford 3.0 L) used in this investigation.

INTRODUCTION

The use of magnesium alloys in elevated temperature automotive applications has been slow due to the lack of published data from component tests and from test specimens sectioned from component castings. Therefore, to address the continuing need to reduce vehicle weight, material suppliers and component manufacturers have teamed together to develop the necessary design expertise and produce the necessary design data [1,2]. Dead Sea Magnesium [3], Hydro Magnesium [4] and Noranda [5,6] have each developed new, high temperature, magnesium alloys. In this study, material provided by Dead Sea Magnesium and Noranda was evaluated. The evaluation consisted of casting a front engine cover by high pressure die casting in a production facility, machining of the castings, performance of Ford ES tests on the castings and the performance of standard tensile tests on sections cut from the castings. Figure 1 shows the front engine cover used in this investigation. The production aluminum front engine cover weighed an average of 4221 grams fully machined and the magnesium front engine covers weighed an average of 2693 grams fully machined, which represented a weight savings of 36%.

CASTING PROCEDURES

Casting trials were conducted at INTERMET's Hannibal Missouri magnesium casting plant. The machine used for these trials was a fully automated Prince 1446CCM die casting cell. The metal for each of the casting trials was prepared in accordance with supplier recommendations the day prior to the trials. The casting tool used for the trials was a "retired" production die from one of our aluminum casting facilities. The gates and runners were not modified from the aluminum configuration for the casting trials.

For each magnesium alloy, a process trial period of two to three hours was used to adjust the process parameters and program robotic extraction and lubricant systems tailored to the part and the specific material. Immediately following the process trial period, a three-hour "completely automatic" production run was commenced. During these automatic runs, all casting parameters and production setups remained unchanged. Production personnel and management from INTERMET Hannibal conducted the trials with assistance from INTERMET's Technical Services group.

All three magnesium alloys, AZ91D, MRI-153M and AJ62x, were successful during the three-hour automatic production run, although the casting parameters were considerably different from alloy to alloy. The three-hour casting runs averaged approximately 200 castings per alloy.

Aluminum castings used in the evaluation and testing were produced at INTERMET Racine Casting during normal production operations.

INTERMET currently can see no impediment to the use of either MRI-153M or AJ62x high temperature magnesium alloys in high volume production based on the production trials conducted at the Hannibal plant.

EVALUATION PROCEDURES

All of the castings produced for this study were tested in the as-cast (F temper) condition. Three standard Ford Motor Company ES tests were performed at room and elevated temperatures (bracket thread integrity test, bracket distortion test and fastener/attachment point acceptability test). Standard tensile tests were also performed at room temperature on specimens sectioned from ten to eleven locations on the castings.

TENSILE. ASTM, standard, sub-size, flat tensile specimens with a 25.4 mm (1.0") gage length were tested at room temperature in accordance with ASTM E8 on a MTS Model 810 servohydraulic tension/compression machine. A minimum of 10 specimens were tested and the results averaged. All samples had machined surfaces (only enough material to produce flat, parallel surfaces was machined off of the samples). A 25.4 mm (1.0") gage length extensometer was used to measure strain up to 1%. The initial strain rate was 0.8% per minute for the magnesium alloys and 0.6% per minute for aluminum alloy 380. At 1% strain, the test was paused, the extensometer removed and the test continued until failure at a crosshead speed of 2.54 mm/min (0.1" per min). Ultimate tensile strength, 0.2% offset yield strength, % elongation at fracture and Young's modulus were then determined.

BRACKET THREAD INTEGRITY. Samples were tested in accordance with Ford ES-1X4E-19E708*-AA, REV-0, Section III.E, which states that the fastener must fail before the threaded hole. Spacers were fabricated to simulate the actual connection point of each component to the front engine cover. New production fasteners (M6 or M8) were used for each test. The fasteners were inserted through the spacers and threaded into the front engine cover. Each fastener was slowly tightened until failure occurred. The failure mode for each location tested was recorded. Three aluminum alloy 380 front engine covers were tested and two of each magnesium alloy. Figure 2 shows the test locations on the front engine cover.

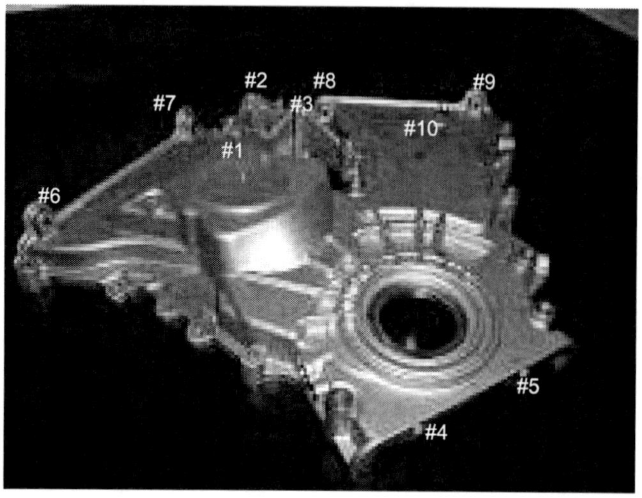

Figure 2. Bracket thread integrity test locations on front engine cover.

BRACKET DISTORTION TEST. Samples were tested in accordance with Ford ES-1X4E-19E708*-AA, REV-0, Section III.F, which states that pulley misalignment may be no greater than 0.08°. Front engine covers were mounted in accordance with Ford engine assembly drawing ILYF1E-030002-E0543F, frames 16A and 68. Each front engine cover was bolted to a rigid mounting plate, which simulated the engine configuration, with sixteen M8 fasteners initially tightened to 5% of the design torque. A production steering pump was then mounted to the front engine cover with bolts tightened to the specified design torque. Three dial indicators were positioned against the power steering pump pulley in the Z-axis as shown in Figure 3. The sixteen M8 front engine cover fasteners were then tightened to the maximum design torque and the dial indicator deflections recorded. The diameter of the power steering pump pulley was 142.88 mm (5.625").

Figure 3. Bracket deflection test setup.

FASTENER / ATTACHMENT POINT ACCEPTABILITY TEST. Samples were tested in accordance with Ford ES-1X4E-19E708*-AA, REV-0, Section III.D, which states that 90% of the initial load must be retained after five thermal cycles at 121°C (250°F). Samples were bolted to a plain carbon steel mounting fixture with new production M8 fasteners (Ford P/N W701525 for joint numbers 2 and 6 or P/N W701581 for joint number 4). Load washers (Lebow model no. 3718-112-8MM high temperature bolt force sensor) were positioned between the bolt head and the bracket with a hardened precision washer contacting the bracket. The contact area for the washers was 86.8 sq mm (0.1346 sq in) for boss numbers 2 and 6 and 138.6 sq mm (0.2148 sq in) for boss number 4. At room temperature, each bolt was tightened to the maximum clamp load (21200 N, 4766 lbf) as specified in Table III.D.2 of the Ford engineering specification. Samples were placed in an oven set at 121°C (250°F) for a minimum of five and one-half hours. A thermocouple was attached to one sample and the actual heat-up time determined for each run (about 90 minutes). Therefore, the samples were "at-temperature" for the specified minimum of four hours. Samples were then removed from the oven and allowed to cool to room temperature for at least four hours. This cycle was repeated a total of five times; bolt load readings were recorded at the end of each cycle. Two samples of aluminum alloy 380 and a minimum of five samples of each magnesium alloy were tested and the results averaged. The fastener/attachment point acceptability test setup and boss locations are shown in Figure 4.

Figure 4. Fastener/attachment point acceptability test setup.

RESULTS AND DISCUSSION

TENSILE. All of the magnesium alloys tested had lower ultimate tensile strength, yield strength and Young's modulus compared to aluminum alloy 380. However, magnesium alloys AJ62x and MRI-153M had greater elongation at fracture and all three of the magnesium alloys tested showed less variation in properties (smaller standard deviations) compared to aluminum alloy 380. The room temperature, as-cast, tensile properties for all of the materials tested are shown in Tables I and II.

Table I. Average room temperature, as-cast, tensile properties of specimens sectioned from front engine covers (average of a minimum of 10 samples cut from different locations on one casting).

Alloy	UTS, MPa	YS, MPa	EL, %	E, GPa
380 Al	260±23	145±6	2.6±1.0	75±1
AZ91D	201±21	139±5	2.3±1.2	41±1
AJ62x	214±18	126±3	4.5±1.4	43±1
MRI-153M	208±12	137±4	3.1±1.1	44±1

Table II. "Best bar" room temperature, as-cast, tensile properties of specimens sectioned from front engine covers (10 to 11 samples were cut from different locations on each casting).

Alloy	UTS, MPa	YS, MPa	EL, %
380 Al	291	153	4.4
AZ91D	226	142	4.2
AJ62x	236	127	7.3
MRI-153M	217	135	4.6

BRACKET THREAD INTEGRITY. None of the magnesium alloys tested passed the bracket thread integrity test at all locations. Front engine covers made from AZ91D exhibited thread failure in at least one of two tests at locations 1-5, which uses an M8 bolt. Front engine covers made from AJ62x or MRI-153M performed similarly with both materials exhibiting thread failure in both tests at locations 4 and 5 and in one of two tests at location 2. Therefore, there may have been a common design issue at locations 4 and 5 that caused the Mg alloy front engine cover threads to fail. All of the Mg front engine cover thread failures were at locations requiring the larger and stronger M8 fastener. The length of thread engagement was less at locations 4 and 5 (15.2 mm, 0.6") compared to locations 1-3 (18.0 mm, 0.7"). The recommended L/D ratio for magnesium alloys is 2.3 [7]. Thus, a M8 fastener should have a minimum thread engagement of 18.4 mm (0.72"). Therefore, the thread integrity test failures noted for this component may be easily corrected. All of the Mg front engine covers passed the thread integrity test at locations requiring the smaller M6 fastener. Front engine covers made from aluminum alloy 380 passed the bracket thread integrity test at all locations. The bracket thread integrity test data for all of the materials tested are shown in Table III.

Table III. Bracket thread integrity test data for all of the materials tested.

		Alloy			
Description	Location	380 Al	AZ91D	AJ62x	MRI-153M
PS pump (M8)	1	BF,BF,BF*	BF,TF**	BF,BF	BF,BF
	2	BF,BF,BF	BF,TF	TF,BF	BF,TF
	3	BF,BF,BF	TF,TF	BF,BF	BF,BF
Oil pan (M8)	4	BF,BF,BF	TF,TF	TF,TF	TF,TF
	5	BF,BF,BF	TF,TF	TF,TF	TF,TF
Cam cover (M6)	6	BF,BF,BF	BF,BF	BF,BF	BF,BF
	7	BF,BF,BF	BF,BF	BF,BF	BF,BF
	8	BF,BF,BF	BF,BF	BF,BF	BF,BF
	9	BF,BF,BF	BF,BF	BF,BF	BF,BF
Sensor (M6)	10	BF,BF,BF	BF,BF	BF,BF	BF,BF

* BF = bolt failure
** TF = thread failure

BRACKET DISTORTION. All of the magnesium alloys tested passed the bracket distortion test. Three aluminum front engine covers were tested to determine the expected range of test values. Material did not have an effect on the results of this test. The bracket distortion test data for all of the materials tested are shown in Table IV.

Table IV. Bracket distortion test data for all of the materials tested.

Alloy	Max. pulley misalignment
380 Al	0.020°, 0.031°, 0.061°
AZ91D	0.015°
AJ62x	0.046°
MRI-153M	0.031°
specification	Less than 0.080°

FASTENER / ATTACHMENT POINT ACCEPTABILITY. Neither aluminum alloy 380 nor the magnesium alloys tested passed the fastener/attachment point acceptability test. Further, none of the magnesium alloys tested performed as-good or better than aluminum alloy 380. However, magnesium alloys AJ62x and MRI-153M performed better than AZ91D at all locations tested. The measured retained load was determined to be a function of boss length not washer contact area. Boss 4 was significantly longer than boss 2 or boss 6. Magnesium alloys AJ62x and MRI-153M had comparable performance. The fastener/attachment point acceptability data for all of the materials tested are shown in Table V.

Table V. Fastener/attachment point acceptability test data for all of the materials tested.

	Retained Load, %		
Alloy	Boss 2	Boss 4	Boss 6
380 Al	83.7	80.5	78.3
AZ91D	50.3	38.0	53.2
AJ62x	59.6	43.5	56.6
MRI-153M	57.8	46.9	55.9
specification	>90	>90	>90
boss length, mm	27.4	47.2	27.4

ONGOING STUDIES

Resonant survey and shaker durability testing are currently in-progress.

CONCLUSIONS

1. None of the magnesium alloys tested performed as as-good or better than aluminum alloy 380. Thread integrity and fastener/attachment point acceptability were the areas where the magnesium components did not meet the current engineering requirements.
2. The thread integrity test failures may be easily resolved by increasing the thread engagement length to the recommended L/D of 2.3.
3. Neither aluminum alloy 380 nor the magnesium alloys tested met the fastener/attachment point acceptability criteria as currently written.
4. INTERMET currently sees no impediment to the casting of either MRI-153M or AJ62x high temperature magnesium alloys in high volume production.
5. The use of magnesium could result in a weight savings of as much as 36% if the thread integrity and fastener/attachment point acceptability requirements could be met.

ACKNOWLEDGMENTS

The authors would like to acknowledge the assistance of Nathan Sochor (component testing engineer) and Dale tenPas (component testing technician) for conducting the various engineering tests reported in this paper, Donald Argo of Noranda and Nick Fantetti, Nir Moscovitch and Dr. Eli Aghion of Dead Sea Magnesium for their assistance during the castings trials and the personnel of INTERMET Hannibal, including Greg

Addison and Kevin Bichsel, for their time and effort during the casting trials. We would also like to thank Al Dimmitt and Joe McNeill from INTERMET Missouri Operations for their help setting up the casting trials.

REFERENCES

1. Druschitz, et al, "Evaluation of Structural and High-Temperature Magnesium Alloys," SAE Technical Paper 2002-01-0080, Society of Automotive Engineers, Warrendale, PA (2002).
2. Druschitz, et al, "Bolt Load Compressive Stress Retention Testing of Magnesium Alloys," SAE Technical Paper 2003-01-0187, Society of Automotive Engineers, Warrendale, PA (2003).
3. Aghion, et al, "A Comparative Study of New Magnesium Alloys Developed for Elevated Temperature Applications in Automotive Industry," SAE Technical Paper 2003-01-0191, Society of Automotive Engineers, Warrendale, PA (2003).
4. Bakke, et al, "Die Cast Magnesium Alloys for High Temperature Applications," SAE Technical Paper 2003-01-0189, Society of Automotive Engineers, Warrendale, PA (2003).
5. LaBelle, et al, "Comparative Mechanical Properties of AE42 and AJ52x High-Temperature Diecast Magnesium Alloys for Elevated Temperature Applications," SAE Technical Paper 2003-01-0188, Society of Automotive Engineers, Warrendale, PA (2003).
6. Pekguleryuz, et al, "Magnesium Die Casting Alloy AJ62x with Superior Creep Resistance, Ductility and Die Castability," SAE Technical Paper 2003-01-0190, Society of Automotive Engineers, Warrendale, PA (2003).
7. Cameron, "Use of Threaded Fasteners in Die Cast Magnesium," SAE Technical Paper 860289, Society of Automotive Engineers, Warrendale, PA (1986).

CONTACT

Alan P. Druschitz received his PhD in Metallurgical Engineering in 1982 from the Illinois Institute of Technology, Chicago, IL. He is currently the Director of Materials Research and Development for INTERMET Corporation. He is located at the INTERMET Technical Center, Lynchburg, VA 24502. He can be reached at adruschitz@intermet.com or (434) 237-8749. Before joining INTERMET Corporation in 1996, he was a staff research engineer for General Motors Corporation for fourteen years. He has been a member of the American Foundry Society for fifteen years, the Society of Automotive Engineers for twenty-two years and ASM International for twenty-seven years. He is currently the President of the Ductile Iron Society, a member of the Industrial Advisory Board for the Central Virginia Governors School, a Member of the Governors Board of Transportation Safety for the Commonwealth of Virginia and a SAE Fellow

Eric R. Showalter received his Masters Degree in Business Administration in 2003 from the University of Michigan (Ann Arbor) and his Bachelors Degree in Mechanical Engineering in 1993 from Bradley University, Peoria, IL. He is currently the Director of Advanced Manufacturing Engineering for INTERMET Corporation. He is located at the INTERMET Corporate Offices in Troy, MI 48098 and can be reached at (248) 267-5107 or eshowalter@intermet.com. Before coming to INTERMET with the acquisition of Ganton Technologies, where he was Director of Engineering, he was Engineering Manager for the J.L. French Corporation in Sheboygan, WI. He has also held various positions in engineering, sales and manufacturing. He is currently a member of the North American Die Casting Association and INTERMET's representative to the International Magnesium Association. He has been involved in the manufacture of automotive parts in aluminum and magnesium using high-pressure die casting techniques for over ten years.

2004-01-0658

Comparative Evaluation of Automotive Oil Pans Fabricated by Creep Resistant Magnesium Alloy and Aluminum Alloy

Peter Vert
Tesma Engine Technologies

Xiaoping Niu
Toral Cast Light Metal Technologies

Alexander Stickler and Wieslaw Zaton
Magna Advanced Technologies

Eli Aghion
Dead Sea Magnesium Ltd.

ABSTRACT

A series of comparative tests were conducted on two geometrically-identical automotive oil pans, one produced from A380 aluminum – pulled directly from its high-volume production line – and a second produced in a casting trial using DSM's MRI 153M creep-resistant magnesium alloy.

While the theoretical differences in mechanical behavior of cast aluminum parts versus cast magnesium parts are relatively simple to determine, very few quantified, experimental comparisons exist. Where these comparisons do exist, they are often performed on simple shapes. Therefore, the objective of this project was to compare the stiffness, modal behavior, natural frequencies, and damping ratios of the two materials in a complex, functional geometry. Stiffness, natural frequencies, and flexural modes were determined using holographic laser interferometry and finite element analysis. Natural frequencies, modal behavior, and damping ratios were determined via hammer-impact excitations.

The push to reduce vehicle weight by applying magnesium to larger, automotive powertrain components requires the usage of a cost-effective alloy that exhibits creep properties similar to the incumbent material, A380 aluminum. Consequently, the creep behavior of the MRI 153M alloy was compared to A380 via determinations of the bolt load retention (BLR) on the sealing face of the oil pans. BLR was determined on-vehicle using ultrasonic measurement methods.

INTRODUCTION

Magnesium alloys offer an attractive advantage in weight savings for automotive applications. However, the benefit has not been realized in powertrain components used at elevated temperatures, due mainly to poor high temperature creep properties of the common die-casting alloys, such as AZ91D, as well as the high cost and poor castability of the currently available creep resistant alloys containing rare earth additions. Therefore, recent research has been focused on the development of magnesium alloys with a combination of high temperature creep resistance, comparable to A380 aluminum, and die castability that is similar to, or better than, AZ91D magnesium.

To date, aluminum alloy is the dominant material used for producing large automotive powertrain components. High-pressure die-casting (HPDC) is the preferred manufacturing process, as it provides high productivity at relatively low cost. To replace aluminum for the purpose of reducing vehicle weight, magnesium alloys must possess similar mechanical, creep and corrosion properties to the incumbent aluminum. During the last decade, several creep resistant magnesium alloys have been developed for die casting, based on existing magnesium alloys with additions of Ca, Sr and Ce-based mishmetal [1-5]. The die castability of the developed alloys was studied on several powertrain components, such as gearboxes [6], clutch housings [6], oil pans [7-8] and covers [8]. The performance of the developed alloys was evaluated mainly based on metallurgical analyses of simply shaped testing specimens [9]. Very few quantified experimental comparisons between aluminum and magnesium via real functional testing of complex components have been reported.

This paper presents recent research results achieved in a collaborative project between Teams/Magna International and Dead Sea Magnesium (DSM). The objective of this work is to compare the static and dynamic properties between aluminum and magnesium in a complex and functional geometry. A series of comparative tests were conducted on two geometrically identical automotive oil pans, one produced from A380 aluminum – pulled directly from its high-volume production line – and a second produced in a casting trial using Dead Sea Magnesium's MRI 153M creep-resistant magnesium alloy. Stiffness and flexural modes were determined using holographic laser interferometry. Natural frequencies, modal behavior, and damping ratios were analyzed via hammer-impact excitations. The creep behavior of the MRI 153M alloy was compared to A380 via ultrasonic measurement of the bolt load retention (BLR) on the sealing faces of the oil pans, which were installed on two road test vehicles.

EXPERIMENTAL PROCEDURES

MATERIALS AND DIE CASTING – A creep resistant magnesium alloy, MRI 153M developed by DSM, was used in this study. The alloy was melted in a Striko-Dynarad furnace protected with a mixture gas of SF_6, CO_2 and dry air. The melting temperature was controlled at 670 °C. The oil pan was cast on a Buehler 1800-ton SC die-casting machine. After casting, selected magnesium oil pans were machined and assembled for further tests.

OIL PAN ASSEMBLY – The oil pan chosen for this trial was for a 2003 model year 4-cylinder engine; due to confidentiality obligations neither the nameplate of the vehicle used for the road test nor the engine name can be disclosed here, however it suffices to describe the vehicle as a compact car.

In several of the tests described below, the oil pans were assembled to their bedplates (aluminum casting comprising the lower half of the crankshaft journals) using the same components as on the production Al oil pan. As shown in Figure 1, the pans were attached to the bedplate with 15-M8x1.25 threaded bolts with a head diameter of 13mm. The bolts passed through 8mm holes in the oil pan and threaded directly into the mating aluminum bedplate. The steel bolts used in both the aluminum and magnesium assemblies were exactly as supplied for usage on the production aluminum oil pan, namely with a zinc phosphate organic coating. No other considerations were made for the prevention of galvanic corrosion between the oil pan and the steel bolts. Additionally, a gasket was formed between the oil pans and their bedplates using a silicone paste that cures at room temperature; again, the silicone sealant used was exactly as specified by the vehicle OEM for usage on the production assembly line. Additionally, when assembled in the vehicle, the oil pan was attached to a steel bracket, as seen in Figure 1b. In this case, the oil pan was separated from the steel bracket via the 'E-Coat' typically employed to protect automotive steel components from their operational environment; here, too, the bolts were treated with a zinc phosphate coating. It is important to note that, at this attachment location, the steel bolts threaded directly into the magnesium or aluminum oil pan.

To summarize, the magnesium oil pan was predominantly separated from the steel bolts with a zinc phosphate organic coating on the bolts and separated from the aluminum bedplate with a silicone sealant, just as used for the production aluminum pan. At the location where steel bolts were threaded directly into the magnesium oil pan, the bolts were zinc phosphate coated and the steel bracket was separated from the magnesium via E-Coat.

a

b

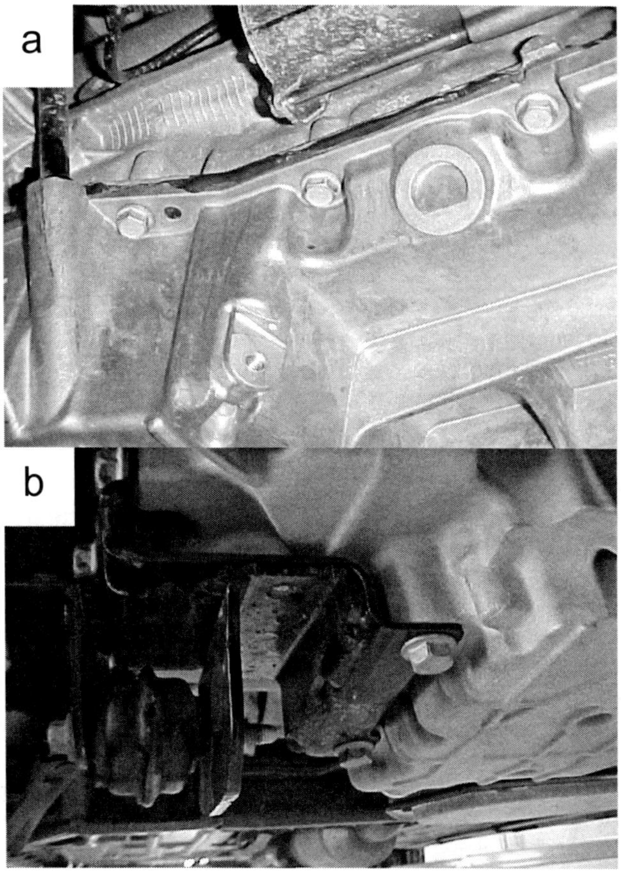

Figure 1: Oil pan installation showing (a) attachment to bedplate via M8 bolts and silicone sealant, and (b) attachment to steel bracket with bolts threaded into magnesium pan.

LASER INTERFEROMETRIC TEST – The comparison of rigidities and natural frequencies between A380 aluminium and MRI 153M magnesium oil pans was studied by laser interferometer testing under varying levels of internal vacuum (762-5 mm Hg). The test was conducted on a continuous-wave laser holographic interferometer system with a Newport Verdi-V Neodymium Vanadate DPSS Class 4 Laser (532 nm, 5 W). Figure 2 presents the holographic interferometry set-up for a test oil pan. Holometry makes use of the three dimensional images recorded as holograms in the following way: after a hologram of the initial undisturbed state of the oil pan has been taken, the real image of the vacuum-deformed oil pan is superimposed and an interferogram is observed. Displacements or movements anywhere on the oil pan's front surface, resulting from the internal vacuum, generate an interferogram with patterns of bright and dark lines called fringes. Fringes depict areas of equal normal displacement – topographic isolines of stress-induced front surface displacement.

A Microtrack 7000 laser displacement sensor was used for displacement measurements. The device has two independent channels for remote displacement measurement in the range of ±5 mm with an accuracy of ±0.001 mm. During testing, the heads were facing the oil pan front surface at different locations to trace the absolute displacement for the whole range of applied vacuum.

FINITE ELEMENT ANALYSIS – A comparative structural analysis was conducted on the oil pans under vacuum. Vacuum was simulated by applying by a uniform, inward acting pressure (0.1 N/mm^2) on the inside surfaces of the part. The finite element model was generated using the original CAD data set. The analysis focused on determining the displacement distribution within the oil pan constructed using two different materials: aluminum alloy A380 and magnesium alloy MRI 153M. The analytical result was compared to laser interferometric result.

DYNAMIC MODAL ANALYSIS – Dynamic properties, such as natural frequencies, vibrational modes and damping ratios of Mg and Al pans were determined using hammer impact excitation. The test was conducted in a free state (oil pan alone) and in an assembled state with the bedplate, gasket, and fasteners. The test set-up is shown in Figure 3. A miniature, instrumented hammer was used to apply impact excitation sequentially to a series of points in such a way that the impact was directed perpendicularly to the flange plane. The acceleration response was monitored with accelerometers (sensors), as shown in Figure 3. During testing, Modal Monitor of Cada-X was used to extract the natural frequencies, damping ratios and mode shapes of several lowest-frequency modes from the collected frequency response functions (FRFs).

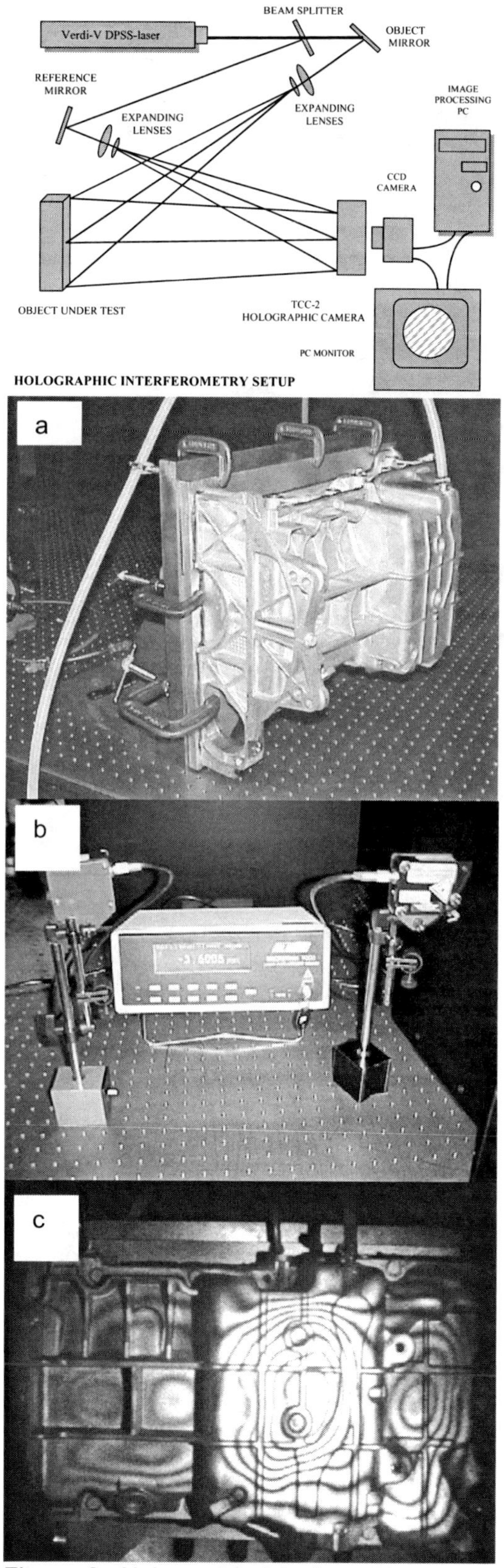

Figure 2: Holographic interferometry set-up; a test oil pan on the holographic bench under vacuum (a), a

Microtrack 7000 laser displacement sensor (b), and an example of an interferogram of the oil pan under test (c).

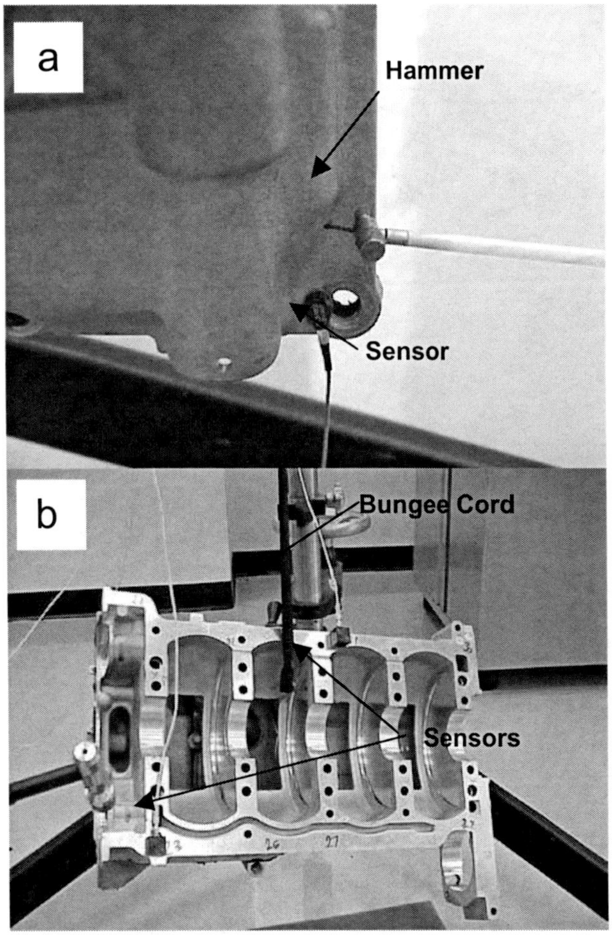

Figure 3: Modal test set-up: (a) oil pan alone (free state), (b) oil pan with bedplate (assembled state).

IN-VEHICLE BOLT LOAD RETENTION TEST – In order to examine the magnesium oil pan performance due to thermo-cycling and galvanic corrosion, two identical 2003 vehicles were fitted: one with an Mg (MRI 153M) oil pan and the other with a new Al (A380) oil pan. An ultrasonic non-destructive testing (NDT) apparatus was employed to examine the relaxation of the bolt load. Ultrasonic NDT relies on measuring the echo return time of a signal passed into a component. As a load is applied to the component, it deforms. In this case, bolts were initially loaded under tension, causing them to stretch. As the mating component (Mg or Al) begins to creep out from underneath the bolt head, the bolt shortens. The contracted bolt has a shorter echo return time than it did under its higher preload. Each echo return time can then be correlated to a specific axial load in the bolt. Before installation, all bolts (Zn-coated) were prepared and calibrated. The calibration was performed by measuring the echo return time for each unloaded bolt and for several load levels, up to the level corresponding to the maximum assembly torque of 25 ±2 Nm. The echo return time for the unloaded bolt was then subtracted from the measurements taken at the varying load levels, thus providing a proper calibration factor, relating the echo delay time to the bolt load level. The vehicles were driven regularly for several months in harsh, wintertime conditions and the remaining bolt preload was checked at chosen intervals. The temperature outside the pan was monitored by a thermocouple. Galvanic corrosion around the mounting bolts was examined visually. Figure 4 shows the installed magnesium oil pan and in-vehicle testing.

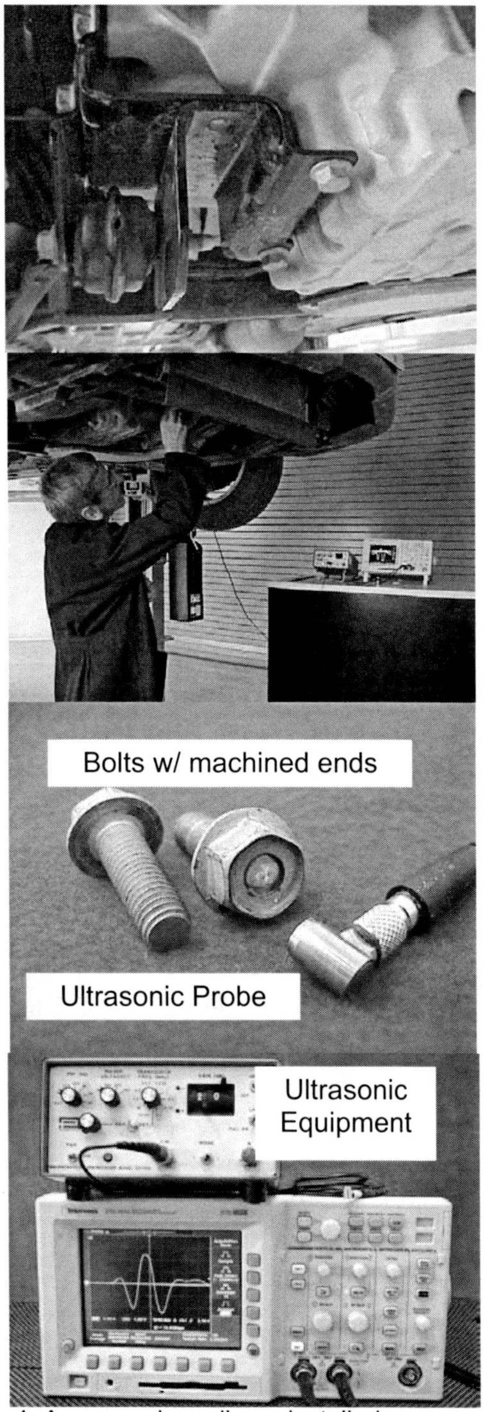

Figure 4: A magnesium oil pan installed on a car and the bolt load being checked on vehicle with an ultrasound apparatus.

RESULTS AND DISCUSSION

LASER INTERFEROMETRIC TEST – Typical interferometric fringe patterns obtained for the two test oil pans are shown in Figures 5 and 6. It is seen that the fringe density increases with decreasing internal pressure. While they have similar shapes and fringe distributions, the aluminum and magnesium oil pans differ in overall fringes numbers, i.e. the magnitudes of displacement. The magnesium prototype oil pan experienced more deformation than the original aluminum oil pan, as more fringes are present.

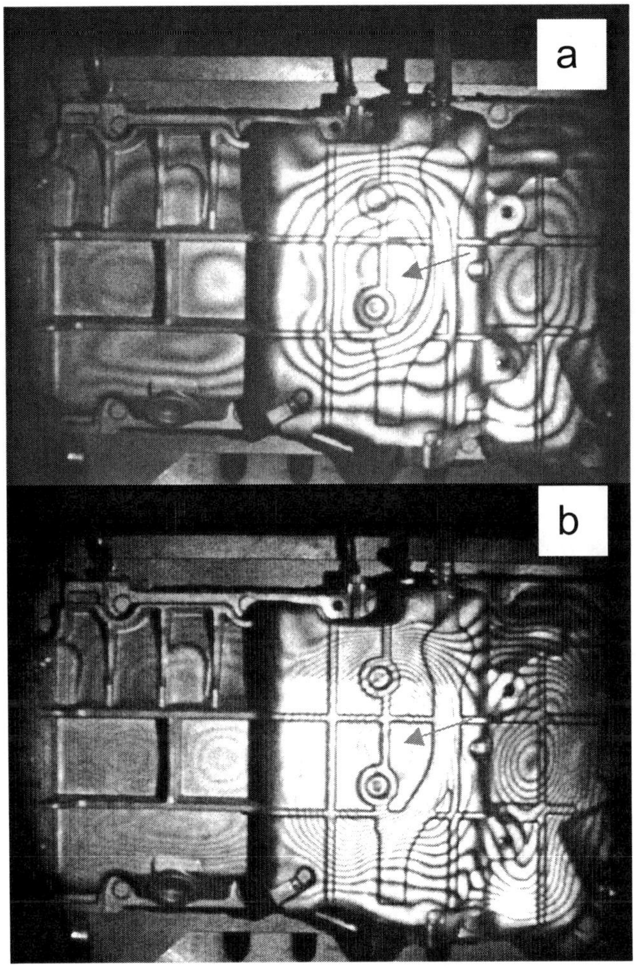

Figure 5: Fringe pattern on aluminium (a, c) oil pans. Internal pressure was changed from atmospheric (762 mm Hg) to 750 mm Hg (a), and from 525 mm Hg to 475 mm Hg (b). Arrows indicate the maximum displacement location.

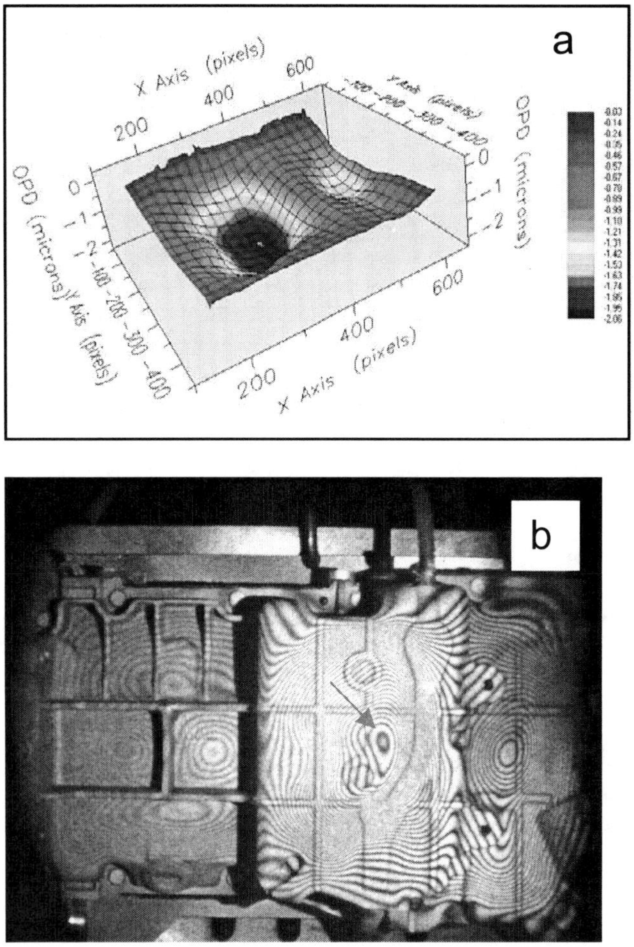

Figure 6: A digitized mapping of a typical surface deformation (a), fringe pattern on magnesium (b) oil pan with internal pressure changed from 525 mm Hg to 475 mm Hg (b). Arrow indicates the maximum displacement location.

The absolute displacement for the whole range of internal pressure was measured. Figure 7 shows the results for absolute displacement at the pre-determined point of maximum displacement. Both tested Al and Mg oil pans showed similar displacement patterns but with different magnitudes. Maximum deformation determined by interferometry was 88 μm for the aluminium oil pan and 142 μm for the magnesium oil pan at 0.1 N/mm^2 internal pressure. This reduced rigidity (54 μm more deformation) can be attributed to the lower stiffness of MRI 153M magnesium (Young's modulus: 45 GPa for MRI 153M and 71 GPa for A380).

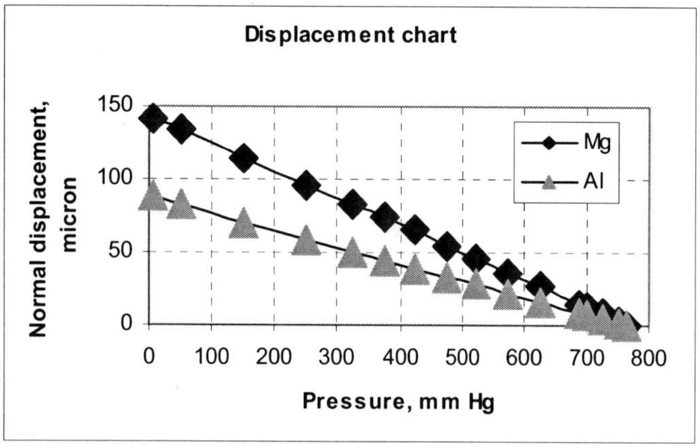

Figure 7: Displacement vs. internal pressure at point of maximum displacement

FINITE ELEMENT ANALYSIS – The oil pan was considered to be rigidly attached to the engine block at the bolt locations. Consequently, all bolt holes were considered to be fixed, while the mating surface with the engine block was constrained in out of plane direction (Fig. 8a). The load was applied through a pre-calculated, uniformly distributed partial vacuum at the inside surfaces the oil pan. Considering the linear behavior of the material, only the vacuum pressure of 0.05 and 0.1 N/mm^2 was applied to the oil pan (Fig. 8b). It should be recognized that the conditions of the FEA runs for the aluminum and magnesium pans varied only by the respective material properties of the alloys being tested.

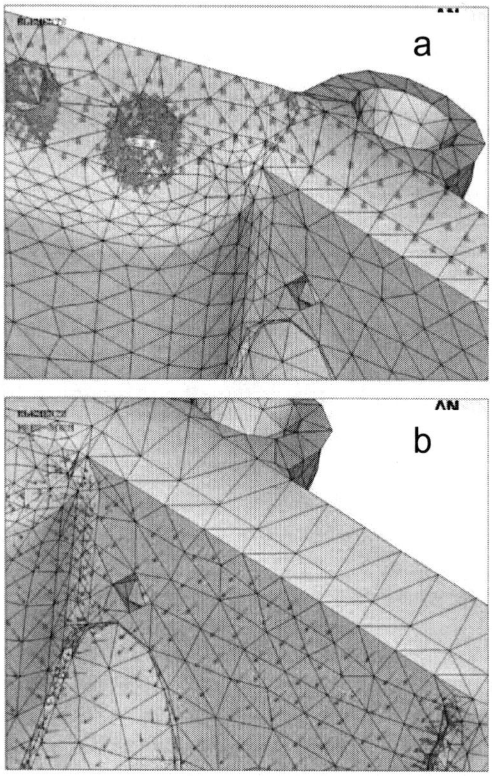

Figure 8: FEA boundary conditions: (a) detail of restraints; (b) detail of applied pressures.

Figure 9 presents the FEA results for aluminum and magnesium oil pans under a vacuum pressure of 0.1 N/mm^2. The aluminum oil pan exhibited a maximum displacement of 93 μm in the out of plane ("z") direction. The magnesium oil pan showed a similar displacement pattern to the aluminum oil pan, but with more deformation: 145 μm in the out of plane ("z") direction. The Mg/Al displacement ratio is 1.57. The FEA results agreed well with the laser interferometric test: comparing the measured displacements Figure 6 with the displacements calculated by the FEA showed outcomes that agreed to within 5 μm.

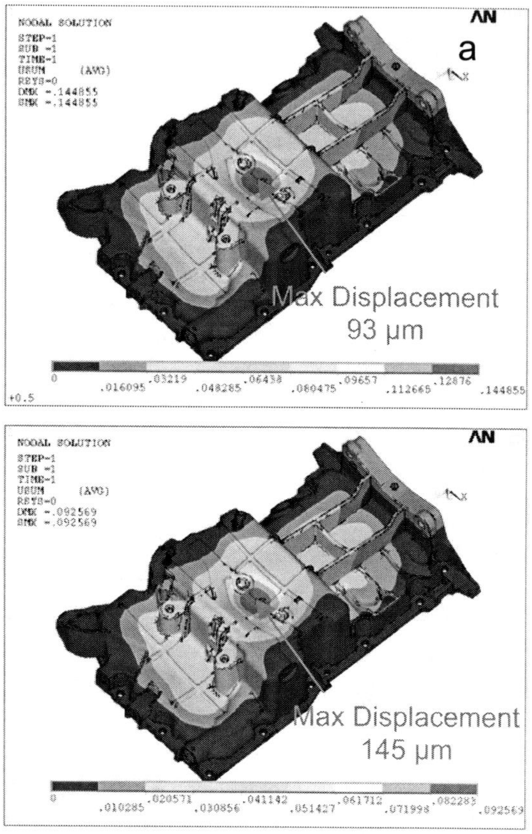

Figure 9: FEA out of plane displacement due to internal vacuum (@ 9.9 mm Hg): a. aluminum oil pan, b. magnesium oil pan

DYNAMIC MODAL ANALYSIS – Table 1 presents the dynamic modal analysis results for the oil pan alone and the assembly with the bedplate, respectively. The table summarizes the natural frequencies identified for the first few vibration modes, natural frequency ratio between Mg and Al, percent critical damping, and the damping ratio between Mg and Al.

In the free state (oil pan alone), the natural frequencies of the Mg oil pan were consistently lower than their Al counterparts. In fact, the ratio of the respective natural frequencies hovers close to 0.94. The magnesium oil pan exhibited better damping properties than the aluminum pan. Similar to the oil pans alone, the identified natural frequencies were lower for the assembly with magnesium oil pan. However, a wider spread in the natural frequency ratios was observed in the assembled state. Separate modes involved different participation of the bedplate and oil pan in the overall deformation pattern of the assembly. The damping ratios were significantly higher for the assembly than for the isolated oil pan, underscoring the role structural joints play in the vibration energy dissipation. Some of the damping ratios are higher for the assembly with the aluminum, while others are higher for the assembly with the magnesium oil pan. Numerous resonant modes were identified over a range of frequencies. Figure 10 shows two of vibrational modes identified in the free and assembled states at two frequencies, 548 and 1337 Hz.

The modal analysis results indicate that Mg MRI 153M pan does have improved damping characteristics over A380 aluminum. In assemblies, though, magnesium damping may not play as significant a role since the structural joints appear to dissipate the majority of the vibration energy.

Table 1: Summary of modal test

mode no.	natural frequency (Hz) Mg	natural frequency (Hz) Al	damping ratio (% critical) Mg	damping ratio (% critical) Al	nat. freq. ratio Mg/Al	damp. ratio Mg/Al
Free state (Oil pan alone)						
1	174.9	185.2	0.046	0.012	0.945	3.8
2	514.5	548.4	0.091	0.047	0.938	1.9
3	568.7	605.6	0.132	0.068	0.939	1.9
4	720.2	761.3	0.053	0.046	0.946	1.2
5	859.3	909.0	0.068	0.043	0.945	1.6
Assembly (oil pan-bedplate-gasket & fasteners)						
1	647.9	679.0	0.12	0.18	0.954	0.67
2	955.2	992.8	0.11	0.29	0.962	0.38
3	1002.4	1046.0	0.28	0.16	0.959	1.8
4	1103.3	1181.3	0.27	0.21	0.934	1.3
5	1139.0	1227.4	0.35	0.29	0.928	1.2
6	1261.2	1336.7	0.36	0.36	0.944	1.0

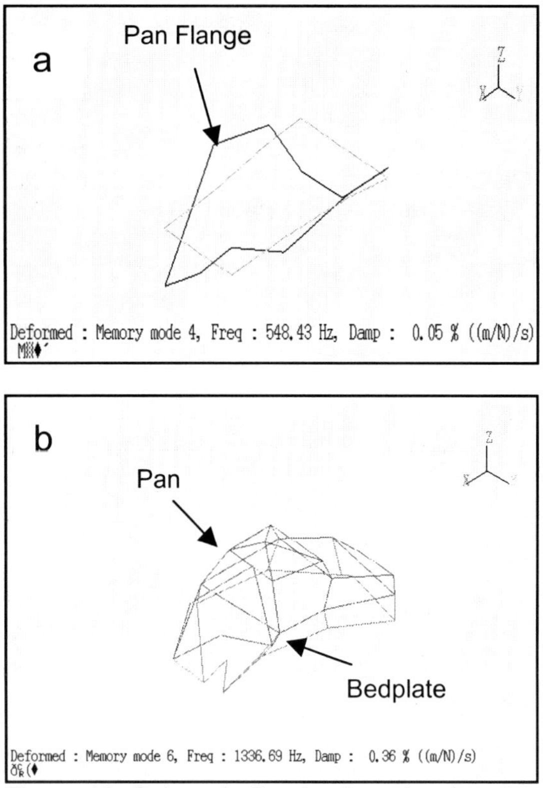

Figure 10: Selected vibrational modes for oil pan alone (a), and assembly (b)

IN-VEHICLE BOLT LOAD RETENTION TEST – Figure 11 shows axial bolt load as a function of distance driven after both cars were run for approximately 7000 km in wintertime conditions. Though neither closely tracked nor quantified during the testing period, December through April, the Southern Ontario winter of 2002-2003 was considered harsh and included many days with snow and slushy road conditions. Under such conditions, roads were typically salted to melt the ice formed on the roadways. Most days during the test period hovered around the freezing point (0° C). Along with the high humidity typical in this area, the test conditions were ideal for the purpose of monitoring the magnesium and aluminum assemblies' corrosion behavior.

It was observed that the preloaded bolts did relax in both the Al and Mg oil pans. The MRI 153M oil pan did creep more than A380 Al pan. The bolt axial load retained, from an equal initial preload, was about 77%-89% in the Al oil pan and 68%-79% in the Mg oil pan over the test period. Further analysis of the averaged data by linear regression analysis indicates that that long-term bolt load stabilization will occur for both the Al and Mg oil pans. On the assumption that a vehicle is driven for 10,000 miles (16,000 km) per year, one can predict from Figure 11d that after 10 years of service that the bolt load would drop to approximately 17 kN for the aluminum oil pan and 13 kN for the magnesium oil pan

During the road test, temperatures observed on the outside of the pan near the mounting flange ranged between 75-80 °C. Minimal galvanic corrosion was observed in the Mg oil pan and no oil leakage was observed; pictures of the components are shown in Figure 12. Some degradation of the surface was seen on both the aluminum and magnesium oil pans, while the areas underneath the bolt locations were completely free of any signs of galvanic corrosion, pointing to the effectiveness of the bolt coating. Independently of the on-vehicle testing, additional oil pans from the casting trial are being analyzed in a more quantifiable manner using a conventional salt spray chamber; at the time of writing, this testing has not been completed.

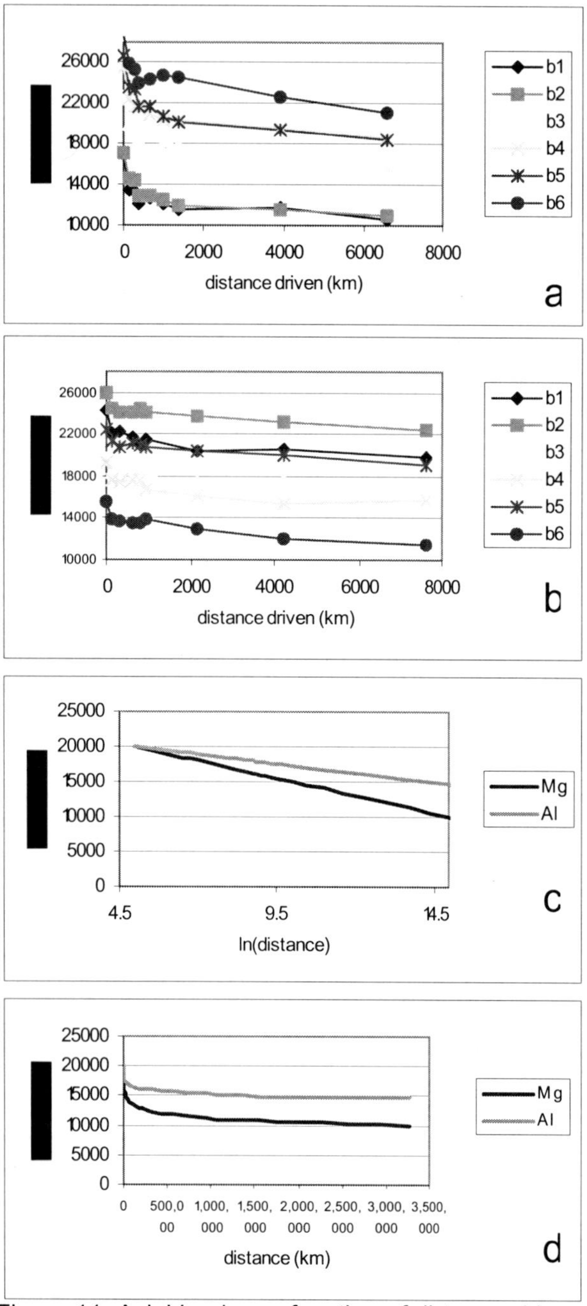

Figure 11: Axial load as a function of distance driven: (a) Al oil pan, (b) Mg oil pan, (c) average bolt load vs. natural log distance driven, (d) predicted load stabilization averaged over all bolts recorded.

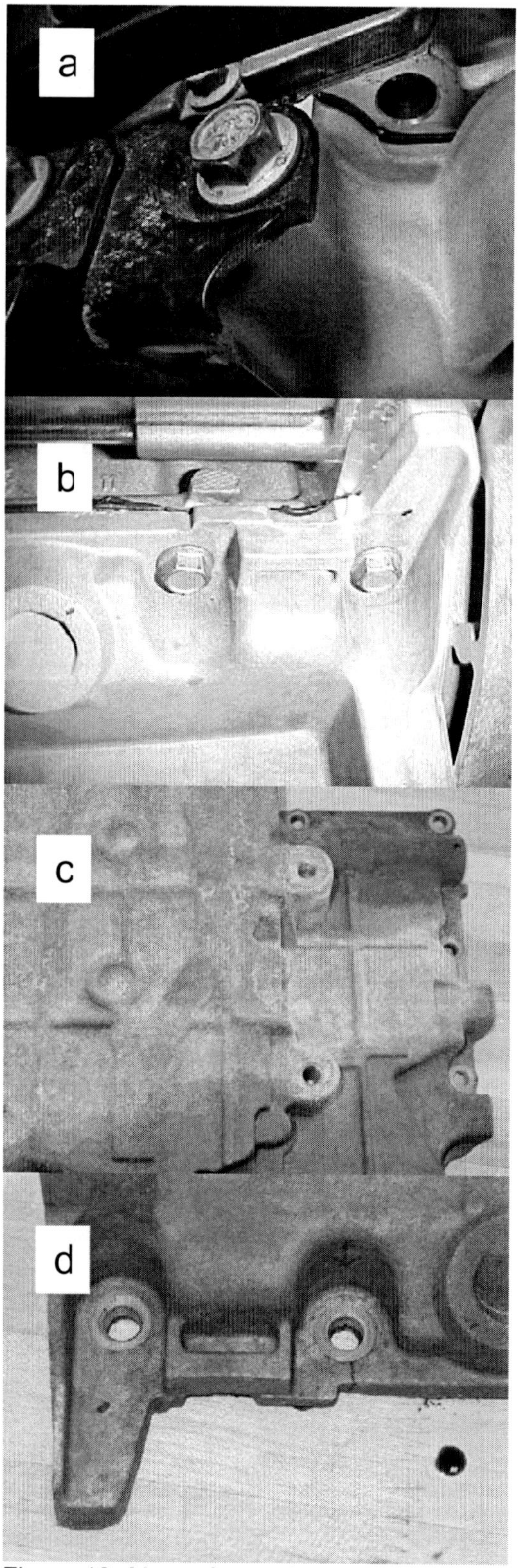

Figure 12: Mg surfaces shown at test start (a, b) and a test conclusion (c, d). Surface shown in (c) indicates a heavy build-up of road salt (light area) and oil stains from oil spilled during pan removal (dark area). Bright areas around bolt holes (d) indicate good isolation from environment by zinc-phosphate coated bolt heads.

CONCLUSIONS

1. Automotive oil pans fabricated by A380 aluminum and DSM 153M magnesium were evaluated by using advanced testing technologies, such as laser interferometry, dynamic modal analysis and ultrasound non-destructive testing.
2. Laser interferometric and FEA study revealed that under a vacuum pressure, the MRI 153M magnesium oil pan showed more (145μm vs. 92μm) but similar deformation behavior to the A380 oil pan. The maximum displacement ratio between them is 1.57. The MRI 153M magnesium oil pan exhibited improved damping characteristics over A380 aluminum.
3. The BLR tests in-vehicle indicated that the Mg oil pan, beginning with the same initial preload as the production aluminum pan, could maintain reasonable level of bolt axial load (about 68%-79%) over the test period, although it was lower than that in the aluminum oil pan (about 77%-89%). Linear regression analysis of the data indicates that long-term bolt load stabilization occurs for both the Al and Mg oil pans. Only minimal galvanic corrosion and no oil leakage were observed in the Mg oil pan and during the road test conducted over three winter months.
4. The successful completion of the road test (no leaks anywhere on the part), in which the Mg oil pan performed as well as the Al pan, indicates that MRI 153M magnesium is a promising and suitable alloy for oil pan applications. The developed testing methodology is suitable for testing magnesium powertrain components. The obtained data and information are very important for the design of magnesium powertrain components.

ACKNOWLEDGEMENTS

Magna Advanced Technologies wishes to thank Chris Ostrowski and Vitaliy Grabovsky for their efforts during the project. Tesma would like to thank Romulus Crisan for conducting the FEA, and Toral Cast's team for their efforts during the magnesium die casting trial. The work was done with the support of the Canada-Israel R&D Foundation under the guidance of Mr. Henri Rothschild.

REFERENCES

1. E. Aghion, B. Bronfin, F. Von Buch, S, Schumann and H Friedrich, " Dead Sea Magnesium Alloys Newly Developed For High Temperature Applications" Magnesium Technologies 2003 (2003), TMS, pp177-182
2. K. Pettersen et al, "Creep Resistant Magnesium alloy Development" Magnesium alloys and Their Applications (2000), Weinhein, Germany, pp 29-34
3. B.R. Powell et al, "The Die Castability of Calcium-containing magnesium alloys: Thin-Wall Computer Case", Magnesium Technology 2002 (2002), TMS, pp 123-130
4. S. Koike, et al, " Development of Lightweight Oil Pans Made of a Heat-Resistant Magnesium alloy for Hybrid Engine", SAE 2000 World Congress (2000), Paper 2000-01-1117
5. M. Pekguleryuz and E. Baril, "Development of Creep Resistant Mg-Al-Sr Alloys", Magnesium Technologies 2001 (2001), pp 119-126
6. F. Von Buch, et al, " New Die Casting Alloys MRI 153M for Powertrain Applications" Magnesium Technologies 2002 (2002), TMS, pp 61-68
7. A. Luo and T. Shinoda, "A new Magnesium Alloy for Automotive Powertrain Applications" SAE Congress 1998 (1998), Paper 980086
8. M. Pekguleryuz, P. Labelle, et al "Magnesium Die Casting Alloy AJ62X With Superior Creep Resistance, Ductility and Die Castability", Magnesium Technologies 2003 (2003), pp 201-206
9. A. P. Druschitz, et al, " Evaluation of Structural and High-Temperature Magnesium alloys", SAE 2002 World Congress (2002), paper 2002-01-008

2004-01-0659

AJ (Mg-Al-Sr) Alloy System Used for New Engine Block

Éric Baril, Pierre Labelle
Noranda Inc.- Montréal, Canada

Andreas Fischersworring-Bunk
BMW Group – Munich, Germany

ABSTRACT

AJ alloy is used with a new Aluminum-Magnesium Composite Design, which is an innovative approach to lightweight crankcase technology. The component is manufactured using high pressure die cast process. A wide range of chemical compositions was used to develop a good understanding of the behavior of this alloy system (castability, thermophysical, mechanical, microstructure). The basic mechanical properties were determined from separately die cast samples and also from samples machined out from high pressure die cast components. Tensile, creep, bolt load retention/relaxation and high cycle fatigue properties were established and analyzed using multivariate analysis and statistical approach. This methodology was used to select the optimal chemical composition to match the requirements. The sensitivity of the alloy to heat exposure was investigated for both mechanical properties and microstructure. The patented AJ alloy system offers a wide range of chemistry to accommodate a large variety of design and process requirements. For this specific application, AJ62x was selected because of the optimal set of mechanical properties in combination with process capability.

INTRODUCTION

The recent announcement of Aluminum-Magnesium Composite engine block occurs after an intensive research program lead by BMW Group, where standard magnesium AZ and high temperature alloys candidates was tested [1,2]. The initial testing program included the AJ52x alloy. Despite the excellent creep resistance obtained, the casting constraints related to such intricate design led to the early conclusion that AJ62x might be more suitable for this application. Due to limited information available on this alloy at that time, an intensive test program was initiated. The program focussed on the optimization of the mechanical properties (especially retained stress in bolted assembly) and castability of AJ62x alloy. This paper summarizes a three years program, which included casting and testing campaigns, and presents the approach used to optimize the AJ alloy for this new application.

DESIGN ASPECT OF THE ALUMINUM-MAGNESIUM COMPOSITE CRANKCASE

A step change in weight reduction within the power train can be achieved by usage of magnesium, particularly with the engine individual heaviest component, i.e. the cylinder crankcase. The design concept is driven mainly by the material capabilities of advanced magnesium alloys, i.e. creep resistance as key mechanical property next to the static and fatigue properties, the corrosion resistance and castability in the high pressure die casting process on the other side. For the engine crankcase design a hybrid material concept was chosen (see figure 1). This allows the best compromise on magnesium material capabilities of available advanced magnesium and the geometrical package constraints.

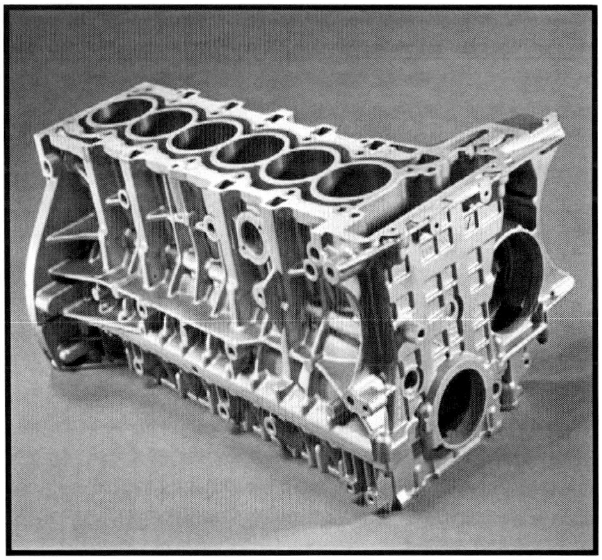

Figure 1 : Aluminum-Magnesium composite crankcase

The insert of a hypereutectic aluminum alloy avoids the requirement of additional liner technology and allows to incorporate highly loaded bolted joints for both the cylinder head and the main bearing. The insert also includes the water jacket, avoiding the still persistent problem of coolant fluid corrosion. The magnesium housing surrounding the aluminum insert, in turn, serves primarily for the oil ducts and the connection of ancillary units. The lower section of the crankcase, in turn, also made of high pressure-cast magnesium, comprises cast-in sintered steel inserts for the crankshaft mounts. The use of aluminum screws for all threaded joints in magnesium and consistent implementation of corrosion protection measures are additionally required.

The development and selection of a suitable high-temperature resistant magnesium alloy, process-validated control of the high-pressure die casting process and a suitable heat treatment are significant elements as regards the implementation in standard production.

On the whole, this advanced approach leads to a considerable reduction in weight of the cylinder crankcase.

METALLURGICAL ASPECT OF THE AJ ALLOY SYSTEM

The microstructure of Mg-Al-Sr alloys is characterized by primary α(Mg) dendrites and one or more second phases at the grain boundaries and in the interdendritic regions. The second-phase morphologies are lamellar, divorced eutectic, or massive. The aluminum and strontium contents of the alloy have a strong influence on the relative volume fraction of the various phases [3]. Based on the microstructure, the alloy system can be divided into three groups.

The Mg-5Al-2Sr (AJ52x), the first-generation alloy formulation, has shown a superior creep resistance and excellent high-temperature properties [4,5]. It was proposed that the creep resistance of AJ52x is related to the low aluminum supersaturation of primary magnesium and the absence of the $Mg_{17}Al_{12}$ phase due to the precipitation of high-melting-point Al_4Sr and $Mg_{13}Al_3Sr$ phases (see figure 2) [6].

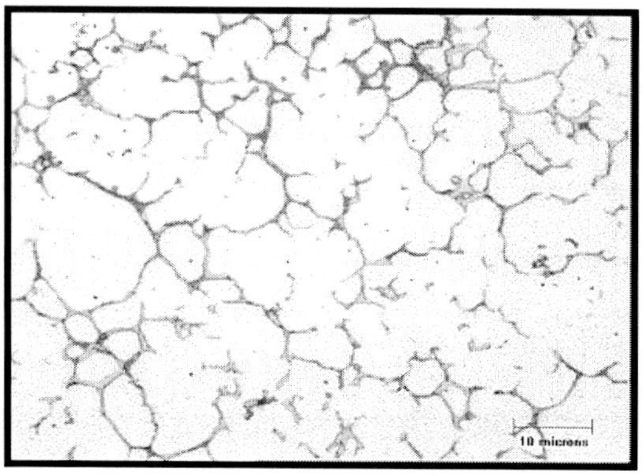

Figure 2 : Optical micrograph of a die cast AJ52x alloy.

The Mg-6Al-2Sr (AJ62x) offers more advantages: improved die castability without significant degradation of the creep resistance and high-temperature properties [5]. Aluminum was added to improve castability and room-temperature tensile properties. The strontium content was also raised to avoid any precipitation of $Mg_{17}Al_{12}$. The Al_4Sr is found in a well define lamellar-eutectic morphology (see figure 3). As for AJ52x, the aluminum supersaturation is low, which offers a good creep resistance.

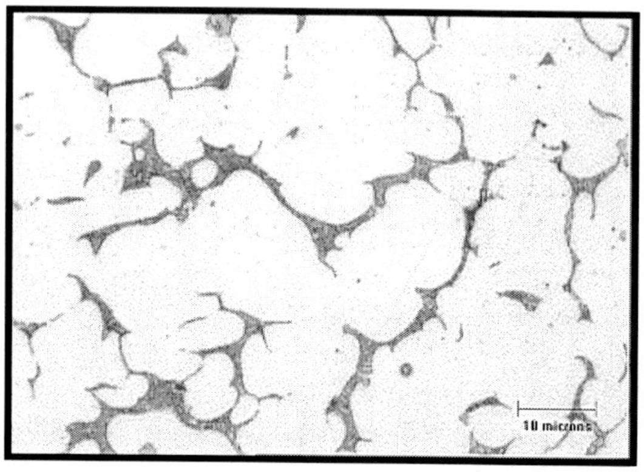

Figure 3 : Optical micrograph of a die cast AJ62x alloy.

A low strontium version (AJ62Lx) was formulated to slightly increase the dissolved aluminum and, therefore, increase the tensile properties. The excess aluminum leads to the precipitation of $Mg_{17}Al_{12}$. However, the presence of Al_4Sr-α(Mg) eutectic phase offers a significant creep resistance up to 150° C(see figure 4).

The AJ alloys are characterized mainly by a low Al-supersaturation in the primary magnesium dendrites and the precipitation of a lamellar Al_4Sr-α(Mg) eutectic in the interdentritic/grain boundary region. The effect of the aluminum and strontium contents on the microstructure is clearly visible in figures 2 to 4.

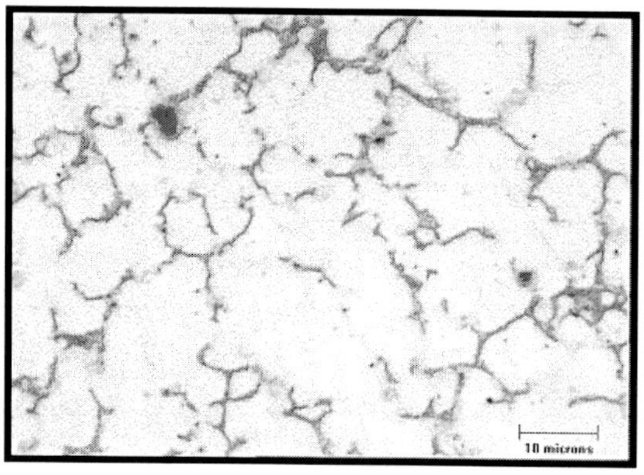

Figure 4 : Optical micrograph of a die cast AJ62Lx alloy.

EXPERIMENTAL PROCEDURES

All test samples required for the mechanical properties determination were cold-chamber high pressure die cast (HPDC). The samples were produced using a mold for ASTM samples or machined to ASTM standard from crankcases (see appendix). The samples were produced at ITM, Performag and EDC Inc. in Canada using a 600 tons Prince machine. The engine blocks were cast by BMW Group LITZ, in Landshut, Germany. The HPDC trials were done with optimized conditions with various melt compositions. The aluminum and strontium contents were ranging between 4.5-7.5 wt.% and 1.0-3.0 wt.% respectively. Table I presents some of the melt compositions selected to support the results presented in this paper.

Table I. Melt composition of samples

	Al(%)	Sr(%)
AJ52x	5.35	2.25
AJ53x	5.41	3.10
AJ62x-1	5.70	2.15
AJ62x-2	6.01	2.38
AJ62Lx	6.46	2.14
AJ73x	7.15	2.82

Except for the bolt load retention (BLR) test, mechanical properties were determined according to ASTM standards and the tests were carried out at Noranda Inc. - Technology and/or by accredited laboratories. The tensile tests were done according to B557 "Test Method of Tension Wrought and Cast Aluminum and Magnesium Alloy Product" and ASTM E8 M "Testing Method for Tension Testing of Metallic Material". The creep tests were performed according to ASTM E139 "Test Method for Conducting Creep, Creep-Rupture, and Stress Rupture Tests of Metallic Material". The fatigue tests fulfill ASTM E466-96 "Standard Practice for Conducting Force Controlled Constant Amplitude Axial Fatigue Tests of Metallic Materials". Other tests, such as multilevel compressive creep and tensile relaxation, were done according to internal standards.

ALLOY OPTIMIZATION APPROACH

The main task of the program was to fulfill the Al-Mg composite engine block design requirements by the optimization of the chemical range of the alloy with the constraint of the HPDC casting process. This approach is schematized in figure 5. The goal is to specify the limits in strontium and aluminum, which offer (1) a robust operating window for HPDC of the component and (2) high temperature performances above the limits specified for the mechanical design. The approach requires a good understanding of the impact of chemical composition on the mechanical properties, mainly BLR as this is close to the design of bolted joints.

As mentioned before, the Mg-Al-Sr metallurgical system shows three groups of microstructure. For a giv manufacturing process (e.g. pressure die casting), the aluminum and strontium contents are the governing variables that influence the microstructure. In turn, the microstructure, temperature and loads influenced the mechanical properties of the alloy: BLR, tensile properties and creep resistance. The most reasonable way to determine these relationships with the amount of parameters involved is using multivariate analysis.

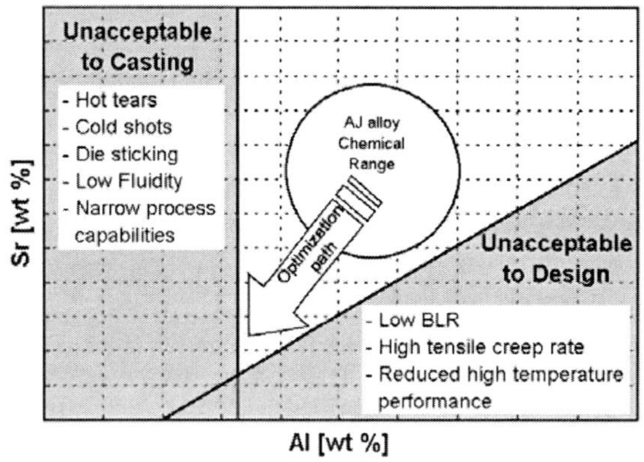

Figure 5 : Schematic of the alloy optimization approach for the Al/Mg composite crankcase design.

The effect of multiple factors (or variables-*X*) on the material response (property-*Y*) can be understood using multivariate analytical tool, namely partial least squares projections to latent structures, or PLS. PLS is a regression technique for modeling the association between X and Y, but it is also a tool to deal with complicated and approximate relationships. The major advantage of PLS analysis is the ability to work with poorly structured dataset not generated by design of experiments (DOE) techniques.

In this paper, the results of the modeling exercise for BLR, tensile creep and, tensile properties are reported. Samples with more than 20 different chemical compositions were used to generate the datasets for PLS analyses. In addition, various testing conditions (initial load and temperature) were used. The PLS models include the results of more than 50 observations.

In a second step, in order to quantify the risk of missing the design targets, a risk analysis using the PLS model was done. The risk is evaluated using a statistical distribution of the chemical composition (for both aluminum and strontium) based on the convoluted process capabilities (variation in ingot composition and component composition). The alloy chemistry is then adjusted to minimize the risk to an acceptable level.

As an example, the risk analysis approach used for bolt load retention is presented in the coming paragraphs along with the PLS analysis results.

BOLT LOAD RETENTION ANALYSIS

Since there is not standard available, the bolt load retention test was done using the arrangement as describe in reference [4]. It can be summarized as a cylinder machined from separately die cast sample or machined from component of various chemistries. The dimensions are 18 mm long, 18 mm diameter with 10 mm central hole. Using an M10 instrumented steel bolt, the initial loads were adjusted between 40 and 80 MPa to obtain the desired peak stress (from 50 to 90 MPa) due to thermal shock after immersion in the oil bath.

The tests were monitored for a period up to 1000 hours. The BLR curve can be described by four parameters (see figure 6): (1) the peak stress, (2) the initial rapid drop of the load, which is expressed by the ratio (k_o) of the extrapolated stress (σ') and the peak stress (σ_o), (3) the stress reduction rate (slope) associated to the long term relaxation and, (4) the final retained stress after 1000 hours. A similar approach was used by other researchers [7] to describe the BLR curves.

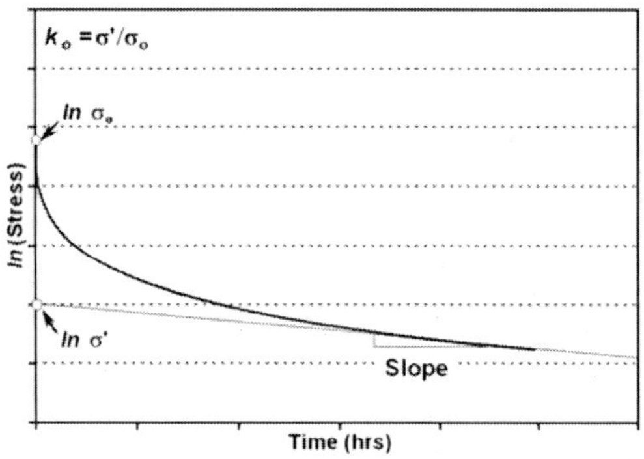

Figure 6 : Parameters of the BLR curve model.

The effect of the temperature on the BLR done at 60 MPa initial stress with AJ62x is illustrated in figure 7. The effect of thermal shock is less severe at the lower temperature (120° C). As a result, the peak stress is lower. The initial drop and the stress reduction rate are larger at higher temperatures. Consequently, the final retained stress decreases with increased temperature.

The effect of initial stress on the BLR done at 150°C with the AJ62x is illustrated in figure 8. The increased initial load (and resulting peak load) increases the initial drop in load. The stress reduction rate is similar for both 40 and 60 MPa initial stress and higher for the 80 MPa. As a result, the final retained stress increases with increased initial stress. But, for longer testing period, the 80 MPa condition may result in lower retained stress than the 60 MPa initial load.

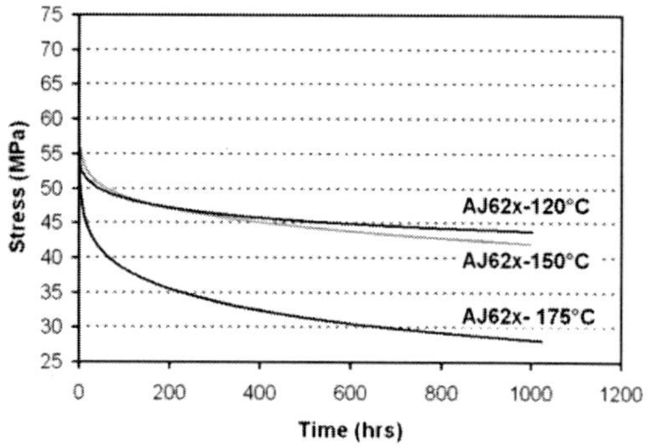

Figure 7 : Effect of the testing temperature on the retained stress with 60 MPa initial stress.

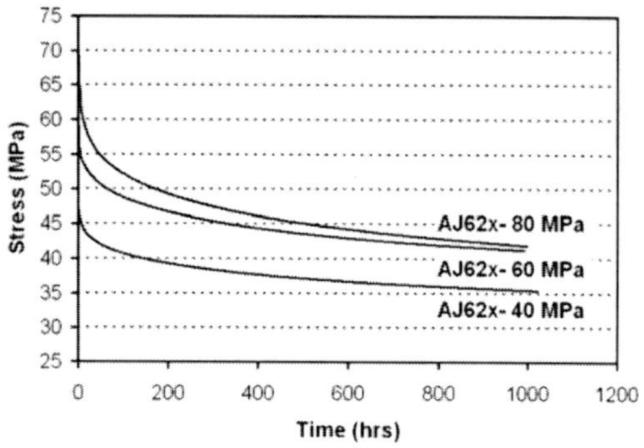

Figure 8 : Effect of the initial stress on the retained stress at 150° C.

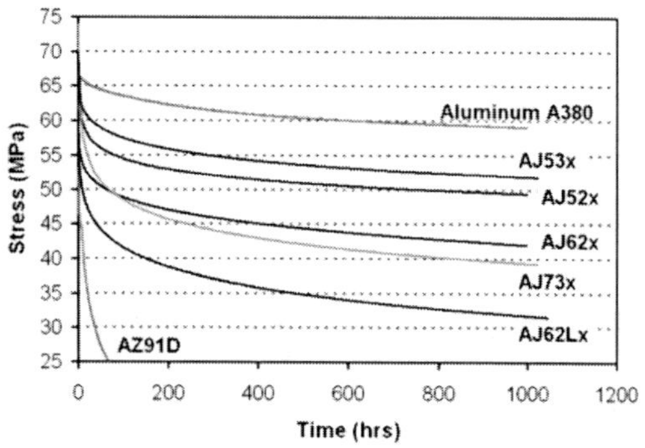

Figure 9 : Effect of the alloy composition on the retained stress at 150° C and 60 MPa initial stress.

The effect of the alloy composition is illustrated in figure 9. The alloys listed in table 1 were tested with 60 MPa initial stress and 150° C. The AJ53x and AJ52x alloys show the highest retained stress and, therefore, the lowest initial load drop and stress reduction rate. The

AJ62Lx shows the lowest retained stress, the highest stress reduction rate and highest initial load drop. The AJ62x and AJ73x show an intermediate behavior with slight difference in the progress the BLR test. The AJ73x shows a larger stress reduction rate than AJ62x. On the other hand, the initial load drop is lower for AJ73x than for AJ62x. In comparison, the aluminum A380 alloy show lower stress reduction rate and a lower initial load drop than any of the AJ alloy tested in the program. As additional reference, AZ91D is given.

As shown in the previous description of a limited set of results, the systematic interpretation of the BLR curves is relatively complex and support from analytical tool is needed. In order to conduct a methodical analysis, the BLR curves are first quantified using the parameters described previously (see figure 6). The quantitative results of the BLR analysis on selected alloys compositions (table I) are given in table II. These results are then used as a dataset for the multivariate analysis.

Table II. Effect of the composition, initial stress and temperature on the BLR results

	Initial Stress (MPa)	Temperature (° C)	Retained Stress after 1000 hrs (MPa)	k_o	Slope
AJ52x	60	150	49.4	0.717	-2.143e-5
AJ53x	60	150	51.9	0.779	-2.336e-5
AJ62x-2	60	120	43.8	0.676	-2.686e-5
AJ62x-2	60	150	42.0	0.701	-4.151e-5
AJ62x-2	60	175	28.2	0.443	-7.804e-5
AJ62Lx	60	150	31.8	0.535	-6.612e-5
AJ73x	60	150	37.5	0.635	-6.127e-5
AJ62x-1	40	150	35.5	0.768	-3.231e-5
AJ62x-1	60	150	41.4	0.642	-3.305e-5
AJ62x-1	80	150	41.9	0.525	-5.241e-5

The PLS model was built using 58 individual BLR experiments. The model used, as *X*'s, (a) Al content, (b) Sr content, (c) initial stress and, (d) temperature. The *Y*'s were (i) retained stress after 1000 hours, (ii) k_0 factor and, (iii) the slope of the retained stress as a function of the time after 1000 hours. The variation of the responses was explained at a level of 82% by the PLS model (cumulative R^2=0.819). Using cross-validation technique, 74% of the variation in BLR results can be predicted by the model (Q^2=0.738). This is considered as an excellent PLS model. In order to compare the contribution of the *X*'s on the *Y*'s, the scaled and centered coefficients of the model are given in figure 10.

The initial stress is the largest single effect on both the retained stress after 1000 hours and the initial stress loss (k_o). An increased initial stress leads to increased retained stress and reduced k_o. The initial stress has a smaller impact on the stress reduction rate (slope) where increased initial stress leads to larger negative slope (or larger stress reduction rate). On the other hand, the temperature and Al-content show equivalent relative effect. An increased Al-content and/or temperature gave reduced retained stress, k_o and larger stress reduction rate. The Sr-content has a smaller but positive effect on the three responses. Larger Sr-content gives higher retained stress, higher k_o and smaller stress reduction rate.

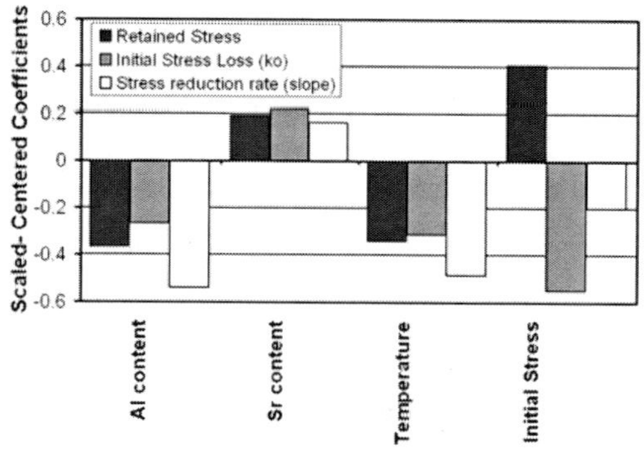

Figure 10 : Scaled and centered coefficients of the multivariate (PLS) model.

The effect of aluminum and strontium contents on the retained stress after 1000 hours at 150° C with 40 MPa initial stress is shown with figure 11. As described previously, the retained stress increases with increased strontium and decreases with increased aluminum. This prediction of the retained stress is in good agreement with the measured values. As an example, the PLS model predicts, for AJ62x-1 (5.7 wt.% Al and 2.15 wt.% Sr), 33.7 MPa of retained stress after 1000 hours while the measured value is 35.5 MPa. It should be noticed that 40 MPa initial stress gives 50 MPa of peak stress.

The PLS model of the retained stress shows good predictability within the range of interest of the AJ alloy system. The alloy model was then used in conjunction with synthetic prediction sets to assess the risk of having a retained stress lower than given limits related to the aluminum-magnesium composite crankcase design. The synthetic prediction sets were generated using the convolution of the AJ ingot casting process capabilities and the crankcase HPDC process capabilities.

The AJ ingot casting process capability was determined during three weeks of production of 8 kg ingots with an industrial casting line. During this production campaign, the chemical composition of each batch was recorded and, then, the average and standard deviation were calculated. The process followed a normal distribution for both aluminum and strontium.

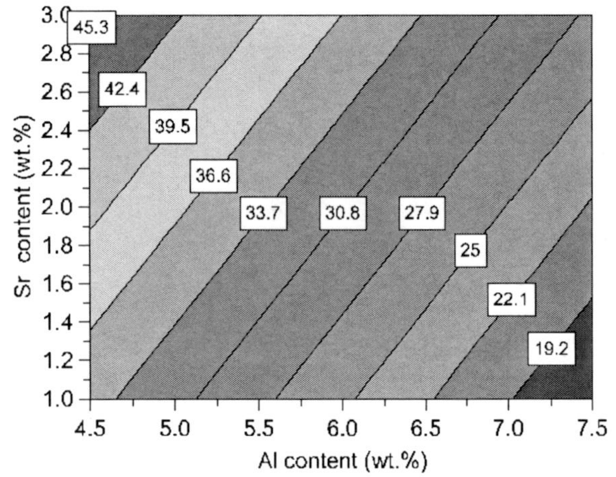

Figure 11 : Prediction of the effect of the alloy composition on retained stress after 1000 hours at 150° C with 40 MPa initial stress (50 MPa peak stress).

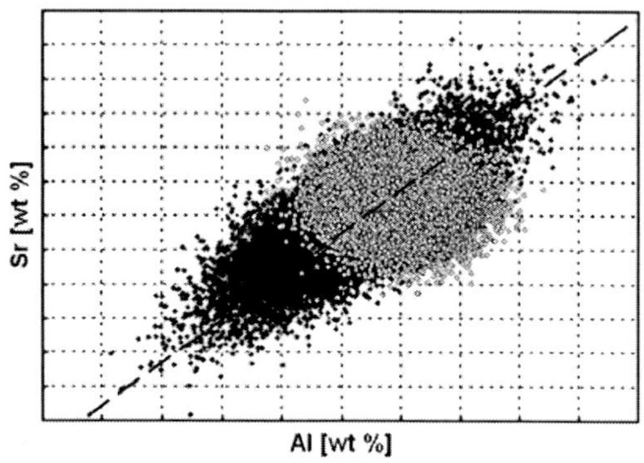

Figure 12 : Schematic of the alloy chemistry composition as generated using a standard ingot casting process capability (white circles) convoluted with the segregation in the engine block (black circles).

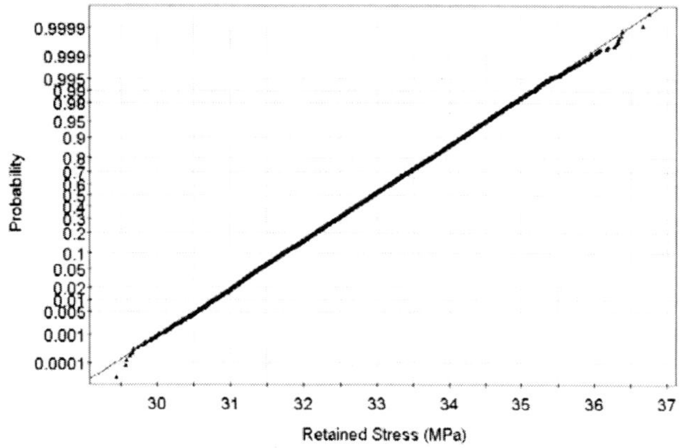

Figure 13 : Normal probability plot of the retained stress, after 1000 hours at 150° C with 40 MPa initial stress, using a convoluted chemical distribution.

The crankcase HPDC process capabilities were determined using chemical composition mappings. The composition was determined at various locations of the component. The average of several crankcases led to a good evaluation of the segregation within the part. Due to the intricate design, the aluminum and strontium tend to be different from one location to the other. It was noticed that the distributions of aluminum and strontium are strongly correlated. Basically, the ratio of strontium over aluminum (in wt.%) is identical to the Sr/Al ratio measured in Al_4Sr phase. This observation can be related to the variation of the volume fraction (or weight fraction) of Al_4Sr phase in a matrix of primary α-Mg. Since the Al and Sr contents of the α-Mg is much smaller than in Al_4Sr, the change in Al and Sr of the bulk alloy is directly related to the Al_4Sr content in the analyzed volume. Therefore, the ratio of Sr/Al should be close to the one of Al_4Sr (between 0.57 and 0.75 according to [6]).

The prediction set is used to calculate the retained stress after 1000 hours using the PLS model. The results are then illustrated in a normal probability plot and the risk to be below a specified limit can be evaluated. The figure 13 shows the normal probability plot of the retained stress after 1000 hours with 40 MPa initial stress at 150° C. As an example, the risk to be below 30 MPa retained stress is 0.001 (0.1%).

If the prediction set gives high probability to be below the target retained stress limit, the center point of the ingot distribution is modified to reduce the probability to acceptable level. This process was iterated until the probability reached the acceptable level and, then, the alloy composition was considered optimized.

Next to the BLR tests, which are closest to the design aspect of bolted joints into Mg, other tests have been performed. Pure tensile relaxation tests performed for a variety of initial load and temperature conditions and chemical compositions fit into the prediction model of the BLR tests.

PLS PREDICTION OF OTHER MECHANICAL PROPERTIES

CREEP PROPERTIES – The creep test specimens used in the program were both separately cast to defined chemical compositions and coming from crankcases with variation in chemical compositions. Sets of different creep test types were conducted within the program. It included tensile and compressive single and multilevel load testing plus multilevel temperature tests, which were performed with variation in load, temperature and hold time. The creep strain rates associated with these tests were analyzed using the same statistical modeling approach. An example is given in figure 14, predicting the tensile creep strain rate depending on alloy composition for an initial load equivalent to 70 MPa at 150° C.

The trend observed in the creep behavior is supporting the trend, i.e. effect of Al, Sr, load and temperature, as observed in the BLR testing.

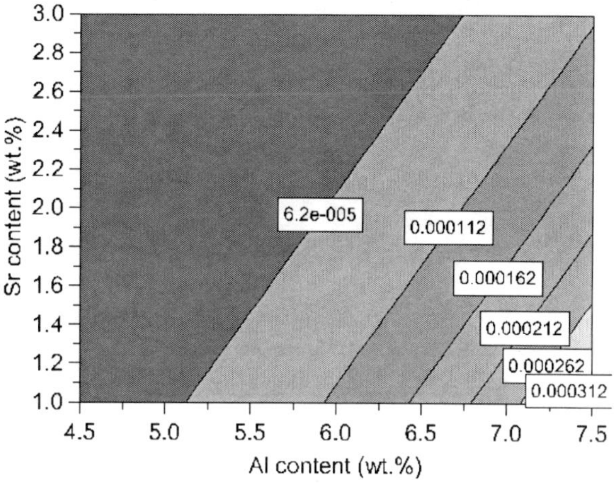

Figure 14 : Prediction of the effect of the alloy composition on the tensile creep strain rate (%/hr) after 500 hours at 150° C and 50 MPa.

TENSILE PROPERTIES – The tensile properties were determined for a large set of alloy compositions using standard test specimens from separately HPDC and machined from components. Around 480 tests results were compiled for multivariate analysis. The resulting PLS model shows excellent characteristics. The variation of the responses was explained at a level of 76% by the PLS model (cumulative R^2=0.763). Using cross-validation technique, 74% of the variation in BLR results can be predicted by the model (Q^2=0.74). The prediction of the effect of the alloy composition on the tensile properties at room temperature is illustrated in figures 15, 16 and 17. The ultimate tensile strength (UTS) and the elongation increase with increased aluminum content and reduce with increased strontium content. The high UTS and elongation correspond to the AJ62Lx alloy composition. As described previously [3], this alloy composition shows increased dissolved aluminum and precipitation of $Mg_{17}Al_{12}$ phase.

On the other hand, the elongation is strongly correlated with the tensile creep strain rate and BLR (see figures 11, 14 and 16). Increased elongation corresponds to higher creep strain rate and lower retained stress after 1000 hours.

The predicted yield strength (YS) is shown in figure 17. In this case, the YS increases with increased Al and/or Sr contents. This effect corresponds to the increased volume fraction of secondary phase particles [3].

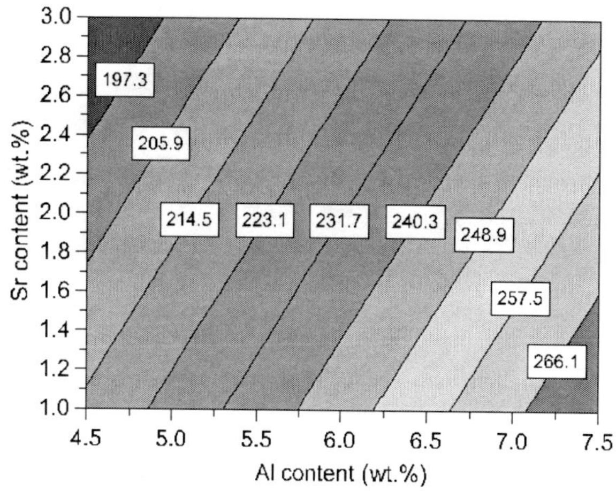

Figure 15 : Prediction of the effect of the alloy composition on the room temperature ultimate tensile strength (MPa).

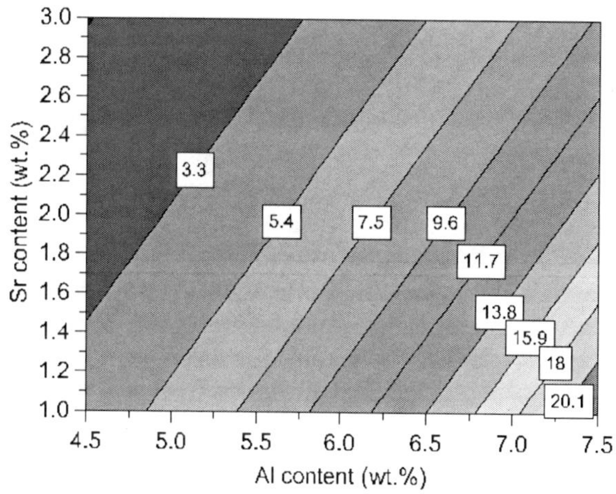

Figure 16 : Prediction of the effect of the alloy composition on the room temperature elongation (%).

The tensile properties of the AJ alloys are mainly controlled by the level of dissolved Al (solid solution strengthening), volume fraction of second phase particles (grain boundary strengthening) and the precipitation of $Mg_{17}Al_{12}$ phase (increased ductility and lower creep resistance). The effect of Al and Sr contents on the mechanical properties can be largely explained by the evolution the AJ alloy system microstructure within the range studied in this project.

The tensile properties of the optimized alloy, according to the PLS approach for BLR, meet the specified target set by the design. The alloy composition was not further optimized using the PLS model of the tensile properties.

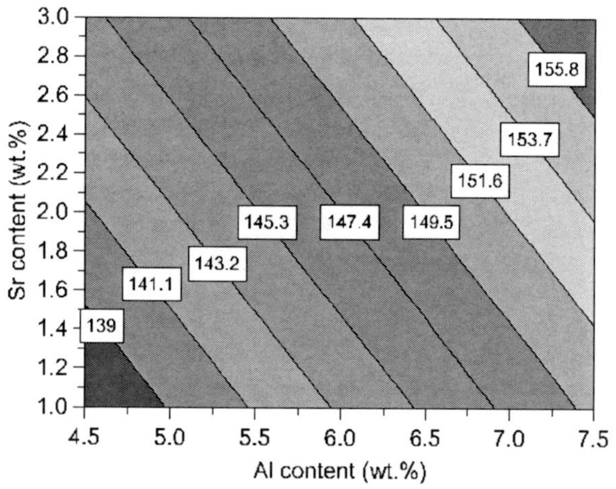

Figure 17 : Prediction of the effect of the alloy composition on the room temperature yield strength (MPa).

FATIGUE PROPERTIES

Axial fatigue tests were carried out on more than 120 AJ62x samples at room temperature, 120° C, and 150° C. Stress ratios (R) were 0.1, -0.5 and –1.0. Fatigue limits were determined using the S-N curves for 50% probability of failure. The fatigue limits are reported on a Goodman Diagram to validate the predictability of such approximation of the fatigue limit [8].

At room temperature, the fatigue limit of AJ62x is higher than the fatigue limits observed for AZ, AM [9] and AJ52x [10] alloys. It is proposed that the absence of brittle intermetallics such $Mg_{13}Al_3Sr$ and $Mg_{17}Al_{12}$ improves significantly the fatigue behavior of the AJ62x system which gives similar results than the AJ51x alloy system [11]. It should be noticed that the Goodman approximation overestimates the fatigue limit at a mean stress of 56 MPa (see figure 18). However, the more conservative Soderberg approximation [8] agrees with the experimental data for AJ62x.

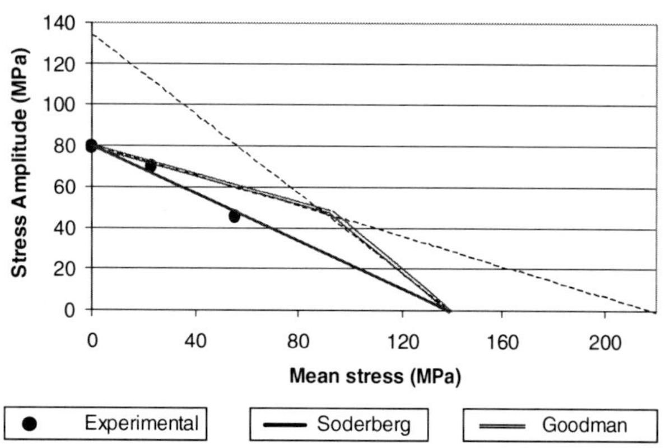

Figure 18 : Fatigue limit diagram of AJ62x for axial stress at room temperature.

For 120° C, the increased temperature leads to a reduction of the fatigue limit at R= -1. The fatigue limit is reduced by 25% when compared to room temperature data while the reduction in YS and UTS are close to 15%. In contrast with room temperature data, at 120° C, the Goodman approximation predicts accurately the fatigue limit (see figure 19).

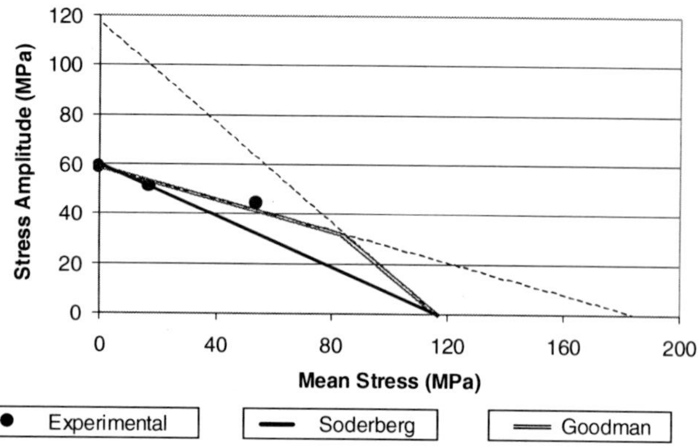

Figure 19 : Fatigue limit diagram of AJ62x for axial stress at 120° C.

At 150° C, the increased temperature reduces the fatigue limit at R=-1 by 40% when compared to room temperature data. This reduction is larger than the reduction observed for both UTS and YS. However, the Goodman approximation overestimates the fatigue limit at 39 MPa mean stress (see figure 20). On the other hand, the Soderberg approximation offers a better predictability.

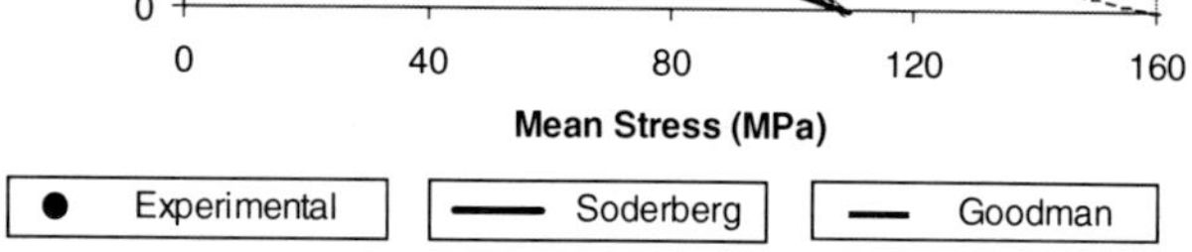

Figure 20 : Fatigue limit diagram of AJ62x for axial stress at 150° C.

The fatigue limit determined using Goodman diagram is overestimated for AJ62x alloy. The fatigue limit of AJ62x alloy is generally closer to the Soderberg prediction line. This phenomenon was observed for aluminum-zinc magnesium alloys [12]. When compared to fatigue behavior of steel or aluminum, it was proposed that magnesium can failed in fatigue after general yielding due to limited number of slipping planes in hexagonal structure of magnesium.

In addition, the estimation of the fatigue limit assuming only the reduction of the YS related to the increased temperature is inaccurate. The ratio between YS at high and room temperatures cannot be used to estimate the equivalent fatigue limit reduction. However, fatigue limit of AJ alloy can be predicted accurately using standard fatigue theories applicable to other metallic material (Soderberg line), when using fatigue limit at R= -1 and Yield Strength at the desired temperature.

HEAT EXPOSURE SUSCEPTIBILITY

The effect of heat exposure on the mechanical properties and microstructure was evaluated for selected alloy compositions. Samples were artificially aged at temperatures between 120° C to 350° C for 2 to 500 hours. The mechanical performance of the AJ alloy system is not significantly affected by heat exposure prior to tests, minor improvement of elongation was observed after high temperature exposure. Small modification of the second phase particles shape was observed at the grain boundaries of AJ62x. Therefore, the effect of heat exposure was not considered in the mechanical properties PLS modeling exercise.

CONCLUSION

An alloy optimization approach was successfully used in this program. It includes the multivariate analysis of large set of mechanical results with many variables, such chemical composition and testing conditions. It also includes a risk analysis using a probabilistic approach based on the process capabilities.

For this specific application, AJ62x was selected because of the optimal set of mechanical properties in combination with process capability. The optimized AJ alloy shows:

- Adequate Bolt Load Retention performances,
- Higher fatigue limits than AJ52 or other die cast magnesium alloys,
- Adequate overall mechanical performances.

ACKNOWLEDGMENTS

The authors would like to acknowledge BMW Group and Noranda Magnesium Inc., for sponsoring this project and for permission to publish this paper. The authors thank all the engineers and technologists from Noranda, BMW Group and external laboratories who were involved in die casting campaigns and performing the mechanical testing required by this program.

REFERENCES

1. C. Landerl, R. Jooss, A. Fischersworring-Bunk, J. Wolf, A. Fent, S. Jagodzinski, "Aluminium-Magnesium Verbundkonstruktion - ein innovativer Ansatz fur Leichtbau im Kurbelgehause. *(Aluminum-Magnesium Composite Design – an innovative Approach to Lightweight Crankcase Technology)*" Aachener Colloquium October 6-8th,2003
2. S. Birch, "BMW's secrets revealed", *Automotive Engineering International*, October (Detroit, MI: SAE 2000).
3. E.Baril, P. Labelle, M. O. Pekguleryuz, "Elevated Temperature Mg-Al-Sr: Creep Resistance, Mechanical Properties, and Microstructure", *JOM*, Nov. (2003), pp. 34-39.
4. P. Labelle, M. Pekguleryuz, M. Lefebvre, R. Bouchard, "New Aspects of Temperature Behavior of AJ52x, Creep Resistant Magnesium Alloy", SAE 2002-01-0079 (Detroit, MI: SAE 2002).
5. M. Pekguleryuz, P. Labelle, D.Argo, "Magnesium Die Casting Alloy AJ62x with Superior Creep Resistance, Ductility and Die Castability", SAE 2003-01-0190 (Detroit, MI: SAE 2003).
6. M.O. Pekguleryuz, E. Baril, "Creep Resistant Magnesium Diecasting Alloys Based on Alkaline Earth Elements", *Met. And Matls. Trans.*, 42 (2001), pp. 1258-1267.
7. K.Y. Sohn, J.W. Jones, J.E. Allison, "The Effect of Calcium on the Creep and Bolt Load Retention Behavior of Die-Cast AM50 Alloy", *Magnesium Technology 2000*, ed. H. Kaplan, J. Hryn, and B. Clow (Warrendale, PA: TMS, 2000), pp. 271-278.
8. R. Juvinall, *Stress, Strain, and Strength* (New York: McGraw-Hill, 1967).
9. P.D.D. Rodrigo, M. Murray, H. Mao, V. Chandrasekar, Y. Kisioglu, A.Deshpande, J. Brevick, C. Mobley, R.Esdaile, "Fatigue Properties of Die cast Magnesium Alloys", SAE 2000-01-1122 (Detroit, MI: SAE 2000).
10. P. Labelle, D. Argo, M. Pekguleryuz, Y. Fasoyinu, R. Bouchard, M. Sahoo, "Comparative Mechanical Properties of AE42 and AJ52x High-Temperature Diecast Magnesium Alloys for Elevated Temperature Applications", SAE 2003-01-0188 (Detroit, MI: SAE 2003).
11. P. Labelle, M. Pekguleryuz, D. Argo, M. Dierks, T. - Sparks, T. Waltemate, "Heat Resistant Magnesium Alloys for Power-Train Applications", SAE 2001-01-0424 (Detroit, MI: SAE 2001).
12. E.F. Emley , *Principles of Magnesium Technologies* (Oxford: Pergamon Press 1966).

APPENDIX

Tensile Die Cast Samples ASTM B557 Standard Test Methods of tension Testing Wrought and Cast Aluminum- and Magnesium- Alloy Products, Fig 13 Standard die cast specimen for die casting

Tensile Machined Samples ASTM B557 Standard Test Methods of tension Testing Wrought and Cast Aluminum- and Magnesium- Alloy Products, Fig 8, Pin-Loaded Tension Test Specimen with 2-in. Gage Length, using Small-Size Specimens Proportional to Standard with 0.25 in. diameter

Fatigue Machined Samples ASTM E466 Standard Practice for Conducting Force Controlled Constant Amplitude Axial Fatigue Tests of Metallic Materials, Fig 1, Specimen with Tangentially Blending Fillets Between the Test Section and the Ends, 0.25 in. diameter.

2003-01-0183

Die Casting of Magnesium Alloys – The Importance of Controlling Die Filling and Solidification

Stian Sannes, Haavard Gjestland and Håkon Westengen
Magnesium Materials Technology, Norsk Hydro

Hans I. Laukli and Otto Lohne
Norwegian University of Science and Technology

ABSTRACT

High pressure die casting is characterised by rapid die filling and subsequent rapid cooling and solidification of the metal in the die. These characteristics are favourable for the mechanical properties of magnesium die casting alloys. Since the filling pattern and the cooling rate of the metal is highly dependent on both process parameters and geometry of the part, there is a natural variation in mechanical properties. Variations in filling pattern can be caused by differences in the filling conditions set up by the gating system, pre-solidification in the shot sleeve and during filling as well as variations in the timing of the pressure intensification. In the present work the effects of solidification during filling are discussed with emphasis on the resulting microstructures and the correlation with mechanical properties.

INTRODUCTION

An ideal flow pattern for a casting is characterised by a unidirectional flow of metal with a plane flow front from the gate towards the end of the casting where the overflows are located. In a multi-gated casting, all gates should start to feed metal into the cavity at the same time. The overflows should trap the metal in the flow front to bring it out of the casting. Oxidation of the metal in the flow fronts should also be avoided. All solidification shrinkage should have adequate feeding to avoid porosity. If all these conditions where fulfilled, the microstructure in the casting should be determined by the local solidification conditions set up by the die. The mechanical properties should consequently be given by the grain size and dendrite arm spacing of the microstructure [1].

These ideal conditions are, however, difficult to achieve in practice. Meeting flow fronts from several gates from an unbalanced runner system, multidirectional flow and stagnant back eddies due to complex geometries, fragmentation and oxidation of the flow front are all circumstances that disrupt the ideal flow pattern. If we add the temperature aspects such as generation of floating crystals in the shot sleeve, solidification along the die wall during filling, segregation and lack of feeding of local hot spots, the prediction and understanding of the final microstructure starts to get complicated. In spite of all these conditions, that are considered as negative with respect to the local mechanical properties, magnesium die-castings show excellent mechanical performance meeting the requirements in new and challenging areas especially in the automotive industry. However, increased understanding of the process will therefore be a valuable tool in further optimising of the performance of HPDC magnesium applications.

EXPERIMENTAL

All castings were made of the AM60 B alloy. The solidification characteristics of this alloy are shown in Figure 1. According to the phase diagram the alloy starts to solidify at 619°C and is completely solidified at 433°C. In the early stages of solidification, the formation of fraction solid is very high as indicated by the broken lines.

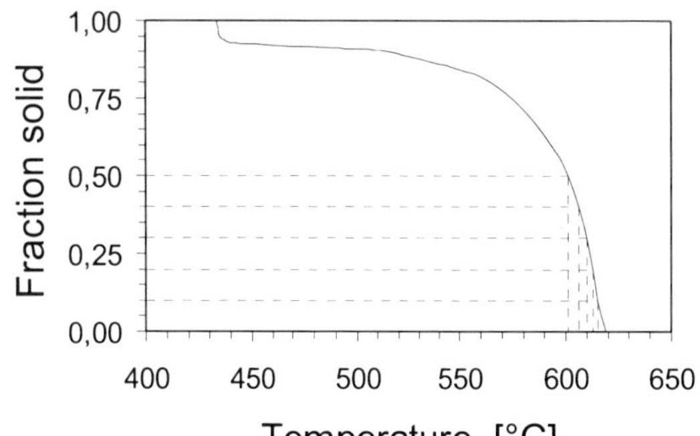

Figure 1 The solidification characteristics of the AM60 alloy as described by the fraction solid versus temperature curve. At 613°C the fraction solid is 20% while 50% of the metal is solidified at 601°C.

The relation between fraction solid and temperature was based on thermal analysis of cooling curves with a cooling rate of 1°C/s [2]. These solidification conditions are far from equilibrium and correspond to a solidification path described by the Scheil equation [3].

The casting experiments were done on a Bühler SC42D cold chamber die casting machine equipped with the DATASPEED system for process control and monitoring. In the process, the metal was transferred from the furnace to the shot sleeve by a gas displacement pump [4]. The shot sleeve diameter was 60 mm while the length was 345 mm. The metal temperature at the outlet of the pump was 680°C during the experiments and the metal was dosed to the shot sleeve in 1.25 s. Immediately after filling, the plunger starts to move in the 1st phase. The last part of the shot profile is shown in Figure 2. The time scale refers to the start of piston movement after shot sleeve filling. Metal enters the runner system after about 1.15 s after end of shot sleeve filling and the 2nd phase starts at about 1.2 s. This means that the total time from start of shot sleeve filling until the metal reaches the runner system is about 2.4 s while metal enters the cavity after about 2.45 s.

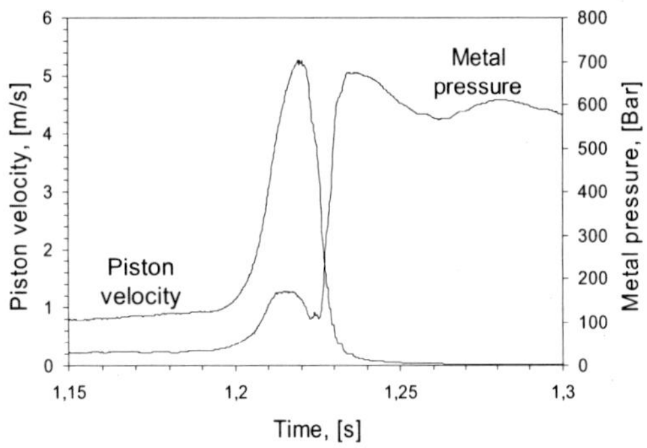

Figure 2 Typical shot profile for the castings. The time scale refers to the start of the piston movement.

For the flow and solidification simulations the MAGMAsoft program, with its special module for high pressure die casting MAGMAhpdc, was used.

GATE VARIATIONS ON A MULTI CAVITY DIE

As a basis for runner and gate variations, an arrangement of test plates and test bars was designed as shown in Figure 3a. The dimensions of the two test plates were 3mmx100mmx165mm. The long tensile test bar in the middle was dimensioned according to ASTM B557. The challenge with this configuration of cavities was to design a gating system giving a balanced filling of all the cavities. Three different gate designs were studied as shown in Figure 3b, c and d. Gate design B had a 25% increase in cross sections compared with gate design A. In addition, the fan gate design for the plates was redesigned to give a linear sided gate, compared to the concave fan design of gate design A.

Gate design C was made with the same cross sections as gate design B, but with an asymmetric runner arrangement with the runner to the test bars forming a Y-junction with the runner to one of the plates.

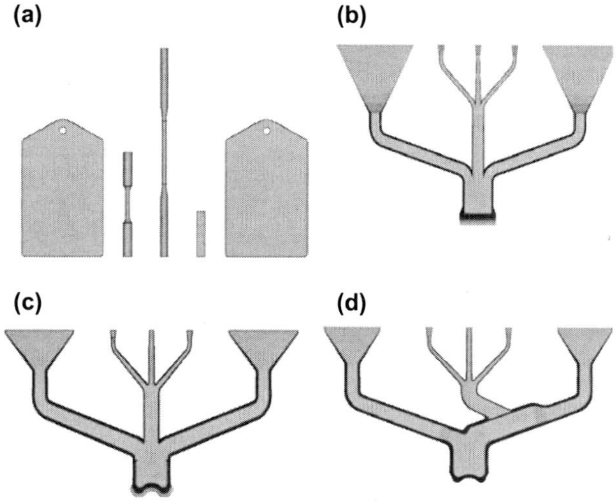

Figure 3 (a) The test plate and test bar casting and (b-d) the three different gate designs: (b) Design A, (c) Design B and (d) Design C.

RESULTS

MECHANICAL PROPERTIES AND MICROSTRUCTURE

In the following, the work was focused on the long tensile bar in the middle of the casting. Typical microstructures in the middle of the gage length area of the tensile test bars are shown in Figure 4.

The microstructure for design A and C possess similar appearance with a fine grained structure from the die surface towards the centre. In the centre a core of coarse grains can be seen. The cross section of design B shows a characteristic layer appearance [5, 6]. The outer skin has a fine microstructure, while the inner core shows a more coarse grain structure.

Typical mechanical properties for the tensile test bar with the different gate designs are given in Table 1. The yield stress was about the same for all three gate designs. Gate design C gave the highest fracture strength, closely followed by gate design A, while design B gave considerably lower values. The elongation with gate design A and C showed the same average value of 13.3%, while the elongation with gate design B was only 7.4%. The tensile strength values show the same pattern. If we look at the scatter in the data, the variation for gate design C is very low compared with the other two. Gate design A generated the largest scatter in properties.

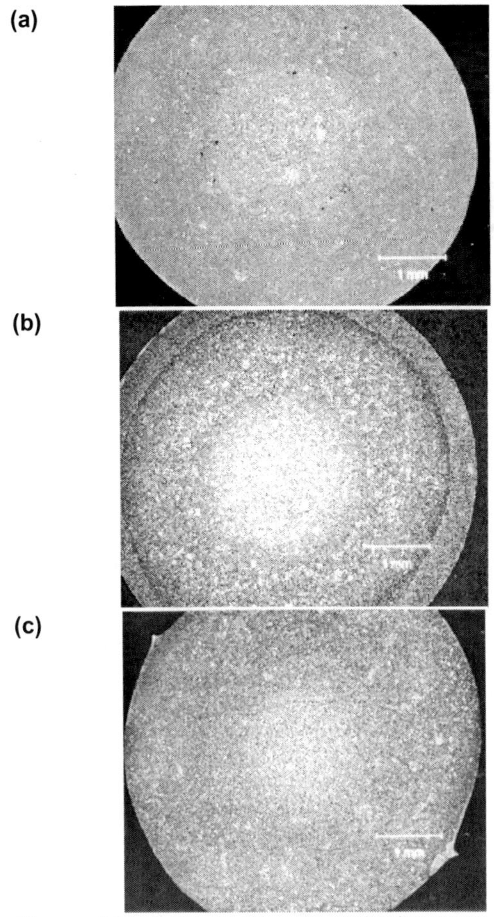

Figure 4 Typical microstructures in the cross section of the long tensile bar for the three runner and gate designs: (a) Design A, (b) Design B and (c) Design C. The diameter of the tensile test bars is 6 mm.

Table 1 Mechanical properties for the long tensile bar with the various gate designs shown in Figure 2.

Gate design	$R_{p0.2}$ [Mpa]	R_m [Mpa]	A [%]
A	120±5	235±14	13.3±3
B	120±2	216±14	7.4±2
C	122±1	245±2	13.3±0.4
*	123	247	12

* Typical tensile properties for die cast AM60 [7].

FLOW PATTERN

In order to understand the variations in microstructure and mechanical properties, flow simulations were made to reveal the general flow pattern of the tensile test bar. Figure 5 shows the filling pattern at three different stages: 1) When the metal has reached the gate of the tensile bar (t=0), 2) when the tensile bar is filled and metal has reached the gate to the overflow and finally 3) when the entire multi cavity die is filled. The time refers to the point when the metal has reached the gate of the tensile bar. All simulations were run with start of the fast filling phase at the same average height in the runner system.

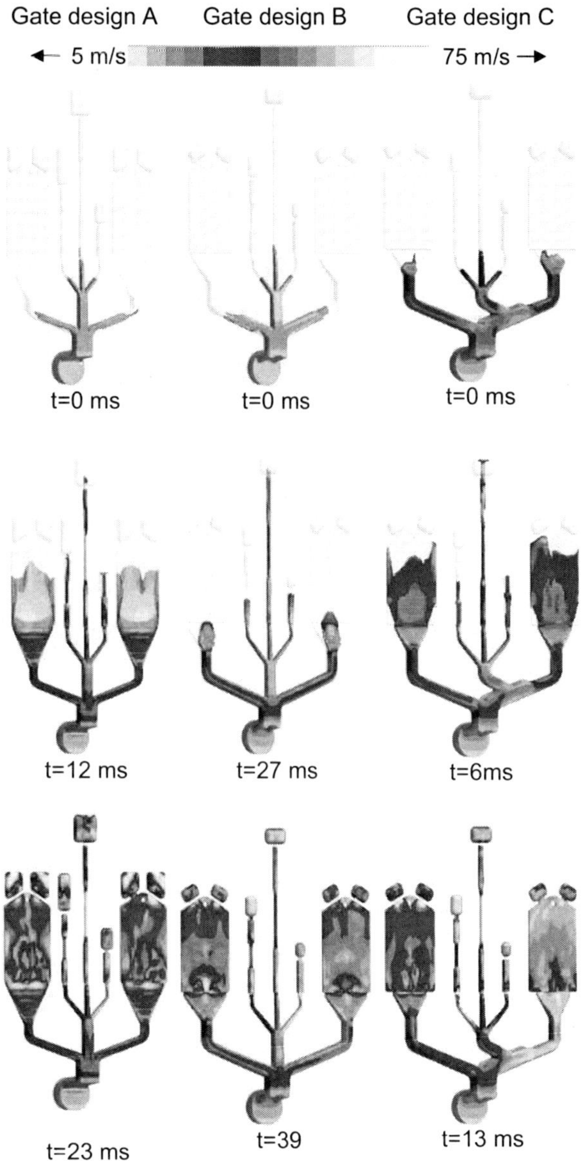

Figure 5 The simulated flow pattern for the test specimen die at three different stages. The color scale indicates the velocity distribution of the flow pattern.

As seen from the three stages the three different gate designs produce different filling sequences of the tensile test bar. Metal reaches the overflow after 12 ms with gate design A. With gate design B metal reaches the overflow 15 ms later, that is after 27 ms. Design C gives, however, the largest difference. Here metal reaches the overflow after just 6 ms. The casting is completely filled after 23, 39 and 13 ms for the designs A, B and C, respectively.

The difference in flow pattern is caused by the difference in balancing of the runner and gating systems. Design C shows the best balancing. When the fast filling phase has started, metal reaches the gates of the two plates at about the same time as the gate of the tensile test bar. After 6 ms the tensile bar is filled and both plates also show a high degree of fraction filled so the remaining filling of the cavity and the overflows is completed within 7 ms. For both design A and B, metal enters the tensile bar already in the slow filling phase. Design A shows, however, reasonable balancing and about half the volume of the plates is filled when the tensile bar is filled after 12 ms. The time to fill the rest of the plates and the overflows is therefore increased to 11 s compared with design C. Design B shows poor performance with respect to balanced filling of the cavities. Because metal already enters the tensile bar before the fast filling phase has started and because the volume of the runner to the test plates is large, it takes 27 ms before metal has reached the overflow. At this stage metal has not yet entered the plate cavities. Due to the small cross section of the entrance to the overflow of the tensile bar, a counter pressure will occur and the metal becomes more or less stagnant in a period of 12 ms as long as the test plates are being filled. Therefore, the final filling of the overflows does not take place until the metal has reached the overflows of the test plates. This filling sequence then gives a total filling time for the tensile bar of 39 ms.

SKIN FORMATION DURING FILLING

Modeling

In the simulated flow patterns in Figure 5, a standard coarse meshing with a free slip boundary condition was used. A constant dynamic viscosity of 10^{-6} m^2/s was used with no restrictions on the fluidity. In order to study the impact of an abrupt stop in the filling sequence on the solidification conditions, a simplified model was set up. In this model only the long tensile bar with overflow was simulated as illustrated in Figure 6. The overflow volume is about half the volume of the tensile bar. The element size was refined to approximately 170 μm. A non-slip boundary condition was used at the die wall.

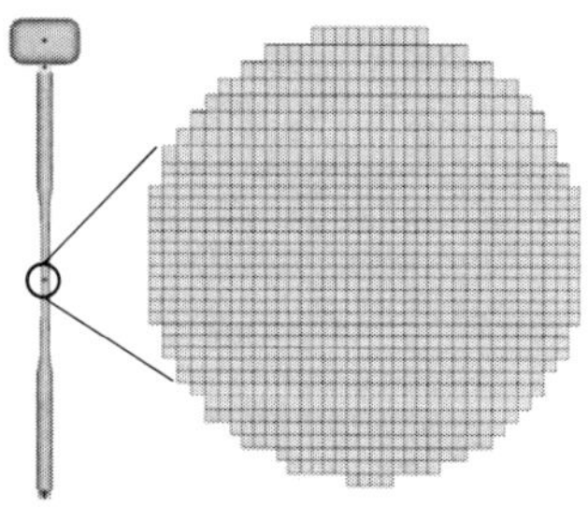

Figure 6 Detailed modeling of the flow pattern in the tensile bar with overflow.

The heat transfer between the casting and the die surface was described by a model where the heat transfer coefficient is constant at a high value of 20 000 W/m^2 above the liquidus temperature and reduces linearly down to 10 000 W/m^2 at the solidus temperature of the alloy [8]. In practice this means that the heat transfer coefficient is close to 20 000 W/m^2 during the filling phase.

If a thin layer of melt sticks to the die wall, the growth of the solidified layer will depend on the balance between the removal of heat from the solidifying skin to the die surface and the input of heat from the convective flow in the core. This layer can have a thickness varying from the order of the typical grain size (1-20 μm) up to several layers of grains. Defining a critical fraction of solid for flow through an element in the simulations is difficult without micro modeling. The dendrite network starts to build up resistance for flow when dendrite coherency occurs, that is, when the dendrite arms start to interact to form a continuous solid skeleton. The dendrite coherency and the following strength development is very much dependant on the grain size and the dendrite morphology which again are given by the alloy composition, grain refinement, cooling rate etc. For aluminum alloys dendrite coherency have been reported to occur in the fraction solid range 15-50% [9]. For the AM50 alloy, dendrite coherency has been reported at a fraction solid of 20% [10].

Another aspect is the mesh size relative to the skin thickness. Even though a solidified skin is formed on the die wall it will be thin compared to the mesh size used in the model in Figure 6. Based on these considerations it was assumed that the metal in an element became stagnant if the average solid fraction in the element became higher than 20%. This corresponds to a drop in temperature in the element of 6°C below the liquidus temperature, that is T_{Stop}=613°C as illustrated in Figure 1. Recovery of flow through the element was assumed if temperature again increased above 613 °C.

Two cases were simulated. In the first case a constant flux giving a total filling time of 30 ms was applied as shown in Figure 7. In the second case, shown in Figure 8, a constant flux was applied in 18 ms. This corresponds to complete filling of the tensile bar. After a rapid reduction, the metal flux was kept at zero for 10 ms. The metal flux was then increased again to the same level as before the drop. In the final filling stage, the overflow was filled through the tensile bar in 10 ms. This flux profile gave a total filling time of 42 ms. In both cases the metal temperature of the incoming flux was held constant at 620°C, that is 1°C above the liquidus temperature.

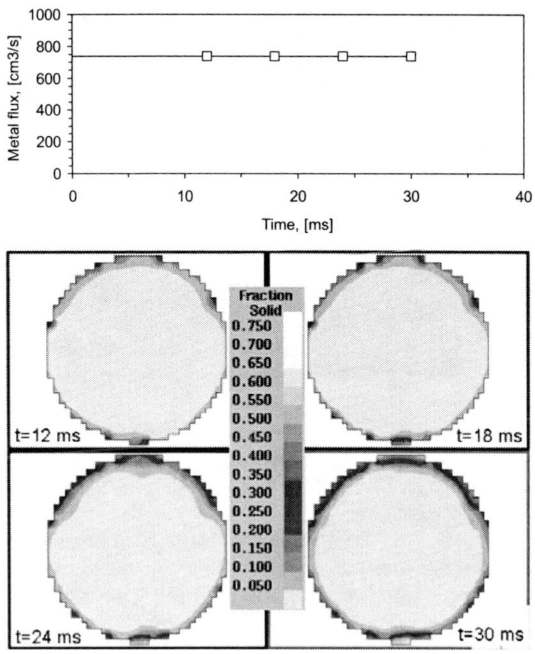

Figure 7 Filling of the tensile bar with constant flux. The simulation shows the fraction solid distribution in the cross section of the tensile bar during filling at four different stages as illustrated with squares in the metal flux curve above.

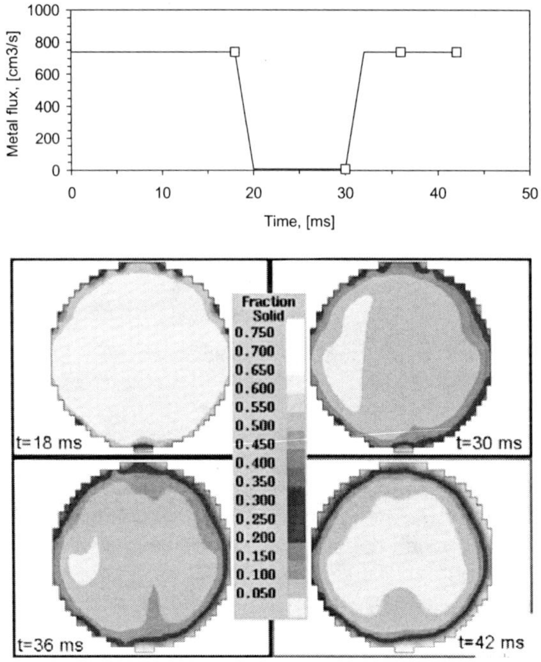

Figure 8 Filling of the tensile bar with a stagnant period of 10 ms. The simulation shows the fraction solid distribution in the cross section of the tensile bar during filling at four different stages as illustrated with squares in the metal flux curve above.

In the case of constant flow the fraction solid pattern shows only small changes during the simulated filling sequence. Already after 12 ms some of the surface elements have developed a fraction of solid more than 20% and then become stagnant with the applied criteria. The stagnant elements continue to cool down, but few new elements become stagnant even after 30 ms. This indicates that a quasi stationary heat balance has been established. The heat flux removed from the solidifying skin is balanced by the convective heat input from the metal flux through the core.

If we look at the second case in Figure 8, the change in flow pattern has a significant impact on the solidification conditions along the die wall. Within the first 18 ms of the filling sequence, the two cases are identical. At 30 ms, after 10 ms with no flux of hot metal into the tensile bar, the temperature has dropped in the cross section and the corresponding fraction solid distribution shows that several element layers have become stagnant. When the flux after 36 ms is back at the same level as before the drop, a flux of hot metal is again flowing into the tensile bar. This metal has, however, not yet reached the cross section at the middle of the tensile bar. At the very end of the filling sequence (42 ms), the flux of hot metal is again flowing through the core of the cross section. The surface elements that became stagnant during the period of no flux have not been re-melted but have cooled down even further. A continuous skin of stagnant metal with a fraction solid in the range 40-70% is then covering the surface of the tensile test bar after die filling.

Comparison between modeling and experiments

The formation of the outer skin with gate design B can then be explained as a result of unbalanced filling. In the simulations the surface thickness with a fraction solid 20% or higher is about 2 elements. This corresponds to a thickness of about 342 μm. In Figure 4 the skin thickness is about 430 μm with gate design B.

The total filling time of the tensile bar with gate design A was 23 ms. Even though the metal flux through the tensile bar in this case showed variations, it must be characterised as continuous without stagnant periods. Compared with the simulation case in Figure 7 after 24 ms, a thick outer skin should therefore not be expected.

In the case of gate design C, the tensile bar was filled with a close to constant flux within 12 ms and no surface skin should be expected.

One condition to emphasize is that even though the metal flux into the tensile test bar is delayed in 10 ms in the simulation case shown in Figure 8, the flow will continue due to inertia effects. In real back eddies in a complex part, the flow can cease completely after changes in the flow pattern. This means that the solidifying skin effect can be even more pronounced than the apparent effect seen in Figure 7 and 8.

Another effect is changes in the heat transfer coefficients. If a solidifying skin is formed during a stagnant period, the metal will shrink. In areas where the solidifying skin is shrinking away from the die surface, the heat transfer will be reduced. On core surfaces and other areas where the skin is shrinking on the heat transfer will be maintained at a high level. These effects have not been considered in the simulations.

PRE-SOLIDIFICATION IN THE SHOT CHAMBER

In the case studies in Figure 7 and 8 it was assumed that the metal flux had a constant temperature of 1°C above the liquidus temperature (620°C) during the whole filling sequence. In order to verify this assumption the filling sequence of the shot sleeve was simulated. The model is illustrated in Figure 9. In this case 104 elements were used to describe the shot sleeve diameter of 60 mm, that is each element was about 577X577 μm. The shot sleeve was filled in 1.25 s with metal at pouring temperature of 680°C. The simulated temperature and fraction solid distributions are shown in Figure 10.

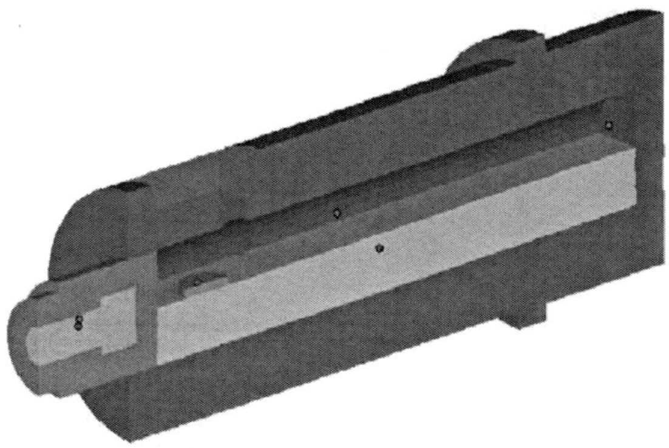

Figure 9 Cut through the centerline of the model for the simulation of shot sleeve filling. Except for the mesh size, the material model and simulation conditions were the same as for the model presented in Figure 6.

As seen from Figure 10 the metal starts immediately to solidify on the cold walls of the shot sleeve. When the piston starts to move, the metal to be injected into the die cavity therefore has a non-isothermal temperature distribution. In the front of the shot sleeve close to the start of the runner system, the temperature in the hottest area has dropped to about 635°C. About 1.15 s later, that is, 2.4 s after start of shot sleeve filling, the temperature had dropped even further down to about 629°C in the same area. This corresponds to the time in the shot profile where metal starts to enter the runner system. In the model it was, however, not possible to simulate the piston movement with the corresponding metal flow during the slow filling phase. The calculations show, however, that the assumption about a drop in metal temperature down to about 620°C before the metal reach the tensile bar is reasonable.

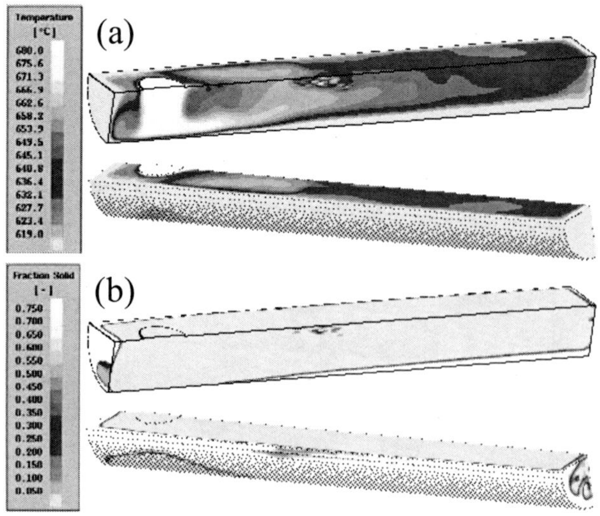

Figure 10 (a) Temperature and (b) corresponding fraction solid distribution after filling of the shot sleeve. The pouring whole is to the left in the pictures, while the right end is connected to the start of the runner system.

DISCUSSION

The microstructure in magnesium die castings is normally showing a duplex structure through the cross section. Close to the surface the microstructure is characterized by an outer skin of fine grained structure with little or no porosity and good mechanical properties. The inner core shows typically a mixture of fine and coarser grains as well as a tendency for some porosity. The mechanical properties of this inner core are therefore normally somewhat lower than in the surface skin. As mentioned in the introduction, magnesium die castings with this typical microstructure appearance do show excellent mechanical performance. Key factors for further improvements will therefore be to avoid defects in the surface skin and to improve the properties of the inner core.

MICROSTRUCTURES IN MAGNESIUM DIE CASTINGS

The layered structure is caused during the filling and solidification processes. If a cross section is filled without any solidification on the die wall, the layered structure is caused by movement of the inner core relative to the solidified skin during feeding of shrinkage in the pressure intensification phase [5, 6]. In this case a thicker skin can be expected than those observed in this work. Several layers can be caused by subsequent burst feeding events. In the following, skin formation during filling will be discussed. Since the time scale for the die filling process typically is in the range 20-90 ms, a skin formed during the filling stage is expected to be thin compared to layers formed after the filling stage.

If we combine the two mechanisms for a solidifying skin and formation of coarse pre-solidified crystals in the shot chamber a model can be proposed as illustrated in Figure 11. The illustration shows the combined effect of a skin solidifying from the die wall and flow of pre-solidified fragments in the centre at different conditions.

It is reasonable to expect that the metal that originally was close to the runner end of the shot cylinder will enter the cavity first [11, 12, 13]. If the metal that enters the die cavity first is still above the liquidus temperature it will be without pre-solidified crystals. This metal is expected to end up at the surface and in positions far away from the gate locations. The microstructure of this metal will then be given by the high cooling rate in the thin walled die cavity, that is, a fine microstructure through the whole cross section is expected in this case. With a solidifying skin close to the die wall the situation will be as illustrated in case II in Figure 11. As filling proceeds, more coarse fragments will follow the metal flow into the die cavity. At the end of the filling stage, the metal with longest dwell time in the shot sleeve will enter the die cavity. This metal will consequently have the highest fraction of pre-solidified coarse crystals and will end up close to the gate as in case I. This mechanism will give a declining gradient of coarse crystals from the gate towards the end of the flow path.

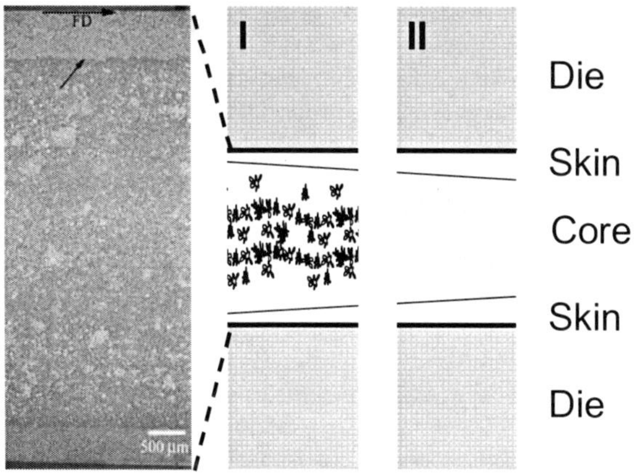

Figure 11 Model for skin formation and distribution of pre-solidified coarse dendrites along the flow path. The micrograph to the left shows a cross section in the middle of the tensile bar (gate design B) normal to the flow direction (FD).

The thickness of the outer skin will be given by the balance between heat removed through the die surface and the heat flowing through the core of the cross section. Close to the gates, where the metal flux normally is high, a thin skin is expected. Further away from the gate and in areas with less metal flux, the surface skin is expected to become thicker. This characteristic feature can also be seen from the micrograph in Figure 11 and from the simulations in Figure 8. The tensile test bar cavity was placed symmetrical in both the cover and the ejector die half. The gating to the tensile bar was, however, placed in the ejector half of the die. This means that the metal flux was not completely centred in the tensile bar cross section. The highest metal flux was therefore in the ejector half of the cross section. In the micrograph in Figure 11 the ejector side is the lower side of the micrograph. As seen from the micrograph, the lower surface skin on the ejector side is thinner (approx. 450 µm) compared to the upper surface skin (approx. 600 µm) on the cover side of the die. In the simulation in Figure 8 the ejector side is the left side of the cross section.

Similar microstructure appearance that can be explained by the model in Figure 11 has been reported in a more complex box casting for an AlSi7Mg alloy [12]. Further work is in progress to investigate the generation and distribution of floating crystals in the shot sleeve filling and in the further filling of the cavity.

GENERATION OF SURFACE DEFECTS

If splashing or meeting flow fronts have formed weld lines or cold shuts during the initial filling stage, these defects cannot be repaired until pressure is applied. This situation is illustrated in Figure 12.

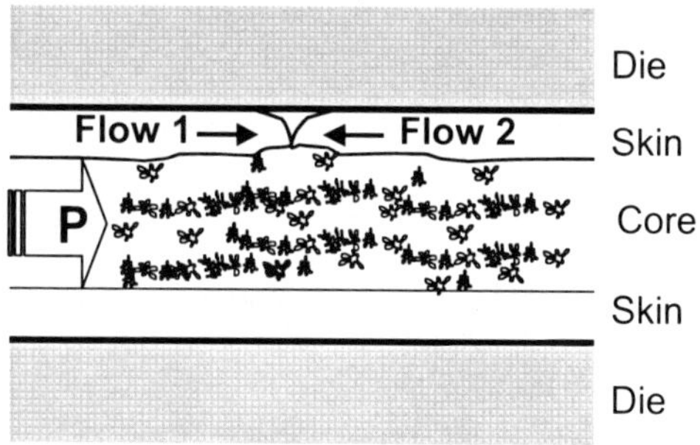

Figure 12 Formation of surface defects such as weld lines and cold shuts when two metal flows are meeting. The final characteristics of this possible surface defect will depend on the transferred pressure P.

The final characteristics of such surface defects will then be given by the pressure intensification conditions in the final solidification process after end of filling. This surface defect can then be completely merged, filled with molten metal from the core or left as an open surface groove in the form of a cold shut or incomplete filling. An example of such a surface defect in the tensile test bar with gate design B is shown in Figure 13.

The tensile test bars cast with gate design B showed in general more surface defects than test bars cast with gate design A and C. In a load condition such as uniaxial tensile stress, these type of surface defects will be determinant for the ductility. Tensile test bars cast with gate design B also showed a higher fraction of pre-solidified crystals in the core. These effects explain both the lower mechanical properties and the high scatter in the data.

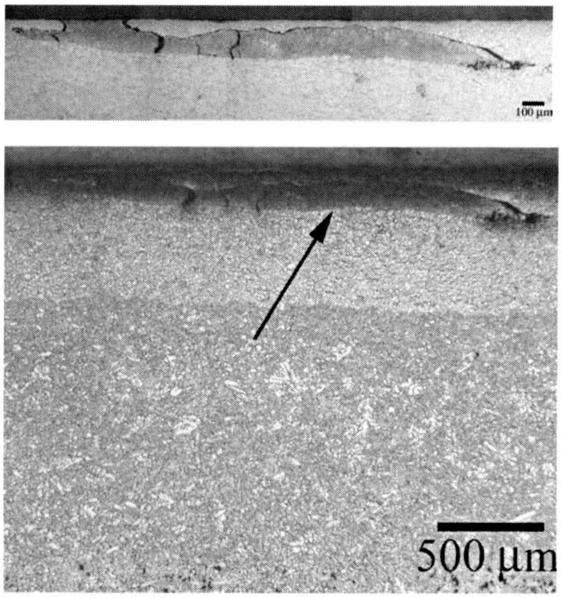

Figure 13 Surface defect in the tensile test bar cast with gate design B. The lower micrograph shows some of the inner core and the surface skin. The upper micrograph shows in more detail the surface defect.

GATE AND PART DESIGN

Optimised metal flow in large thin walled castings is considered a key to obtain good and uniform mechanical properties in the part. Gate and part design is therefore very important in order to promote cavity filling with an ideal flow pattern. Magnesium alloys for die casting are known for having good die filling and fluidity properties. Flow distances of 1 meter or even more in cavities with wall thickness of less than 2 mm is possible under the right conditions. One condition that needs to be fulfilled is to maintain a high and constant velocity in the flow during the whole filling sequence. Pressure drops in the flow due to unbalanced filling or back eddies will promote skin formation during filling. In the meantime, when other parts of the die become filled, solidification proceeds from the die walls. At the time when the pressure again is raising, the flow through the cross section can cease or be significantly redirected due to reduced effective cross section. As seen in this work these conditions can give significant reduction in mechanical properties. Careful analysis should therefore be done in the design process to avoid filling patterns that locally gives more or less stagnant periods in the filling of the part. If areas that have experienced a stagnant period during filling are coinciding with high load areas in crash requirements, the actual part or the gating system should be redesigned.

Another aspect is the feeding efficiency through the gates. If the model of an increasing fraction of pre-solidified coarse crystals during filling pertains, this will give a high fraction of solid close to the gate before the casting becomes pressurized after filling. Depending on the general temperature distribution in the die and the wall thickness distribution of the part, the following feeding of the solidification shrinkage can be blocked at an early stage giving generally high porosity levels in the casting.

The proposed models and mechanisms in this work was based on experiments and simulations of about 50% filling in a 60 mm diameter shot sleeve. The effects and their relative importance will be dependent on shot sleeve dimensions and shot sleeve filling as well as shot sleeve temperature and metal temperature [14].

In order to directly predict the pre-solidification effects that are discussed in this work in flow and solidification simulations, more sophisticated models and criteria need to be developed. Proper analysis of flow and solidification simulations will, however, give a clear indication on flow induced defects. Flow and solidification simulations are therefore important tools to optimise the flow and solidification patterns in complex, thin walled magnesium die castings.

CONCLUSIONS

- The flow pattern has a significant impact on the mechanical properties in magnesium die castings.
- Good and consistent mechanical properties were obtained in a tensile bar when it was filled with a constant flow of metal through the tensile bar. A flow pattern of early filling followed by a more or less stagnant period before further flow gave poor mechanical properties with a large scatter.
- The microstructure in tensile test bars that had experienced a more or less stagnant period showed a typical layered structure. Characteristics in the microstructure were a distinct outer shell of fine microstructure and an inner core with a significant fraction of coarser dendrite fragments.
- The main variations in microstructure appearance can be explained by differences in solidification conditions initiated in the filling phase: The fine microstructure is formed under high cooling rates inside the cavity, while the coarser microstructure is formed under lower cooling rates in the shot sleeve.
- Die design should be carried out to give a die filling pattern without large local variations in the metal flux during the filling sequence.

REFERENCES

1. Gjestland H. et al, High pressure die casting – Getting the best out of magnesium, 10th Magnesium Automotive and End User Seminar, Aahlen, Germany, 2002
2. Taminnen J., Thermal analysis for investigation of solidification mechanisms in metals and alloys, Ph.D. thesis, Dept. of Structural Chemistry, University of Stockholm, 1988.

3. Flemings, Solidification Processing, McGraw-Hill Inc., London, 1974.
4. Automatic Transfer & Metering Equipment for Liquid Mg Alloys, Brochure, Hydro Magnesium, 1999.
5. Sannes S. and Westengen H., The influence of Process Conditions on the Microstructure and Mechanical Properties of Magnesium Die Castings, Magnesium Alloys and Their Applications, ed. B.L. Mordike and K.U. Kainer, pp. 223-228, 1998, Frankfurt, Germany
6. Dahle A. K. et al, Formation of defect bands in high pressure die cast magnesium alloys, Journal of LightMetals 1, pp 99-103, 2000
7. Die Cast Magnesium Alloys, Data sheet, Hydro Magnesium, 2001.
8. Schmidt P., Heat transfer in Permanent mould casting, Ph.D thesis, Dept. of Materials Processing, The Royal Institute of Technology, Sweden, 1994.
9. Dahle A. K., Mushy zone properties and castability of Aluminium foundry alloys, Ph.D thesis, Dept. of Metallurgy, The Norwegian Institute of Technology, Norway, 1996.
10. Rølland T. K., Arnberg L., Solidification of Mg-alloys, SINTEF Materials Technology, internal report, Trondheim, Norway, 1992.
11. Garber L. W., Theoretical analysis and experimental observation of air entrapment during cold chamber filling, Die Casting Engineer, May-June 1982, pp 14-22.
12. Laukli H. I. et al., The effect of solidification of metal prior to injection in HPDC on the grain size distribution in a complex die casting, NADCA 2002
13. Wang L. et al, Simulation of Flow Pattern and Temperature Profile in the Shot Sleeve of a High Pressure die Casting Process, NADCA, 2001.
14. Paliani P. et al, Alloy Cooling And Solidification in Horizontal Cold Chamber Die Casting Shot Sleeves, NADCA 1995.

2003-01-0187

Bolt Load Compressive Stress Retention Testing of Magnesium Alloys

Alan P. Druschitz and Eric R. Showalter
INTERMET Corporation

ABSTRACT

Elevated temperature bolt load compressive stress retention testing of four high temperature magnesium alloys (AJ50X, AJ52X, AS21X and AE42), two structural magnesium alloys (AM50A and AM60B), one aluminum alloy (383) and one gray iron alloy were performed at the INTERMET Technical Center over a period of about one year. Artificial aging of some of these alloys during testing was observed. The effect of a heat treatment designed to thermally stabilize the microstructure was evaluated and determined to significantly improve magnesium performance and degrade aluminum performance. This paper documents the test procedure and the test results.

INTRODUCTION

The automobile and light truck industries are continuing to apply magnesium castings to structural applications (seat backs, seat risers, seat pans, brake pedal adapter brackets, steering wheel armatures, lock housings, valve covers, engine cover assemblies and instrument panels) but has been slow to use magnesium in demanding applications, such as oil pans, transmission cases and cylinder blocks. For many of these demanding applications, high temperature (up to 175°C) tensile and compressive strength and corrosion resistance are critical. A few high temperature alloys (AS21 and AE42) have been or currently are in use. The original AS21 alloy was developed and used in the Volkswagen Beetle for the transmission case, cylinder heads and air-cooled cylinder block. However, because of the combination of casting issues, performance and cost, AS21 and other magnesium alloys do not satisfy the needs of today, and thus, many new magnesium alloys are being developed. In this study, two well known structural magnesium alloys (AM50A and AM60B), which have been reported as "...for production of automotive wheels and other parts requiring good elongation and toughness combined with reasonable yield and tensile properties" [1], two high-temperature magnesium alloys (AE42 and AS21X), both of which have been characterized as having "good creep properties up to about 150°C" [2], two new high temperature magnesium alloys (Noranda AJ50X, Noranda AJ52X), one aluminum alloy (383) and gray iron were simultaneously evaluated and compared. Aluminum alloy 383 was chosen as one of the baseline materials due to its widespread use for non-structural, elevated temperature applications, such as valve covers and engine cover assemblies and gray iron was chosen as another baseline material due to its extensive use in engine applications, such as cylinder blocks, cylinder heads and bedplates.

EXPERIMENTAL METHODS

An experimental test casting containing a fluidity snake, a tensile bar, a bending fatigue bar, an impact specimen and a bolt boss plate was designed and approximately 50 castings of each non-ferrous material produced. The casting trials were performed on a horizontally parted, horizontal shot, cold chamber high-pressure die casting machine at the INTERMET Monroe City plant. The geometry of the test casting is shown in Figure 1.

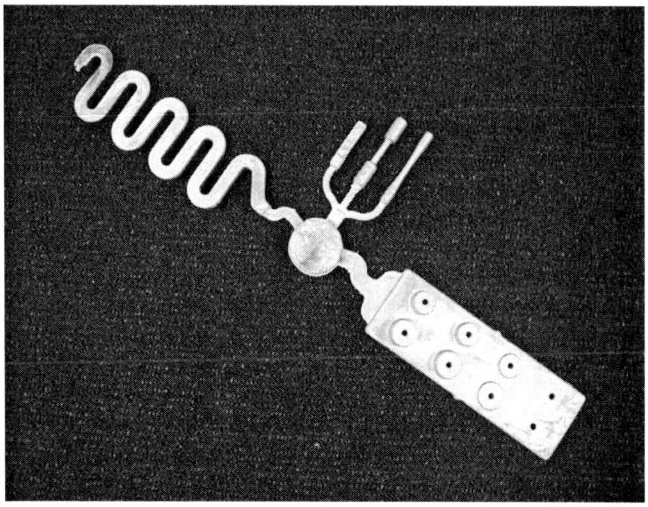

Figure 1. Geometry of Complete Test Casting

Melting was performed in two, 227 kg capacity, portable melting crucibles. Before use, each crucible was either previously unused or thoroughly cleaned and steamed to remove any residue. Great care was taken to be certain that new alloy melts were not contaminated with the previously melted material. Material coupons were taken at the beginning and end of the runs and tested by the material suppliers for correct chemical composition. Cover gas consisted of a mixture of dry air, SF_6 and CO_2. On the days of the casting trials, the ambient temperature, dew point and atmospheric pressure were recorded. No attempt was made to optimize the casting conditions for any alloy. Details of the casting conditions were previously reported [3].

The stabilizing heat treatment evaluated in this study consisted of aging at 225°C (437°F) for 6 hours. This produced a thermally stable, over-aged microstructure.

Samples of gray iron were cut from a production bedplate casting produced at the INTERMET Radford Shell foundry.

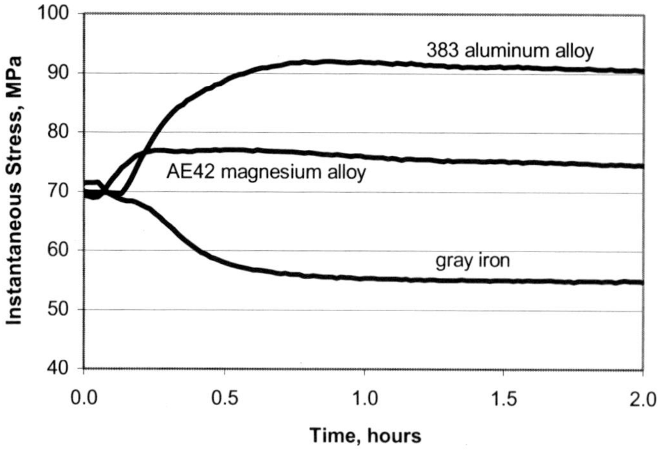

Figure 2a. Typical Bolt Load Compressive Stress Retention Curves for Magnesium, Aluminum and Gray Iron During the Beginning of a Test.

BOLT LOAD COMPRESSIVE STRESS RETENTION TEST

The castings produced for this study were tested in both the as-cast and heat treated condition. A comprehensive analysis (tensile, impact, fatigue, creep, corrosion, microstructure) was performed on these materials in the as-cast condition and previously reported [3,4].

Prior bolt load compressive stress retention studies have measured bolt stretch [5], bolt torque [6,7] or bolt load [8] before and after elevated temperature exposure. None of these studies discussed the increase or decrease in applied stress that would occur upon heating and cooling due to the differences in the coefficient of thermal expansion of the test specimen (Mg, Al or gray iron) and the bolt (steel). In this study, a technique was developed that allowed this change in stress to be observed and measured. Typical curves for magnesium, aluminum and gray iron during the beginning of a test are shown in Figure 2a and over the entire test are shown in Figure 2b.

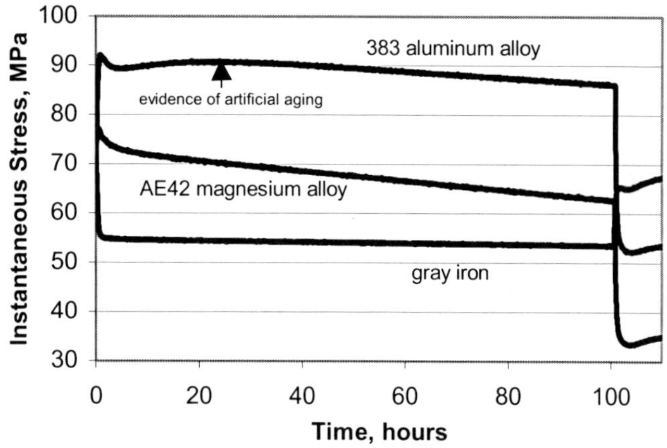

Figure 2b. Typical Bolt Load Compressive Stress Retention Curves for Magnesium, Aluminum and Gray Iron Covering the Entire Test.

Aluminum showed the greatest increase in applied stress during heat-up, which was surprising since the magnesium alloy had a larger coefficient of thermal expansion. Therefore, it was surmised that the magnesium alloy was yielding during this time. Gray iron showed a decrease in applied stress since its coefficient of thermal expansion was less than the steel bolt.

The initial increase or decrease in applied stress was visible along with the final increase or decrease in applied stress during cooling in Figure 2b. Evidence of artificial aging for the aluminum alloy was readily apparent. The differences in the material behavior were readily revealed by this technique, i.e., the applied stress decreases then holds constant for gray iron, the applied stress increases and then decreases at a relatively constant rate for magnesium and the applied stress increases significantly and then decreases slowly for the aluminum alloy.

Bolt load compressive stress retention testing used a bolt boss specimen, two small hardened 440C stainless steel washers (58-60 Rc) that had an 8.3 mm inside diameter and 14.3 mm outside diameter, a Lebow 3718-112-8MM high temperature bolt force sensor, a special large hardened 440C stainless steel washer (58-60 Rc) and a M8 Grade 10.9 bolt (7.8 mm nominal diameter) and nut. The hardness of selected bolts was determined in the as-received condition and after testing to confirm that the properties of the bolts did not change during testing. The large washer was slipped onto the bolt followed by the bolt force sensor and a small washer. The bolt was then passed through the bolt boss sample, a second small washer slipped onto the bolt followed by the nut. To provide consistent clearance, the hole in the 14.9 mm thick bolt boss specimen was drilled out to 7.9 mm. Various views of the assembled test specimen are shown in Figures 3a-c.

Figure 3a. Overall View of Bolt Load Compressive Stress Retention Test Specimen.

Figure 3b. Close-Up of Bolt Load Compressive Stress Retention Test Specimen Showing Bolt Force Sensor and Upper Washers.

Figure 3c. Close-Up View of Underside of Bolt Load Compressive Stress Retention Test Specimen.

The bolt force sensors were connected to Measurements Group Model 2120B signal conditioner/amplifiers, zeroed and balanced. The output from the signal conditioner/amplifiers was captured by a Measurements Computing PCI-DAS64/M2/16 data acquisition board. One sample of each material was prepared in this manner and the data acquisition system, which used SoftWIRE, was started. The load versus strain calibration curves for each bolt force sensor were programmed into the data acquisition system so that applied stress could be displayed during the test. The samples were then loaded to 70 or 105 +/- 1 MPa at room temperature. The samples were then placed in a pre-heated Grieve air-circulating oven at 85°C, 125°C,

150°C, or 175°C and held for 100 hours. Temperature was controlled to +/- 1°C. The bolt force sensor's strain output and calculated stress were continuously recorded and displayed. After 100 hours at temperature, the samples were removed and allowed to cool to room temperature. The samples were then allowed to stabilize at room temperature for a minimum of two hours before the final (retained) stress was measured and recorded. After the final stress measurement, the bolts were loosened and the "zero" behavior of the bolt force sensor was noted. If the bolt force sensor did not return to zero, the final stress measurement was corrected.

Unfortunately, the Lebow bolt force sensors were not temperature compensated. Therefore, apparent strain vs temperature data was collected for each sensor and the measured strain vs time curves were corrected. Improved sensors would be a valuable development for this type of testing.

RESULTS and DISCUSSION

The bolt load compressive stress retention data showed that none of the magnesium alloys investigated in the as-cast condition performed as-good or better than aluminum alloy 383 or gray iron. However, in the heat treated condition and at low initial stress (70 MPa), AE42 and AJ52X had much closer performance to aluminum alloy 383 but did not perform as well as gray iron. In general, the magnesium alloys showed as-good or better performance after heat treatment.

After 100 hours at room temperature, gray iron showed a loss of ≤0.8% of the initial applied stress, aluminum alloy 383 showed a loss of 1.6-2.3% and the magnesium alloys showed losses of up to 5.2%. For comparison, gray iron had losses of ≤3.6% after 100 hours at 175°C. Some permanent plastic deformation was expected in the magnesium alloys at room temperature since the applied stresses were greater than the material's elastic limit.

For the as-cast condition at low applied stress (70 MPa), AE42, AJ52X and AJ50X had similar performance up to 85°C. At 125°C and 175°C, AE42 and AJ52X had comparable performance, which was significantly better than the other magnesium alloys. At 150°C, AJ52X had better performance than AE42; this was apparently due to artificial aging. Surprisingly, AJ52X showed very little change between 125°C and 150°C. AJ50X and AS21X had comparable performance up to 175°C. At 175°C, AM50A and AM60B had comparable "poor" performance.

For the heat treated condition at low applied stress (70 MPa), the magnesium alloys showed as-good or better performance at all temperatures except for AJ52X and AJ50X at 150°C. Apparently, the heat treatment suppressed the artificial aging that occurred during testing for these alloys. In the heat treated condition, AS21X performed better than AJ50X. The bolt load compressive stress retention data for 70 MPa initial stress is summarized in Table I and Figures 4a-c.

Table I. Bolt Load Compressive Stress Retention Test Data (70 MPa initial stress, average of 3 tests)

		Stress Retained at 70 MPa Initial Stress after 100 Hours at Temperature				
Alloy	Condition	RT	85°C	125°C	150°C	175°C
Gray Iron	AC	99.2%		97.8%	97.1%	96.6%
	HT			97.5%	97.5%	95.7%
383 Al	AC	98.4%	92.0%	84.8%	96.9%	81.0%
	HT			84.6%	81.0%	68.7%
AJ52X	AC	96.6%	87.6%	74.1%	73.2%	48.6%
	HT			77.4%	67.5%	66.1%
AE42	AC	96.5%	87.6%	72.0%	64.9%	49.6%
	HT			76.7%	72.9%	63.9%
AS21X	AC	95.4%	82.6%	62.0%	54.6%	29.3%
	HT			70.8%	55.4%	41.5%
AJ50X	AC	96.3%	86.6%	63.8%	56.1%	27.7%
	HT			64.6%	49.1%	33.2%
AM50A	AC	96.6%	83.5%	49.7%	32.9%	7.4%
	HT			47.0%	34.0%	13.2%
AM60B	AC	97.6%	79.6%	49.0%	20.5%	5.2%
	HT			54.0%	38.0%	9.9%

AC = As Cast
HT = Heat Treated

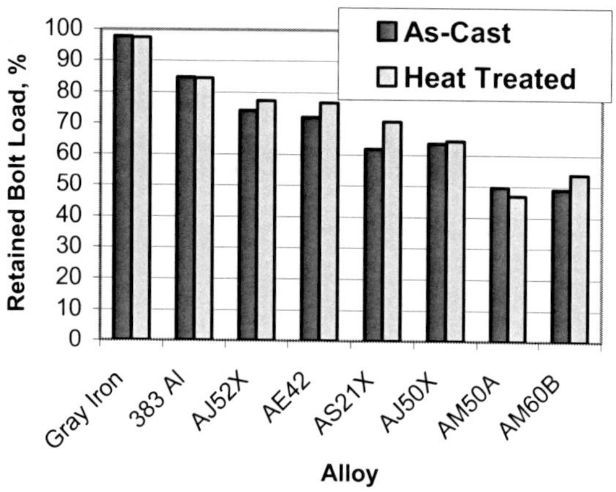

Figure 4a. Effect of Heat Treatment on the Bolt Load Compressive Stress Retention Test Data at 70 MPa Initial Stress and 125°C.

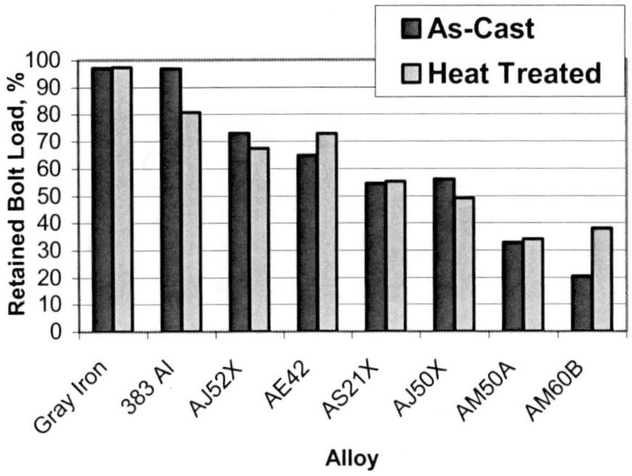

Figure 4b. Effect of Heat Treatment on the Bolt Load Compressive Stress Retention Test Data at 70 MPa Initial Stress and 150°C.

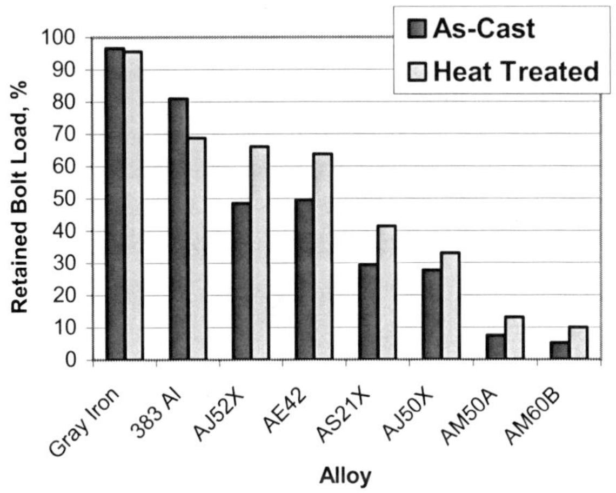

Figure 4c. Effect of Heat Treatment on the Bolt Load Compressive Stress Retention Test Data at 70 MPa Initial Stress and 175°C.

For the as-cast condition at high applied stress (105 MPa), AE42, AJ52X and AJ50X had similar performance up to 85°C. At 125°C, AE42 and AJ52X had comparable performance, which was slightly better than the other magnesium alloys. At 150°C and higher, the performance of AE42 and AJ52X were comparable and much better than the other magnesium alloys. At 175°C, AJ50X had better performance than AS21X. As before, at 175°C, AM50A and AM60B had comparable "poor" performance.

For the heat treated condition at high applied stress (105 MPa) and at 175°C, the performance of AS21X was significantly improved relative to the as-cast condition but the performance of all of the other magnesium alloys was essentially unchanged. The bolt load compressive stress retention data for 105 MPa initial stress is summarized in Table II and Figure 5.

Table II. Bolt Load Compressive Stress Retention Test Data (105 MPa initial stress, average of 3 tests)

		Stress Retained at 105 MPa Initial Stress after 100 Hours at Temperature				
Alloy	Condition	RT	85°C	125°C	150°C	175°C
Gray Iron	AC	99.3%		97.0%	97.0%	96.4%
	HT					94.5%
383 Al	AC	97.7%	92.1%	84.5%	89.3%	71.0%
	HT					68.1%
AJ52X	AC	96.4%	85.2%	73.0%	67.3%	47.2%
	HT					46.6%
AE42	AC	95.3%	84.8%	68.9%	61.7%	47.7%
	HT					46.3%
AS21X	AC	94.8%	78.3%	62.2%	44.4%	19.7%
	HT					29.0%
AJ50X	AC	96.6%	83.8%	65.2%	48.0%	26.6%
	HT					28.2%
AM50A	AC	97.2%	74.4%	46.4%		6.5%
	HT					7.7%
AM60B	AC	95.8%	72.3%	46.8%		5.8%
	HT					5.5%

AC = As Cast
HT = Heat Treated

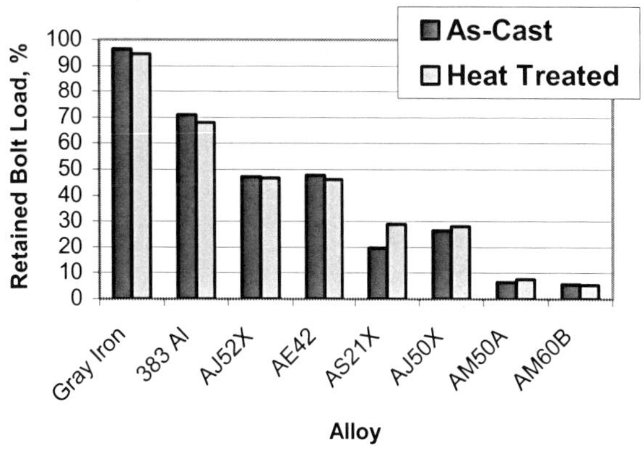

Figure 5. Effect of Heat Treatment on the Bolt Load Compressive Stress Retention Test Data at 105 MPa Initial Stress and 175°C.

CONCLUSIONS

1. None of the magnesium alloys investigated had as-good or better bolt load compressive stress retention properties as aluminum alloy 383 or gray iron. However, in the heat treated condition at low applied stress (70 MPa), AE42 and AJ52X had much closer performance to aluminum alloy 383.

2. At low initial applied stress (70 MPa), heat treatment improved the performance of all magnesium alloys except AJ52X and AJ50X at 150°C. The reduction in performance of these two materials at 150°C was due to the suppression of artificial aging during testing.

3. At high initial applied stress (105 MPa) and at 175°C, heat treatment only improved the performance of AS21X.

4. Heat treatment decreased the performance of aluminum alloy 383 for test temperatures >125°C. Again, the reduction in performance was due to the suppression of artificial aging during testing.

ACKNOWLEDGEMENTS

The authors would like to acknowledge the assistance of Dale tenPas and Ed Druschitz (summer intern), INTERMET Technical Center, Lynchburg, VA for performing the bolt load compressive stress retention testing. Noranda, Hydro Magnesium and Magretech are also acknowledged for providing the materials evaluated in this study.

REFERENCES

1. Magnesium and Magnesium Alloys, Avedesian, M.M. and Baker, H. eds., ASM International, Materials Park OH (1999) p. 226.
2. Hydro Magnesium website, www.hydromagnesium.com
3. Druschitz, A.P., et al, "Evaluation of Structural and High-Temperature Magnesium Alloys", SAE Paper 2002-01-0080, Society of Automotive Engineers, Warrendale, PA (2002).
4. Druschitz, A.P., et al, "Evaluation of Structural and High-Temperature Magnesium Alloys at Room Temperature," in Magnesium Technology 2002, H.I. Kaplan ed., TMS, Warrendale, PA (2002).
5. Sieracki, E.G., Velazquez, J.J. and Kabiri, K., "Compressive Stress Retention Characteristics of High Pressure Die Casting Magnesium Alloys", SAE Paper 960421, Society of Automotive Engineers, Warrendale, PA (1996).
6. Pekguleryuz, M.O., "Development of Creep Resistant Magnesium Diecasting Alloys—An Overview", Noranda website, www.norandamagnesium.com.
7. Pekguleryuz, M.O., "High Temperature Magnesium Alloy Development", Noranda website, www.norandamagnesium.com.
8. Ford Engineering Specification ES-19E708, Test No. IIID, "Fastener/Attachment Point Acceptability".

CONTACT

Dr. Alan P. Druschitz received his PhD in Metallurgical Engineering in 1982 from the Illinois Institute of Technology, Chicago, IL. He is currently the Director of Materials Research and Development for INTERMET Corporation. He is located at the INTERMET Technical Center, Lynchburg, VA 24502. He can be reached at adruschitz@intermet.com or (434) 237-8749. Before joining INTERMET Corporation in 1996, he was a staff research engineer for General Motors Corporation for fourteen years. He has been a member of the American Foundry Society for fourteen years, the Society of Automotive Engineers for twenty-one years and ASM International for twenty-six years. He is currently the Vice President of the Ductile Iron Society, a member of the Industrial Advisory Board for the Central Virginia Governors School and a Member of the Governors Board of Transportation Safety for the Commonwealth of Virginia.

Eric R. Showalter received his Bachelors Degree in Mechanical Engineering in 1993 from Bradley University, Peoria, IL. He is currently the Director of Advanced Manufacturing Engineering for INTERMET Corporation. He is located at the INTERMET Corporate Offices in Troy, MI 48098 and can be reached at (248) 267-5107 or eshowalter@intermet.com. Before coming to INTERMET with the acquisition of Ganton, where he was Director of Engineering, he was Engineering Manager for the J.L. French Corporation in Sheboygan, WI. He has also held various positions in engineering, sales and manufacturing. He is currently a member of the North American Die Casting Association and INTERMET's representative to the International Magnesium Association. He has been involved in the manufacture of automotive parts in aluminum and magnesium using high-pressure die casting techniques for over eight years.

2003-01-0189

Die Cast Magnesium Alloys for High Temperature Applications

Per Bakke, Håkon Westengen and Ketil Pettersen
Norsk Hydro ASA, Magnesium Materials Technology

Darryl Albright
Magnesium Market Development, Hydro Light Metals

ABSTRACT

The development of creep resistant alloys for automotive drive train components has proven to be metallurgically challenging. This paper discusses the principles of high temperature alloy development, featuring metallurgical, microstructural and processing aspects of some alloys, relative to their high temperature performance. The creep resistant alloys within the Mg-Al base system obtain their creep resistance by a relatively low content of Al, and addition of elements that form stable intermetallic phases within the grain mantle. Various third elements affect the high temperature performance differently. The results demonstrate that rare earth elements (RE) show a remarkable potential as the third element(s).

INTRODUCTION

Magnesium applications require different critical properties, such as ductility and energy absorption in safety parts, high yield strength and fatigue strength in structural parts, and creep resistance for automotive drive train components. Especially the latter has attracted considerable interest and has proven to be metallurgically challenging. There is a large research activity worldwide, and numerous combinations of elements have been tested. It has become common practice for new entrants in the magnesium industry to present proprietary 'new' alloy compositions for elevated temperature applications. Exceptionally good properties are claimed, and numerous presentations are made without revealing the chemical composition of the alloys. When the possible alloy candidates are examined in detail, it turns out that they are all based on the same alloying principles, which are known from basic metallurgy [1].

SOLID SOLUTION HARDENING

The α-magnesium needs to be stiffened by solid solution hardening. Potent elements include aluminum, zinc and rare earth elements. Since creep occurs preferentially in the grain boundary regions where the eutectic components are located, it is important that the eutectic temperature is not too low. Mg-Zn has a eutectic temperature around 340°C and ternary elements are required to increase the eutetic temperature. Aluminum is a compromise with eutectic temperature 437°C [2].

GRAIN BOUNDARY HARDENING

Die casting is characterized by rapid cooling. Depending on the wall thickness, cooling rate will typically be in the range 10-1000°C/s and the resulting grain size is very fine, typically 5-100 micrometer. This leads to significant hardening by the Hall-Petch effect [3]. The fine grain size is harmful for the creep properties, and it is known that finegrained magnesium alloys deform super-plastically at elevated temperatures [4].

DISPERSION HARDENING

Adding alloying elements that form a dense distribution of intermetallic particles during solidification, play a key role in providing creep resistance of die casting alloys. Such particle distributions may have a high dispersion level f/r, where f is the volume fraction of particles and r the mean particle radius. These particle dispersions become increasingly efficient to reduce the creep rate when they form 3-dimensional networks along the grain boundaries capable of reducing creep by thermal activation. An additional positive effect is observed when the dispersoid forming element combine with the solid solution hardening element to reduce the amount of low melting eutectic along the grain boundaries.

AGE-HARDENING

Age hardening in the usual sense of performing a full T6 heat treatment is not compatible with high volume production of die cast parts. There are several reasons for this: During solution treatment, the loss in strength due to grain coarsening would be larger than the increase caused by subsequent age hardening. Also, formation of blisters due to expansion of gas porosity would be a problem, unless special precautions were taken. Age hardening is commonly used for sand cast parts. An example is provided by WE43 where age

Table 1
Some high temperature resistant magnesium die casting alloys, with patents and relevant literature.

Alloy design.	*Al*	*Zn*	*Si*	*Ca*	*Sr*	*RE*	*References*
AS41 (DOW/VW)	4		1				6, 7, 8, 9, 10, 11, 12
AS21 (DOW/VW)	2		1				7, 11, 12, 13
ASE210 (Hydro)	2		1			0.1-0.2	14, 15, 16
AE42 [1] (DOW)	4					2-3	17, 18
AX81[2] (Hydro)	8			1			19, 20
AXJ530 [2] (GM)	5			3			21, 22, 23
AEX522 [2] (Honda)	5			2		2	24, 25
AJ52 (Noranda)	5				2		26, 27, 28
ZA [3]	4	12					11, 29, 30, 31, 32, 33
EZ30 (MEL)						3	34, 35, 36

[1]) AE alloys were also discussed in the early 1970's [8,30,37]
[2]) Mg-Al alloys with Ca patented already in 1959 by Otto Fuchs [38], and later described by Mann [39]. Mg-Al-Ca alloys were rediscovered [40, 41, 42], and re-patented [43, 44, 45] in the 1990's.
[3]) Originally proposed compositions were ZA12-4, AZ88, AZ55, ZA102+Ca.

hardening precipitates are formed in extremely high density during a final age hardening [5]. This particle dispersion provides both strength and creep resistance, due to the high thermal stability of the particles formed. This explains the paradox that while magnesium die casting alloys show lower creep resistance than aluminum alloys, the special magnesium sand casting alloys are superior to any aluminum alloy.

OVERVIEW OF EXISTING ALLOYS

Some die casting alloys with known chemical composition are summarized in Table 1 [6-45]. Most of the alloys contain aluminum as a main alloying element. The dispersoid forming elements Si, RE, Sr and Ca are added in amounts up to several wt%. Exceptions are EZ30, which is very soft due to the absence of solid solution hardening elements. Another exception is the high Zn alloys (ZA) where solid solution hardening is provided by both Zn and Al, and creep properties are obtained by a network of $Mg_{32}(Al,Zn)_{49}$ and MgZn phases rather than the brittle $Mg_{17}Al_{12}$ [31,32].

In addition to the alloys listed in Table 1 several heat resistant die casting alloys have been patented, including AZE and ZX alloys (Toyota) [46,47], AEX alloys (UBE/Nissan) [48], EX alloys (Kobe Steel) [49] and ZS alloys (UBE) [50]. However, so far none of these alloys have been considered for mass production and industrial application.

While the solubility of Si, RE, Sr and Ca in molten magnesium is quite high for each element separately, the liquid solubility may fall dramatically if several elements are added. This is why only very limited amounts of RE elements can be dissolved in molten AS21, to produce ASE210 [16].

The challenge in composing a creep resistant die casting alloy is to keep enough hardening elements to give adequate solid solution strengthening, maximize the volume fraction, and optimize the spatial distribution of dispersoids without introducing brittleness. Castability must be adequate and other properties like corrosion resistance etc. must conform to standards. Recycability is also an important issue [1, 51].

For high power engines and automatic transmissions, the operating temperatures may exceed 150°C and the moderate creep resistance of AZ91D becomes critical. This is in line with the experience with the VW Beetle engine that was initially produced in AZ81, and later changed to AS41 and further to AS21 as the engine power was increased [12]. Unfortunately, none of these alloys which all show excellent castability performance, are suitable for service at temperatures of 150°C or above. For applications requiring <150°C and moderate loads, an inexpensive, castable and recyclable alloy like ASE210 may be considered. For more demanding applications AEX522, AXJ530 and AJ52 may withstand temperatures up to 175°C, but poor castability and recycling problems (incompatibility with salt as Ca and Sr in the alloys will react with $MgCl_2$ in the salt [52]) are strong arguments against them. The creep properties of AE42 may be at the limit, but within the AE system further optimization is possible.

AE ALLOYS

The beneficial effect of rare earth elements (Ce) to Mg was discovered in the 1930's [53,54]. Sand and gravity casting alloys i.e. EZ33 and ZE41 grain refined with small additions of Zr, and subsequently heat treated, have been in use for some time [55,56]. Die casting alloys of the AE system have not received the attention they probably deserve from property and metallurgical points of view.

Recent work comparing the performance of various creep resistant alloys rank AE42 as the alloy having the best combination of properties at room temperature [57] and die castability similar to AZ91D for large automotive (power train) components [58]. With respect to elevated temperature properties and corrosion, AE42 is the current benchmark.

EXPERIMENTAL

Test bars and plates of different experimental AE alloys were die cast on a 420 ton Bühler SC42D Evolution machine, with a five cavity plate/bar die (version 0), Fig. 1. In addition, specimens of the alloys A380 (aluminum die casting alloy), AZ91D, ASE210, AE42, AJ52 and AXJ530 were prepared and cast.

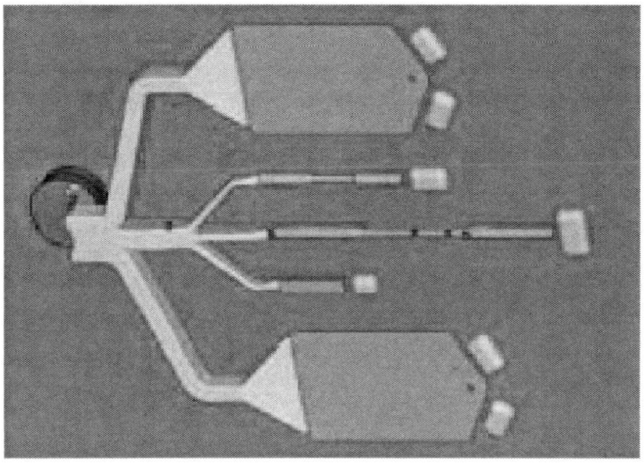

Figure 1 The five cavity plate/bar die for the Bühler SC42D Evolution die casting machine.

Permanent mould cast medallions were withdrawn from each alloy and analyzed by ICP-AES.

Standard 6 mm test bars from each composition were tensile tested at a 10 kN Instron machine at temperatures ranging from room temperature and up to 210°C in accordance with standard procedure ISO 6892. Strain rates 1.5/10 mm/min were used throughout.

Two specimens from each were tensile creep tested for minimum 150 hrs in accordance with internal procedure MgMT-05-02 (standard). Different loads and temperatures (150°C/50 MPa, 175°C/40 MPa and 175°C/75 MPa) were applied. Tensile creep testing at 175°C/40 MPa and 175°C/75 MPa were also performed on samples heat treated at 175°C for 500 hrs. This because microstructural changes (phase transformations) possibly taking place during prolonged exposure at elevated temperatures could negatively affect the creep properties. This information would be highly relevant for e.g. engine blocks in service.

Die cast test plates were subjected to corrosion testing in accordance with ASTM B117 (salt spray test) after pre-treatment according to HRE MAT A38 pkt.3 (standard). Reference plates of AZ91, AM60 and AE42 (5 plates each) were tested in the same chamber.

RESULTS AND DISCUSSIONS

TENSILE TESTING

Existing alloys

Tensile yield strength as a function of temperature for high temperature magnesium alloys and aluminum alloy A380 are shown in Fig. 2. It is seen that AXJ530 has the highest TYS among the Mg alloys tested. It should be emphasized that tensile testing is a high strain rate test, and there is no clear connection between tensile yield strength and creep resistance. A clear example is AZ91D, which perform very well at high strain rates, but due to supersaturation of Al precipitation processes will be active under creep conditions. (The solid solubility of Al in Mg at 150 °C is 1.8 % [2])

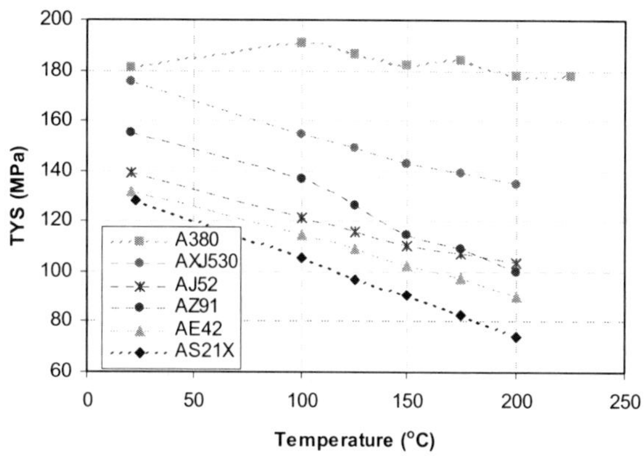

Figure 2 Tensile yield strength as a function of temperature for high temperature magnesium alloys and aluminum alloy A380.

The tensile yield strength is an absolute upper limit for the load that can be applied in a creep test. If creep tests are carried out at high loads/low temperatures where the softest alloys, like ASE210 and EZ30 are stressed above the yield strength, the results can be misleading. In such cases the alloys with highest yield stresses, like AZ91 will perform comparatively well.

It should also be noted that the value of the yield stress is very dependent on the method of measurement, and direct comparison of values from different sources may be misleading. Therefore we have chosen to use only our own data in Fig. 2.

Experimental AE alloys

Tensile yield strength as a function of temperature for some experimental AE alloys is shown in Fig. 3. AE42 and AXJ530 are shown for comparison. It is seen that TYS for the experimental AE alloys is well above that of AE42. Specifically, at 200°C, TYS for four of the AE alloys is in the range of 105 -115 MPa. For AE42, TYS at 200°C is measured to 90 MPa. On the other hand, TYS for AXJ530 at 200°C is as high as 135 MPa.

Ultimate tensile strength for the same alloys is shown in Fig. 4. It is seen that at room temperature all AE alloys tested exhibit UTS in the 205-250 MPa range. All the experimental AE alloys perform better than AE42. At room temperature UTS for AXJ530 is found to be only about 190 MPa. At 200°C, UTS of the AE alloys is in the range of 110-135 MPa, while for AXJ530 it is 160 MPa.

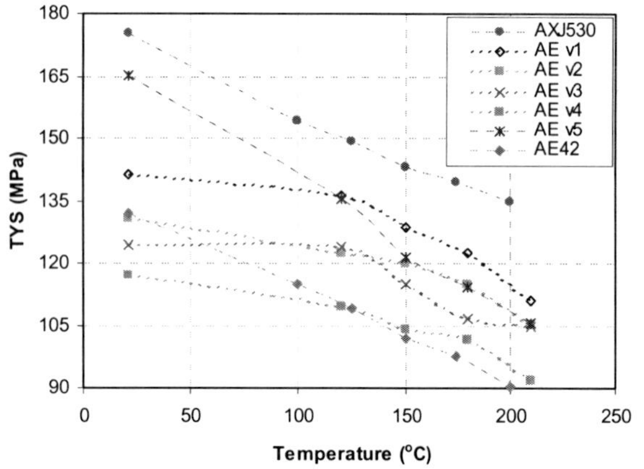

Figure 3 Tensile yield strength as a function of temperature for some experimental AE alloys, AE42 and AXJ530.

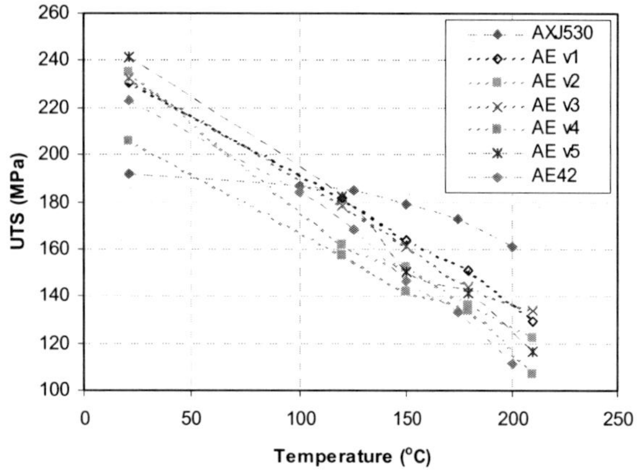

Figure 4 Ultimate tensile strength as a function of temperature for some experimental AE alloys, AE42 and AXJ530.

The range of alloying elements in AE type alloys is quite wide due to the extensive liquid solubility. Compared to other dispersoid forming alloying elements like Sr and Ca, RE elements have a less harmful effect on ductility/fracture toughness.

TENSILE CREEP TESTING

As cast condition

Fig 5 shows the steady state creep rates at 100 hrs of various die cast AE-alloys compared with those of AZ91, AS21 and AE42 at 150°C. All samples were tested as cast. It can be seen that all AE alloys, as well as AJ52, have better creep properties at 50 MPa than AE42. Furthermore, Fig. 5 illustrates the shortcomings of AZ91D, at 50 MPa, for example, the creep rate is about 100 times larger than that of the AE alloys, and roughly 50 times larger than AS21.

Average creep rates at 100 hrs (average between 80 and 120 hrs) are used as estimates for minimum creep rates during phase II creep. This estimate is acceptable as long as the onset of phase III creep occur at >>120 hrs. Also, phase I creep should be terminated at << 80 hrs. However, for many creep resistant alloys minimum creep rate will occur at >>100 hrs. Thus, the average creep rate at 100 hrs will slightly overestimate the minimum creep rate. The resolution of the tensile creep measurements is at the level of 5.10^{-7} hr^{-1}.

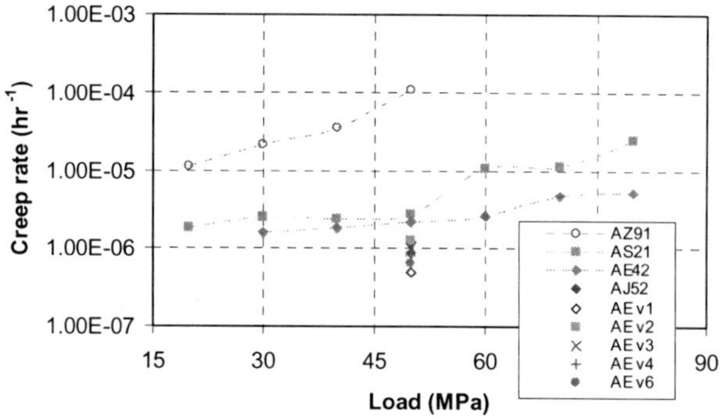

Fig. 5 Average (secondary) creep rate at 100 hrs as a function of load at 150°C.

Fig. 6 shows the steady state creep rates at 100 hrs of various die cast AE-alloys compared with those of AS21 and AE42 at 175°C Again, all samples were tested as cast. In the diagram we have included a single point for A380, at load 75 MPa. This shows that significant improvements can be obtained with the AE alloys, relative to AE42. It is remarkable that the AE alloys can match the creep resistance of A380 at such high temperatures and loads. This is very encouraging for a variety of new applications

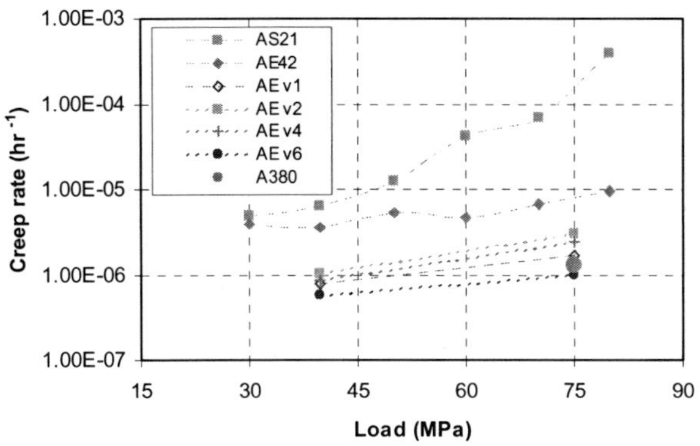

Fig. 6 Average (secondary) creep rate at 100 hrs as a function of load at 175°C.

Aged specimens

As mentioned earlier, specimens were heat treated for 500 hrs at the highest likely test (and operating) temperature 175°C before creep testing. In Fig. 7 the steady state creep rates at 100 hrs (at 175°C) of the heat treated (HT) AE samples are indicated. These are compared with the results of the as cast specimens as well as with as cast AE42. It is seen that the creep rates of the as cast samples and the samples aged at 175°C are almost indistinguishable. This demonstrates that an AE alloy in service, for example in an engine block, will exhibit the same and predictable creep properties regardless of operating time.

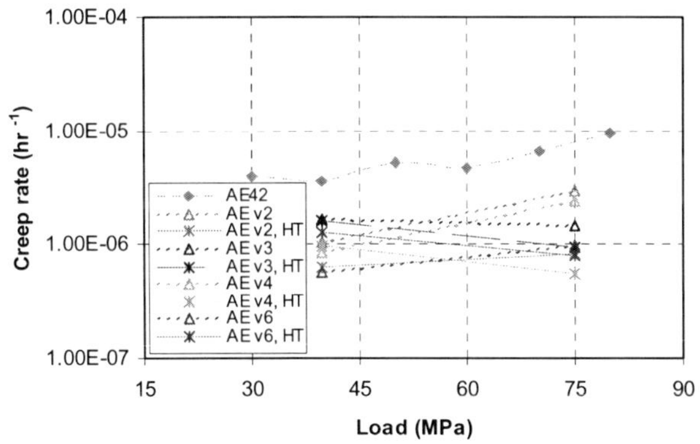

Fig. 7 Average (secondary) creep rate at 100 hrs as a function of load at 175°C. Results from heat treated samples are indicated.

CORROSION PROPERTIES

Fig. 8 shows results of the corrosion testing. MCD values ($mg/cm^2/day$) are plotted as function of Al and RE contents. The corrosion rates are low and consistent, in the range of 0.024-0.069 $mg/cm^2/day$ within the composition range tested. It is seen that the corrosion rates decrease slightly with increasing Al content. Results of reference samples AZ91, AM60 and AE42 tested alongside the AE alloys confirm the corrosion rates of the AE alloys.

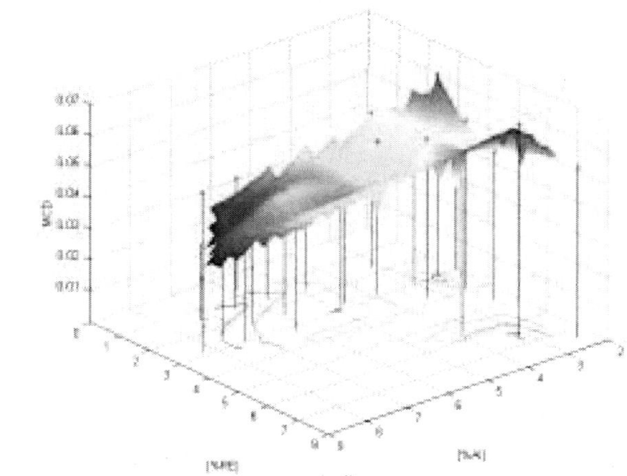

Fig. 8 Corrosion rates ($mg/cm^2/day$) of AE alloys as a function of Al and RE contents.

It must be stressed that corrosion values at this level can only be obtained as long as the contents of the heavy metal impurities Fe, Ni and Cu are kept below the maximum limits specified in the High Purity standards.

METALLOGRAPHY

The microstructure as illustrated by the SEM micrographs in Figs. 9-11, partly explain the ductility and the creep behavior of AE alloys.

Creep

AE42 contains 2-3 wt% Cerium-rich Mischmetal (RE), which forms $Al_{11}RE_3$. The intermetallic phases are distributed in the grain boundary zone, and their effect on creep is primarily to inhibit grain boundary sliding. The formation of $Al_{11}RE_3$ (or Al_2RE) also reduces the effective amount of Al, which otherwise would lead to Al-enrichment in the grain boundary zone. The grain interior is not significantly influenced by the addition of RE [51]. Initial SEM-EDS analyses may indicate that at a certain RE/Al ratio, the grain boundary phase formed in the die cast specimens is Al_2RE, rather than $Al_{11}RE_3$. This will be studied in greater detail in a later paper [59]. It has been reported that sharply decreasing creep strength of AE42 above 150°C is attributed to a partial decomposition of $Al_{11}RE_3$ into Al_2RE and Al, subsequently forming $Mg_{17}Al_{12}$ [60]. With a higher RE/Al ratio than that of AE42, possibly producing Al_2RE during solidification, this unwanted phase transformation no longer occurs. This is also supported by the creep test results of the heat treated samples (at 175°C for 500 hrs, Fig. 7), where all samples with a higher RE/Al ratio than that of AE42 were unaffected by the heat treatment.

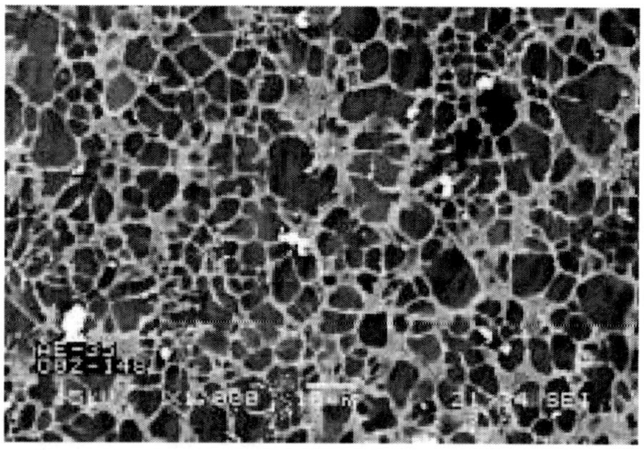

Fig. 9. SEM micrograph of alloy AE35.

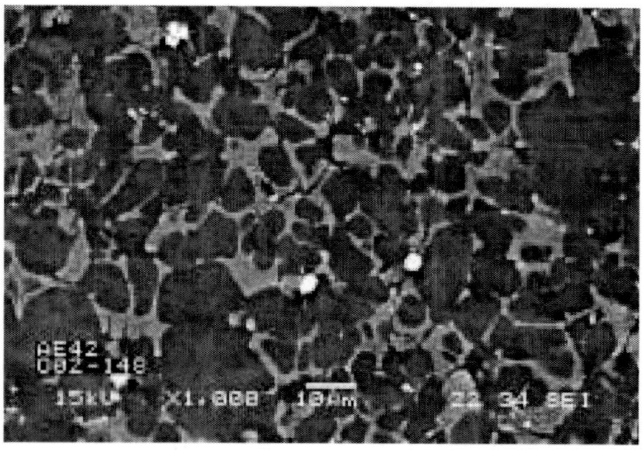

Fig. 10. SEM micrograph of alloy AE42.

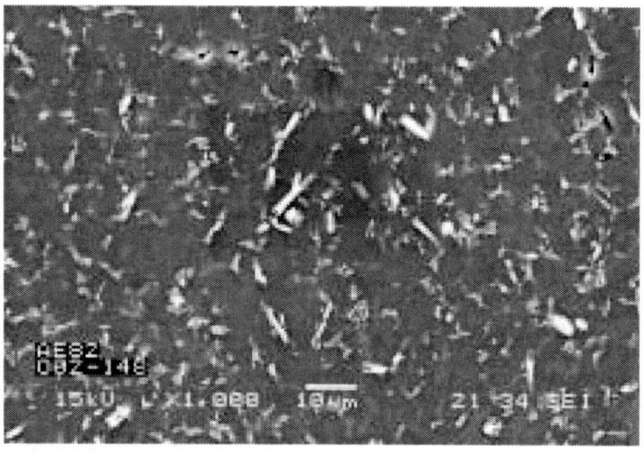

Fig. 11. SEM micrograph of alloy AE82.

<u>Ductility</u>

In alloys with a RE/Al ratio corresponding to AE42, the $Al_{11}RE_3$ phase has a lamellar structure with a relatively large extension from the grain boundaries. With increasing Al content the $Al_{11}RE_3$ phases are coarsening, and more concentrated close to the grain boundaries. The formation of $Al_{11}RE_3$ suppresses the formation of $Mg_{17}Al_{12}$ by binding up Al. At higher RE/Al ratios the ductility is reduced due to formation of a continuous skeleton of an intermetallic phase decorating the grain boundaries. This is illustrated in Fig. 9 where the continuous phase on the grain boundaries is seen in alloy AE35, which is a fairly brittle alloy with elongation at fracture typically in the 2-4% range. In alloy AE42 (Fig. 10) the network is starting to disintegrate, and the elongation at fracture of die cast AE42 typically is between 8 and 12%. In Fig 11 (AE82), the continuous network is absent, and the second phase ($Al_{11}RE_3$) appears like isolated particles. The ductility of this alloy can be expected to be at the AE42 level [61].

CONCLUSION

By combining well known principles of alloy design, with the characteristics of the die casting process, magnesium alloys can be tailored to specific property requirements. Although the mechanical strength and creep properties do not completely reach the levels of competing aluminum alloys, the differences are minor, and can be compensated for by proper design.

The tensile-, creep- and corrosion results demonstrate that AE alloys show a remarkable potential among the Mg-based alloys. If we include recyclability, the AE alloys should become a very attractive light weight solution for the automotive industry.

ACKNOWLEDGMENTS

Sten Bjørneboe, Jan Gjerstad and Vidar Sjøberg are acknowledged for alloy preparation and die-casting. Per R. Gjertsen is acknowledged for tensile and creep testing. Eva Thorjussen and Jan Ivar Skar are acknowledged for corrosion testing. Finally, Kari Straume Andersen is acknowledged for microscopy and SEM analyses.

REFERENCES

1. H. Westengen, and P. Bakke, "Magnesium die casting alloys for use in applications exposed to elevated temperatures: Can they compete with aluminium?", <u>PSTAM</u>, Osaka, Japan, January 2003
2. A.A. Nayeb-Hashemi, and J.B. Clark, <u>Phase Diagrams of Binary Magnesium Alloys</u>, ASM International, Metals Park, Ohio, U.S.A. 1988
3. H.Gjestland, and H.Westengen, "High pressure die casting – Getting the best out of magnesium", <u>10th Magnesium Abnehmer Seminar</u>, Aalen, Germany, Sept. 2002
4. H. Watanabe, H. Tsutsui, T. Mukai, K. Ishikawa, Y. Okanda, M. Kohzu, and K. Higashi, "Superplastic Behaviour in Commercial Wrought Magnesium Alloys", <u>Magnesium Alloys 2000</u>, Materials Science Forum Vols. 350-351, Trans Tech. Publ., Switzerland 2000, pp.171-176
5. J.F. King, "Development of practical high temperature magnesium casting alloys", <u>Magnesium Alloys and Their Applications</u>, ed K.U.Kainer, Wiley-VCH, Weinheim, Germany, 2000, pp. 14-22

6. G.S. Foerster, Norwegian Patent No. 121.753, Filed 19-02-1968
7. G.S. Foerster, French Patent No. 1.555.251, Filed 29-02-1968
8. K.E. Nelson, "Magnesium die casting alloys", Proc. 6th SDCE, Paper no. 13, 1971
9. S.L. Couling, "New magnesium alloys", Metals Eng. Quarterly, May 1972, pp.7-13
10. E.W. Balzereit, "Neuentwiklung von Magnesium-legierungen fur den Drückguss", Metallwissenschaft und Technik, 26, March 1972, pp.245-250
11. G.S. Foerster, P.C.J. Gallagher, D.L. Hawke, and E.N. Aqua, "Research in Magnesium", Die casting Engineer, Jan.-Feb. 1977, pp 12–18
12. F. Hollrigl-Rosta, E.Just, J.Kohler, and H.-J.Melzer, "Magnesium in Volkswagenwerk", Metall, 34 J, H12, Dec. 1980, pp.1138-1141
13. S.L. Couling, US. Patent No. 529.885, Filed 05-12-1974
14. K. Pettersen, J.I.Skar, and M.Videm, Intl. patent application WO0102614, Filed 25-10-1999
15. K. Pettersen, H. Westengen, J.I. Skar, M. Videm, and L.-Y. Wei, "Creep resistant alloy development", Magnesium Alloys and their Applications, Ed. K.U.Kainer, Wiley-VCH, Weinheim, Germany, 2000, pp.29-34
16. K. Pettersen, P. Bakke, J.I.Skar, C. Tian, and D.Albright, "Magnesium alloy development for automotive drive-train applications", NADCA 2001, Cincinnati OH, Oct.29.-Nov.1, 2001, pp.193-201
17. W.E. Mercer II, "Magnesium Die Cast alloys for Elevated temperature applications", SAE Technical Paper No. 900788, 1990; J.S. Waltrip, "Fresh look at some old magnesium die casting alloys for elevated temperature applications", Proc. 47th IMA Conf., Cannes, 1990, pp.124-129
18. T.Aune, H.Westengen, and T.Ruden, "The effects of varying aluminium and rare-earth content on the mechanical properties of die cast magnesium alloys", SAE Technical Paper No. 94077, 1994
19. Norsk Hydro, British Patent GB1,163,200, Filed 30-01-1967
20. W. Schmidt, "Magnesium-Gusswerkstoffe mit erhöhter Warmfestigkeit", Giesserei, Vol. 57, 1970, pp.535-538
21. B. Powell, V. Rezhets, A. Luo, and B. Tiwari, European Patent Application, EP 1 048 743 A1, 2000
22. A. Luo, M.P. Balogh, and B.R. Powell, "Tensile creep and microstructure of magnesium-aluminium-calcium based alloys for power train applications – Part 2 of 2", SAE Technical Paper 2001-01-0423, 2001; A.Luo, B. Powell, "Tensile and compressive creep of Magnesium-Aluminium-Calcium based alloys", Magnesium Technology 2001, Ed.: J.Hryn, TMS-AIME, 2001, pp.137-144
23. B.Powell, A.Luo,B.L.Tiwari, and V.Rezhets, "The die castability of calcium-containing magnesium alloys: Thin-walled computer case", Magnesium Technology 2002, Ed. H.I.Kaplan, TMS-AIME, 2002
24. T. Baba, and S. Wako-shi, European Patent Application EP 0 791 662 A1, Filed 20-02-1997
25. S. Koike, K. Washizu, S. Tanaka, T. Baba, and K. Kikawa, "Development of lightweight oil pans made of a heat resistant magnesium alloy for hybrid engines", SAE Technical Paper No. 2000-01-1117, 2000
26. M. Pekguleryuz, and P. Labelle, Intl. Patent application PCT/CA00/01527, Filed 14-12-2000
27. M. Pekguleryuz, and E. Baril, "Development of creep resistant Mg-Al-Sr alloys", Magnesium Technology 2001, Ed.: J. Hryn, TMS-AIME, 2001, pp.119-125
28. D.Argo, M.Pekguleryuz, P.Labelle, P.Vermette, R.Bouchard, and M.Lefebvre, "Process parameters and die casting of Noranda's AJ52 high temperature Mg-Al-Sr alloy", Magnesium Technology 2002, Ed. H.I. Kaplan, TMS-AIME, 2002
29. G.S. Foerster, Norwegian Patent 132492, Priority date 01.10.1973; and G.S. Foerster, Norwegian Patent 132493, Priority date 01.10.1973
30. G.S. Foerster, "Improved Magnesium Die Casting Alloys", 8th SDCE Intl. Die Casting Exposition & Congress, Paper no G-T75-112, 1975
31. Z. Zhang, A. Couture, and A. Luo, "The Creep Properties of Mg-Zn-Al Alloys", Magnesium Alloys and their Applications, Werkstoff-Informations-gesellschaft, Frankfurt, 1998, pp 289–294; Z. Zhang, A. Couture, and A. Luo, "An Investigation of the Properties of Mg-Zn-Al Alloys", Scripta Materialia, Vol 39, No 1, 1998, pp 45–53
32. M. Vogel, O. Kraft, G. Dehm, and E. Arzt, "Quasi-crystalline grain boundary phase in the magnesium die cast alloy ZA85", Scripta Materiala, Vo.45, 2001. pp.517-524; M. Vogel, O. Kraft, E. Arzt, E.D. Reese, and R. Rauh, "Creep of Mg-Zn-Al-Alloys", Magnesium Alloys and their Applications, Ed. K.U.Kainer, Wiley-VCH, Weinheim, Germany. 2000, pp 692-698
33. I.A. Anyanwu, T. Honda, S.Kamado, Y.Kojima, S.Takeda, and T.Ishida, "Heat and Corrosion Resistance of Mg-Zn-Al-Ca Alloys", Magnesium Alloys and their Applications, Ed. K.U. Kainer, Wiley-VCH, Weinheim, Germany, 2000, pp 110 – 115
34. P. Lyon, J.F. King, and K. Nuttall, "A new magnesium HPDC alloy for elevated temperature use", Proc. 3rd Intl. Magnesium Conf., Ed. G.W. Lorimer, The Institute of Materials, London, UK, 1996, pp.99-108
35. I.P. Moreno, K.Y. Sohn, J.W. Jones, and J.E. Allison, SAE Technical Paper 2001-01-0425, 2001
36. I.P. Moreno, T.K. Nandy, J.W. Jones, J.E. Allison, and T.M. Pollock, "Microstructure and creep behavior of a die cast magnesium-rare earth alloy", Magnesium Technology 2002, Ed. H.I.Kaplan, TMS-AIME, 2002
37. G.S. Foerster, "Designing die casting alloys", Proc. 7th SDCE, Paper no. 9372, 1973
38. Otto Fuchs, Norwegian Patent 99341, Filed 22-01-1959

39. K.E. Mann, "Über die Warmfestigkeit der Magnesiumlegierungen", Metall Vol.14, 1960, pp. 206-213
40. R.Ninomiya, T.Ojiro, and K.Kubota, "Improved heat resistance of Mg-Al alloys by the Ca addition", Acta Metall. Mater., Vol.43, No.2, 1995. pp.669-674
41. M. Pekguleryuz, and J. Renaud, "Creep resistance in Mg-Al-Ca casitng alloys", Magnesium Technology 2000, Eds.: H.I. Kaplan, J.N. Hryn, B.B. Clow, TMS-AIME, 2000. pp.279-284
42. F. von Buch, S. Schumann, E. Aghion, B. Bronfin, B.L. Mordike, and M. Bramberger, "Development of a low-cost, temperature- and creep-resistant Magnesium Die casting Alloy", Magnesium Alloys and their Applications, Ed. K.U.Kainer, Wiley-VCH, Weinheim, Germany, 2000. pp 23-26
43. M. Pekguleryuz, and A.A. Luo International Patent Application WO9625529, Filed 02-14-1996
44. A. Luo, and T. Shinoda, European Patent Application EP 0 879 898 A1, Filed 08-04-1998
45. B. Bronfin, E. Aghion, S. Schumann, P. Bohling, and K.U. Kainer, UK Patent application GB 2 340 129 A, Filed 30-07-1999; US patent Application 6 139 651, Filed 04-08-1999
46. H. Iba, and C. Maeda, European Patent Application 92112699.1, Filed 24-07-1992
47. K. Makino, T. Kawada, K. Kanemitsu, K. Watanabe, M. Matsunaga, and M. Sayashi, Japanese Patent Application JP7011374, Filed 28-06-1993; US Patent 5681403, 1997-10-28
48. T. Shintani, and H. Uchida, Japanese Patent Application JP8027533, Filed 12-07-1994
49. T. Horie, H. Iwabori, Y. Awano, Y. Suzuki, T. Takeuchi, H. Iba, and A. Matsui, Japanese Patent Application JP7018364, Filed 30-06-1993
50. K. Makino, K. Kanemitsu, N. Miyamoto, Norwegian Patent Application 930127, Filed 04-05-1993
51. K. Pettersen, P. Bakke, and D. Albright, "Magnesium Die Casting Alloy Design", Magnesium Technology 2002, Ed.: H.I. Kaplan, TMS-AIME, 2002
52. H.S. Tathgar, P. Bakke, and T.A. Engh, "Impurities in Magnesium and Magnesium Based Alloys and their removal", Magnesium Alloys and their Applications, Ed. K.U.Kainer, Wiley-VCH, Weinheim, Germany, 2000. pp 767-779
53. A. Beck, The Technology of Magnesium and its Alloys, F.A. Hughes & Co. Ltd, London UK, 1940. pp.250-252 *(Translation from German)*
54. A. Beck, Ibid. p.286
55. E.F. Emley, Principles of Magnesium Technology, Pergamon Press, London UK, 1966. pp.308-341
56. W. Unsworth, "The role of rare earth elements in the development of magnesium base alloys", Intl. J. of Materials and Product Technology, Vol.4, no.4, 1989, pp.359-378
57. A.P. Druschitz, E.R.Showalter, J.B.McNeill, and D.L. White, "Evaluation of structural and high-temperature magnesium alloys at room temperature", Magnesium Technology 2002, Ed. H.I.Kaplan, TMS-AIME, 2002
58. J.J. Berkmortel, H. Hu, J.E. Kearns, and J.E. Allison, "Die castability assessment of magnesium alloys for high temperature applications", SAE Technical paper 2000-01-1119, 2000
59. P. Bakke et al., To be published 2003
60. B. Powell, V Rezhets, M.P. Balogh, and R.A. Waldo, "The relationship between microstructure and creep behavior in AE42 magnesium die casting alloy", Magnesium Technology 2001, Ed.: J. Hryn, TMS-AIME, 2001. pp.175-181
61. P. Bakke, K. Pettersen, and H. Westengen, "Development of magnesium die casting alloys with emphasis on strength – ductility", Magnesium Technology 2003, Ed. H.I. Kaplan, TMS-AIME, 2003

2003-01-0190

Magnesium Die Casting Alloy AJ62x with Superior Creep Resistance, Ductility and Die Castability

Mihriban Pekguleryuz, Pierre Labelle and Donald Argo
Noranda Inc. - Technology

ABSTRACT

Magnesium die casting alloys for elevated temperature applications are coming of age. Several research centers and companies have been working on alloy systems based on alkaline earth and rare earth alloying additions to push the limits for the creep performance of Mg-based die casting alloys.

Noranda's Mg-Al-Sr based alloys have shown superior creep performance and high-temperature performance at temperatures as high as 150-175C and stress levels of 50MPa - 70MPa. The most recent alloy formulation AJ62x (Mg-6Al-2Sr) has in addition shown excellent castability, and superior hot-tear resistance. Based on these attributes AJ62x is positioned well for applications such as transmission cases and oil pans.

In this paper, the mechanical properties (creep and tensile) of AJ62x are presented. The high ductility of the AJ62Lx version is an added advantage for this alloy. Lab-scale evaluation of AJ62x shows that the alloy is more resistant to hot-tearing and cracking than all other magnesium alloys and the A380 aluminum alloy and industrial trials indicate that the alloy is highly castable.

INTRODUCTION

The mass of the vehicle has an effect on fuel consumption by affecting rolling resistance, resistance to climbing and acceleration. A fuel consumption target of 3 to 4 L /100Km is being pursued in various countries, in addition to raising customers expectation in terms of performance, comfort and safety. To achieve these diverse targets, the magnesium alloys with a density of 1.8 g/cm^3 are suitable materials. New part applications are primarily responsible for the increasing use of magnesium in the automotive industry. But the growth has been limited to components that used standard die cast magnesium alloys such AZ91D, AM50 and AM60B and are mainly located in interior applications. There are also distinct advantages to using light-weight magnesium in power-train components where large die cast components are used. However, the service temperatures of power-train components are in the 150°C range, which exceed the service temperature of standard magnesium die casting alloys, as previously mentioned. Several research centers and companies have been working on developing magnesium die casting alloys with improved high-temperature performance. Alloy systems based on alkaline earth and rare earth alloying additions have tried to push the limits for the creep performance of Mg-based die casting alloys. Several types of heat-resistant magnesium alloys have been developed to date, including AS, AE, Mg-Al-Ca, Mg-Al-Ca-RE, Mg-Al-Ca-Sr-RE and Mg-Al-Ca-Sr alloys. Unfortunately, many of these alloys suffer from inferior die-castability (Mg-Al-Ca) or are limited in terms of marginal performance improvement (AS, Mg-Al-Ca-RE, Mg-Al-Ca-Sr-RE) and high material cost (AE, Mg-Al-Ca-RE). Mg-Al-Sr alloys offer a good combination of die-castability, creep at 150°C and 175°C, and cost.

Noranda's Mg-Al-Sr based alloys are clearly positioned at the top performance level with superior creep resistance and a first generation alloy formulation Mg-5Al-2Sr (AJ52x) has also shown good castability when the correct process parameters are used. A new alloy formulation with Mg-6Al-2Sr (AJ62x) offers more advantages, that of

1. Superior creep performance
2. High ductility (AJ62Lx version)
3. Very good diecastability and hot-tear resistance
4. Excellent corrosion resistance.

In this paper, mechanical (creep, tensile) corrosion (salt-spray) properties and the castability of the alloy AJ62x will be discussed.

BACKGROUND

In selecting alloys for a particular application the user needs to consider different criteria such as performance, manufacturability, and cost. Clearly an engine block that sees service temperatures as high as 200 C needs an alloy that can offer superior creep performance, while an oil pan or an engine cradle may utilize a more moderate creep performance. Having an alloy that has the highest creep resistance possible at high temperatures and stress conditions as well as good-to-excellent tensile properties and ductility, castability, cost and recycling is the current challenge of the magnesium industry. Often we have alloys that have good castability but moderate creep resistance (1), or good creep resistance but very challenging castability (Mg-Al-Ca alloys). The user is now faced with electing a few candidate alloys to eventually develop a Mg-based power train component. Having a number of different alloys, however, will create the need for the establishment of alloy technology (database, recycling, casting) for many different alloy systems.

Noranda's Mg-Al-Sr alloys give the automotive user the versatility to provide alloy formulations that meet the demands of the various powertrain components from one single alloy system. The AJ52x (Mg-5Al-2Sr) formulation has creep resistance that is significantly better than all other Mg alloys at the 175 C at 50 MPa. Its castability has been industrially evaluated in powertrain components and has been found to be castable with simple change of die and melt temperature (2, 3).

Recently an AJ62x (Mg-6Al-2Sr) composition has also been tried and has been found to have the same superior creep performance as the AJ52x but with the added advantage of better castability. AJ62Lx version of this alloy gives excellent ductility as well. The composition ranges of these alloy formulations are given in Table I below. Noranda AJ62x (as well as AJ52x) are recommended for applications such as transmission cases and engine blocks that see high stresses at high service temperatures. On the other hand AJ62Lx is recommended for oil pan, engine cradle and other components requiring moderate creep performance. Alloys still under development such as AJ7xx, AJ8xx compositions will target components requiring slight improvement in creep resistance over the AM and AZ alloys.

TABLE I. AJ ALLOY COMPOSITIONS

ALLOY	Al%	Sr	Comments
AJ62x	5.5 - 6.5	2.1-2.6	Superior creep resistance, and good castability, excellent corrosion resistance
AJ62Lx	5.5 - 6.5	1.6-2.1	High creep resistance, good castability, high ductility, excellent corrosion resistance
AJ52x	4.5 - 5.5	1.6-2.3	Superior creep resistance, acceptable castability, excellent corrosion resistance (first formulation)
AJ7/8xx	x	x	Target for moderate creep resistance and exceptional castability, high ductility and strength (future formulation),

In order to rationalize the alloy use and minimize the number of alloys Noranda's aim is to commercialize only two of the various developmental formulations. These currently are the AJ52x and AJ62x (both of superior creep performance). Among options is also AJ62Lx (high creep performance and high ductility). An AJx (7-8%Al) is also under evaluation to target the applications needing only a small improvement in creep performance. This way one alloy system (Mg-Al-Sr) can meet the demands of the whole power train. This would be similar to the use of the AM alloys based on varying ductility requirements of the applications. In this paper the properties of the AJ62x and AJ62Lx alloys will be presented.

EXPERIMENTAL PROCEDURES

Mechanical and Corrosion Test Specimens

All die-cast test specimens required for the mechanical property determination were cold-chamber die cast. AJ62Lx and AJ62x alloys were die cast at Lunt manufacturing Co. and Meridian Technologies Inc. AJ52x and all other magnesium alloys and the aluminum alloy A380 were cold-chamber die cast at Performag (Montreal).

Mechanical and Corrosion Testing

Creep testing and tensile testing were carried out at Westmoreland and Noranda according to ASTM B557 and ASTM E8. Salt-spray corrosion test was conducted at Noranda Technology Center according to ASTM B117.

Hot-Tear Resistance Testing

Hot-tear resistance testing was carried out at Noranda according to a test method developed by Noranda Technology Centre (4, 5) that is a modification of a method used for aluminum alloys (6). The test involves the casting of the alloys into a mold that has cavities to produce four constrained rods of varying lengths (Fig. 1). Cracks are induced due to the constraints of the rod ends during solidification. A rating system based on the severity and the location of the crack determines the hot-tear susceptibility of the alloy. For the tests, all alloys were cast with a 100°C superheat above their liquidus temperatures.

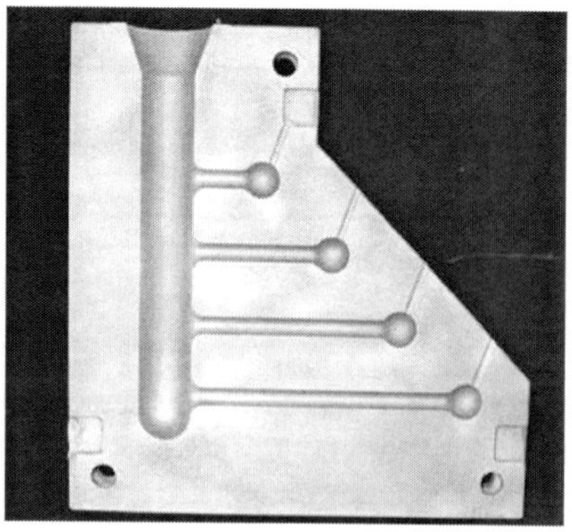

Figure 1. Hot-tear resistance-test mold

Industrial Casting trials

Industrial casting trials were carried out at various die casters in North America and Europe. The die-castability was evaluated qualitatively and compared to other alloys cast with the same machines and dies.

PROPERTIES OF AJ ALLOYS

Creep

The creep resistance of the various alloys is shown in Table II. Tensile creep and compressive creep rankings of the alloys show different trends. At 150 C and a 50 MPa stress the tensile creep deformation after 200 hours can be quite comparable in die cast AJ, AE, AS and even the aluminum 380 alloy. At 175 C and 50 MPa the differences in creep resistance of the alloys become more clear. AJ52x and AJ62x alloys are close contenders to the aluminum A380 with about 0.1% creep deformation while the AE42 alloy shows four times more creep deformation. Tensile creep data at 70 MPa/175C have been presented by Showalter et al (1) and is also provided in Table II for comparison with other alloys.

TABLE II. TENSILE CREEP OF DIECAST ALLOYS

ALLOY	% Creep @ 50 MPa, 200 hrs		% Creep @ 50 MPa, 500 hrs		% Creep @ 70 MPa,200 hrs (1)
	150 C	175 C	150 C	175 C	175 C
AJ52x	0.04	0.05	0.03	0.09	0.14
AJ62x	0.05	0.05	-	-	-
AJ62Lx	0.08	0.29	-	-	-
AE42	0.06	0.33	0.08	0.44	0.18
AS41	0.05	2.48	0.07	-	-
AS21x	0.19	1.27	-	-	8.95
MR153	-	-	-	-	4.01
AZ91D	2.7	*	6.35	-	-
A380	0.08	0.04	0.10	0.05	0.22**

*failed after 80 hrs **A383 alloy

The compressive creep mode, which is more representative than tensile creep, of the stress conditions in bolted assemblies of power train components, aluminum 380 alloy has the highest resistance to deformation with only 0.03%. AJ alloys outperform all other magnesium alloys such as AE42, AS41, AS21x and the AZ91D. Clearly AJ alloys AJ62x and AJ52x demonstrate significantly better creep performance than all the other magnesium die casting alloys. Bolt-load retention data on alloys presented by Showalter et al (1) is also given for comparison in the table.

TABLE III COMPRESSIVE CREEP IN DIECAST ALLOYS

ALLOY	% Compressive Creep @ 150 C, 70 MPa, 200 hrs	% Bolt-Load Retention 70 MPa, 100 hours (Ref 1)	
		150 C	175 C
AJ52x	0.24	73%	49%
AJ62x	1.73	-	-
AJ62Lx	-	-	-
AE42	2.16	65%	50%
AS41	6.13	-	-
AS21x	3.97	55%	29%
MR153	-	42%	11%
AZ91D	21	-	-
A380	0.03	-	81% *

*A383 alloy

Tensile Properties

Room temperature and high temperature tensile properties of the AJ alloys are given in Table IV. All AJ alloys have UTS and YS that are higher than all other alloys at all temperatures. The MR153 alloy has slightly higher yield strength at the higher temperatures. The AJ62Lx has the UTS and the elongation. The versatility of the AJ alloys is seen in this trend as well. The same alloy system with a slight modification in composition is able to provide a high ductility alloy in addition to improved creep performance.

TABLE IV. TENSILE PROPERTIES at 20 C

Tensile Properties at 20C			
ALLOY	**UTS, MPa**	**YS, MPa**	**%E**
AJ62Lx	**270**	**147**	**11**
AJ62x	**239**	**142**	**8**
AJ52x	212	134	6
AE42	220 (230)	128 (145)	10 (11)
AS41	197 (215)	136 (140)	4 (6)
AS21x*	210	121	5.5
AZ91D	204 (240)	139 (160)	3 (3)
MR153*	197 (250)	157 (165)	2.2 (5)
A380	290 (290)	155 (180)	3.2 (3)
Tensile Properties at 150 C			
ALLOY	**UTS, MPa**	**YS, MPa**	**%E**
AJ62Lx	**166**	**116**	**27**
AJ62x	**163**	**108**	**19**
AJ52x	**163**	**110**	**12**
AE42	140 (157)	88 (100)	23 (20)
AS41	153 (155)	94 (95)	17 (24)
AS21x	130	87	20
AZ91D	169 (157)	105 (108)	16 (23)
A380	251 (257)	148 (159)	6.4 (7)
Tensile Properties at 175 C			
ALLOY	**UTS, MPa**	**YS, MPa**	**%E**
AJ62Lx	**142**	**109**	**27**
AJ62x	**143**	**103**	**19**
AJ52x	**141**	**100**	**18**
AE42	121 (130)	81 (91)	23 (23)
AS41	127 (127)	85 (85)	19 (25)
AS21x	110	78	23
AZ91D	138	89	21
MR153*	139	113 (110)	3.4
A380	248 (228)	154 (148)	7.1 (10)

* data from Ref 1 **A383 values in brackets are IMA, AA or supplier

Corrosion

Salt-spray corrosion resistance of the alloys is given in Table V. AJ alloys have exceptional corrosion resistance. Tolerance levels for the Fe, Ni, Cu contaminants of the AJ alloys are similar to AM alloys as shown in Table V.

TABLE V. CORROSION RESISTANCE

ALLOY	Salt-Spray Corrosion Rate (mg/cm²/day)
AJ62Lx	0.04
AJ62x	-
AJ52x	0.09
AE42	0.21
AS41	0.16
AZ91D	0.10
A380	0.34

CASTABILITY

Castability is influenced by a number of factors that are either alloy dependent or process dependent. Alloy dependent parameters include feeding characteristics (i.e. fluidity), susceptibility to hot-tearing and cracking, freezing range, shrinkage and thermal properties (heat capacity, thermal conductivity) while process dependent factors are comprised of casting parameters and die design considerations. A castability index is under development by Noranda for assessing "a priori" some of the alloy dependent factors and is discussed elsewhere (4,5). In this paper hot-tear resistance of the alloys is discussed.

Hot-Tear Resistance

The hot-tear resistance of the magnesium alloys evaluated is shown in Table VI. The highest hot-tear susceptibility is observed in the AE42 alloy and the lowest in the AJ62x alloys. Fig. 2 shows a hot-tear test casting made with an AJ62x alloy and Fig.3 shows hot-tears produced in an AS41 alloy casting.

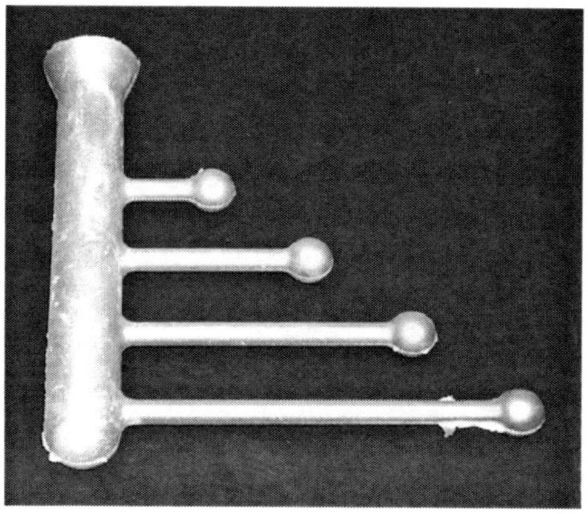

Figure 2. Casting produced with an alloy having high resistance hot-tearing,

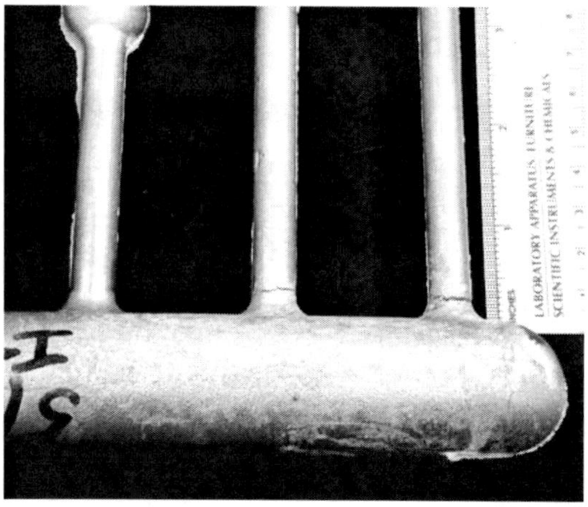

Figure 3. Tears in a casting produced with an alloy having low resistance to hot-tearing.

Table VI. Hot-Tear Ratings of Various Alloys

Alloy	Average Hot Tear Rating*
A380	2
AZ91D	20
AZ91E	14
AM50A	44
ITM alloy (0.8%Ca)	61
AJ50x (0.2%Ca, 0.5%Sr)	45-60
AS21	69
AS21x	27
AS41B	82
AE42	94
AJ51x	35-50
AJ52x	24
AJ53x	5
AJ62x	0
AJ62Lx	0

*A lower number means less susceptibility to hot tearing

Industrial Die Casting Trials

Several die casting component trials have been conducted in North America and Europe with the AJ alloys. The first generation AJ52x alloys were die cast as valve covers, an oil pan and other automotive components weighing from 2 to 12kg. The valve cover in Figure 4a and the oil pan in Figure 4b were die cast at Spartan Light Metal Products using a cold chamber machine.

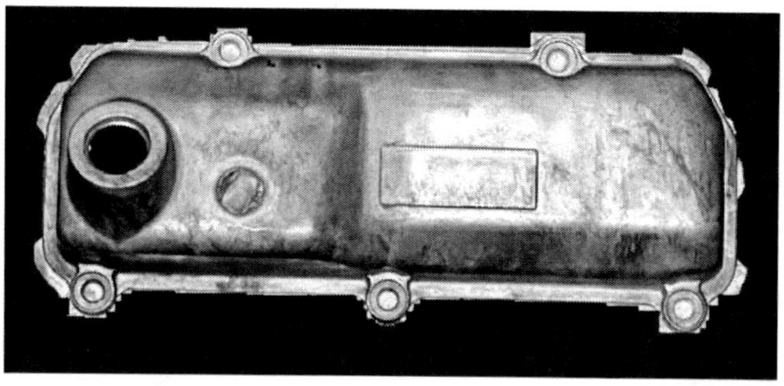

Figure 4a. Valve-cover cast in AJ52x at Spartan Light Metal Products with very good results.

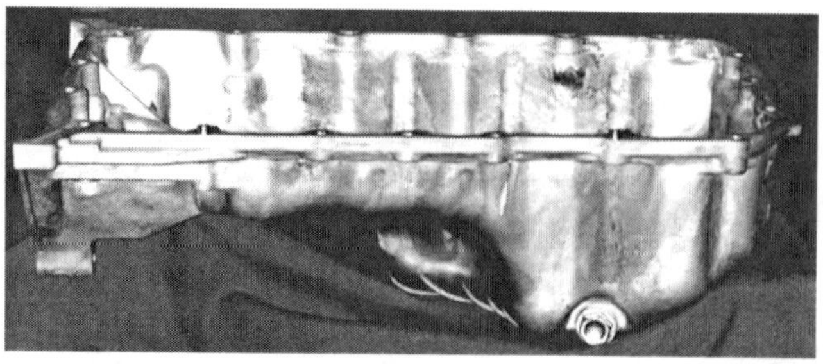

Figure 4b. Oil pan cast in AJ52x at Spartan Light Metal Products with very good results.

A medium-walled powertrain component was diecast in Europe using AJ52x. After some optimization of the diecasting process (see table VII), excellent casting soundness (no tears, no feeding problems or internal shrinkage) was obtained and confirmed through fluoroscopic analysis of 50 consecutively cast parts. For confidentiality reasons the part cannot be identified.

Differences in the castability of AJ52x and AJ62x were evaluated at Lunt Manufacturing. The die cast part, called a "electronics cover", used for the testing is a thin-walled casting produced in an unheated die. The unheated die gave an opportunity to compare castability under the adverse condition of a cold die surface. Because of the cold die surface the AJ52x experienced certain difficulties that could have been remedied when it were possible to heat the die. The main difficulties were die sticking and cracking. The AJ62Lx alloy, however, demonstrated very good castability using similar process parameters.

Figure 5. Electronics cover cast with AJ62Lx at Lunt Manufacturing.

AJ62Lx and AJ62x alloy die casting trials were also carried out on powertrain components in North America and Europe. Very good castability was observed in these alloys with no die-sticking or cracking tendencies, good fluidity and hence good die filling. (Table. VIII). Recommended casting parameters for AJ62x alloys are given in Table VII. Many power-train development trials are currently underway with the AJ alloys. Among those that can be mentioned are the EUCAR Mg-Engine and the USCAR power-train programs.

TABLE VII. DIECASTING GUIDELINES FOR AJ ALLOYS

PARAMETER	RECOMMENDED
Die Surface temperature	**AJ52x** : 300-350 °C **AJ62x** : 250-300 °C
Metal temp.	710-720°C
Gate area	30-50% greater than Al
Gate velocity	Low end of North American Die Casting Association (NADCA) recommendations
Fill time	Low end of NADCA recommendations
Cycle time	Short to maintain die temperature
Shot sleeve	Heated/high fill percentage, short dwell time before injection
Draft	Generous where possible
Ejection	As early as possible to allow free contraction

TABLE VIII. CASTABILITY RATING OF AJ ALLOYS

ALLOY	Die Sticking	Cracking	Fluidity	Overall Rating
AJ51x	low	high	moderate	fair
AJ52x	moderate	moderate	good	Good. If recommended parameters are used very good castability can be obtained.
AJ62Lx	Very low	Very low	Similar to AM60	Very good
AJ62x	Very low	Very low	Similar to AM60	Very good

CONCLUSIONS

1. AJ alloys AJ52x and AJ62x have superior creep performance surpassing all other magnesium alloys at high temperatures (150-175 C) and high stresses (50-70 MPa).
2. AJ62x alloys have the highest tensile properties of all magnesium alloys tested (both room temperature and elevated temperature). AJ62Lx alloy has very substantial ductility.
3. Corrosion resistance of AJ alloys is excellent. The alloys show salt-spay corrosion resistance equal to or better than AZ91D.
4. AJ62x alloys have exceptional resistance to hot-tearing and cracking. The hot-tear tendency is even lower than that of aluminum A380 alloy.
5. The diecastability of AJ62x alloys is very good, and superior to AJ52x alloy.
6. AJ62x alloys are positioned extremely well for use in automotive power train components.

REFERENCES

1. A. P. Druschitz, E. R. Showalter, J.B. McNeill, D. L. White, "Evaluation of Structural and High-temperature Magnesium Alloys", **SAE Paper 2002-01-0080.**
2. P. Labelle, M. Pekguleryuz, D. Ago, M. Dierks, T. Sparks, T. Waltamatte, "Heat Resistant Magnesium Alloys for Power-Train Applications" **SAE Technical Paper 2001-01-0424.**
3. Mihriban O. Pekguleryuz, Pierre Labelle, Don Argo. Mike Dierks, Todd Sparks, Ted Waltematte, "Creep Resistant Magnesium Alloys for Power-Train Applications," **Proc. 57th Ann. World Magnesium Conf**., IMA, Brussels, May, 2001, pp. 27-33.
4. D. Argo, M. Pekguleryuz, P. Labelle, P. Vermette, R. Bouchard and M. Lefebvre "Process Parameters And Diecasting of Noranda's Aj52 High Temperature Mg-Al-Sr Alloy," **2002 Magnesium Technology**, H. Kaplan, Ed., TMS, Seattle, Feb 2002 pp. 87-93.
5. M. O. Pekguleryuz, D. Argo, Michel Lefebvre, Pierre Vermette, "Diecastability And Hot-Tear Susceptibility Of Mg-Al-Sr Based Alloys," Aalen Conference, 9th Proc. of Automotrive End User Seminar, FH Aalen, Germany, Dec 2001 (CD ROM)
6. S. Lin, C. Aliravci and M. Pekgüleryüz "A Study of Hot-Tearing Tendency in Aluminum Alloys," **Proc. Light Metals, Proc Int.Symp**, C. M. Bickert and R.I.L. Guthrie, Eds., Canadian Inst. Mining & Metallurgy (CIM), pp. 451-461,1997.

2003-01-0191

A Comparative Study of New Magnesium Alloys Developed for Elevated Temperature Applications in Automotive Industry

E. Aghion and B. Bronfin
Magnesium Research Institute, Dead Sea Magnesium

F. Von Buch, S. Schumann and H. Friedrich
Volkswagen AG

ABSTRACT

Recently several new magnesium alloys for high temperature applications have been developed with the aim to obtain an optimal combination of die castability, creep resistance, mechanical properties, corrosion performance and affordable cost. Unfortunately, it is very difficult to achieve an adequate combination of properties and in fact, most of the new alloys can only partially meet the required performance and cost.
This paper aims at evaluating the current status of the newly developed alloys for powertrain applications. The paper also addresses the complexity of magnesium alloy development and illustrates the effect of alloying elements on properties and cost. In addition, the paper presents an attempt to set the position of each alloy in the integrated space of combined properties and cost

INTRODUCTION

Magnesium alloys are the lightest structural materials and hence are very suitable for car industry where vehicle weight and consequently fuel economy are becoming the most important factor.
Magnesium alloys are usually produced by various casting processes. The most applicable methods are high-pressure die casting and gravity casting; particularly sand and permanent mold casting.

A. DIE CASTING ALLOYS

To date die casting is the dominant form of casting providing high productivity at relatively low cost. It is widely accepted by general consensus that commercial magnesium alloys AZ91D, AM60B and AM50A offer a good balance between key properties such as:

- Good combination of mechanical and physical properties at ambient temperature.
- Excellent die castability.
- Good corrosion resistance.
- Ease of handling.
- Relatively low cost.

The last factor is very important for comparison with aluminum alloys when magnesium is considered as substitution for aluminum.
Although the above mentioned alloys of AZ and AM series account for the major consumption of magnesium die casting alloys, there is an important and material consumable area where improved elevated temperature performance is required.
This area includes drive train components such as gearbox housing, intake manifold, oil pan, transfer case, crankcase, oil pump housing and others. These parts are operating at various elevated temperatures and therefore improved creep resistance and bolt load retention properties are critical issues for the alloy to be used for production of such components.
The insufficient creep strength of alloys used for manufacturing various housings can result in a clamping load reduction of bolted joints that causes poor bearing-housing contact, leading to an oil leak and increased noise and vibration.
Unfortunately, existing AZ and AM series alloys are not suitable for manufacturing drive train parts operating at temperatures higher than 130°C. In addition, the creep resistant alloys developed in the 70's (AS21 and AS41) and in the late 80's (AE42) also did not find their place in power train applications due to different reasons (poor castability and corrosion resistance, increased cost, low strength, etc.).
It is evident that the most of drive train components for regular cars have to be produced by high-pressure die casting – a high volume production technique. For this method, alloy cost represents a significant proportion of the component cost. Hence, alloy price becomes a major restriction in the alloy development. It means that one of principal requirements to new alloys is their price competitive ability with existing magnesium and aluminum commercial alloys. This requirement

combined with die castability issues reduces possible options to alloy systems containing Al or Zn as major alloying elements, with using Mn, Si, Ca, Sr and Ce-based mishmetal as relatively small additions.

The increased activities in the development of creep resistant die casting alloys have been indicated during last decade [1-7].

Hydro Magnesium [4] announced development of a creep resistant automotive die cast magnesium alloy named AS21X and based on veteran AS21 alloy but with improved corrosion resistance due to small addition of rare earth mishmetal combined with reduced level of Mn. However, it seems that other disadvantages of AS21 alloy such as low yield strength and inadequate castability for complex parts remained unchanged also for the new alloy.

Recently Noranda Inc. [6] developed a new alloy based on AM50 alloy with addition of 1.7-2% Sr. It was shown that this alloy named AJ52X has good creep resistance in the temperature range of 150°C - 175°C. However, in order to be die castable, the alloy requires very specific casting conditions such as casting temperature of 720°C and die surface temperature of 300-350°C that makes very difficult to die cast AJ52X using the existing equipment. Even the above conditions are fulfilled, nevertheless it is very difficult to cast AJ52X alloy into thin wall component (with less than 2 mm wall thickness) [9]. In addition, it should be pointed out that the price of Sr containing master alloys is considerable higher than that of Ca containing master alloys and Ce-based mishmetal. This fact also could be certainly an obstacle for mass production of alloys containing high Sr concentrations.

Another approach to new creep resistance alloy development was demonstrated by Honda R&D center [3] and General Motors R&D center [5].

The alloy ACM522 developed by Honda is based on AM50 alloy with additions of 2.5% Ce-based mish metal and 2% Ca. This alloy is already in small-scale industrial use for production of oil pan for Honda's 1-liter engine of the "Insight" hybrid car with fuel consumption of 35 km per liter. It was demonstrated that this alloy exhibit creep strength in the range of 150-200°C similar to that of aluminum alloy A384. However, ACM522 alloy, in addition to rather high cost, has very low ductility (2-3%) and impact strength (4-5 J) that can limit its potential application. Furthermore, the presence of 2% Ca in this alloy can lead to hot cracking problems in components with variable wall thickness and more complicated shape than specially designed oil pan with wall thickness in the range of 2.5-3 mm.

The new AXJ alloy developed by General Motors R&D Center is based on AM50 alloy with addition of 1.7 to 3.3% Ca and up to 0.2% Sr. This alloy exhibits excellent creep resistance. However, the alloy castability is very sensitive to Ca content. In general, the alloy castability and hot cracking susceptibility was regarded as lower than that of AM50A alloy [10].

Magnesium Electron [2], with long experience in developing aluminum free gravity casting alloys decided to move away from Mg-Al system also in the development of a new alloy for die casting applications. The alloy named MEZ includes 2.5% RE, 0.35% Zn and 0.3% Mn with small addition of Zr or Ca. This alloy exhibits creep resistance and bolt load retention better than AE42 at temperatures within the range of 150-175°C. However, MEZ alloy has very low ductility (1-2%) and low tensile yield strength (95-110 MPa) at room temperature. Other limitations of MEZ alloy relate to castability and molten metal handling problems as well as relatively high cost.

B. GRAVITY CASTING ALLOYS

Although die casting is the dominant process used for manufacturing magnesium alloy components, the low density and other advantages of magnesium alloys are especially important for large and heavy parts produced by sand or permanent mold casting and mostly operating at elevated temperatures [11].

All high-temperature resistant, gravity casting magnesium alloys are aluminum free and contain Zirconium, which is a very powerful magnesium grain refiner [12]. However, the use of Zirconium for refining requires saturation of the molten metal with soluble Zr at high temperature range of 770-800°C, which leads to oxidation and subsequently loss of magnesium and expensive elements such as Y, Nd, Gd, Dy, Yb etc. In addition, the existing alloy preparation technology can be carried out only in small crucibles up to 300kg. The process efficiency is relatively low due to the need to retain a considerable heel in the crucible. This allows preventing entrapment of settled inclusions and excess Zr compounds in the castings.

All of these measures as well as the use of expensive alloying elements result in relatively high cost of such alloys both in ingot form and in final castings. Nevertheless, this situation can be accepted in the production of high quality complicated castings for aerospace industry, and luxury and racing cars, where performance is more important factor than cost.

In contrast to the increased activities in the field of developing new creep resistant die casting alloys, no new gravity casting alloy emerged since 80's when WE43 and WE54 alloys were developed by Magnesium Electron. These alloys combine both excellent creep and corrosion resistance and are considered as the state-of-the-art for high temperature Mg alloys. However, extremely high cost of WE alloys restricts their use only for very specific applications.

Currently the largest portion of gravity casting applications is covered by ZE41 alloy, which has moderate strength and creep resistance, combined with good castability. Although this alloy exhibits very poor corrosion resistance it is still most widely used as preferred material in various helicopter transmission applications.

The present paper will highlight the results achieved by the Magnesium Research Institute (MRI) of Dead Sea Magnesium (DSM) and Volkswagen AG in the development of new die cast alloys and gravity casting alloys. The DSM-VW joint research activities are of prime interest as they synergistically combine the physical metallurgy capabilities of DSM with the casting process capabilities and end user requirements of VW, as a car manufacture company

DSM-VW NEWLY DEVELOPED MAGNESIUM DIE CASTING ALLOYS

As was indicated in the introduction, the existing alloys do not exhibit an adequate combination of castability, creep performance, corrosion resistance, ambient strength and reasonable cost in order to meet requirements, which are necessary for manufacturing power train components.
In order to address these specific issues DSM and Volkswagen AG together with other partners from the academic world started a comprehensive alloy development program aiming at developing creep resistant and cost effective die casting magnesium alloys. As an outcome of this work two new Mg alloys namely MRI 153M and MRI 230D have been developed. MRI 153M is a Beryllium free, low cost creep resistant alloy with the capability of long operation at temperatures up to 150°C under high stresses. MRI 153M alloy exhibits die castability similar to that of AZ91D alloy. This was demonstrated and confirmed by die casting trials of complicated components such as transmission housings for VW/Audi vehicles, for example Polo, Golf, Passat, Audi 3, Audi TT and Audi 6 [1,7,8].

Table 1. Mechanical properties of MRI 153M & MRI 230D

Properties	MRI 153M	MRI 230D	AZ91D	AE42	AS21
Tensile Yield Strength (MPa)					
At 20 °C	170	180	160	135	125
At 150 °C	135	150	105	100	87
Ultimate Tensile Strength (MPa)					
At 20 °C	250	235	260	240	230
At 150 °C	190	205	160	160	120
Elongation (%)					
At 20 °C	6	5	6	12	16
At 150 °C	17	16	18	22	27
Compression Yield Strength (MPa)					
At 20 °C	170	180	160	115	110
At 150 °C	135	150	105	85	80
Impact strength (J)	8	6	8	12	14
Fatigue strength (MPa) (1)	120	110	100	80	75
Corrosion Rate (2) (mg/cm2/day)	0.09	0.10	0.11	0.12	0.34

(1) Rotating bending 5x 10^7 Cycles
(2) 200 hr. Salt Spray Test (ASTM Standard B-117)

Various properties of MRI 153M alloy in comparison with commercial alloys AZ91D, AE42 and AS21 are summarized in Table 1 and Figures 1-3. It is evident that MRI 153M has corrosion resistance and mechanical properties at room temperature similar or better than AZ91D and other commercial alloys.

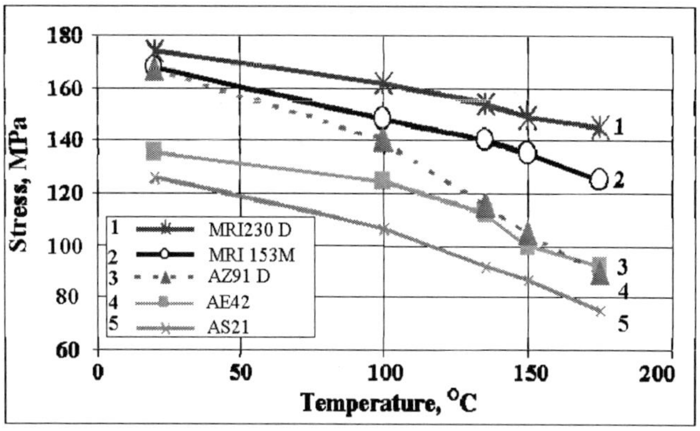

Fig. 1. Tensile yield strength vs. temperature for MRI 153M & MRI 230D in comparison with conventional alloys

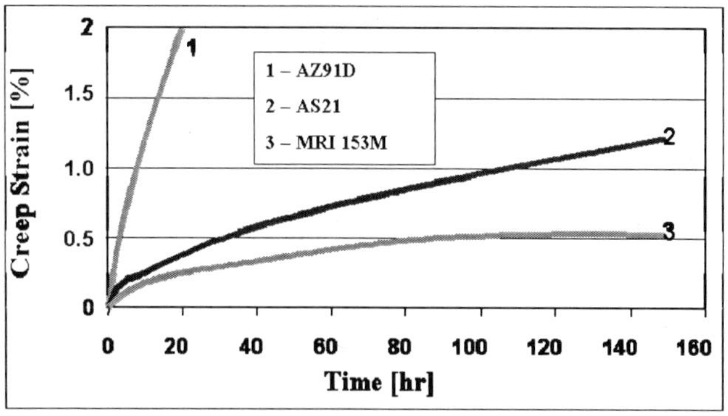

Fig. 2. Creep resistance of MRI 153M in comparison with conventional alloys at 135°C under stress of 85 MPa

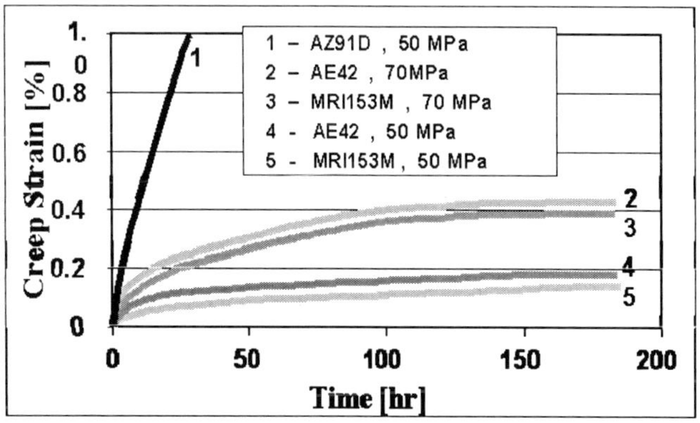

Fig. 3. Creep resistance of MRI 153M at 150°C in comparison with conventional alloys.

The elevated temperature strength and creep resistance of MRI 153M, at 130-150°C under stresses of 50-85 MPa, are also substantially superior to these commercial alloys. This can make MRI 153M a superior choice for automotive applications such as transmission housings, oil pans, valve covers, intake manifolds, etc.

MRI 230D is a die casting alloy, which was developed to address automotive engine blocks operating at higher temperatures up to 190°C. The alloy has excellent creep resistance combined with good castability, high strength and superior corrosion behavior.

The properties of MRI 230D alloy are summarized in Table 1 and in Figures 1, 4 and 5 in comparison with commercial magnesium alloys and aluminum alloy A380. The results obtained show that MRI 230D and A380 exhibit similar tensile creep behavior at 150-175°C under stress of 70 MPa.

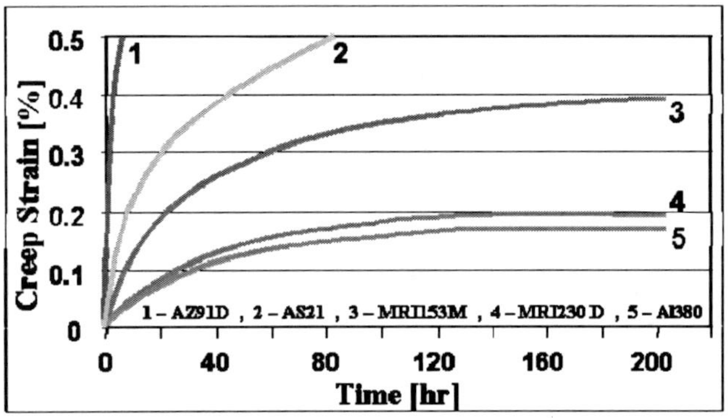

Fig. 4. Creep resistance of MRI 230D and MRI 153M alloys in comparison with commercial magnesium and aluminum alloys at 150 °C / 70 MPa

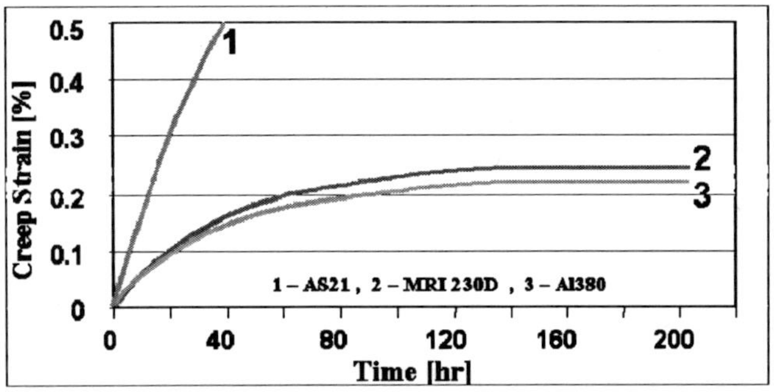

Fig. 5. Creep resistance of MRI alloys in comparison with conventional alloys at 175 °C / 70 MPa

DSM-VW NEW MAGNESIUM ALLOYS FOR GRAVITY CASTING

As indicated in the introduction, two gravity casting alloys ZE41 and WE43 are being mostly used to date for different applications. However, there is a big gap between high performance and extremely high cost of WE43 alloy and moderate cost but limited creep resistance and corrosion properties of ZE41 alloy.

In order to address this gap, two new alloys designated MRI 201S and MRI 202S were developed by DSM and VW. The alloy MRI 201S in the T-6 conditions exhibits excellent mechanical properties and creep resistance combined with outstanding castability, pressure tightness, weldability, dimension stability and corrosion resistance. Comparison data presented in Table 2 shows that in accordance with its performance the alloy MRI 201S can be considered as a very cost effective substitution for WE43 based on it's alloy composition and production procedure.

The alloy MRI 202S is high purity corrosion resistant gravity casting alloy, with moderate strength and excellent creep performance (Table 2 and Figure. 6). Creep and corrosion performance of this alloy are significantly superior to those of ZE41 alloy.

In general, it is believed that various specific automotive and even aerospace components such as large engine block, cylinder head covers, helicopter transmission housings, etc. can benefit from the MRI 201S and MRI 202S performance.

Table 2. Mechanical properties of MRI 201S & MRI 202S alloys

Properties	MRI 201S-T6	MRI 202S-T6	WE43-T6	ZE41-T5
Tensile Yield Strength (MPa)				
at 20 °C	170	150	180	140
at 150 °C	170	145	175	120
Ultimate Tensile Strength (MPa)				
at 20 °C	260	250	260	220
at 150 °C	245	220	210	170
Elongation (%)				
at 20 °C	6	7	6	5
at 150 °C	11	15	7	22
Compression Yield Strength (MPa)				
at 20 °C	190	145	190	140
at 150 °C	190	140	185	115
Fatigue strength (MPa) (1) at 20 °C	110	85	100	95
Corrosion Rate (2) (mg/cm2/day)	0.10	0.12	0.10	3.1
Stress [MPa] to produce 0.2% creep strain				
at 200 °C	160	100	160	50
at 250 °C	75	40	60	20

(1) Rotating bending 5x 10^7 Cycles
(2) 200 hr. Salt Spray Test (ASTM Standard B-117)

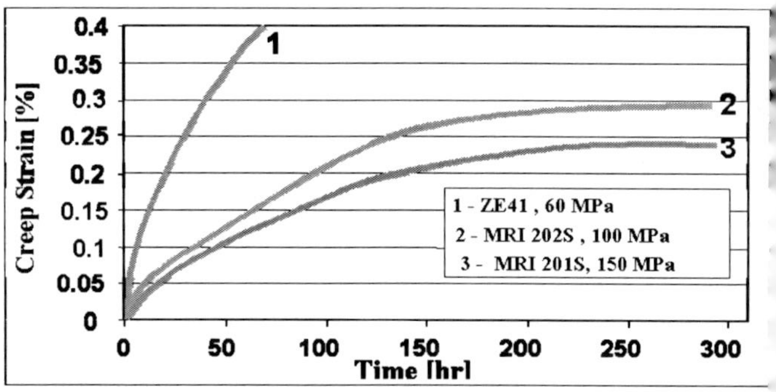

Fig. 6. Creep resistance of MRI 201S & MRI 202S in comparison with ZE41 alloy at 200°C.

DISCUSSION

Most of the above reviewed die cast Mg alloys contain various concentrations of Al, Ca, Sr, and rare earth (RE) mishmetal. These alloying elements were used to improve the creep resistance while Al was kept to provide adequate castability. Of course the final cost of the new alloy certainly depends on the concentration of the alloying elements and their price. In fact, it was clearly evident that the use of different concentrations of the above alloying elements resulted in positive modification of some properties while other properties including cost were jeopardized.

This statement can be schematically illustrated by two dimension diagrams showing the effect of Al and alloying elements on castability, properties and cost of magnesium based die casting alloys (Figure 7).

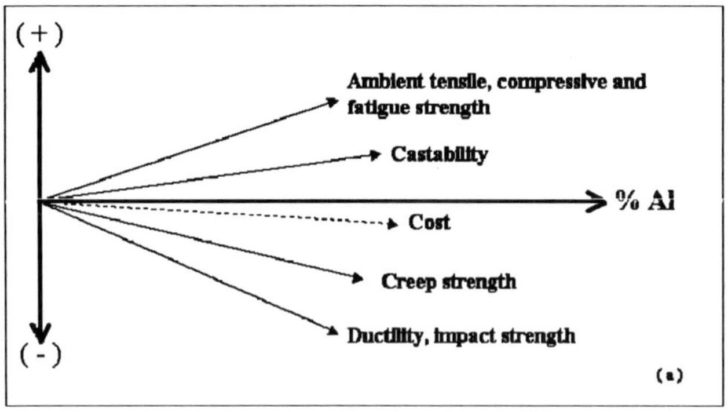

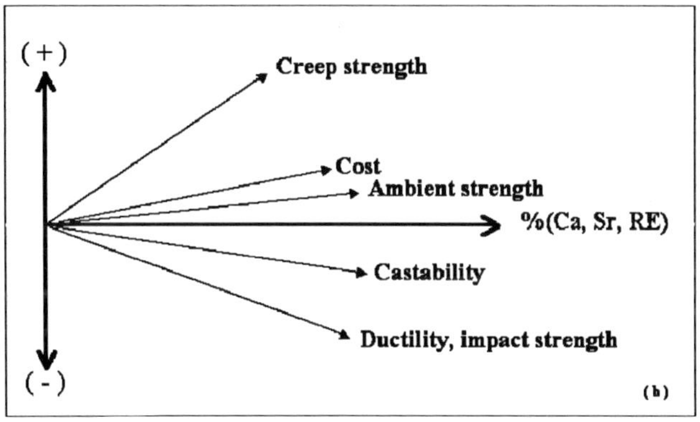

Fig. 7. Schematic illustration of effect of Al (a) and alloying elements (b) on properties and cost of magnesium die casting alloys.

It is well known and documented that increasing Al content in magnesium alloys results in better castability and an increase in room temperature tensile, compressive and fatigue strength as well as improving the corrosion resistance. On the other hand, creep properties, ductility and impact strength are significantly deteriorated (Figure7a). Alloying with Al practically does not affect an alloy cost. In contrast to this, alloying with alkaline earth elements (Ca, Sr) and rare earth mishmetal (RE) leads to deterioration of castabillity, ductility and impact strength as well as increasing cost. The benefit of this alloying is mainly associated with significant improvement of creep resistance followed by moderate increase of room temperature strength (Figure 7b).

Alloying with Ca provides maximum increase of both ambient tensile and compressive yield strength, and especially creep strength for equal weight unit compared to Sr and RE mishmetal. As Calcium is a relatively inexpensive element and has atomic weight which is half than that of Strontium and less than third than the atomic weight of RE elements, it means that alloying with Ca results in maximum increase of the above properties for each dollar of cost increase.

On the other hand, it also evident that alloying with Ca in concentration more than 0.3% causes castability and ductility deterioration. Hence, it is a challenge for alloy developers to select optimum combination of alloying ingredients (Al, Ca, Sr, RE) to provide the best service properties combined with affordable cost.

In principle, if one would like to make a short list of the main properties to be addressed for high temperature applications, this may include creep resistance, castability and cost. This can be illustrated by a two dimensional diagram where creep resistance and castability form the XY axes and cost is represented by the relative size of a circle (small circle for low cost and large circle for increased cost). Using this diagram one can display any new alloy in the combined space of properties. As an example Figure 8 shows such diagram of combined properties with the positioning of each of the reviewed new alloys. It is clearly evident that each alloy presents a relatively different space of properties

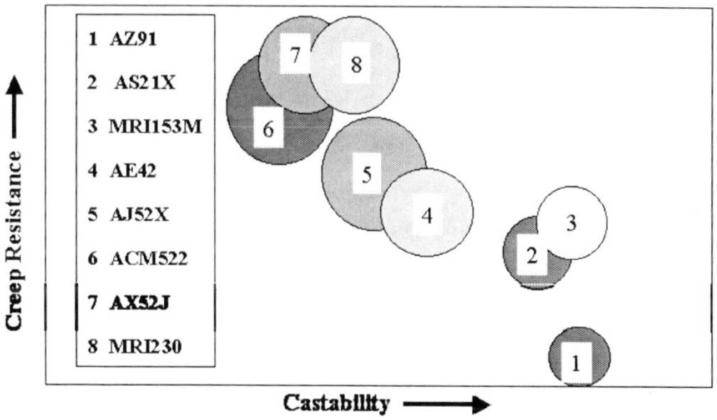

Fig. 8. Creep resistance vs. castability and cost (represented by relative size of circle, which increases with higher cost) of reviewed Mg alloys.

Although Figure 8 illustrates a general comparison of the short list of properties (i.e. creep, castability and cost) this is not enough for the design engineer to make a practical decision on alloy selection. Hence, additional comprehensive information is required with which the designer can tailor his requirements according to specific alloy properties and performance.

Such information is illustrated in Tables 3 and 4 and Figures 9 and 10.

Table 3. Combined properties rating of new Mg alloys for applications at 150 °C

		Castability	Creep Resistance	Tensile Yield Strength at 150 C	Fatigue Strength	Ductility	Corrosion Resistance	Total Score	% of Max Score
Alloy	Weight	10	10	7	7	3	7		
MRI -153 M	Score	9	8	8	10	5	10		
	W x S	90	80	56	70	15	70	381	100 %
MRI -230 D	Score	6	10	10	9	3	8		
	W x S	60	100	70	63	9	56	358	94 %
AX 52 J	Score	5	10	10	8	3	8		
	W x S	50	100	70	56	9	56	341	90 %
ACM 522	Score	4	10	9	8	3	9		
	W x S	40	100	63	56	9	63	331	87 %
AJ 52 X	Score	6	9	6	6	5	8		
	W x S	60	90	42	42	15	56	305	80 %
AS 21 X	Score	9	7	3	5	10	8		
	W x S	90	70	21	35	30	56	302	79 %
AE 42	Score	7	8	4	5	9	8		
	W x S	70	80	28	35	27	56	296	78 %
AZ 91 D	Score	10	1	5	8	5	9		
	W x S	100	10	35	56	15	63	279	73 %

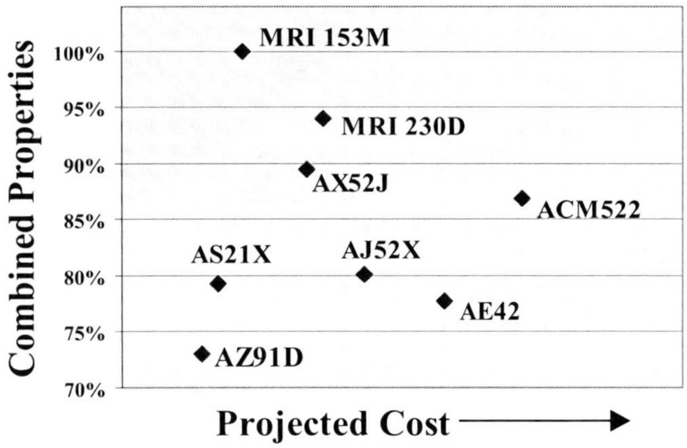

Fig. 9. Combined properties rating vs. projected cost of new Mg Alloys designated for applications at 150 °C

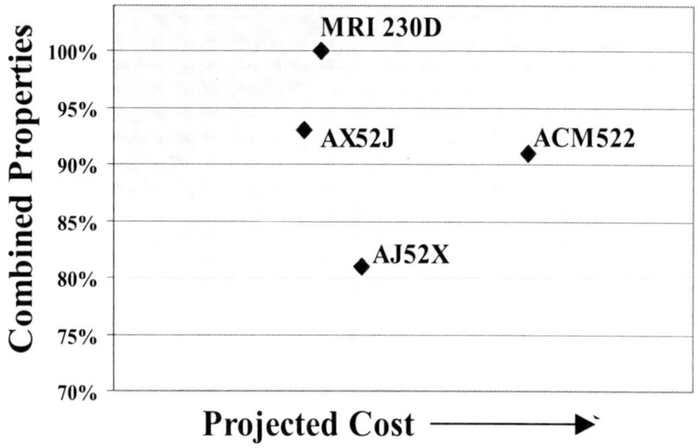

Fig. 10. Combined properties rating vs. projected cost of new Mg alloys Designated for applications at 180 °C

Table 4. Combined properties rating of new Mg alloys for applications at 180 °C

		Castability	Creep Resistance	Tensile Yield Strength	Fatigue Strength	Ductility	Corrosion Resistance	Total Score	% of Max Score
Alloy	Weight	10	10	7	7	3	7		
MRI -230 D	Score	6	10	10	9	4	9		
	W x S	60	100	70	63	12	63	368	100 %
AX 52 J	Score	5	10	10	8	4	8		
	W x S	50	100	70	56	12	56	344	93 %
ACM 522	Score	4	10	9	8	4	9		
	W x S	40	100	63	56	12	63	334	91 %
AJ52X	Score	6	8	6	6	6	8		
	W x S	60	80	42	42	18	56	298	81 %

The presented data relate to two temperatures namely 150°C and 180°C, which represent two typical service conditions of advanced power train applications. In general, according to the component's requirements the design engineer can modify the weight given to specific properties in order to meet the service conditions of the components. Using modified weight of properties and cost target a preferred alloy can be selected for optimal cost and service performance.

CONCLUSIONS

In conclusion, it is clearly evident that it is impossible to distinguish between "good" or "bad" Mg alloy. Each alloy has different metallurgical properties and cost of production and processing, which put it in a different position in the combined property diagrams.
Selection of a new alloy for a practical component should be done based on the criteria of technical requirements and targeted cost. In reality, this selection process is complicated and very much depends on the relative weight given to a specific property and final targeted cost. This task is becoming even more complicated if alternative material systems such as aluminum alloys are considered for the same applications.

REFERENCES

1. H. Friedrich and S. Schumann, "The second age of magnesium research strategies to bring the automotive industry's vision to reality", Proceedings of the Second Israeli International Conference on Magnesium Science & Technology, Ed. by E. Aghion and D. Eliezer, Dead Sea, Israel, 22-24 February, 2000, 9-18.
2. P. Lyon, J.F. King, and K. Nuttal "A New magnesium HPDC alloy for elevated temperature use" (Proceedings of the Third International Magnesium Conference, ed. G. W. Lorimer, Manchester, UK, 10-12, April 1996), 99.
3. S. Koike et al., "Development of lightweight oil pans made of a heat-resistant magnesium alloy for hybrid engines", SAE Technical Paper 2000-01-1117, Detroit, MI, 2000.
4. K. Pettersen et al., "Creep-Resistant Magnesium Alloy Development", Magnesium Alloys and Their Applications, ed. K.U. Kainer (Weinheim, Germany: Wiley-VCH Verlag GmbH, 2000), pp.29-34.
5. B. Powell et al., "Development of Creep-Resistant Magnesium Alloys for Powertrain Applications", Light Metal Applications for the Automotive Industry: Aluminum and Magnesium, SAE Special Publication SP-1622, paper number 2001-01-0422, (Warrendale, PA: SAE, 2001).
6. M. Pekguleryuz and E. Baril, "Development of Creep Resistant Mg-Al-Sr Alloys", Magnesium Technology 2001, ed. J.N. Hryn (Warrendale, PA: TMS, 2001), pp. 119-126.
7. B. Bronfin et al., "Die Casting Magnesium Alloys for Elevated Temperature Applications", Magnesium Technology 2001, ed. J.N. Hryn (Warrendale, PA: TMS, 2001), pp. 127-130.
8. Von Buch et al., "New Die Casting Alloy MRI 153 for Power Train Applications", Magnesium Technology 2002, ed. H.I. Kaplan (Warrendale, PA: TMS, 2002), pp.61-68.
9. D. Argo et al., "Process Parameters and Die Casting of Noranda's AJ52 High Temperature Mg-Al-Sr Alloy", Magnesium Technology, 2002, ed. H.I. Kaplan, (Warrendale, PA: TMS, 2002), pp. 87-94.
10. B.R. Powell et al., "The Die Castability of Calcium-Containing Magnesium Alloys: Thin-Wall Computer Case", Magnesium Technology 2002, ed. H.I. Kaplan (Warrendale, PA: TMS, 2002), pp. 123-130.
11. I.J. Polmear, Metallurgy of Light Alloys, Halsted Press, New York, 1996, 1
96-247.
12. J.F. King, "Development of Practical High Temperature Magnesium Casting Alloys", Magnesium alloys and their applications, ed. K.U. Kainer (Weinheim, Germany: Wiley-VCH Verlag GmbH, 2000), pp.14-22.

2003-01-1365

AMC-SC1: A New Magnesium Alloy Suitable for Powertrain Applications

C. J. Bettles, C. T. Forwood, D. S. Jones and J. R. Griffiths
CRC Cast Metals Manufacturing (CAST), CSIRO Manufacturing and Infrastructure Technology

M. T. Frost
CRC Cast Metals Manufacturing (CAST), Australian Magnesium Corporation

D. H. St. John, Ma Qian and G-L. Song
CRC Cast Metals Manufacturing (CAST), Division of Materials, School of Engineering, University of Queensland

J. F. Nie
CRC Cast Metals Manufacturing (CAST), School of Physics and Materials Engineering, Monash University

ABSTRACT

A new sand cast magnesium alloy, AMC-SC1, suitable for use in powertrain applications has been developed by the Australian Magnesium Corporation (AMC). The alloy is based on the Mg-rare earth system, and it is intended that it be used in a fully heat-treated T6 condition. The creep and bolt load retention behaviours are comparable to those of aluminium A319 in the 150°C – 177°C temperature range. The alloying process has been selected to obtain a cost competitive high temperature magnesium alloy. This paper reports on the physical and mechanical properties of the alloy, and relates these to the properties required for a specific engine block design.

INTRODUCTION

With the increasing needs to control international fuel consumption and reduce harmful emissions into the atmosphere, automobile manufacturers are being pressured into developing more fuel efficient vehicles. Reducing the overall weight of the vehicles is the key to achieving this goal. A major contributor to the weight of any vehicle is the engine itself, and the most significant component of the engine is the block, which makes up 20 –25% of the total engine weight. In the past significant weight savings were made by introducing an aluminium block to replace the traditional grey iron block, and further reductions of the order of 40% could be achieved if a magnesium alloy that could withstand the temperatures and stresses generated during the engine operation was used. However, the development of such an alloy, which combines the desired elevated temperature mechanical properties with a cost effective production process, is necessary before a viable magnesium engine block manufacturing line can considered. In recent years, the search for an elevated temperature magnesium alloy has focused primarily on a high pressure die casting (HPDC) processing route and several alloys have been developed [1]. HPDC was considered to be the best option for achieving the high productivity rates required to counteract the probable high cost of the base magnesium alloy. However, HPDC is not necessarily the best process for the manufacture of an engine block and, in reality, the majority of blocks are still sand cast. The use of magnesium alloys will require not only an appropriate material but also innovative engine design and gravity or low pressure sand casting production processes can more readily accommodate any design variations that may be required for a particular engine design. Therefore, in the search for a magnesium alloy suitable for use in engine blocks, it was thought appropriate to develop a heat treatable sand casting alloy.

In setting the specification for a magnesium engine block alloy it was thought appropriate to consider the particular thermal and stress requirements for a diesel engine, being more severe than those associated with a petrol engine, and base the selection criteria on these conditions. On the assumptions that local reinforcement would be required in the crankcase, for the main bearing, and that, as is also the case with aluminium blocks, a cylinder liner insert would be required, a property specification for the alloy was prepared. This is given in Table 1.

Table 1: Critical property specifications for a sand cast magnesium engine block alloy.

	Room Temperature	150°C	177°C
0.2% Proof Stress (MPa)	120		110
Creep Strength (MPa)*		110	90
Fatigue Limit**(MPa)	50		
Thermal conductivity	115 W/Km		

*stress to give 0.1% creep strain after 100 hours. This target was set to be equivalent to the performance of A319.

** R = -1, n = $5x10^7$ cycles on polished specimens loaded axially in fully reversed tension-compression.

Merely meeting the property requirements will not however ensure immediate acceptance by the automobile manufacturers. The alloy must also be cost competitive and, whilst the cost effectiveness of a magnesium engine block is dependent upon more than just the simple cost of the alloy, the target for economic viability was set at a slight premium to the current price of AZ91D.

BACKGROUND

There are two major classes of magnesium sand casting alloys:

1. -Those based on the magnesium- aluminium binary system, often with small additions of zinc for improved strength and castability. These alloys have adequate room temperature mechanical properties, but do not perform well at elevated temperatures and are inappropriate at temperatures in excess of 125°C [4]. These alloys do not contain expensive alloying elements and are widely used in areas where high temperature strength is not a requirement.

2. Those able to be grain refined by the addition of zirconium. The major alloying elements in this group are zinc, the rare-earth elements, yttrium and silver. With the right choice of alloying additions, alloys in this group can have excellent room and elevated temperature mechanical properties. However, with the exception of zinc, the alloying additions within this group, including the grain refiner, are expensive and the alloys are generally restricted to aeronautical applications. Cost savings can be achieved by the use of misch metals rather than the elemental rare earths and yttrium, but this is often at the expense of the mechanical properties. A summary of the properties of the currently available sand casting alloys is given in Table 2

It is clear from Table 2 that QE22 is the only alloy that meets all the physical and mechanical property criteria, although the WE-type alloys may also be suitable as there is some flexibility with the thermal conductivity specification, in that a slightly lower value may be accommodated through novel design. However, all three of these alloys contain expensive alloying additions at relatively high levels which make them commercially unattractive as potential engine block materials.

Considering Table 2 carefully, several trends can be noted:

1. A high aluminium content has a deleterious effect on the elevated temperature creep properties.

2. High zinc contents are detrimental to the creep behaviour, even in the presence of rare earth elements.

3. Rare earth elements alloyed with either silver or yttrium give superior room and elevated temperature properties.

Clearly any new alloy would need to contain some rare earth additions to achieve the desired elevated temperature properties, and AMC-SC1 has been developed on this basis.

EXPERIMENTAL PROCEDURE

Samples were prepared from gravity cast stepped plates, as shown in Figure 1. The mechanical properties presented here were determined from samples cut from the 15mm step, where the grain size achieved was approximately 40μm. Standard melt handling procedures and controlled environment heat treatment conditions were used throughout the preparation of the cast plates.

MICROSTRUCTURE - Samples for metallographic examination were polished with diamond pastes to 1 μm followed by 0.05 μm colloidal silica. Etching was carried out in a solution of nitric acid in ethylene glycol and water for approximately 12 seconds.

Table 2: Comparison of the mechanical and physical properties of several commercially available magnesium sand casting alloys, highlighting those properties which meet the specification for an engine block,

Alloy	AZ91C# T6	EZ33 T5	ZE41 T5	ZE63 T6*	QE22 T6	ZK51 T5	WE54 T6	WE43 T6
UTS (MPa) / 0.2% Proof Strength (MPa) / Elongation								
20°C	275/**130/** 5	160/105/ 3	205/**140/** 5	290/**175/-**	275/**205/** 4	275/**165/** 8	280/**205/** 5.7	270/**186/** 7.2
150°C	185/**115/** 31	150/95/ 10	165/**115/** 15	185/110/-	205/**185/-**	160/**115/** 14	260/**195/** 7.9	260/**180/** 14.3
200°C	140/95/ 33	145/85/ 20	130/95/ 29	130/95/-	185/**165/-**	115/90/ 17	242/**185/** 10	242/**170/** 15.7
Creep Stress (MPa)**								
150°C	15		86		**140**	65		**212**
200°C	5	66	43		75	40	**160**	**148**
Fatigue Limit (MPa) 20°C	**~65**	**70**	**~100**		**100**			
Thermal Conductivity (W/Km)	56.2	99.5	**123.1**	109	**113**	110	52	51.3
Corrosion Resistance ($mg/cm^2/d$) ***	<0.63						0.2	0.2

AZ91C is not an elevated temperature alloy and is listed for comparison purposes only. * T6 condition includes solution treatment in a hydrogen atmosphere ** stress to give 0.1% creep strain after 100 hours at temperature ***ASTM B117 salt spray test

Figure 1: Gravity cast stepped plate from which all test specimens were prepared.

TENSION AND COMPRESSION TESTS - The tensile properties were measured in accordance with ASTM E8M - 96 at 20, 100, 150 and 177 °C in air using an Instron Testing Machine. Samples were held at temperature for 10 minutes prior to testing. The test specimens had a rectangular cross section (6 mm x 3 mm), with a gauge length of 25 mm (figure 2(a)). The compressive yield strength was determined in accordance with ASTM E9-89(a) (1995) at the same temperatures using cylindrical samples 15 mm in diameter and 30 mm long. The elastic modulus of the alloy was determined at room and elevated temperatures using a Piezoelectric Ultrasonic Composite Oscillator Technique (PUCOT) [3].

CREEP TESTS - The creep behaviour was determined on constant load machines at temperatures of 150 and 177 °C and stresses of 46, 60, 75 and 90 MPa, in temperature controlled silicone oil baths according to ASTM E139-95. The test samples were the same geometry as those used in the tensile testing, and the extension during creep was measured directly from the gauge lengths of the samples.

FATIGUE TESTS - The fatigue strengths at 10^6 and 10^7 cycles were determined at 25 and 120 °C in air. The specimens had a circular cross-section, 5mm in diameter and a 10mm gauge length (figure 2(b), polished to 1μm finish which corresponds approximately to the surface finish at the main bearing – the most highly stressed part of the engine block. Specimens were loaded axially in fully reversed tension-compression (i.e. at zero mean stress) and the test frequency was 60 Hz, corresponding to nominal service conditions. There are several procedures for assessing the fatigue strength at a given life and here the staircase method was used (BS 3518 Part 5).

(a

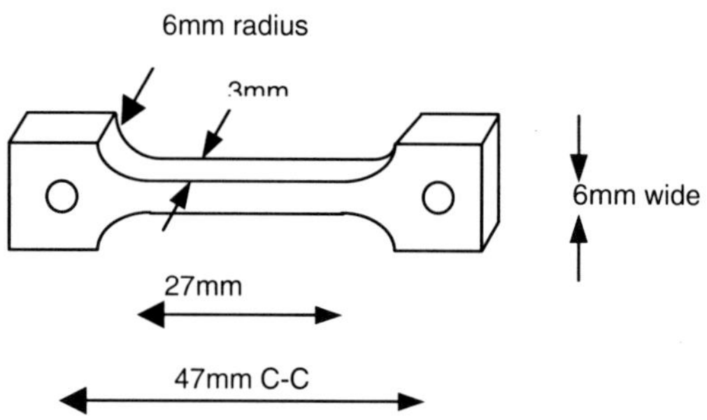

(b

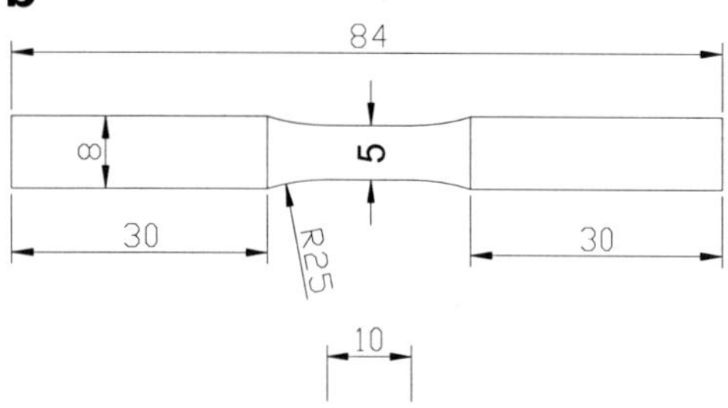

Figure 2: Specimen geometries for (a) tensile and creep and (b) fatigue testing.

BOLT LOAD RETENTION (BLR) TESTS - Bolt load retention testing can be used to simulate the relaxation that may occur in service under a compressive loading. The test method [4] involves applying an initial load (in this case 8 kN, which corresponds to approximately 55 MPa in the magnesium assembly) through an assembly consisting of two identical bosses, 15mm thick, 16mm outside diameter and 8.3mm inside diameter, made of the test material and a high strength M8 bolt instrumented with strain gauges (figure 3). AISI 4340 steel washers, 1.5mm thick, 16mm outside diameter and 8.5mm inside diameter, are used. The change in load over 100 h at an elevated temperature (150 and 177°C) is measured continuously. Three BLR tests are carried out for each alloy at both temperatures. The two significant loads, in terms of defining the BLR behaviour, are the initial load at ambient temperature, P_I, and the load at the completion of the test after returning to ambient conditions, P_F. The ratio of these two values (P_F/P_I) is a measure of the bolt load retention behaviour of an alloy. There is often an initial increase in load as the bolted assembly is heated to the test temperature. This is the result of the combined thermal expansion of the bolted assembly and the yield deformation in the alloy bosses.

THERMAL CONDUCTIVITY - The thermal conductivity was measured on samples 30mm in diameter and 30mm long.

CORROSION RESISTANCE - The corrosion resistance of AMC SC1 was compared to that of diecast AZ91D, using ASTM B117 standard salt spraying test. The tests were carried out over a period of seven days in a saline environment (5 wt%NaCl solution) at 35°C. The corrosion products were removed from the test coupons using a chromic acid wash followed by an ethanol rinse.

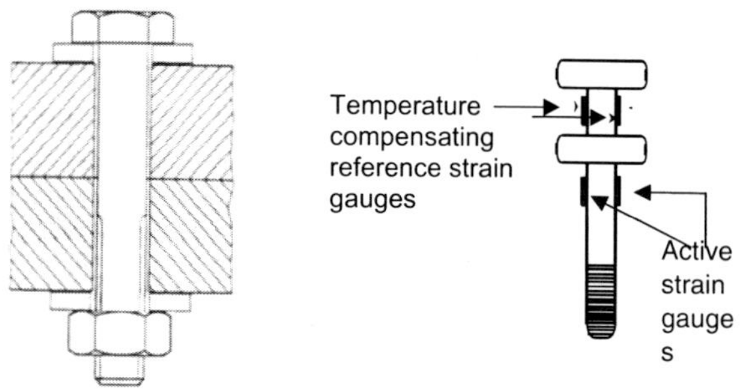

Figure 3: Schematic of the bolt load retention test assembly and the positioning of the reference and active strain gauges.

RESULTS AND DISCUSSION

MICROSTRUCTURE - Being a sand casting alloy, AMC-SC1 requires a T6 treatment (solution heat treatment, cold water quench, and elevated temperature anneal) to fully develop its mechanical properties. The recommended heat treatment regime is a balance between mechanical property requirements and commercially acceptable holding times after casting. The T6 microstructure of AMC-SC1, which is shown in Figure 4, consists of grains of an α-Mg phase (A) locked by a magnesium-rare earth intermetallic phase (B) at grain boundaries and triple points. Clusters of rod-like precipitates (C) are present within the central regions of most grains. The intermetallic phase, B, has a stoichiometry close to $Mg_{12}(La_{0.43}Ce_{0.57})$ [5].

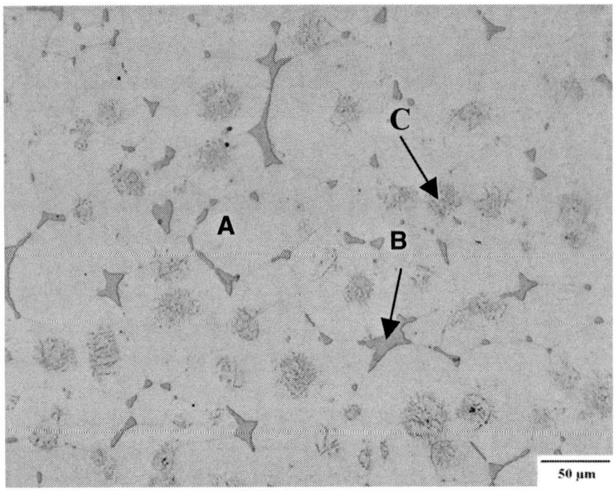

Figure 4: Optical micrograph showing the T6 microstructure of AMC-SC1.

(a

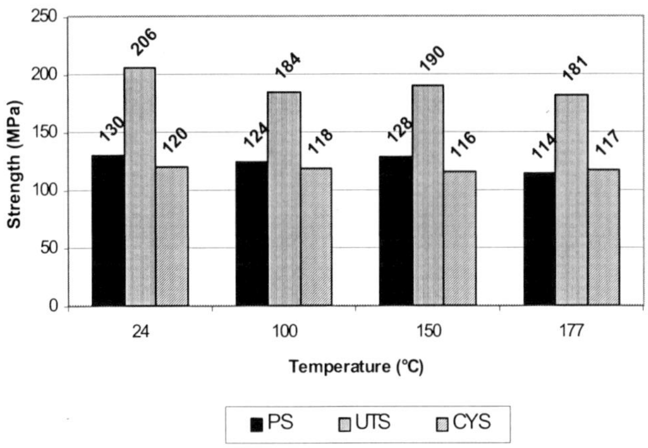

(b

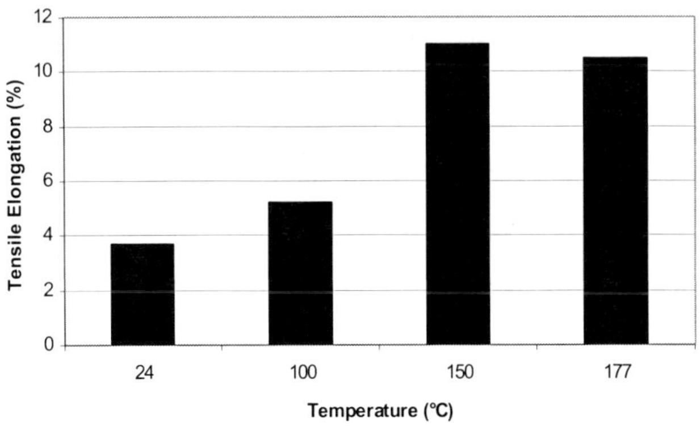

Figure 5: The effect of temperature on (a) the tensile proof strength, compressive yield strength and ultimate tensile strength and (b) the tensile elongation of AMC-SC1.

TENSILE AND COMPRESSIVE STRENGTHS - Figure 5(a) shows both the tensile properties (the 0.2% proof strength and the ultimate tensile strength and the compressive yield strength as a function of temperature. Figure 5(b) shows the tensile elongation, also as a function of temperature. It is significant to note that the mechanical properties of AMC-SC1 are extremely stable at elevated temperatures, with the proof strength in both tension and compression being relatively unchanged between room temperature and 177°C. The room temperature properties of AMC-SC1 are nowhere near as high as most other magnesium sand casting alloys but it is the stability of these properties up to 177°C which makes this alloy particularly attractive for engine block applications.

The results of the elastic modulus determination are shown in Table 3, and it is of note that the elastic modulus shows a drop of less than 10% at 177°C over the room temperature value.

Table 3: Elastic Modulus of AMC-SC1 as determined using a PUCO technique

Young's Modulus (Gpa)		
25°C	100°C	177°C
45.8 ± 0.3	43.9± 0.3	41.9 ± 0.3

CREEP BOLT LOAD RETENTION BEHAVIOUR - The microstructure of AMC-SC1 is extremely stable at temperatures up to 177°C, and this is an important factor, together with the form and distribution of the grain boundary intermetallic phase, in achieving the requisite creep resistance. The use of a creep stress, being the stress to produce a creep strain of 0.1% after 100 hours at temperature, as a measure of creep resistance is an arbitrary one, but it is nonetheless a useful method for comparing alloy behaviour. Using this concept, the behaviour of AMC-SC1 may be compared to that of A319 (Figure 6) and it is clear that the two alloys are very similar in their creep responses in the temperature range 150 to177 °C. More importantly, however, it should be noted that the stresses required to produce a creep strain of 0.1% in AMC-SC1 after 100 hours at both 150 and 177 °C are approaching the tensile yield strengths (0.2% offset) of the material.

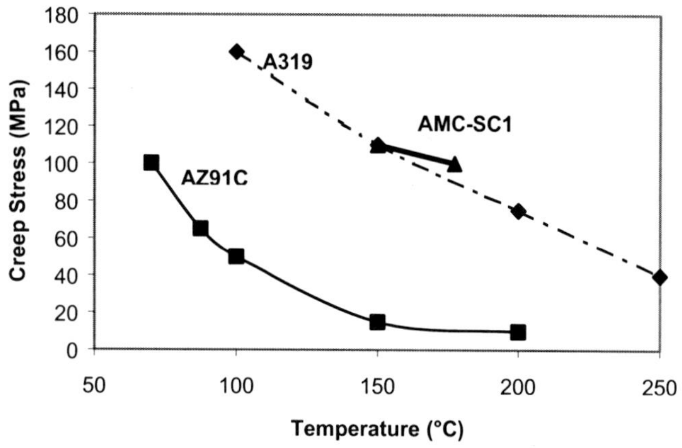

Figure 6: Stress required to produce 0.1% creep strain after 100 hours at temperature for AMC-SC1, A319 and AZ91C.

Typical bolt load retention curves for AMC-SC1, A319 and AE42 at 150°C and 8kN load are shown in Figure 7(a). AMC-SC1 is in the T6 condition, A319 is as sand cast and AE42 is high pressure die cast (ie all three alloys are in their normal operating condition). The increase in load occurring at the commencement of the test is the net result of the thermal expansion of the bolted assembly less the yield deformation in the alloy bosses. Two significant loads are the initial load at ambient temperature, P_I (8kN in this case), and the load at the completion of the test after returning to ambient conditions, P_F. The ratio of these two values is taken as a measure of the bolt load retention behaviour of an alloy, and has been used in this case to compare AMC-SC1 with die cast AE42 at 150 and 177°C (Figure 7(b)). The bolt load retention behaviour at elevated temperatures again reflects the high temperature stability of this alloy and it is clear that AMC-SC1 is as good as the aluminium alloy A319 and superior to AE42 in this respect. In the work referenced [4], the BLR behaviour of die cast AE42 oil filter housings was tested at 150°C and 8 kN using M8 bolts and standard steel washers (ISO 7089). The P_F/P_I value obtained after 100 hours at 150°C was approximately 0.3, compared to approximately 0.39 obtained from this work using two identical AE42 bosses.

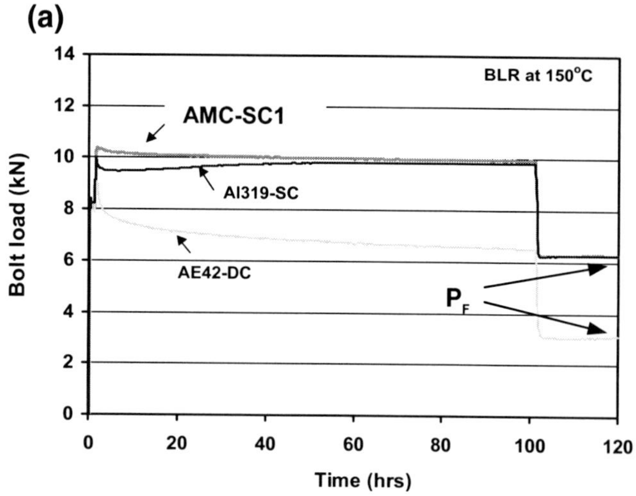

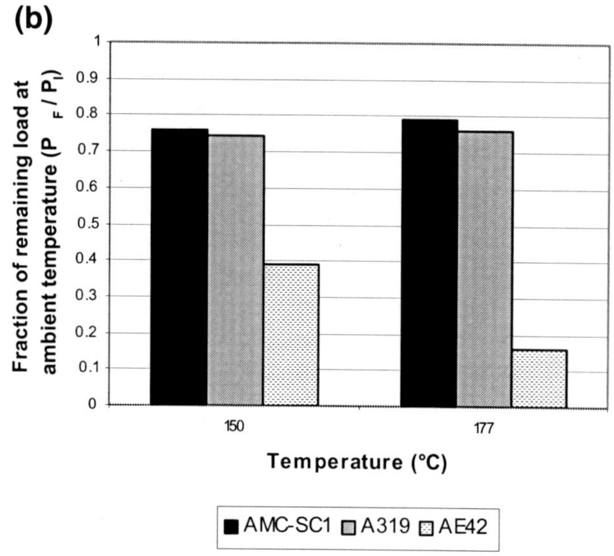

Figure 7: (a) Typical bolt load retention curves at 150°C and 8kN (~55 MPa) for AMC-SC1, AE42 and A319 and (b) a comparison of the bolt load retention behaviours of AMC-SC1, AE42 and A319 at 150°C and 177°C with a load of 8kN.

FATIGUE PROPERTIES - An engine block is continually subjected to cyclic stresses during service and it is necessary, therefore, to ensure that the material chosen for the block can withstand this fatigue loading. The fatigue strengths of AMC-SC1 at 10^6 and 10^7 cycles were determined at both 24 and 120 °C, and the figures quoted in Table 4 are the stresses giving a 50% probability of fracture. The limits represent the stresses for the 10% and 90% probabilities of fracture. It should be noted that these results are for a maximum of 10^7 cycles, rather than the $5x10^7$ specified in the design criteria. Nonetheless, the strengths are sufficiently high for the alloy to be considered to have met the target.

Table 4: Fatigue Strengths of AMC-SC1 at two temperatures (R=-1).

Temperature	Fatigue Strength (MPa)	
	10^6 cycles	10^7 cycles
24°C	~80*	75 ± 18
120°C	74 ± 9	71 ± 7

~ denotes 12 samples only tested, rather than the 15 required by the standard.

CORROSION - The corrosion behaviour of the alloy, both internally and externally, is of paramount importance. Corrosion on the internal surfaces may be controlled by the use of an appropriate engine coolant combined with careful design to ensure compatibility of all the metal components in contact with the coolant liquid. With regard to the external corrosion, there is no one test alone which can determine the corrosion resistance of an alloy in all environments. In this study AMC-SC1 was compared to diecast AZ91D using the ASTM B117 standard salt spraying test. AZ91D was in the as-cast condition and AMC SC1 was T6-heat-treated. The mean weight loss rates over 7 days of salt spraying were found to be 0.68 $mg/cm^2/day$ for AMC-SC1 and 0.27 $mg/cm^2/day$ for diecast AZ91D.

THERMAL CONDUCTIVITY - The thermal conductivity of AMC-SC1 was found to be 102 W/mK, which is slightly less than that originally specified in the design criteria. However, with this information available, it is not difficult to modify the design of an engine block to accommodate this thermal conductivity value.

CONCLUSION

The recently developed sand casting alloy, AMC-SC1, is able to meet the following specifications:

- 0.2% proof strength of 120 MPa at room temperature and 110 MPa at 177°C.
- Creep resistance comparable to that of A319 at temperatures of 150°C and 177°C.
- Fatigue limit in excess of 50 MPa at room temperature.
- Under salt spray testing conditions the corrosion resistance is lower than AZ91E.
- Thermal conductivity is comparable to other High temperature Mg alloy.

This combination of superior elevated temperature mechanical properties and cost effectiveness makes AMC-SC1 an interesting option as an engine block material.

ACKNOWLEDGMENTS

This work was conducted under an alliance between the Australian Magnesium Corporation and the Australian Cooperative Research Centre for Cast Metals Manufacturing (CAST) in association with Hydro Automotive (formerly VAW Aluminium AG), AVL List GMBH and MEL. The research within CAST was carried out at CSIRO Manufacturing and Infrastructure Technology, Monash University and the University of Queensland. The CRC for Cast Metals Manufacturing (CAST) was established under and is supported by the Australian Government's Cooperative Research Centres Scheme.

REFERENCES

1. Druschitz, A.P., Showalter, E.R., McNeill, J.B. and White, D.L. "Evaluation of structural and high temperature magnesium alloys", SAE Paper No 02M-29.

2. ASM Specialty Handbook "Magnesium and Magnesium Alloys" eds. Avedesian, M.M. and Baker, H. The Materials Information Society 1999.

3. Robinson, W.H. and Edgar, A. IEEE transactions on Sonics and Ultrasonics, SU-21 (2) 1974 98-105.

4. Pettersen, K. and Fairchild, S. SAE Technical Paper 970326.

5. Rossouw C.J., Bettles C.J., Davis T.J., Miller P.R. and Venkatesan K. "Zn incorporation within the intermetallic $Mg_{12}(La_xCe_{1-x})$ lattice" Acta Cryst **A57** pp321-332 (2001)

2003-01-0821

Development in Lost Foam Casting of Magnesium

Michael Marlatt
wfv / Roeperwerk

David Weiss
Eck Industries, Inc.

John N. Hryn
Argonne National Laboratory

ABSTRACT

Preliminary work was conducted in the casting of magnesium using the lost foam casting process. The lost foam or expendable pattern casting (EPC) process is capable of making extremely complicated part shapes at acceptable soundness levels and with low manufacturing costs. Standard test shapes were used to determine the ability of the magnesium to fill the mold and to assess the types of defects encountered. This paper will briefly explain how this project evolved including the developmental strategies formed, the products selected, the casting trials performed, and the casting results.

INTRODUCTION

The last 10 years have seen a dramatic increase in the production and utilization of magnesium. This demand is primarily driven by the automobile industry and its desire to increase vehicle fuel efficiency. One way to improve vehicle fuel efficiency is to reduce vehicle weight by using alternative materials in vehicle components. Magnesium has found uses in several automotive components because of its low density and high strength-to-weight ratio as compared to conventional automotive materials. The majority of magnesium components are produced using the high pressure die casting process. This industry has shown significant growth as the automotive industry strives to reduce vehicular weight (Figure 1). Alternatively, low pressure and gravity fed magnesium casting processes have not achieved the same kind of growth in this industry sector. The key element missing is the lack of a technological base for foundry casting of magnesium alloys. To address this need, several industry members met in May 2000 and established a Magnesium Committee within the American Foundry Society (AFS). The committee's primary purpose was to encourage magnesium use by foundries and customers and to move the state-of-the-art forward. The Magnesium Committee has grown rapidly within AFS and was elevated to Division 6 status in June 2002.

Magnesium castings are traditionally produced either by high pressure die casting or precision sand casting. Each of these processes has its advantages as well as disadvantages for producing components. Compared to lost foam casting processes, disadvantages of the conventional processes are that several casting design geometries cannot be produced by the steel die sections and are limited by moving cores of die cast tooling. Other parts cannot be produced by means of high pressure die casting with acceptable levels of porosity. While precision sand casting can produce a wide range of part shapes with high levels of soundness, the process is expensive and generally limited to high value components such as those used in the aerospace industry.

Finding alternative casting processes, such as lost foam casting, to high pressure die casting and precision sand molding are a priority of the AFS Magnesium Division. One major advantage of the lost foam process over any of the other casting techniques is the ability to create one component to replace several fabricated components. Using the lost foam process is a major manufacturing cost savings.

There have been great strides within the magnesium casting industry to address the concerns of cast magnesium component properties. Corrosion resistance, bolt load retention, yield strengths, creep resistance and microporosity of magnesium casting have greatly improved in recent years. Although many foundries still believe that the melting, handling, machining of magnesium is a safety hindrance in the production of magnesium casting components, this also is no longer the case. By following today's recommended magnesium industry standards and procedures, the safety risks to casting facilities and employees working with magnesium is no greater to those casting with other non-ferrous metals.

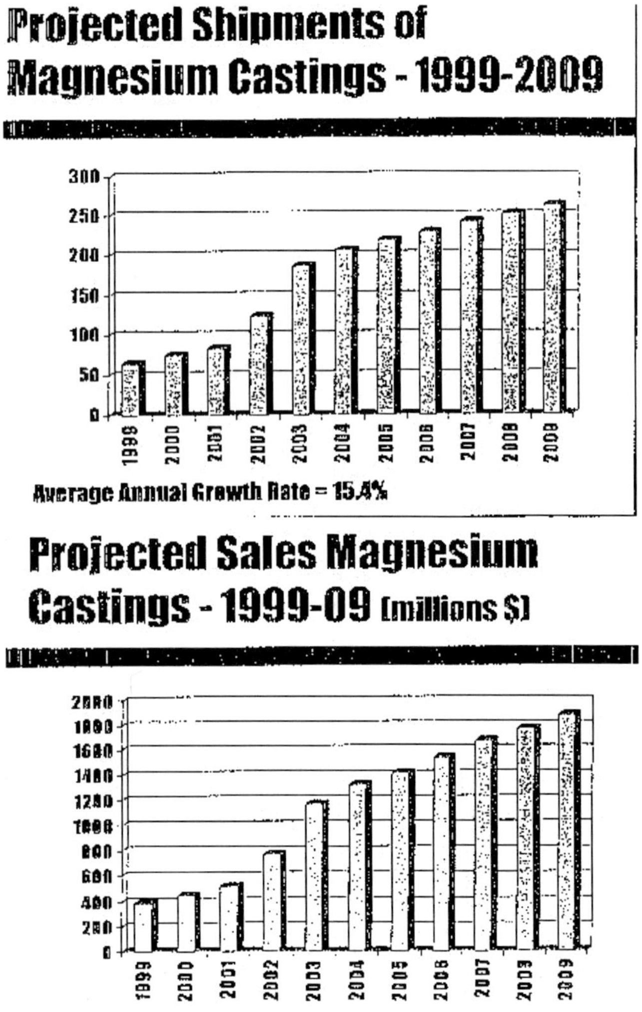

Figure 1. Projected Shipments and Sales of Magnesium Castings.[1]

PROJECT DEVELOPMENT

In June 2002, selected members of the American Foundry Society (AFS) Magnesium Division 6 joined together with several members from the AFS Lost Foam Division 11 to form a committee to work on the lost foam magnesium project. The goal of the project was to produce sound magnesium castings using the lost foam casting process. Major items to be investigated and selected included a foundry location that casts magnesium, a lost foam material and pattern, a pattern coating material, a molding sand type, and an alloy type.

MATERIAL SELECTION

The committee immediately began researching materials to best cast lost foam with magnesium. Although aluminum is often cast using the lost foam casting process, there are a number of differences between aluminum and magnesium that can affect the casting process. The two primary differences are in their thermal conductivity and chemical reactivity. Because of magnesium's low thermal heat content, filling a lost foam mold created the greatest challenge. Comparing the specific heat values of aluminum and magnesium, the k/cal value of aluminum is 0.70 compared to magnesium which has a value of 0.57. While aluminum would require a mold media that has a high conductivity media and low thermal expansion, magnesium would require a low conductivity media with low thermal expansion properties.

The reactivity of magnesium would be controlled by the use of inhibitors, common in the magnesium casting industry. It was also decided to hold the casting trials at a foundry that already casts magnesium, as opposed to a foundry with lost foam casting proficiency but little experience with molten magnesium. Eck Industries was selected for the trials.

Fluidity differences between molten magnesium and aluminum alloys can also affect lost foam casting. Studies by Liu and Hu [2] et al of common aluminum and magnesium casting alloys show that the indicated fluidity of magnesium alloy AZ91 is about one third that of aluminum alloy 319. Those authors speculate that vacuum may be an indispensable parameter in the magnesium EPC process. In order to simplify the initial experiments, vacuum was not used in this initial work.

PATTERN, FOAM, AND COATING SELECTION

As metal would be poured directly onto the foam patterns during magnesium casting, the foam would form decomposition products at the degradation (leading) edge of the advancing metal. Therefore, selecting a low-density foam material would result in less decomposition products present during metal filling. The idea was to maintain as much heat in the metal as it advanced through the foam pattern during casting.

Two kinds of foam patterns were selected for the casting trials. One was a test box pattern while the other was a window test pattern. Both of these patterns are used by General Motors to evaluate the aluminum foam casting process.

The foam patterns using low-density silene lubricant beads with a triblend mixture of three different pentane isomers as the blowing agent were chosen. Two bead grades selected were: T170 and T180C. The T180C grade allows for easier pre-expansion and a lower density compared to the T170 bead. Foam patterns were made with 1.5 and 1.04 densities (specific gravities).

Once the foam patterns and ingates were glued together, the patterns were coated with a mica-based slurry. The mica-based slurry typically used in lost foam casting of iron was selected because of its high thermal insulation and permeability characteristics.

Figure 2. Coated and gated window-test patterns (front) and test-box patterns (right and rear).

DOWNSPRUE, SAND, AND ALLOY SELECTION

Instead of the standard foam downsprues commonly used for aluminum lost foam casting, two sizes (LF4 and LF10) of hollow fiber downsprues were used. The major difference between the two downsprues was length. Prior to use, the fiber downsprues were coated with an inhibitor to reduce the risk of reaction with molten magnesium. The fiber downsprues were then glued to the foam ingate just prior to molding the patterns in the flask.

High-permeability 40 GFN silica sand was selected to hold the patterns in the flask. As an additional safety measure, a mixture of sulfur and potassium fluoroborate inhibitor was mixed in with the sand as an added safety factor and to decrease the possibility of reaction between the sand and molten magnesium.

AZ91E magnesium alloy was chosen for the trials because of its high purity level of all AZ91 alloys, good corrosion resistance properties, and its reputation of being the "workhorse" alloy for magnesium sand casting.

Figure 3. LF10 (left) and LF4 (right) downsprues coated with inhibiter.

LOST FOAM CASTING TRIALS

The casting trials were held at Eck Industries on June 25-26, 2002. Several hundred pounds of AZ91E alloy was melted using standard magnesium melting practices and SF_6 cover gas for melt protection. After bringing the melt to temperature and drossing, hexachloroethane tablets were added to the metal as a degassing agent and grain refiner.

Figure 4. Degassing/refining process.

The pattern and mold information for the seven castings that were made during the two-day trial is shown in Table I. Gating systems for all patterns expect #7 were side-gated with bottom in-gates. Patterns were placed in a flask half full with sand. The flask was vibrated as more sand was added to the flask to pack the sand around the pattern. On the first day of the castings trials, two patterns (#1 test box and #2 test window) were invested in the sand flask and poured at about 1500°F (816°C) with the AZ91E alloy. These pouring temperatures are considered typically on the high side for magnesium alloys; however our main goal was to completely fill these patterns with no signs of cold lap or misruns. Pour times for both pattern molds was approximately 6-8 seconds. After the castings solidified, they were removed from the flask and sand-blasted clean. From visual examination of the first castings, it was evident that successful lost foam casting of magnesium was possible.

Figure 5. Pouring the mold.

Figure 6. Pouring molten magnesium.

On the second day of the casting trial, five additional patterns were cast. Again AZ91E was poured at 1500°F for all the patterns. A minor modification was made to pattern #6. Visual examination of casting #1 (test box) which was cast the previous day suggested the presence of a slight gas or cold lap. In an attempt to correct this defect, some of the coating was scraped off the pattern in the rib section to allow easier escape of foam's gas decomposition products, which were presumed to cause the lap.

The last pattern (#7) to be molded and poured had some major modifications made to it. X-ray examination of the previous day's castings showed some levels of foreign material (presumed to be oxides) in some of the castings. Therefore, the gating system was modified to include a stainless steel wire mesh to the bottom of the fiber downsprue to act as a filter. Using metal mesh filters to the gating systems is a common foundry practice in magnesium sand casting. The window pattern was made with a shortened LF4 downsprue. A foam piece was glued to the bottom of the fiber cup and to the window pattern. The gating system was now top gated directly into the top of the test window.

Table I. Trial Patterns

Mold number and pattern	Foam grade (bead size)	Foam density (specific gravity)	Downsprue (with inhibitor)	Comments
1 – test box (day 1)	T180	1.5	LF4	
2 – window (day 1)	T180	1.04	LF4	
3 – window (day 2)	T180	1.5	LF4	
4 – window (day 2)	T180	1.04	LF10	
5 – test box (day 2)	T170	1.5	LF4	
6 – test box (day 2)	T180	1.5	LF4	Coating scraped off in rib section
7 – window (day 2)	T180	1.04	LF10 - shortened	Steel filter and modified gating

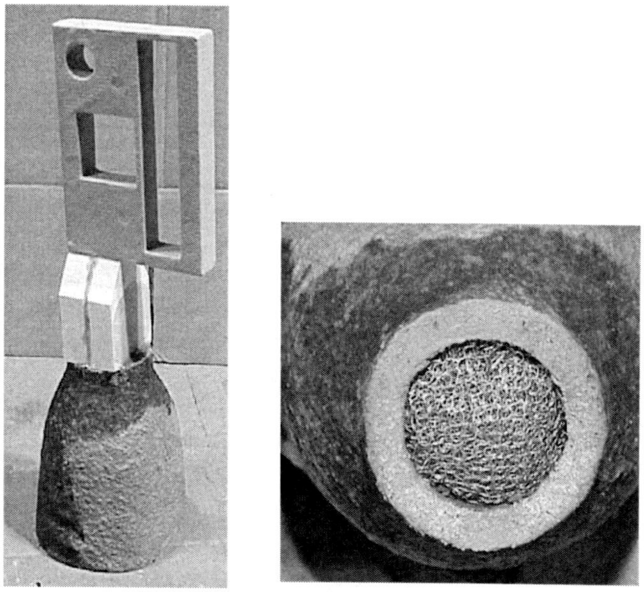

Figure 7a and 7b. Pattern #7 modified with shortened downsprue, steel filter, and top in-gate.

All the molds were poured at 1500°F from a hand held ladle, and all but #7 had pouring times of 6-8 seconds. The pouring time for casting #7 was slightly longer. The pouring cup was filled initially as with the other castings, but the pour was interrupted for a few seconds to allow the metal level to drop. Once the level in the cup dropped, the cup was topped off with additional molten metal. After all the molds had been poured off from each trial day, the castings were blasted clean for visual inspection, followed by x-ray inspection. Following the inspections, some of the castings were heat treated to a T6 condition and sectioned for determination of mechanical properties.

Figure 8a and 8b. Filled mold sprinkled with inhibiter to extinguish magnesium oxidation reaction.

Figure 9. Casting shakeout.

RESULTS

VISUAL OBSERVATIONS

All seven castings filled completely with the exception of casting #1 which showed a slight gas or cold lap.

Figure 10. Castings #1 (rear) and casting #2 (front).

Figure 11. Casting # 7 (left) and Casting #3 (right).

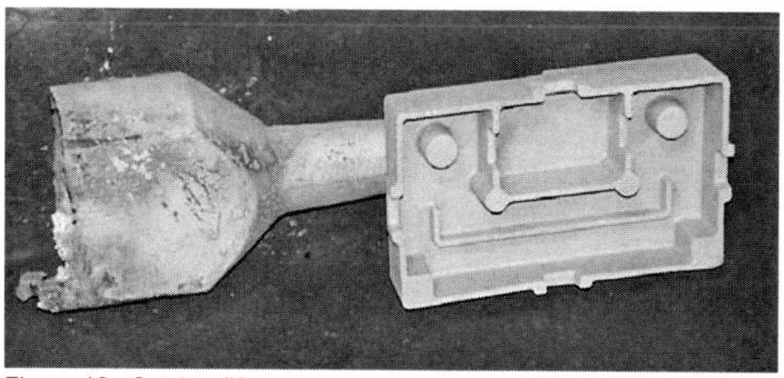

Figure 12. Casting #6.

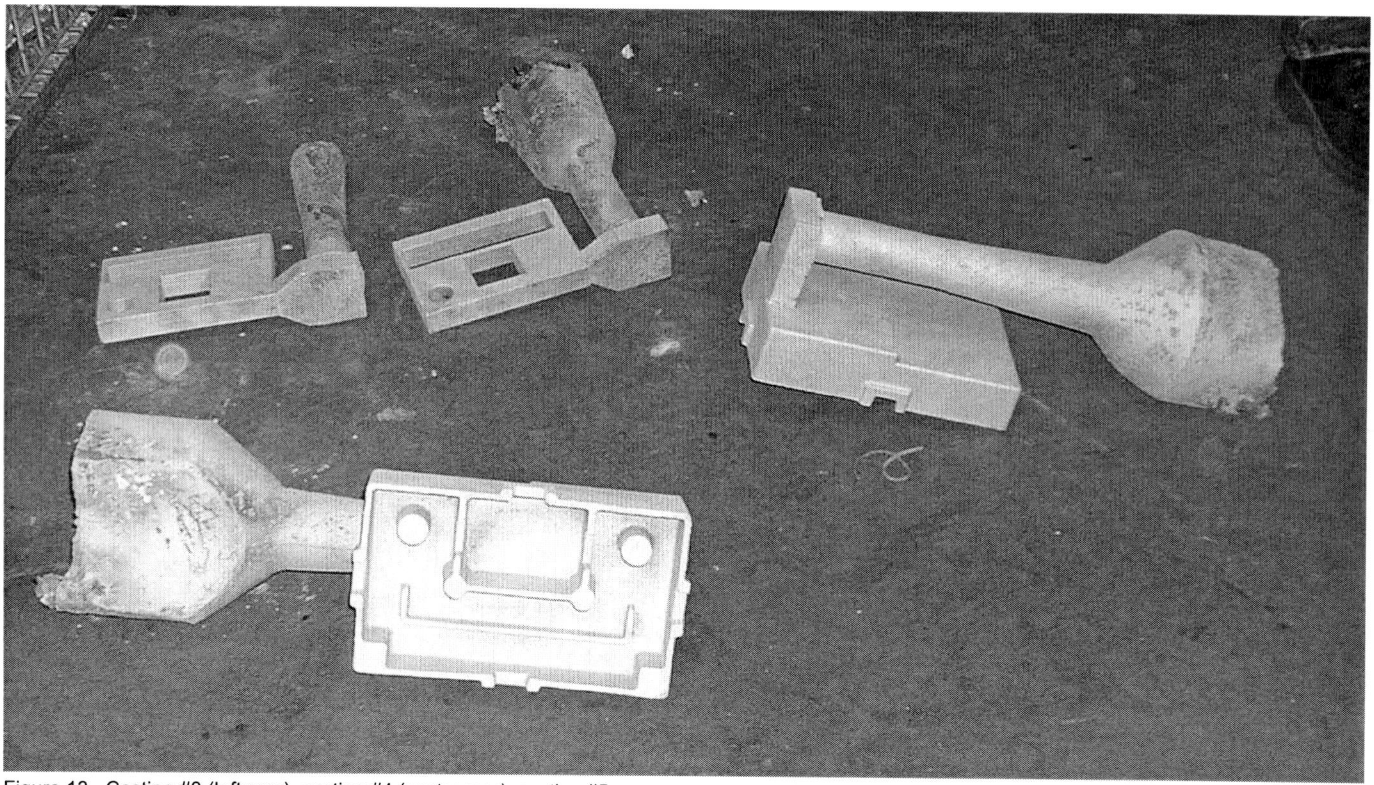
Figure 13. Casting #3 (left rear), casting #4 (center rear), casting #5 (right), and casting #6 (front).

X-RAY QUALITY INSPECTIONS

The castings after visual inspection were sent to x-ray for quality inspection. All castings were confirmed grade C minimum except for casting #7, the test window pattern with the modified gating system and the metal mesh filter. Not only was casting #7 visually a good casting, but the x-ray results confirmed a grade B or higher results. (X-ray grades per Mil-Std-2175).

Figure 14a. X-ray photographs of casting # 7 indicate grade B quality.

Figure 14b. X-ray photographs of casting # 7 indicate grade B quality.

MECHANICAL PROPERTIES

After x-ray inspection was completed the castings went through a T6 heat treatment cycle for the purpose of obtaining mechanical properties. Figure 15 shows the mechanical property data determined for selected castings. The values are comparable to those obtained by precision sand castings.

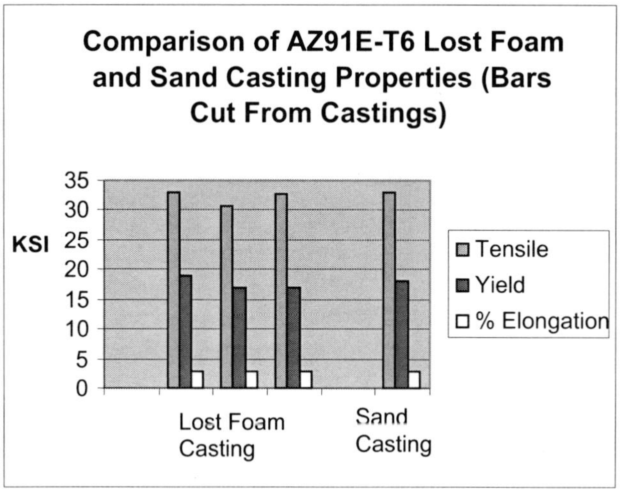

Figure 15. Measured physical property data of magnesium lost foam castings.

CONCLUSION

The trials proved that casting magnesium using the lost foam technique can be done successfully when all good foundry practices are followed closely. This is also true for the safety aspect of casting magnesium in a foundry environment.

ACKNOWLEDGEMENTS

The authors are extremely grateful to the following organizations that participated in the project.

Eck Industries, Manitowoc, Wisconsin, USA
General Motors, R&D Center- Warren, Michigan, USA
GM Powertrain, Milford, New Hampshire, USA
Styrochem, Ft. Worth, Texas, USA
Joy-Mark, Cudahy, Wisconsin, USA
Ashland Chemical, Cleveland, Ohio, USA
American Foundry Society, Des Plaines, Illinois, USA
Argonne National Laboratory, Argonne, Illinois, USA

REFERENCES

1. Overview of magnesium casting market presented by Dave Kanicki at the inaugural AFS Mg Committee Meeting, May 2000, AFS Headquarters, Des Plaines, Illinois, USA
2. A. Z. Liu, J. Hu, W. Ding, Q. Wang, Y.G. Zhu, Y. Lu, W. Chen; Effect of Processing Parameters on Mold Filling in Magnesium Alloy EPC Process, AFS Transactions 01-014, 2001

2002-01-0078

Metal Quality – The Effects on Die Casters and End Users

Per Bakke and Jørild M. Svalestuen
Norsk Hydro Light Metal Research Centre

Darryl Albright
Hydro Magnesium Marketing

ABSTRACT

As more and more magnesium metal is used and recycled, particularly for safety-sensitive components, the maintenance of metal cleanliness remains of critical importance. The term *metal quality* may be defined as 1) chemical composition 2) inclusions and porosity inside the metal, 3) the surface appearance and 4) consistency. This definition may apply to both ingots and cast parts. Chemical composition of ingots will influence on the cast parts, but also process operation at the die casters' will contribute to the chemical composition of the final product. The inclusion contents of a cast component is only indirectly determined from the cleanliness of the ingots from where it origins, emphasising the need for metal cleanliness assessment in the die casting shop. Beside the die casting process itself, the ingot surface appearance and the housekeeping in the die casting furnaces are important for the property of the product. This affects the economy of the die caster operation due to loss of metal and up-time.

INTRODUCTION

Magnesium alloys are being successfully used for increasingly sophisticated, more demanding components /1,2/. Part of the reason for this success is an increasing primary supply base and the utilization of recycling on a broad scale /3-7/. In order to insure high quality components, continued focus on metal quality and its assessment is necessary /8,9/.

The quality of die cast product is to a large extent considered satisfactory if the mechanical properties meet certain criteria, reflected by the intrinsic properties of the alloy (i.e. strength, ductility and corrosion rates) and engineering know-how. The die casting engineer is usually expert in the design of parts and gating systems, and the related operational parameters on the die casting machine. Given the chemical composition of the alloy, the operational parameters will normally determine the properties of the die cast part. Nevertheless, metal quality is considered important and necessary from different standpoints as will be discussed here.

There may be diverging interpretations or perceptions of the word *quality*. However, in a standard dictionary /10/ *quality* is defined as:

- Something having a distinguishing characteristic or attribute
- The basic character or nature of something
- The degree or standard of excellence, especially a high standard

Furthermore, a *quality product* is something having or showing excellence or superiority. According to ISO 8402 /11/ quality is defined: the totality of characteristics of an entity/item that bear on its ability to satisfy stated or implied needs.

How do this relate to metal quality? Traditionally we understand the term *metal quality* as:

1) Chemical composition including both alloying elements and some trace elements
2) Inclusions and porosity inside the metal
3) The surface appearance with associated defects such as discoloring, oxidation including corrosion deposits and moisture, cold shuts and visible porosity
4) Consistency

Items 1) and 2) of ingots have direct influence on the quality of the cast parts. While international standards are established for chemical composition of both ingot metal and die cast parts, also including the most common and undesirable trace elements, such standards exist neither for surface quality nor metal cleanliness. The term *metal cleanliness* is often used when discussing inclusions.

Surface quality, item 3), could affect the dross formation in die caster furnaces. However, items 3) and 4) are also psychological factors, affecting the relationship between the customer and the supplier. Surface appearance and product consistency give the customer a first impression of the supplier and the supplier's ability to meet the customer's expectations.

Consistency, item 4), is here defined as the ability of keeping the same level or standard over time. Consistency can be followed-up on measurable parameters (i.e. chemical composition, weight, packaging, delivery on time etc.) Delivering to a high standard of consistency gives the customer confidence in the supplier. Statistical process control (SPC) is used to ensure consistency in chemical composition and ingot- and bundle weight.

The price of the product is usually not considered to be a part of the quality issue. But it is a common understanding that a "quality product" is associated with higher price, as it may be more costly to produce.

The above definition of metal quality should apply to both ingots and cast parts. However, additional criteria such as hot tears and dimensional stability are also important to die cast parts.

The ISO 9001 /12/ formalize the agreements between supplier and customer on the quality issue, and deals with the most important features of quality. Hydro Magnesium, for example, is certified according to ISO 9001. However, the customer's expectations, reflected by cultural differences, are often difficult to formalize. A close interaction between the customer and supplier is needed to develop a common understanding of these unwritten expectations.

Producing quality is a management issue. The ***T**otal* ***Q**uality* ***M**anagement* (TQM) philosophy focuses on each individual's responsibility for quality of their own work, and the use of teamwork (quality improvement group) to improve process and products. TQM is implemented in several organizations, i.e. Hydro Magnesium.

QUALITY IN PROCESS AND PRODUCT

In the following the Key Process Parameters (KPPs) and Key Quality Parameters (KQPs) of ingots, and the impact of these on the performance of the die casting operations and final die cast products are discussed. Furthermore, it addresses the influence of the KPP's in the die casting shop, and their influence on the quality parameters of the final die cast products. Including transport and storage, this leads to six "arrows" as illustrated in Fig. 1 /8/. Each arrow or connection will be discussed in detail.

PROCESS CONTROL DURING INGOT PRODUCTION VS. QUALITY OF INGOTS - Magnesium alloys are normally prepared batchwise in (induction) furnaces. Here, alloying elements are added to a batch of pure Mg or recycled alloy. In order to meet the specifications of the High Purity alloys, temperature control is a key word, both with respect to hitting target compositions of Mn and Fe,

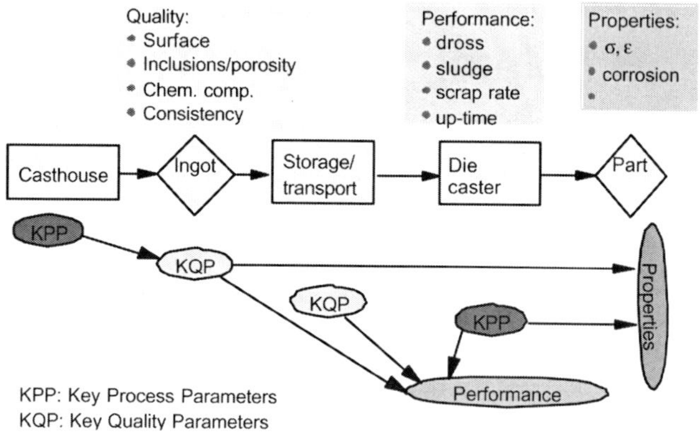

Figure 1. Key process parameters and key quality parameters and their influence on the quality parameters of the die cast product and performance indicators in the die casting operation /8/.

but as much as to have the precipitated intermetallic particles removed. The extremely low tolerance for contaminants such as nickel, put strong restrictions on the choice of materials for processing. This applies of course not only to the alloy producer, but also to the die caster. However, from a maintenance point of view, the steel qualities having the best combination of cost and lifetime are those containing nickel, possibly giving nutrition to conflicts.

From the alloy preparation furnace, the alloy is usually transferred to casting furnace, or a combined holding and casting furnace. Ingots are cast in different ways, from simply pouring of tilting crucibles to advanced metering systems. Depending on the mould filling technique, turbulence and splashing may cause formation of foam and oxides on the ingot. Some producers put effort in skimming off the surface just after completed filling to improve the customer's visual impression of the ingots. This procedure is mainly used by small, manual operations, and might sometimes lead to entrapment of inclusions, despite that the purpose of this operation is to remove surface oxides, foam etc.

Mould coatings that are used for easing the release of the ingots may lead to discoloring and even sometimes attachments to ingots. If the mould coating contains humidity, gas evolution may cause surface defects and even gas porosity in the ingots. Likewise, sulfur sometimes distributed into the mould for extra protection against oxidation, may lead to discoloring and other defects if not distributed properly. Cold shuts, resulting from too low mould surface temperatures or too low flow rates during mould filling, may give some additional surface of the ingots.

Investigations of AZ91D ingots from a large number of producers showed that porosity is always present /13/. It is concentrated along the top center of the ingot, but not uniformly distributed along the ingot's length. This porosity caused by interdendritic flow of low melting point eutectic during the last stages of solidification and forms a network connected to the atmosphere, causing significant additional surface. The finely distributed porosity in Mg alloy ingots is the counterpart to the localized cavity in pure Mg ingots. It is a consequence of ingot casting, and should be considered as a natural part of the ingot.

A magnesium alloy ingot is not a homogeneous piece of metal. Analyses by spectrograph in various positions inside AZ91D ingots (slices) show large and systematic variations in Zn and Al concentrations with strong depletion near the top center /13/. Towards the end and towards the bottom of the ingot, the ingot is fairly homogeneous. The depletion of certain elements is worst in the regions solidifying last. The presence of low-melting point eutectics, combined with pressure gradients set up by solidification shrinkage, will cause macrosegregation. In general, specimens cut from ingots should not be used to certify the chemistry of an entire heat or pallet of ingots; it is strongly preferable to melt the ingot(s) and cast chemistry disks. However, if ingots provide the only access to the metal, specimens for chemical analysis should be selected from near the bottom of the ingot rather than near the top, and towards the ends rather towards the middle.

In Hydro Magnesium's furnaces, inclusions have been measured regularly by the HMIAM technique, Fig. 2 /14/, documenting the consistency of metal cleanliness in the metal to be cast. The chemical composition as well as ingot weights are subjected to strict statistical process control.

QUALITY OF INGOTS AND THEIR INFLUENCE ON PERFORMANCE IN THE DIE CASTING SHOP - The surface quality of the ingots are presumably the most important factor, but also chemical composition (especially Be) may have a significant effect. However, there are probably still arguments around which are based on perceptions and "myths" rather than science. Cultural differences may also play a role as people in different parts of the world may evaluate things differently.

Apart from the porosity, which is a feature of all alloy ingots, the surfaces of ingots are the main contributors to dross. The thickness of the surfaces, possibly varying in the range 5-500 nm depending on the conditions during solidification and cooling, should thus be a key quality parameter. However, evaluations of oxide skin thickness and compositions are complicated, and not easily undertaken. Unfortunately, the "nicest" looking ingots may not be those having the thinnest oxide skin, and vice versa.

Hydro Magnesium's ingot production is mainly automatically operated. Automatically operated processes usually provide more consistency of the product than more manually operated processes.

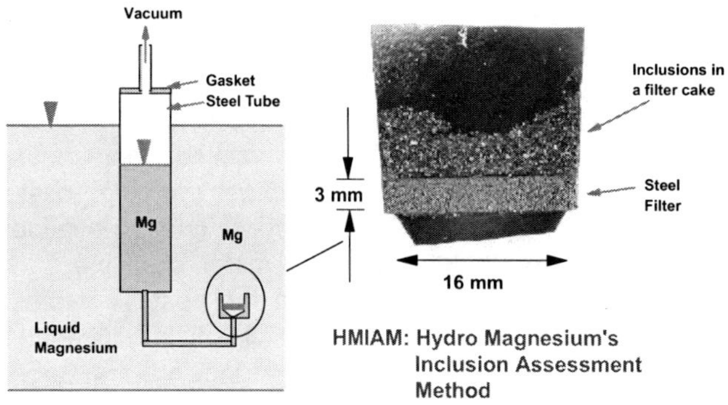

Figure 2 Hydro Magnesium Inclusion Assessment Method.

Some producers put considerable effort on making ingot nice looking. For instance, as mentioned earlier, the ingots may be skimmed before solidification. This operation is supposed to remove superficial oxides. Ingots may also be brushed before packaging. Ingots may even be washed in acid to get bright, shiny surfaces. Although the ingots having gone through such treatment may look impressive at first sight, the effect of acid washing tends to fade after long time storage; acid-treated and non-treated ingots stored for a long time are practically indistinguishable. Furthermore, nobody can document that acid washing has any effect on i.e. dross formation in the die casting shops.

Hydro Magnesium's ingot casting and cooling processes ensure that the surface oxide is consistently of minimum thickness. Provided good operational practice, which will be explained later, this will generate low and predictable dross volumes in the die casting shops. The so-called valve mark, which is a characteristic feature of ingots produced by Hydro Magnesium, is a consequence of the metering process during casting. It is a highly visible but insignificant part of the ingot surface.

QULITY OF INGOTS AND THEIR INFLUENCE ON THE PROPERTIES OF THE DIE CAST PRODUCT -Chemical composition and inclusions of the ingots will have an influence on the properties of the die cast products. Regarding surface appearance, the relation is probably even less quantified than the one discussed in the previous chapter, and unfortunately myths and perceptions possibly reflecting the customer's cultural background, are often decisive whether or not ingots from specific producers are to be purchased or not. In addition to the certified chemical analysis, the first impression based on visual and superficial observations of ingot appearance is considered important by many.

In an attempt to contribute to an understanding of this complex issue, ingots from different sources have been remelted in the laboratory and sampled by the HMIAM technique /14/. Perhaps surprisingly, the results revealed little difference in inclusion contents in the melts of these ingots. Although some ingots did not show particularly good appearance, the resulting molten metal was very clean. Generally, no correlation was found between the surface appearance of the ingots and the inclusion content in a melt coming from the same ingots.

What principally distinguishes the ingots is their ability and potential to form dross when remelted, representing unwanted and unforeseen challenges in die caster operations. It seems that there is no correlation between dross formation (metal loss) and metal cleanliness. In fact, the metal in the test showing the best cleanliness figures was identical to that which clearly produced the largest amounts of dross. This might indicate that dross layers accumulates at rates that may be characteristic for the actual metal source.

In a sample from one of the sources, intermetallic particle content 6-8 times higher than what we normally find in remelted ingots from Hydro Magnesium were found. As the succeeding samples indicated, the intermetallic particles may settle out fairly quickly and thus may not harm either the metering equipment or the die cast part. However, it necessarily accumulates in the die casting furnace causing additional sludge, resulting in additional loss of metal and up-time when the sludge is to be removed.

The ingots were also examined with respect to chemical composition. As cutting specimens for analytical purposes directly from ingots is not recommended due to considerable inhomogenities /13/, permanent mould cast medallions (disk samples) were cast from melted alloy ingots. In one case the certificates of analysis following the ingots were very different from results obtained by spectrograph and ICP-AES analyses of the cast disk samples. Both AM60 and AZ91 from this specific supplier were far out of ASTM spec. on Ni, while on AZ91 Si was off spec. as well. For both alloys the certificate of analysis showed excellent values for Ni as well as for Si. Such mismatches are dangerous, especially when highly critical elements like Ni are involved.

QUALITY OF TRANSPORT AND STORAGE AND THE INFLUENCE ON PERFORMANCE IN THE DIE CASTING SHOP - It must be stressed that an ingot should not be considered as a product as such; it is rather an intermediate and convenient storage of metal. The quality can deteriorate by unfavourable conditions during transport and storage causing pick-up of moisture and corrosion. Recommended conditions for storage include:

- Avoid direct exposure to water either through condensation or rainy weather.
- Magnesium products should be stored in clean temperature controlled warehouses with concrete floors.
- Keep magnesium away from easily combustible materials.
- Avoid storage together with acids, salts and fertilizers.

Transport should take place in closed containers. Due to variable temperature and humidity conditions, pick-up of moisture can occur. When necessary, Hydro Magnesium uses desiccants inside the containers during long sea transports to prevent excessive corrosion.

PROCESS CONTROL IN THE DIE-CASTING SHOP INFLUENCING THE PERFORMANCE - Remelting of ingots in a die casting shop is not necessarily a straightforward operation. Preheating should eliminate possible surface moisture that may represent a safety hazard as well as a source for dross and inclusions. Gentle charging minimizes unnecessary splashing and entrapment of oxides from the molten metal surface.

Although the reasons are not completely understood, it is observed that most of the surface of ingots (and scrap) floats to the surface when the parts melt /15/. This happens in spite of the fact that these impurities should have a density higher than magnesium. Presumably the explanation is that a gas layer attaches to the oxides, thus lowering the mean density. Such gas layers have been detected for Al_2O_3 in aluminum /16/. Perhaps water vapor and dissolved hydrogen play a role in creating the gas layers. The fact that there is a gas layer shows that the particles prefer gas to the magnesium melt, the particles are "wetted" by gas /17/.

The growing layer of oxides on top of the melt puts strong requirements on the housekeeping of the furnace(s), including removal of surface dross and bottom sludge, and cover gas protection. If dross and sludge are not allowed to accumulate extensively, modern transfer and dosing techniques will probably restrict introduction of additional inclusions. However, in cases where hand ladling is still practiced oxides from the molten metal surface (dross) may be fed directly to the die casting machine. If the metal quality is inferior, housekeeping may become notoriously difficult.

As mentioned earlier, it is considered extremely important that the metal is at equilibrium with respect to Mn and Fe when cast to ingots, as well as when it is remelted in the die casting shop /18/. If the ingots are cast under non-equilibrium conditions, or at a considerably higher temperature than the working temperature in the die casting furnace, precipitation of intermetallic particles will occur in the die casting furnace generating sludge, and subsequent loss of up-time due to its removal. This means that the ingot casting temperature ideally should be the same as the metal temperature in the die casting furnace. Presently there are no distinctions between ingots to be used for hot or cold chamber machines; ingot casting temperatures are usually the same. As a result, some additional precipitation of intermetallic particles (sludge formation) may occur in hot chamber operations where the operating temperature maybe is as low as 630°C. In any case, proper temperature control and rapid melting minimize sludge formation.

As settling either to the top surface (dross) or to the bottom (sludge) is the main mechanism for inclusion removal, long and even metal residence time in the die casting furnace is important. If space permits, this can partly be achieved by a separating wall dividing the furnace into a melting and a casting chamber. Preferably, a two furnace system /3/ can be used. When melting in one furnace and casting from a second, excellent temperature stability can be achieved. Temperature gradients generated during charging are thus confined to the melting furnace, and the metal to be cast comes from a constant temperature reservoir. Furthermore, a two-furnace system gives a long and evenly distributed metal residence time, facilitating good and consistent metal quality. The effect of temperature gradients and temperature control on chemical composition and sludge (and dross) formation may not be very well documented.

Measurements of quality are usually based on narrow selections of sample material. However, the only 100% test is melting and processing in the die casting shop. Small quality differences can accumulate to give large and unwanted problems over time.

PROCESS CONTROL IN THE DIE CASTING SHOP AND ITS INFLUENCE ON THE QUALITY OF THE DIE CAST PRODUCT – To a large extent, the quality of a die cast product is considered satisfactory if the mechanical properties meet certain criteria. These are reflected by the intrinsic properties of the alloy (i.e. strength, ductility and corrosion rates) and engineering know-how. The die casting engineer is usually expert in the design of parts and gating systems, and the related operational parameters on the die casting machine. Given the chemical composition of the alloy, the operational parameters will normally fully determine the properties of the die cast part. However, in certain situations, inclusions in the metal may lead to a deterioration of the intrinsic properties /19/.

Die casting furnaces at a number of locations have been sampled with the HMIAM technique. The results show a substantial spread in results ranging from "clean" (similar to contents measured in Hydro Magnesium's casting furnaces) to levels that has a potential to influence negatively on mechanical properties /19/.

Inclusions are also entrapped during the die casting, due to oxidation in the shot sleeve, entrapment of air during casting, and possibly pick-up of lubricants from the die. Moreover, casting defects such as cold shuts, hot tears and porosity will normally dominate over inclusions in determining the properties of a die cast alloy part. However, when the casting defects are minimised, the impact of inclusions is assumed to increase /19/. This is illustrated in Fig. 3. In a casting free from casting defects, the inclusion content and distribution of inclusions are thus assumed to be a critical factor. These considerations suggest a critical inclusion content to be a function of the level and distribution of casting defects.

Another aspect is also the indirect influence of inclusions on the mechanical properties. Inclusions have a significant influence on fluidity in die-casting /19/. Inclusions can then directly indicate cold shuts or other surface defects in the castings and nucleate tears either in the semi solid material or later in the completely solidified metal.

Porosity is common in die cast products, due to entrapped gas and/or shrinkage. Hydrogen, although having a considerable solubility in molten magnesium /20/, will normally not contribute significantly to the porosity.

METAL CLEANLINESS ASSESSMENT IN THE DIE CASTING SHOP

Die casters need a quick and reliable method for checking the metal cleanliness in their die casting furnaces. This can be a part of the process control, although the

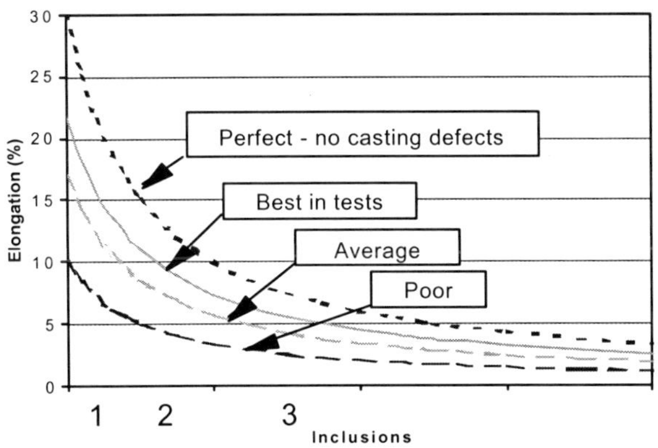

Figure 3. Metal cleanliness vs. mechanical properties. On a "perfect" casting a low level of inclusions (1) will deteriorate the elongation, while on a "poor" casting even a high level of inclusions (3) hardly makes any impact /19/.

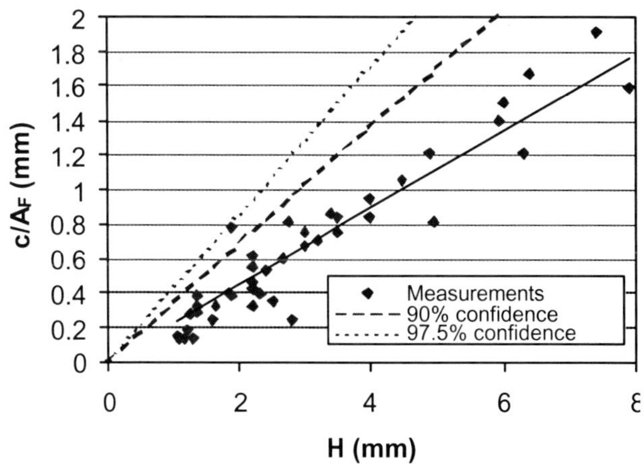

Figure 4. Measurements of gross filter cake thickness vs. oxide content from Fig. 4, with extreme value statistics /9/.

frequency of sampling and the acceptance levels could be an item for negotiation with the end-user of the specific part being cast. Various methods for metal cleanliness assessment have been considered, and an updated survey has recently been published /9/. Nevertheless, there are no instruments/tools for metal cleanliness evaluation capable of generating a quick and reliable response at a reasonable detection level, and at an affordable price, either for magnesium or aluminium.

However, major improvements with respect to the reliability of the HMIAM sampling system, Fig.2, have recently been realised /9/. A two-piece filter cup is press-fitted against the filter and welded from the outside. This ensures a perfect seal of the filter, avoiding bad samples due to leakage. Even more important, charts where gross filter cake thickness is compared to historic metal cleanliness values have been developed, Fig. 4.

By introducing the simplified analysis, the total time for sampling and analysis is reduced by roughly 50%; more importantly, the analysis which is the most difficult part, can be made optional. Provided that the majority of inclusions are oxides, which usually is the case, the gross filter cake thickness approach may be applicable for inclusion contents where gross filter cake thickness $H > 1$ mm. A cake thickness corresponding to 1 mm per kg metal filtered gives a resolution corresponding roughly to 140 ppm MgO (assuming that all the oxides are MgO) or 55 ppm O. This should be sufficient for most practical purposes, making the simplified HMIAM method an attractive tool for rapid metal cleanliness evaluation in the die casting shop.

Measurements of quality are traditionally based on a narrow selection of sample material for instance ingot samples or disk samples. The HMIAM method provides rapid and reliable measurements of metal cleanliness directly from the molten metal in the die casting furnace, bringing the quality evaluation one step closer to the actual die cast part. However, as mentioned earlier, sampling in the molten alloy before die casting does not necessarily reflect the true oxide content in the die cast part, because certain amounts of oxide inclusions are probably generated in the shot cylinder. Furthermore, the extremely high pressures and velocities involved in the high pressure die casting process, preferably causing atomization of liquid metal /21/, surely affect the number size distributions of inclusions. Oxide particles, and oxide films in particular, may be broken up into smaller particles. Despite this, it is believed that the more large particles entering the shot sleeve, the more large particles survive the high forces involved in the die casting process. One may also speculate about (large) particles affecting the liquid metal atomization process, thereby indirectly deteriorating the castablilty as well as the properties.

CONCLUSIONS

Traditionally we understand the term metal quality as:

1) Chemical composition
2) Inclusions and porosity inside the metal
3) Surface appearance
4) Consistency

How a supplier complies with the customer's expectations on these four issues is important for the relationship between the customer and the supplier. The customer may correlate the first impression of the ingot delivery with the quality of the producer, but this may not always be correct, and of minor importance.

Dross formation in the die casting furnaces is strongly influenced by the surface of ingots. Despite having an excellent appearance, new metal sources may represent unwanted and unforeseen challenges in die caster operations, creating a need for increased quality consciousness on ingot purchase, and supplier qualification. There seems to be no clear correlation between metal cleanliness and dross formation (metal loss).

Measurements of quality are traditionally based on a narrow selection of sample material, for instance ingot samples or disk samples. The HMIAM method provides rapid and reliable measurements of metal cleanliness from the molten metal in the die casting furnace, bringing the metal quality evaluation one step closer to the actual die cast part. However, the only 100% test is melting and processing in the die casting shop. Small quality differences can accumulate to give large and unwanted problems over time.

Molten metal handling and furnace housekeeping is crucial both with respect to metal cleanliness in the furnaces which is vital for the performance of the cast parts, and on the up-time of operation; the latter being crucial for the operating cost. If the metal quality is inferior, housekeeping may become notoriously difficult. After all, the die casters' business is to cast parts, not to clean up furnaces.

REFERENCES

1. Friedrich, H., Schumann, S. "The use of magnesium in cars – today and in the future", 56th IMA Conf, Coronado, CA, USA, 1998. pp.1-7
2. Fink, R., "Large thin-walled magnesium die castings", 56th IMA Conf, Coronado, CA, USA, 1998. pp.24-28
3. Øymo, D., Holta, O., Hustoft, O.M., Henriksson, J. "Magnesium recycling in the die casting shop". Proc. of the 1992 conf. on the recycling of metals, Düsseldorf/Neuss, Germany, May. ASM International, 1992. pp.143-150.

4. Leyndecker, T., Rauch, M., Economical process for the in-house remelting in Magnesium die casting shops. In Mordike BL, Kainer KU (eds) Magnesium Alloys and their Applications. Werkstoff-Informationsgesellschaft mbh, Frankfurt, Germany, 1998. pp.703-707
5. Berkmortel, R., Wang, G.G., Bakke, P., "Fluxless in-house recycling of high purity magnesium die cast alloys at Meridian operations". 57th IMA Conference, Vancouver BC, Canada, May. 2000. pp.22-27
6. Galowsky U, Kuhlein M. "A new conti-process for the fluxless recycling of high purity magnesium". In Hryn JN (ed) Magnesium Technology 2001. TMS-AIME, Warrendale PA, USA. 2001. pp49-54
7. Berkmortel, R., Wang, G.G., Bakke, P., "Continuous fluxless recycling of high purity magnesium die cast alloys", CIM conference, Toronto, Canada, Aug. 2001
8. Bakke, P., Svalestuen, J.M., Mellerud, T., "Metal quality – Does it matter for die casters and end users ?". 58th IMA Conference, Brussels, Belgium, May 2001
9. Bakke, P., Wang, G.G., Albright, D., "Metal cleanliness assessment – Recent developments", NADCA 2001, Cincinnati, OH, USA, Oct.-Nov. 2001
10. W.T. McLeod, Editor, The New Collins Dictionary and Thesaurus, William Collins Sons & Co Ltd, London, UK, 1987
11. NS-ISO 8402, Quality management and quality assurance. Vocabulary, 2. Edition, October 1994
12. NS-ISO 9001, Quality management systems. Requirements, 3. Edition, December 2000
13. Bakke, P, Fuerst, C.D., Westengen, H., "Solidification induced inhomogenities in magnesium alloy AZ91", Magnesium Technology 2000, TMS-AIME, 2000, 201-210
14. Bakke, P., Karlsen, D.O., "Inclusion assessment in magnesium and magnesium base alloys", SAE Technical Paper 970330, 1997
15. Hillis, J.E., Mercer II, W.E., "Separation of non-metallic contaminants in fluxless melting and refining of magnesium alloys", SAE Technical Paper 2000-01-1125. 2000.
16. Haugland, E., "Removal of inclusions by gas purging", Dr. Ing. Thesis, NTNU, Trondheim, Norway, 1998
17. Tathgar, H.S., Bakke, P., Engh, T.A., "Impurities in magnesium and magnesium based alloys and their removal", 5th Intl. Magnesium Conference, Munich, Germany, Sept. 26-29, 2000, pp.767-779
18. Holta, O., Westengen, H., Røen, J., "High purity magnesium die casting alloys. Impact of metallurgical principles on industrial practice", 3rd Intl. Magnesium Conf., Manchester, UK, April 10-12, 1996.
19. Bakke, P., Pettersen, K., Guldberg, S., Sannes, S., "The impact of metal cleanliness on mechanical properties of die cast magnesium alloy AM50". In Kainer, K.U. (ed) Magnesium Alloys and their Applications. Wiley VCH Verlag Gmbh, Weinheim, Germany, 2000. pp739-745
20. Øvrelid, E.J., "Hydrogen in magnesium and magnesium alloys", Dr. Ing.-thesis, NTNU, Trondheim, Norway, 1997
21. Hao, S.W., Hu, B.H., Niu, X.P., Pehlke, R.D. "Atomization in high pressure die casting – a problem and a challenge", Die Casting Engineer, 2000, pp.42-56.

2002-01-0079

New Aspects of Temperature Behavior of AJ52x, Creep Resistant Magnesium Alloy

Pierre Labelle, Mihriban Pekgularyuz, Michel Lefebvre and Robert Bouchard
Noranda Inc. - Technology

ABSTRACT

The new patented high temperature alloy AJ52x has the potential to replace the aluminum alloy 380 in power-train applications. The creep properties of AJ52x are similar to the aluminum alloy A380 and overall mechanical performance is superior to other magnesium creep resistant alloys AS41 and AE42. These properties were presented in the SAE paper 2001-01-0424. The present paper reports on further evaluation of the properties of the AJ52x alloy.

Thermal conductivity of AJ52x has been investigated in a range of temperatures from R.T. to 450°C; the thermal conductivity is higher than any die-cast magnesium alloy. So due to the high thermal conductivity, the AJ52x alloy is able to evacuate 75% of energy as the aluminum alloy currently used in the power-train components. These thermal properties have been compared with thermal properties calculated from electrical resistivity at room temperature. The effect of temperature on Young's Modulus and the percent-retention have been found to be similar to standard aluminum alloys. Additional test results show best torque and bolt-load retention for aluminum A380, however AJ52x is superior to standard magnesium die-casting alloys.

I. INTRODUCTION

The mass of the vehicle has an effect on fuel consumption by affecting rolling resistance, resistance to climbing and acceleration. A fuel consumption target of 3 to 4 L /100Km is being pursued in various countries, in addition to rising customers expectation in terms of performances, comfort and safety. To achieve these divergent targets, magnesium with a density of 1.71 g/cm^3 is a suitable material. New part applications are primarily responsible for the increasing use of magnesium in the automotive industry. But the growth has been limited to components that used standard die cast magnesium alloys such AZ91D, AM50 and AM60B and are mainly located in interior applications. Recently, power-train components have been produced by Honda (1) using a high temperature resistant magnesium alloy. The general characteristics of magnesium are suitable for application in power-train components where large die cast components are used. However, the service temperatures of power-train components are in the 150°C range, which exceed service temperature of standard magnesium die-casting alloys previously mentioned. To overcome this temperature limit Noranda developed the Mg-Al-Sr alloy AJ52x (2) which provides good overall mechanical properties without castability problems usually associated with rare-earth or calcium containing alloys.

In addition to the mechanical properties previously investigated (2), designers are concerned with how the alloy behaves with respect to other thermally related properties. To address this specific demand Noranda Inc - Technology has started an investigation under the high temperature alloy development program.

The following properties have been investigated and compared to various standard magnesium alloys:

1) Thermal Properties vs. Service Temperature,

2) Electrical Resistivity,

3) Young's Modulus vs. Service Temperature

4) Bolt-Load Retention under various loads and temperatures using instrumented bolts

II. BACKGROUND

Several types of heat-resistant magnesium alloys have been developed to date, including AS, AE, ZC, Mg-Al-Ca, Mg-Al-Ca-Re and Mg-Al-Ca-Sr alloys. Unfortunately, many of these alloys suffer from inferior die-castability or have disadvantages in terms of marginal performance improvements and high material cost. The AJ52x (Mg-5Al-2Sr) offers a good combination of die-castability, creep at 150°C and 175°C, and mechanical properties. There are other major points that need to be considered when deciding whether to select magnesium high temperature alloy for a specific part. The ability to evacuate heat generated inside the component, the electrical resistivity, the retention of the stiffness over a certain range of temperatures and the bolt-load retention are few of these points.

III. EXPERIMENTAL PROCEDURES

All samples used in this study have been taken from die cast plates, ASTM tensile bars and solid cylinders described in past publication (2). Additional details and data from experiments are given in Appendix A.

A. THERMAL CONDUCTIVITY AND THERMAL DIFFUSIVITY

Three disk samples, each about 12.68 mm diameter and 1.67 mm thick have been sectioned and machined from die cast corrosion plates. The Laser-Flash Method. (ASTM E1461-01) has been used to determine the thermal conductivity and thermal diffussivity (3).

B. THERMAL LINEAR EXPANSION

The coefficient of thermal linear expansion of AJ52x and AE42 was determined (4) using three cylindrical samples, each about 10 mm in length and 6 mm in diameter, machined from a cylindrical castings of 25 mm diameter and 40 mm length. An MMC Quench-Deformation dilatometer has been used to measure the thermal coefficient of linear expansion from room temperature to 200°C.

C. ELECTRICAL RESISTIVITY

The four-points method developed by Hydro-Québec, Montréal, Canada has been used, This method is similar to the ASTM B193 method but uses four points to improve accuracy when measuring electrical resistance of highly electrically conductive materials like metals.

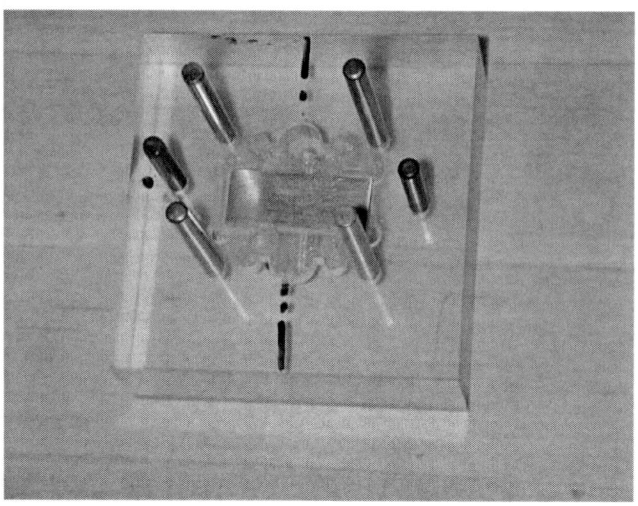

Fig.1. Sample holder for resistivity test.

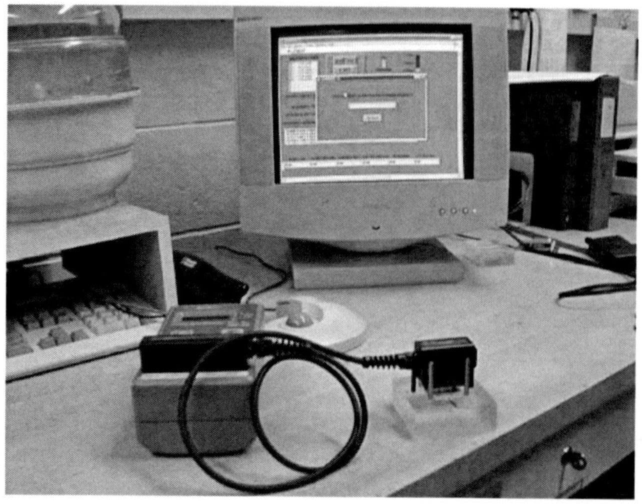

Fig 2. Set-up for resistivity measurements.

The electrical resistivity of AJ52x, AE42, AM60B, AZ91D and aluminum A380.1 has been measured at room temperature using an Ultra-Optec micro ohmmeter. For each alloy, three samples of dimensions 3x1 cm, machined from die cast corrosion plates, have been used. The magnesium samples were cleaned using 4%NaOH + deionized water, 4% H_2SO_4 + deionized water and sanded with 320 grid paper. Samples were held in place using a special fixture (Fig 1) and resistivity was measured using a four-point probe (Fig 2).

D) YOUNG'S MODULUS

The Young's Modulus or the Modulus of Elasticity has been determined for aluminum A380, magnesium AZ91D and magnesium AJ52x at R.T, 150°C and 175°C using a water-cooled extensometer (Fig. 3).

Fig. 3. Sample and water-cooled extensometer.

The test was performed using a Universal Instron testing machine, Model 8502. Three die cast cylindrical specimens of 6,35 mm (0.25 in.) diameter, 50,8 mm (2 in.) gage length, per alloy has been used. The tests were repeated a minimum of two times for each alloy. For tests performed at elevated temperatures, specimens were held in an oven for a period of 30 minutes prior to testing. The modulus has been measured at stress levels corresponding to a 13 - 17 MPa range where the modulus of magnesium is linear. The Young's Modulus was obtained by using the best R^2 fit for the data obtained in this range.

E. BOLT-LOAD RETENTION

Bolt-load retention tests are greatly affected by the sample geometry and tests procedures; no standardization is available to compare materials. The tests procedures used in this study have previously been developed by Ford Motor Co. and University of Michigan (5) but at that time, a special die cast fixture was used. In the case of this study (Fig 4), the sample has been replaced by a simple cylindrical sample previously used by Dow Magnesium Inc. in study on creep for the VW 411 air-cooled engine (6).

The samples were machined from a solid die cast cylinder to the following dimensions: 18 mm diameter,18 mm length and 10 mm hole. Special instrumented M10 (Fig. 4) bolts have been used to create initial clamping forces of 7 / 10.5 / 14 KN which correspond to 40 / 60 / 80 MPa of initial compressive stress on the samples. The assemblies have been heated using a Neslab EX251 silicon oil bath at 125°C, 150°C and 175°C. The results were monitored using a data acquisition system based on a Datatracker Process Instrument, Model 200, linked to a PC (Fig 5).

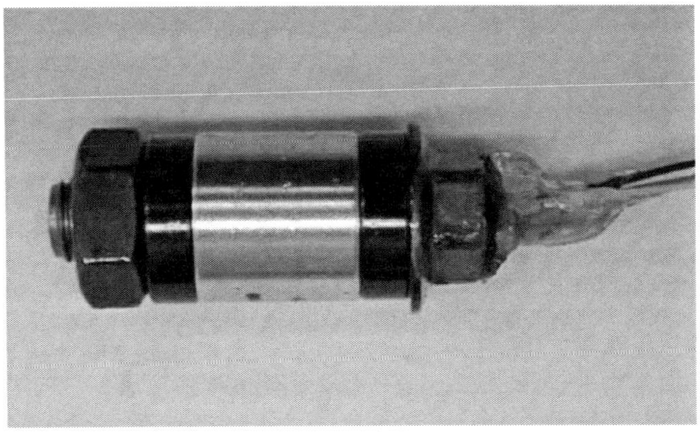

Fig. 4. Instrumented M10 bolt, steel washer.

The tests were monitored continuously for a period of 120 hrs, and two samples per alloy have been used for magnesium alloys. To verify the accuracy of the system a steel sample has been machined and tested at 60 MPa and 150°C. The assembly retained 96% of the initial load.

Fig. 5. Data acquisition system used in the bolt-load retention test.

IV. RESULTS & DISCUSSION

A) THERMAL CONDUCTIVITY

The measured thermal conductivity curves of AJ52x (7), AE42 (7) AM60B, AZ91D and Al 380.1 are presented individually in Figures 6-10. Historically, the thermal conductivity were calculated from electrical resistivity values (8) using Bungardt & Kallenback (9,10) or Weidmann-Franz (11,12) laws and these rules yield similar values to those obtained experimentally in this study for AJ52x and Aluminum A380. However for poorer electrical conductors such AZ91D, AM60B, the calculated values differ by a small percentage from values measured in this study.

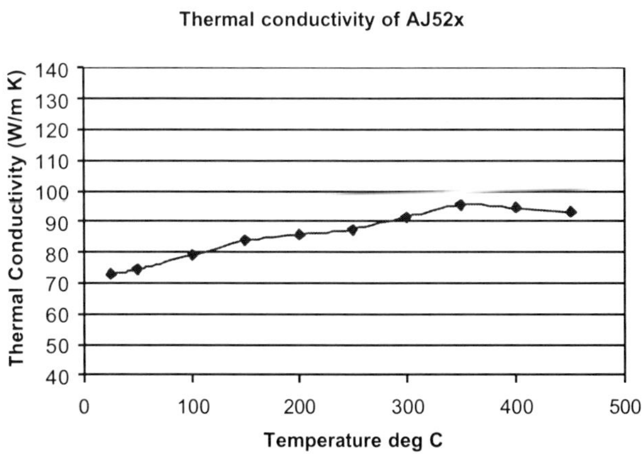

Fig. 6. Thermal conductivity of AJ52x.

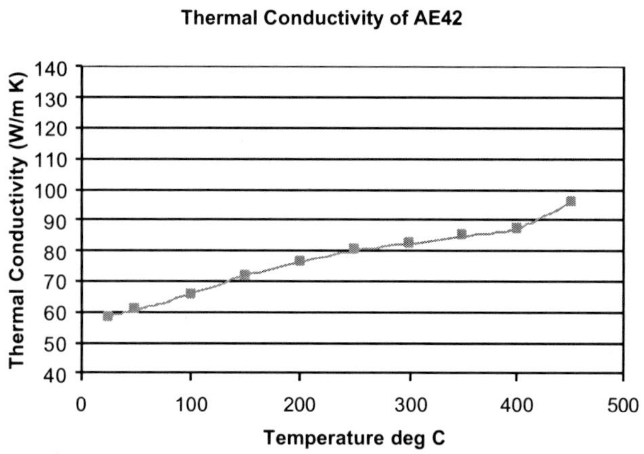

Fig. 7. Thermal conductivity of AE42.

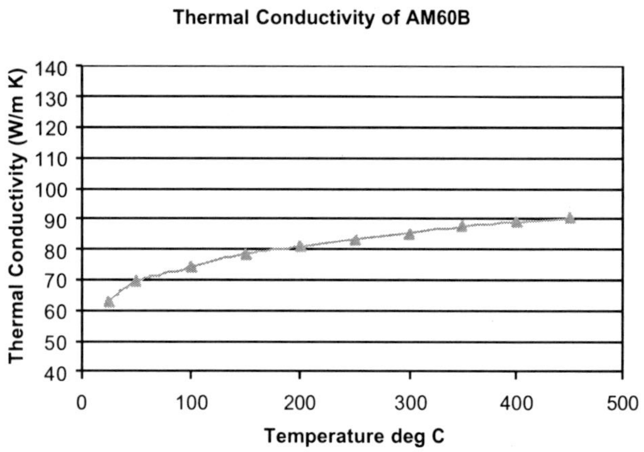

Fig. 8. Thermal conductivity of AM60B.

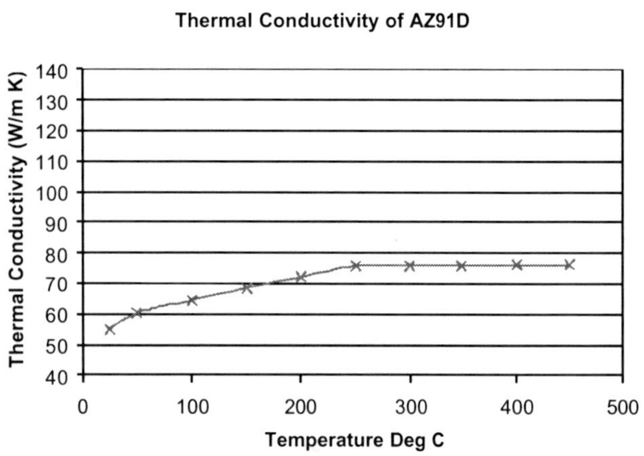

Fig. 9.. Thermal conductivity of AZ91D.

However, thermal conductivity values measured for AE42 are lower than the values calculated using electrical conductivity laws reported in the literature (9,10,11,12). When compared with other magnesium alloys (Fig 11), AJ52x shows the best thermal conductivity and is closer to the range of aluminum alloys than other die-cast magnesium alloys.

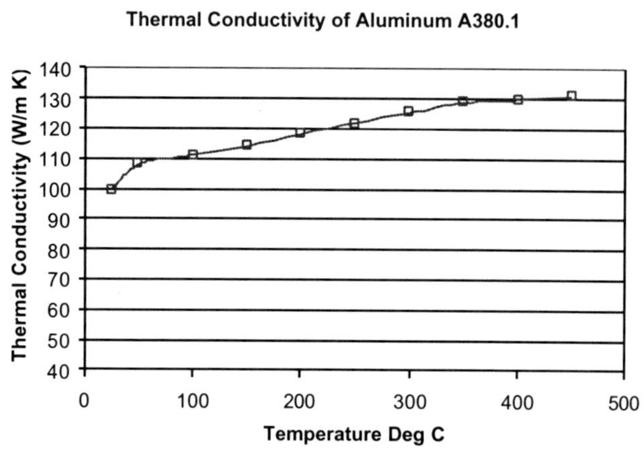

Fig. 10. Thermal conductivity of Aluminum 380.

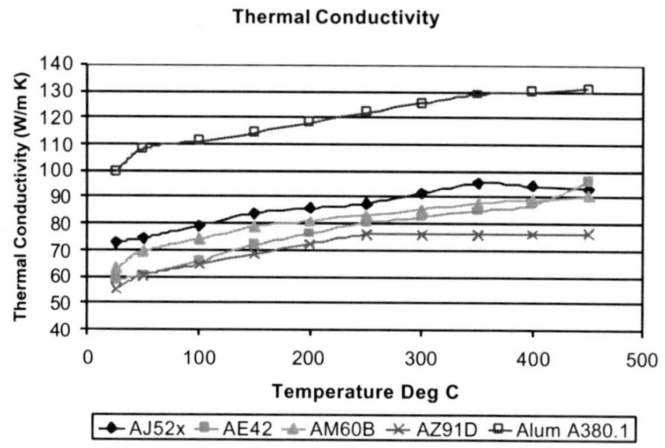

Fig. 11. Comparative thermal conductivity

The high thermal conductivity of AJ52x increases the ability to transfer heat inside the component, compared to AE42 and other creep resistant Mg alloys. A hot spot on the component will be cooler with AJ52x, which will distribute the heat inside the material. This is an advantage for power-train, electrical, and electronic components that need to evacuate heat, to avoid potential heat build-up at usual service temperatures. It should be noted that the value measured at room temperature for A380 (99.6 W/m°K) is slightly higher than the value reported in the literature (13) for 380 (96.2 W/m°K). This difference may be due to the lower iron content in aluminum A380 compared to 380.

B. THERMAL DIFFUSIVITY

The thermal diffusivity, a measure of the rate at which heat will be transfer through a metal is defined as:

$$\alpha = 10000K/(C_p x\ \rho) \qquad \text{Eq. 1}$$

where:

α	= Thermal Diffusivity (cm^2/sec)
K	= Thermal Conductivity (W/m °K)
C_p	= Specific heat (J/Kg °K)
ρ	= Density (Kg/m^3)
10000	= used to convert from m^2/sec to cm^2/sec easier to visualize

Curves of AJ52x, AE42 AM60B, AZ91D and Al 380.1 are presented at Figs. 12 to 16. As indicated in Figure 17, AE42 shows thermal diffusivity close to the aluminum alloy A380. In the case of AJ52x, thermal diffusivity is higher then AM60B and AZ91D. Since thermal diffusivity is high for magnesium and aluminum alloys, parts used for high temperature applications will run cooler, than steel counterparts if there is a heat dissipation mechanism available like a cooling fan.

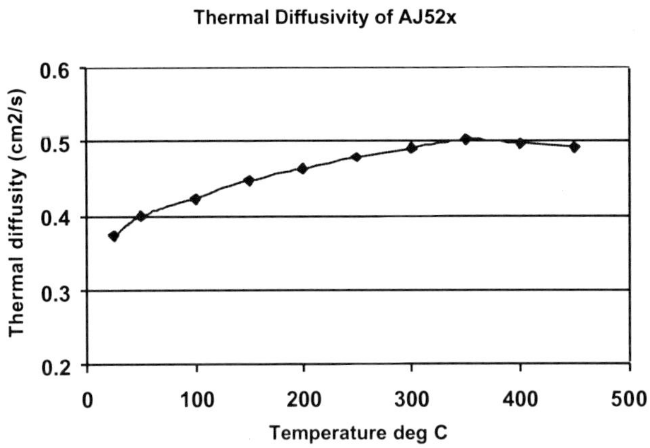

Fig. 12. Thermal diffusivity of AJ52x.

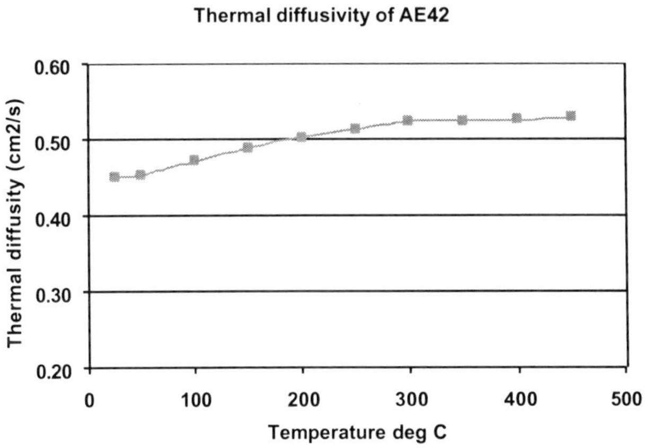

Fig. 13. Thermal diffusivity of AE42.

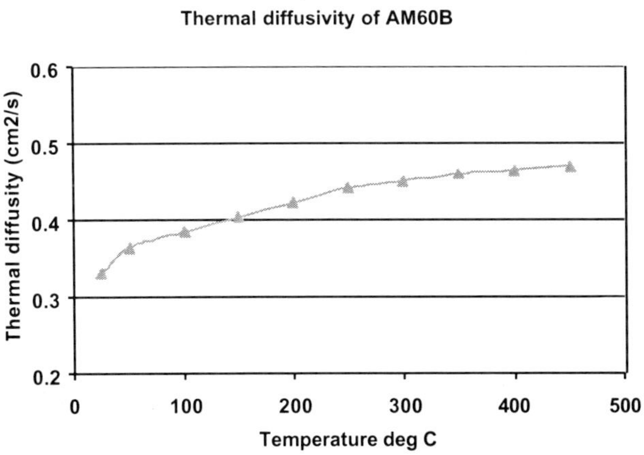

Fig. 14. Thermal diffusivity of AM60B.

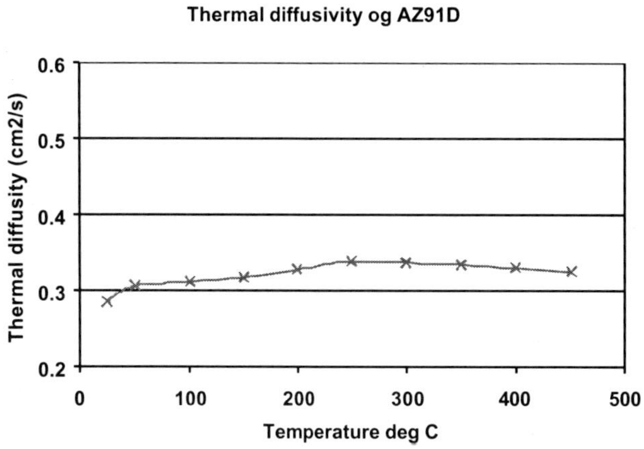

Fig. 15. Thermal diffusivity of AZ91D.

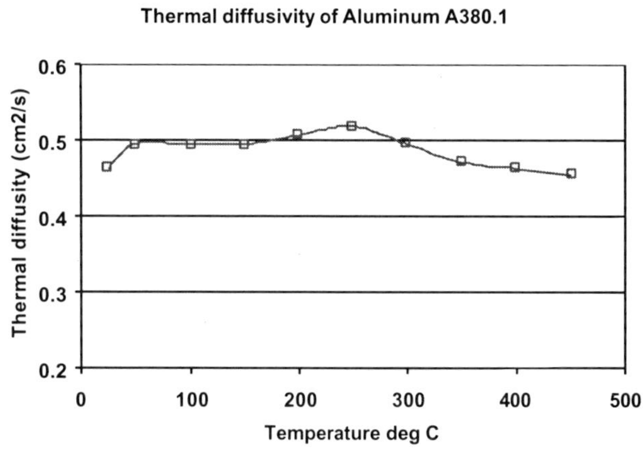

Figure 16. Thermal diffusivity of aluminum 380.

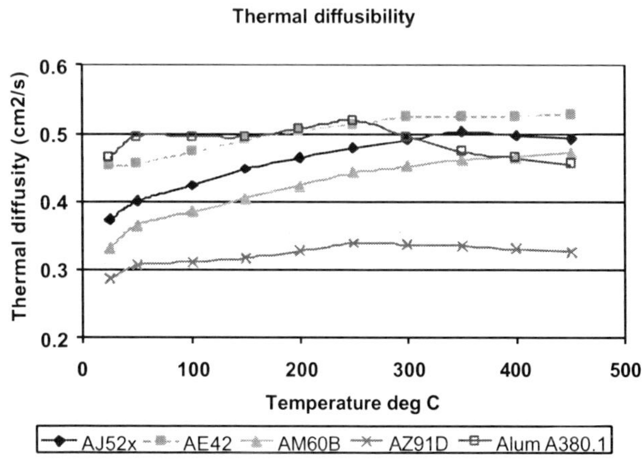

Fig. 17. Comparative thermal diffusivity.

C. THERMAL LINEAR EXPANSION

The thermal linear expansion coefficients of AJ52x and AE42 have been determined (4). And these results have been compared to published results for other die-casting magnesium alloys and the aluminum A380 alloy (Fig.18). The coefficient of AJ52x is similar to AZ91D, so design rules used to calculate stresses induced in press-fitting a steel insert or when interference with other materials exist in an AZ91D component are also applicable to AJ52x.

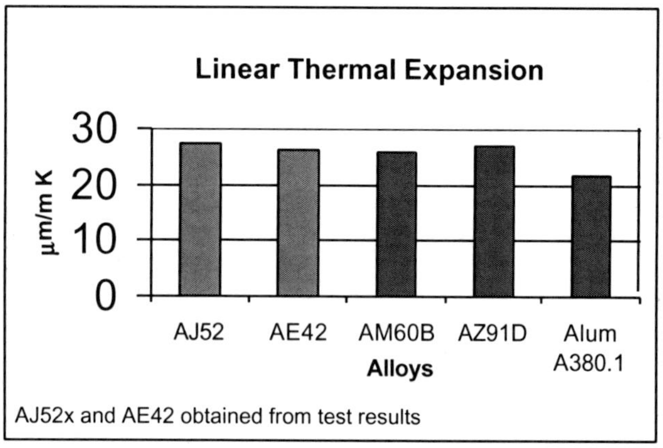

Fig. 18. Linear thermal expansion of diecast alloys.

D. ELECTRICAL RESISTIVITY

Figure 19 compares the results of electrical resistivity obtained in this study and those that are available in the literature (13,14,15) for various alloys.

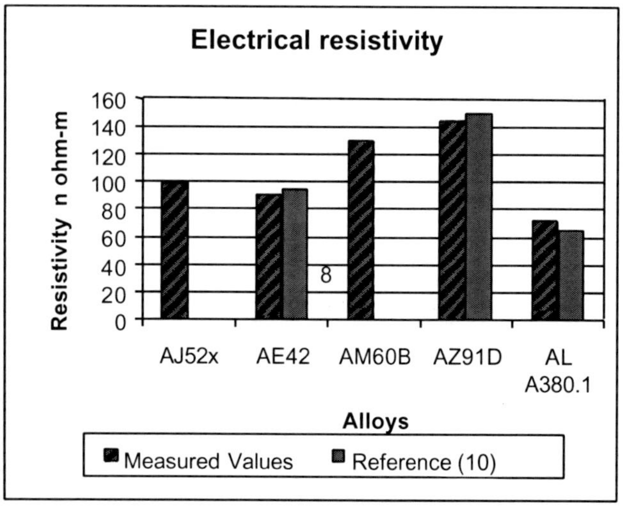

Fig 19. Electrical resistivity of diecast alloys.

Aluminum die cast alloy A380 has the lowest electrical resistivity and the AZ91D alloy is twice as resistive as the aluminum A380. Other magnesium alloys have values ranging between these two alloys. AE42 shows the lowest electrical resistivity of all magnesium alloys tested at room temperature, even though it does not have the highest thermal conductivity. AJ52x has a lower electrical resistivity than current commercially product die-cast magnesium alloys.

The electrical resistivity results obtained do not correlate with the results obtained for thermal conductivity but correlate with the results obtained for thermal diffusivity at room temperature. It is observed that the classification of alloys according to electrical resistivity corresponds to their ranking in thermal diffusivity which is different from the ranking in thermal conductivity. For example if the thermal conductivity (Fig. 11) and electrical resistivity were to correlate (Fig. 19), as is usually the case, the thermal conductivity of the AJ52x alloy should have been between AE42 and AM60B, which is not the case. This may be due to the specific nature of the rare-earth containing alloy which lower the thermal conductivity. This exceptions to the rule have already been noticed for WE54 an Yttrium and Rare Earth containing alloy (9),and it is observed in this study for die cast AE42 alloy.

E. YOUNG'S MODULUS

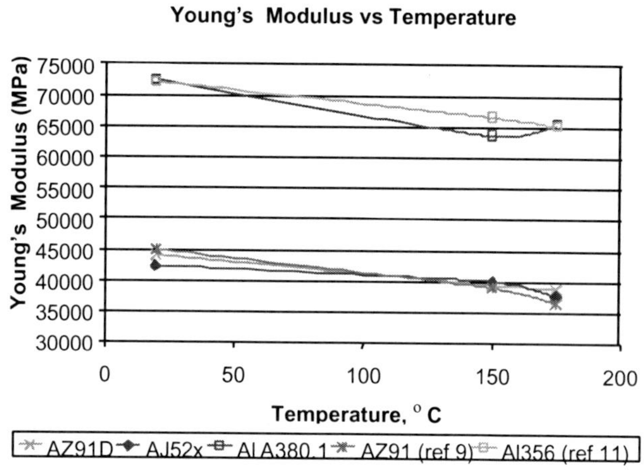

Fig. 20. Young's Modulus vs. temperature.

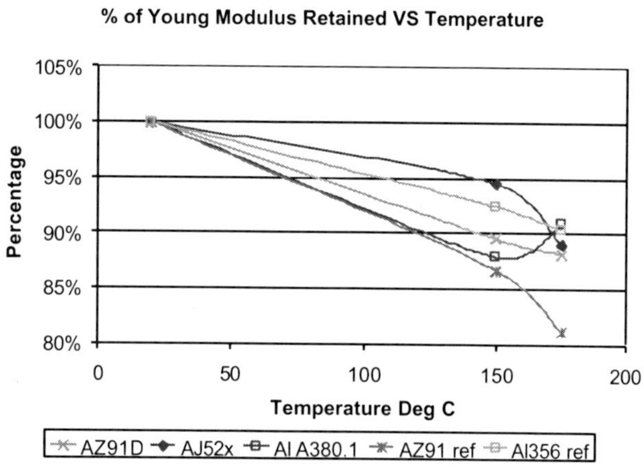

Fig. 21. Percent loss in Young's Modulus

Data obtained in this study (Figs. 20 & 21) were compared to published Young's Modulus values for AZ91D (9) and aluminum permanent casting alloy A356 (16). At room temperature, the Young's Modulus of AJ52x is slightly lower than the modulus of AZ91D (Fig 20). However, the modulus is maintained over a wider temperature range. At 150°C, the percentage of retention of AJ52x is better than any other die-cast commercial magnesium and aluminum alloys measured or found in the literature (Fig 21).

At 175°C, the percent Young's Modulus retained by the AJ52x alloy becomes similar to the aluminum alloy. This means that when temperature increases, a component manufactured using AJ52x will retain stiffness in the same ratio as components made in aluminum alloys. It can be noted that the sudden reduction at 150°C in the Young's Modulus of Al A380 can be due to recrystallization that occurs in this alloy at that temperature. This agrees well with earlier findings (2) that better creep properties were observed at 175°C than at 150 °C for aluminum alloy A380.

F. BOLT-LOAD RETENTION

It is known that the magnesium alloy AZ91D can perform well up to 125°C, even at that level, the residual clamping forces of this alloy were noted to be significantly reduced (Fig 22). Only the application of a high initial load will maintain acceptable clamping forces at 125°C. At 125°C, the load and the temperature affect AZ91D, higher loads retaining proportionally lower percentage of initial clamping forces. But for high temperatures, high initial loads have no effects on the residual compressive stress of AZ91D which stabilizes to 5 MPa at 150°C and to less than 1MPa at 175°C (Fig 23) regardless of the initial load applied on the assembly.

In the case of AJ52x, the alloy retains proportionally at 175°C the same percentage of clamping forces as AZ91D at 125°C. Percentage of loss is a function of temperature only, and less affected by initial stress level. So an AJ52x assembly can maintain clamping forces up to 175°C by providing higher initial clamping forces. Compared to AZ91D, which significantly lost its clamping strength after 125°C, AJ52x, maintained acceptable clamping strength up to 175°C.

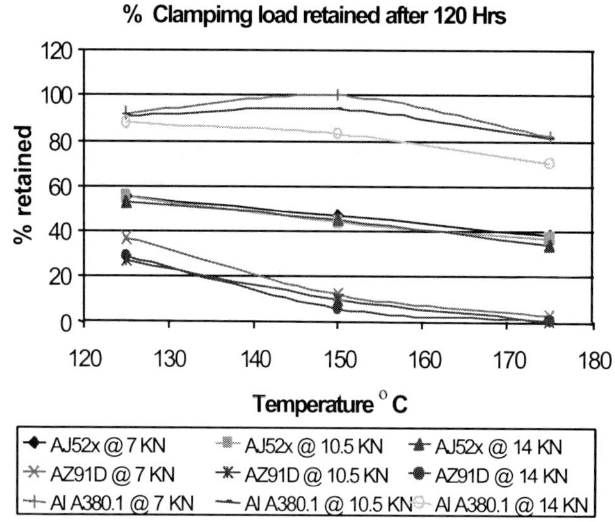

Fig. 22. Percent bolt-load retained.

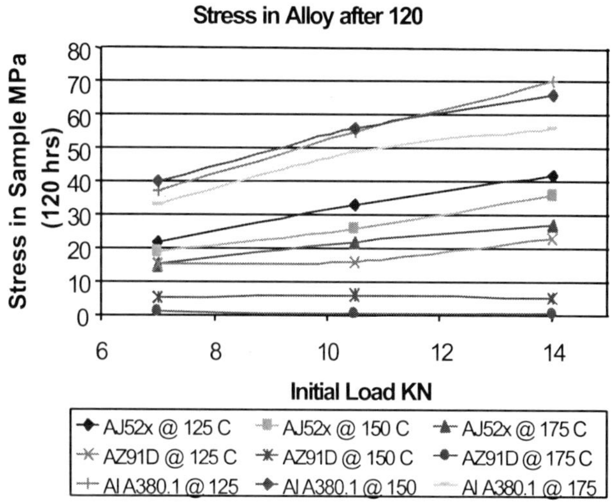

Fig. 23. Stress in the sample after 120 hrs.

As shown in Figs. 22 and 23, aluminum A380 shows the best bolt-load retention results and maintains high percentage of load retention and higher stress in the cylinder than all magnesium alloys tested. Considering that the yield and ultimate stress of AJ52x alloy at 150°C are 30% lower than they are for aluminum A380 (2), one design technique would be to locally modify the geometry of the parts in high stress regions. This would be attained by obtaining clamping forces on gaskets adjusted to approximately 75% of the clamping forces used in aluminum alloy components.

CONCLUSIONS

1) In all temperature ranges, AJ52x shows higher thermal conductivity than other die-cast magnesium alloys. The thermal conductivity of AJ52x is closer to the thermal conductivity of aluminum A380.
2) The thermal conductivity of various magnesium alloys does not follow the electrical resistivity ranking. In this context, it is strongly suggested that care be taken when using thermal conductivity values calculated from electrical conductivity measurements for die-cast magnesium alloys. Thermal properties need to be measured independently; alloying elements such rare earth have a strong influence and change the behavior of the thermal conductivity curves.
3) Thermal diffusivity ranking of magnesium diecasting alloys follows the electrical resistivity ranking.
4) Coefficient of linear thermal expansion of AJ52x is similar to AZ91D; usual design rules developed for AZ91D for press fit and interference stress calculations of AZ91D can be applied to AJ52x.
5) Electrical resistivity of AJ52x is lower than the electrical resistivity of AM60B and AZ91D.
6) At 150°C, the percentage of retention of Young's Modulus in AJ52x is better than AZ91D and aluminum alloys (A380 and A356) measured or found in the literature. However at 175°C, the percent modulus retention is nominally equivalent to that measured for the aluminum alloys and for AZ91D.
7) During bolt-load retention tests, AJ52x maintained acceptable clamping strength up to 175°C, where as AZ91D could give a similar clamping retention only at 125°C.
8) Percent clamping-force retention in AJ52x is independent of the initial load applied, so applying a high initial clamping load to the assembly can increase clamping forces for high temperature service.

APPENDIX A -Numerical values used in various figures.

Table A1 Thermal Conductivity (W/m K)--Ref 4,7

Temp. °C	Thermal conductivity (W/m K)				
	AJ52x	AE42	AM60B	AZ91D	A380.1
25	72.858	58.557	63.122	55.130	99.620
50	74.519	60.630	69.516	60.418	108.077
100	79.224	65.716	74.127	64.502	111.145
150	83.930	71.857	78.739	68.587	114.212
200	85.770	76.162	81.018	72.269	118.135
250	87.611	80.228	83.297	75.951	122.057
300	91.570	82.519	85.495	75.998	125.488
350	95.530	84.945	87.694	76.045	128.918
400	94.405	87.401	89.081	76.250	129.883
450	93.280	96.205	90.467	76.456	130.848

Table A2. Thermal Diffusivity (cm^2/s) (Refs 4,7)

Temp. °C	Thermal diffusivity (cm^2/s)				
	AJ52x	AE42	AM60B	AZ91D	A380.1
25	0.373	0.452	0.331	0.287	0.464
50	0.400	0.455	0.365	0.306	0.496
100	0.424	0.472	0.385	0.312	0.495
150	0.448	0.490	0.405	0.318	0.495
200	0.464	0.503	0.424	0.329	0.507
250	0.479	0.514	0.442	0.339	0.519
300	0.491	0.524	0.451	0.337	0.496
350	0.503	0.525	0.461	0.335	0.473
400	0.498	0.525	0.465	0.331	0.464
450	0.492	0.529	0.470	0.326	0.455

Table A3 Coefficient of Thermal Expansion

Alloy	Coefficient of Linear Thermal Expansion (μm/m°K)	
	RT to 200°C (μ m/ m °K)	Source
AJ52	27.3	Ref . 4
AE42	26.4	Ref. 7
AM60B	26.0	Ref. 15
AZ91D	27.2	Ref. 15
A380.1	21.8 - 22	Ref. 13

Table A4 Electrical Resistivity

Alloys	Electrical Resistivity nΩ - m	
	Measured Values	Reference 14
AJ52x	98.2	N/A
AE42	90.5	94
AM60B	129.1	N/A
AZ91D	144.3	150
A380.1	72.6	65

Table A5 Modulus of Elasticity

Temp. °C	Young's Modulus, MPa				
	AZ91D	AJ52x	A380.1	AZ91D (ref 9)	Al356 (ref 16)
20	44074	42372	72400	45000	72088
150	39454	40046	63610	39000	66618
175	38857	37697	65780	36500	65138

Tale A6. % Retention of Room Temperature Young's Modulus

Young Modulus, % of Retention from room temperature					
Temp. °C	AZ91D	AJ52x	A380.1	AZ91D (ref. 9)	Al356 (ref. 16)
20	100%	100%	100%	100%	100%
150	90%	95%	88%	87%	92%
175	88%	89%	91%	81%	90%

Table A7 Bolt-Load Retention

After 120 hrs immersion in oil bath			% Stress retained in Alloy Cylinder		
Temp., °C	Load, KN	Stress, MPa	AJ52x	AZ91D	A380.1
125	7	40	55% 22MPa	37% 15MPa	92% 37MPa
125	10,5	60	55% 33MPa	27% 16MPa	91% 55MPa
125	14	80	53% 42MPa	29% 23MPa	88% 70MPa
150	7	40	47% 19MPa	12% 5MPa	100% 40MPa
150	10,5	60	44% 26MPa	10% 6MPa	94% 56MPa
150	14	80	45% 36MPa	6% 5MPa	83% 66MPa
175	7	40	38% 15MPa	3% 1MPa	82% 33MPa
175	10,5	60	37% 22MPa	<1% <1MPa	81% 49MPa
175	14	80	34% 27MPa	<1% <1MPa	70% 56MPa

ACKNOWLEDGMENTS

The authors gratefully acknowledge Noranda Magnesium Inc., for sponsoring the alloy development program. The die-casting of test specimens was carried out at Intermag and Performag Inc, Québec Canada. The thermodynamic measurements have been performed by CANMET, Ontario Canada, the AE42 and AM60B results were obtained from a joint Noranda Inc, CANMET project sponsored by PERD, (Panel on Energy Research and Development) under the CLiMRI (Canadian Light Material Research Initiative). Special thanks to Yemi Fasoyinu, CANMET project leader. Electrical Resistivity measurement was performed at NRC (National Research Council of Canada), Institut des Matériaux Industriels, Québec, Canada.

REFERENCES

1. S. Koike, K. Washizu, S. Tanaka, K. Kikawa, T. Baba, "Development of Lightweight Oil Pan Made of Heat Resistant Magnesium Alloy for Hybrid Engines." SAE Technical Paper 2000-01-1117.
2. P. Labelle, M. Pekguleryuz, D. Argo, M. Dierks, T. Sparks, T. Waltamatte, "Heat Resistant Magnesium Alloys for Power-Train Applications" SAE Technical Paper 2001-01-0424
3. "Test Method E1461-01 Standard Test Method for thermal diffusivity of solids by Flash Method" , American Society for Testing and Materials, (ASTM)
4. F. A. Fasoyinu, "Determination of the Thermal Conductivity and Coefficient of Thermal Expansion of a Noranda Experimental Alloy", Report: 2000-68 (CF-L), CANMET, October 16th 2000, (Protected Business Information).
5. F.C. Chen, J.W. Jones, T.A. McGinn, J.E. Kearns, A.J. Neilsen, J.E. Allison, "Bolt Load Retention and Creep of Die-Cast Magnesium Alloys" SAE Technical Paper 970325,
6. B.C. Peters, "Creep Testing of Mg Die Castings", Report No. NEM 20138-Final, Internal Dow Magnesium Report, May 1970.
7. Y. Fasoyinu, R. Bouchard, M. Sahoo, 'Fundamental Properties of Magnesium Die Casting Alloys Used in Automotive Applications", Nov 2000 (unpublished work).
8. "Magnesium Database" International Magnesium Association", Thermal Conductivity, www. intlmag.org / physical/ ThermalCond.
9. R. S. Busk, Magnesium Product Design, Marcel Dekker, Inc, 1987.
10. W.R.S. Kemp, A.K. Sreedas, and G.K. White, Proceedings of Physical Society, vol. 4, pp. 314 1950.
11. N.F. Mott, H. Jones Properties of Metals and Alloys, Oxford University Press, 1936, p. 306
12. R.E. Peierls, Quantum Theory of Solids, Oxford, 1955, p. 121
13. Properties and Selection of Non Ferrous Alloys and Special Purpose Materials, ASM Handbook 10th edition, Vol. 2, ASM International, 1990
14. 'Magnesium Database" International Magnesium Association", Electrical Conductivity www.intlmag.org/physical/ElectrCond
15. Magnesium and Magnesium Alloys, M. Avedesian, H. Baker, Eds. ASM Specialty Handbook, ASM, 1999.
16. Structural Alloys Handbook, J.M. Holt, H. Mindlin, C.Y. Ho, Eds., Vol 3, Cindas/Purdue University, In, USA, 1994.

2002-01-0081

Evaluation of Protective Coatings on Magnesium for Phosphate Process Compatibility and Galvanic Corrosion Prevention

Guilian Gao and Mark Ricketts
Ford Research Laboratory, Materials Science Department, Ford Motor Company

ABSTRACT

Bare magnesium is not compatible with the zinc phosphate process commonly used in automotive manufacturing today. If a magnesium component is assembled onto the vehicle prior to the phosphate process, it must be coated beforehand to prevent magnesium dissolution in the phosphate bath. In addition, for a component that must function in very corrosive environments, galvanic corrosion problems also need to be addressed because the component is most likely to be connected with a more noble metal such as steel fasteners or to another component made of steel or aluminum. In this paper, several coatings were evaluated for their effectiveness in stopping magnesium dissolution during the phosphate process and preventing galvanic corrosion. The results show that a porous anodized coating or a very thin chemical conversion coating does not completely stop magnesium dissolution during the phosphate process. An epoxy E-coat stops magnesium dissolution during the phosphate process but is not very effective in preventing galvanic corrosion. An epoxy powder coat or a properly sealed anodized film not only prevents magnesium dissolution in the phosphate bath but also provides good galvanic corrosion protection.

INTRODUCTION

Magnesium components are finding increasingly wider applications in the automotive industry. Corrosion protection for each component is dictated by several factors: microenvironment where the component is located, alloy composition and casting method, design and assembly, as well as reliability requirement. A component that resides inside the passenger compartment is shielded from liquid water and thus has minimum corrosion concerns. The magnesium instrument panel is a good example. For a die cast component that is assembled onto the vehicle after the paint process but encounters very corrosive environments such as de-icing salts used in the North America Green Belt, general corrosion is not a concern due to the very good corrosion resistance of the die cast skin. Galvanic corrosion is a major concern and the protection solutions normally involve proper design, use of coated fasteners compatible with magnesium, and insulating spacers. The magnesium transfer case for trucks is a good example of high volume application. For a component that is assembled onto the vehicle prior to the phosphate, E-coat and paint processes, the corrosion protection must perform dual functions of preventing magnesium dissolution in the phosphate bath and preventing general and galvanic corrosion in the field. Coating the component is the most cost effective means of preventing magnesium dissolution in the phosphate bath.

This paper examines the corrosion concerns and protective solutions for magnesium components that must go through the phosphate process and serve in corrosive environments. Impact of a zinc phosphate commonly used by the automotive industry on the corrosion performance of magnesium alloys was examined. Several coatings were evaluated for their effectiveness in reducing magnesium dissolution during the phosphate process as well as preventing galvanic corrosion.

MAGNESIUM DISSOLUTION DURING THE PHOSPHATE PROCESS

The zinc phosphate solution commonly used by the automotive industry is moderately acidic (pH 3.0--3.5) and can result in high dissolution rate of magnesium if the part is not properly protected before entering the phosphate bath. Magnesium dissolution in the phosphate bath not only reduces the thickness of the magnesium component, it also introduces Mg^{2+} contamination into the phosphate bath.

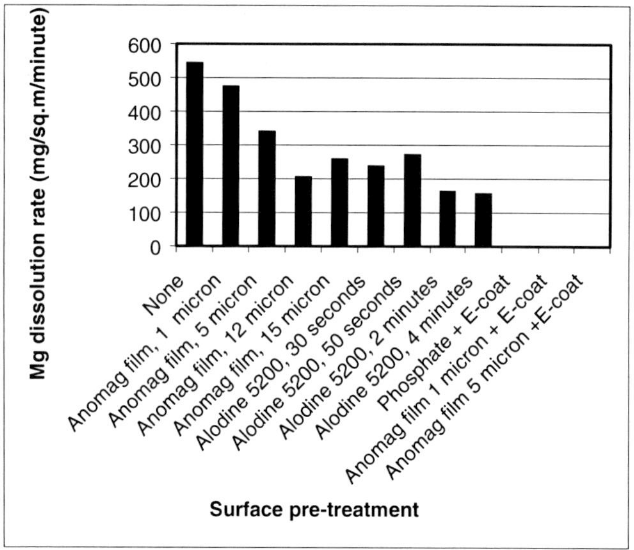

Figure 1: Effect of coatings on the dissolution of magnesium during the phosphate process

Several coatings as listed in Table 1 were evaluated for their effectiveness in preventing magnesium dissolution during the phosphate process. The alloy used for the testing was AZ91. The Anomag film was produced using the "Anomag" process, a patented anodization process developed by Magnesium Technology Ltd of New Zealand. The composition of the film is basically $Mg_3(PO_4)_2$[1]. Alodine 5200 is a patented conversion coating for aluminum and magnesium alloys. It was developed by Henkel Surface Technologies. The E-coat evaluated was an epoxy formulation currently used in production at Ford.

Magnesium dissolution experiments were carried out by PPG INDUSTRIES INC. The phosphate solution used was PPG Chemfos 700. Tests were conducted in a bath containing 250 ml solution. The total surface area of the magnesium test coupon was 0.315 m^2 (0.34 ft^2). Coupons were immersed in the bath for approximately 2 minutes at 49-60°C (120-140°F). Results of the study are given in Figure 1.

Clearly, the most effective means of preventing magnesium dissolution is a non-porous and inert coating such as an E-coat. Either combined with a phosphate treatment or a thin Anomag film, it stopped magnesium dissolution completely.

Both Anomag and Alodine 5200 coatings reduce magnesium dissolution to some degree, but do not completely stop the dissolution. For the Anomag film, the lowest dissolution rate was observed with a film thickness of 12 microns. For Alodine 5200, dissolution rates appear to be higher for samples with process times of 30 seconds and 50 seconds. The rates appear to be approximately constant for samples with 2 minutes and 4 minutes treatment time. Since processing time directly impacts cost and production cycle time, two minutes appears to be the optimal time for corrosion performance and cost. Therefore all the Alodine 5200 samples for subsequent studies used 2 minute processing time.

IMPACT OF THE ZINC PHOSPHATE PROCESS ON GENERAL CORROSION RESISTANCE OF MAGNESIUM

Impact of the chemical pre-treatment such as phosphate on corrosion resistance of magnesium has been reported in the literature in the form of coating creep from a scribe[(2),(3)]. The creepage from scribe reflects the effectiveness of the pre-treatment on coating adhesion. Its direct impact on the general corrosion resistance of the alloys with and without an organic coating has not been reported. This study addressed this issue. Samples were pulled out at different stages of the pre-treatment, cleaning and phosphate process and run through corrosion exposure.

Magnesium die cast plates of AM60 and AM50 were used for the study. Alloy selection was limited by sample availability. The dimensions of samples are 12.5cm x 2.7cm x 2mm. Table 2 lists the samples and various surface treatments tested in this study. Cleaning, phosphate and E-coat was conducted by PPG. The E-coat bake was conducted according to the specification of 177°C (350°F) metal temperature for 20 minutes. No coating blistering or unusual discoloration was observed after the bake. Following the E-coat bake process, one area of the sample was subjected to the crosshatch scribe and peel test, a standard E-coat adhesion test. A scale of 0 to 5 was used to rate the adhesion, with 0 being the best and 5 being the worst. Results of the adhesion test are given in Table 2.

After the scribe and peel test, the sample was exposed to salt spray per ASTM B117 salt spray procedure[(4)] for 240 hours. Corrosion exposure was carried out at Ford Central Laboratory. In most cases, triplicate coupons were used. All samples were weighed prior to corrosion exposure.

Upon completion of corrosion exposure, for samples without E-coat (A, B, C, D), corrosion products were removed by immersion in chromic acid solution (180g/l of CrO_3, 80°C, 2 minutes). These samples were then weighed and % weight loss determined. Optical photographs were also taken before and after the removal of corrosion products. For the samples with E-coat, weight loss measurement was not carried out. Instead, selected samples were cross-sectioned for the determination of maximum depth of corrosion attack. Results are given in Table 2.

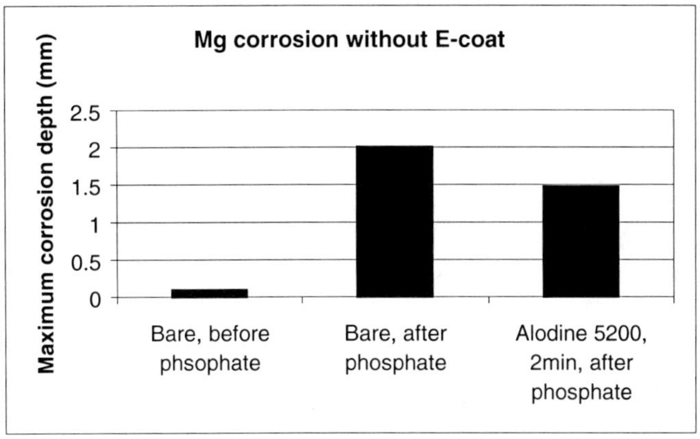

(a)

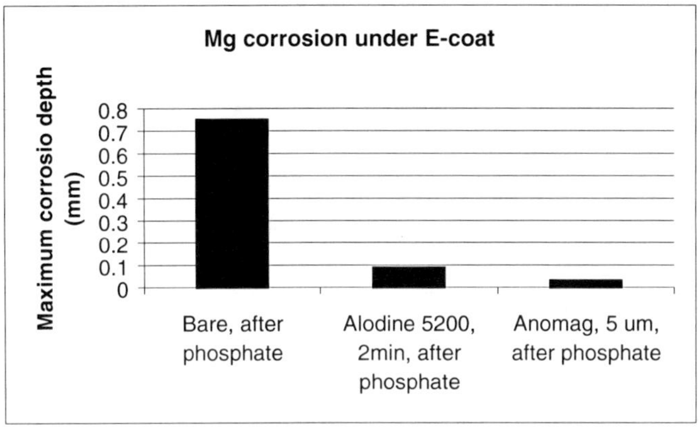

(b)

Figure 2: Maximum corrosion depth of magnesium after 240 hours of salt spray. (a) Without E-coat; (b) Under E-coat.

Figures 2 & 3 show that the zinc phosphate process greatly accelerates corrosion of bare magnesium. After the 240 hour salt spray exposure, an as-cast magnesium coupon showed a maximum corrosion depth of only 91 μm while as the 2mm thick coupon that went through the phosphate process was perforated (Figure 3). SEM EDX analysis of the surface phosphate deposit showed presence of nickel. This can be expected because the phosphate formulation used in this study contained nickel ions. The dramatic increase in magnesium corrosion is probably caused by nickel compounds that deposited out with the zinc phosphate. Hillis and Reichek[(5),(6)] have shown that nickel content in the alloy in excess of 50 ppm greatly increases corrosion rates of high purity magnesium alloys such as AM60 and AZ91. In the case of zinc phosphate treatment, the nickel deposits formed on the surface are likely to be in the form of nickel compounds. However, these compounds could have much higher electrochemical potential than magnesium and can serve as micro cathodes that cause galvanic corrosion of magnesium. Clearly, this particular zinc phosphate formula and bare magnesium are not compatible.

(a)

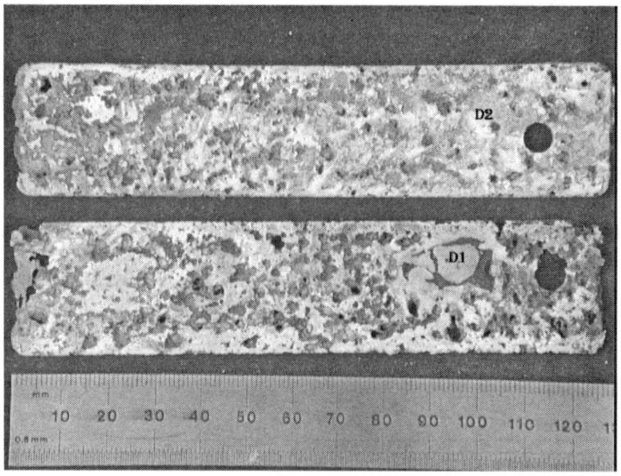

(b)

Figure 3: Photograph of magnesium panels after 240 hours of salt spray exposure. (a) As-received samples; (b) Samples after the cleaning and phosphate processes, showing perforation in many locations.

Samples with Alodine 5200 pre-treatment also showed higher corrosion rate after the phosphate process. Corrosion tends to be localized. Figure 4 shows the surface appearance and cross-section view of a sample with 2 minute processing time. It appears that the conversion coating was not uniform. It is likely that coating in some areas was damaged during the phosphate process, resulting in direct deposition of nickel compounds and zinc phosphate onto the magnesium surface. As mentioned before, the nickel compounds would likely serve as micro cathodes and cause galvanic corrosion of the magnesium.

Both the Alodine 5200 coating and the unsealed Anomag film are conductive and permit deposition of E-coat.

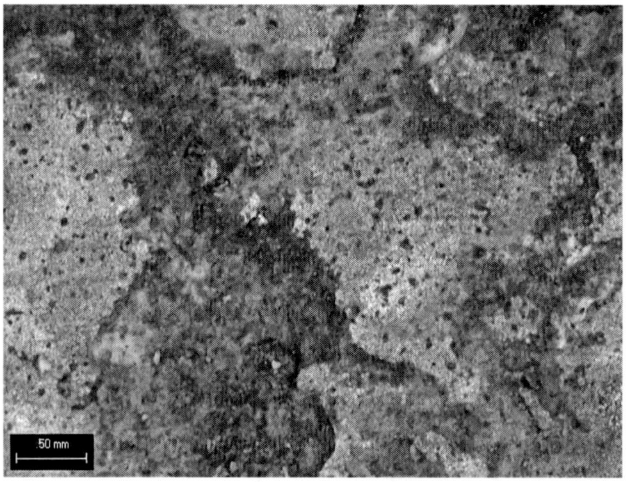

(a)

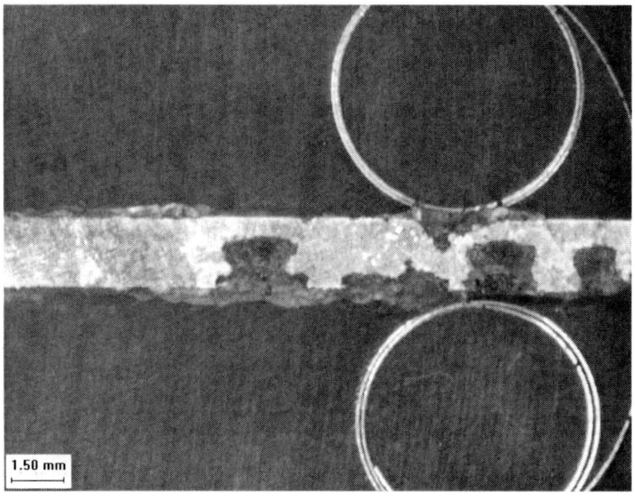

(b)

Figure 4: Magnesium with Alodine 5200 pre-treatment (2 minutes processing time) after 240 hours of salt spray exposure. (a) Surface appearance; (b) Cross section showing localized pitting.

Maximum magnesium corrosion depths under an E-coat or at locations where the E-coat was mechanically damaged are shown in Figure 2b. Both types of coatings substantially reduced the extent of corrosion attack. The Anomag film appears to be more effective than the Alodine 5200 coating in this regard.

EFFECT OF THE COATINGS ON GALVANIC CORROSION

Coupon assemblies simulating materials combinations in production were used for evaluation of effectiveness of several coatings in preventing galvanic corrosion. The test matrix used for this study is given in Table 3. The assembly was held together by a M8 steel bolt coated with JS500 finish. An example of the assembly is given in Figure 6. AZ91 coupons were used in this study. The aluminum spacer used was made of alloy 5183. The tape evaluated was a polyester tape made by 3M. Three coating systems were evaluated: Alodine 5200 alone, sealed Anomag film, and Alodine 5200 + epoxy powder coat. Data on the Alodine 5200 + epoxy powder coat system were generated by Henkel. Samples J & K were included for baseline information.

Figure 5: Appearance of sample O after 240-hour salt spray, showing no corrosion (rated 0).

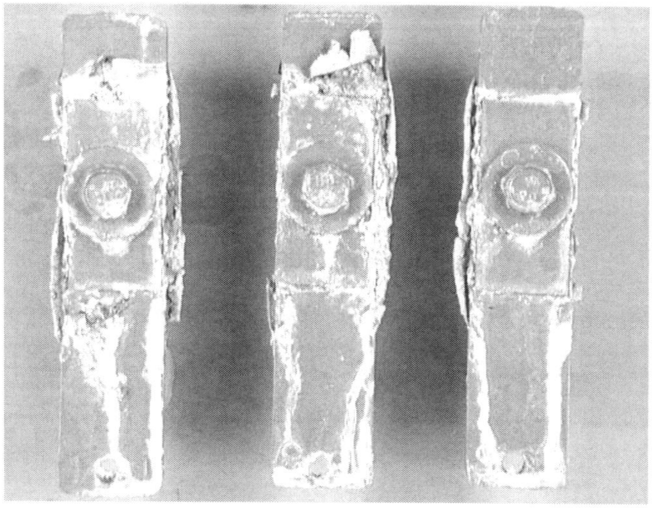

Figure 6: Appearance of sample G after 240-hour salt spray, showing very severe galvanic corrosion (rated 10).

The salt spray tests were conducted for 240 hours. Visual inspection was first conducted on the samples upon completion of the salt spray exposure. A scale of 0 to 10 was used to rate the samples. Samples showing no galvanic corrosion were rated 0. Samples showing very severe galvanic corrosion was rated 10. Figures 5 and 6 show examples of 0 and 10 ratings. Rating results are given in Table 3. After the visual examination, samples showing substantial corrosion were cross-sectioned for measurement of maximum corrosion depth. Results are also given in Table 3. Alodine 5200 alone did not offer adequate galvanic corrosion protection. Sample I showed severe galvanic corrosion after 240 hours of salt spray.

Samples coated with sealed Anomag film performed better. Some corrosion was still observed. Among the three samples with the same coating system, sample A (with a aluminum spacer) performed best, with a rating of 1. Sample D (with the coated magnesium coupon in direct contact with a EG60 galvanized steel) performed worst, with a visual rating of 3 and maximum corrosion depth of 0.79 mm. The results demonstrate effectiveness of a spacer in moderating galvanic corrosion .

Alodine 5200/powder coat showed the best corrosion performance among the three coating systems tested. Even in the absence of an aluminum or plastic spacer, the amount of corrosion observed was minimum.

CONCLUSION

1) The zinc phosphate process commonly used in the automotive industry is not compatible with bare magnesium. It causes substantial magnesium dissolution. In addition, the phosphate deposits formed on the magnesium surface greatly accelerate corrosion of magnesium.
2) Magnesium components that go through the phosphate and E-coat processes must be pre-coated with an inert coating to minimize bath contamination and deterioration of magnesium corrosion properties. E-coat or epoxy powder coat are very effective for this purpose.
3) Alodine 5200 is a suitable surface pre-treatment for magnesium. It substantially reduces magnesium dissolution in the phosphate bath and provides good adhesion of the E-coat to magnesium. However, the coating does not completely stop magnesium dissolution or prevent direct phosphate deposition. It does not offer satisfactory galvanic corrosion protection either.
4) Unsealed Anomag films slow down, but do not stop magnesium dissolution in the phosphate bath. Sealed Anomag films stop magnesium dissolution during the phosphate process and offer good galvanic corrosion protection when used in conjunction with a spacer.
5) Alodine 5200 + epoxy powder coat is the most satisfactory coating system among the coatings evaluated. It stops magnesium dissolution during the phosphate process and offers good galvanic corrosion protection.

ACKNOWLEDGMENTS

The authors like to acknowledge Jason Balzer of Ford Motor Company, Shawn Dolan and Oscar Roberto of Henkel Surface Technologies, Kevin Stewart and David Reed of Meridian, Ronald Zurilla of PPG, John MacCulloch and Philip Ross of Magnesium Technology Licensing Ltd, for their assistance in sample preparation and testing.

REFERENCES

1. V. Tchervyakov, G. Gao, J. Bomback, A.P. Pchelnikov and G. Cole, "Laboratory Evaluation of Corrosion Resistance of Anodized Film on Magnesium", in Magnesium Technology 2000, H.I. Kaplan, J.N. Hyrn and B.B. Clow ed. TMS(200), p143.

2. J.I. Skar, M. Walter and D. Albright, " Non-Chromate Conversion coatings for Magnesium Die Castings", SAE Technical Paper 970324, SAE International, Warrendale, PA, (1997)

3. D. Hawke and K. Gaw, " Effectis of Chemical Surface Treatments on the Performance of an Automotive Paint System on Die Cast Magnesium", SAE Technical paper 920074, , SAE International, Warrendale, PA, (1992)

4. ASTM B117-94, "Standard Practice for Operating Salt Spray (Fog) Testing Apparatus", ASTM, 1994.

5. J.E. Hillis and K.N. Reichek, "High Purity Magnesium AM60 Alloy: the Critical Contaminant Limits and the Salt Water Corrosion Performance", SAE Technical paper 860288, SAE International, Warrendale, PA, (1986)

6. K.N. Reichek, K.J. Clark and J. E. Hillis, " Controlling the Salt Water Corrosion Performance of Magnesium AZ91 Alloy", SAE Technical Paper 850417, SAE International, Warrendale, PA, (1985)

CONTACT

Mark Rickets, Materials Science Department, Ford Research Laboratories, MD 3182, SRL, 2101 Village Rd, Dearborn, MI 48121, tel: 313-322 6726, email: mricket2@ford.com

Guilian Gao, Department of Chemical Engineering, Keck Science Building, Rm 265, Stanford University, Stanford, CA 94305-5025, tel: 650-736 2045, email: ggao@ford.com

Table 1: Coatings evaluated during the magnesium dissolution study.

Surface Coating	Film thickness or process time
None	n/a
Anomag Film	1 μm
Anomag Film	5 μm
Anomag Film	12 μm
Anomag Film	15 μm
Alodine 5200	30 seconds
Alodine 5200	50 seconds
Alodine 5200	120 seconds
Alodine 5200	240 seconds
Phosphate + Epoxy E-coat	Not measured
Anomag Film + Epoxy E-coat	Anomag 1 μm, E-coat not measured
Anomag Film + Epoxy E-coat	Anomag 5μm, E-coat not measured

Table 2: Sample matrix and test results for the magnesium general corrosion study.

ID	Alloy	Surface coating	Clean & phosphate	E-coat	Adhesion rating prior to corrosion exposure	% wt loss after corrosion exposure	Maximum depth of corrosion
A	AM60	None	No	No	N/A	5.57	91μm
B	AM60	None	No	No	N/A	2.63	8.2 μm
C	AM60	None	Yes	No	N/A	32.43	2mm (perforation)
D	AM60	None	Yes	No	N/A	21.35	2mm (perforation)
E	AM60	None	No	Yes	0 to 5	N/A	
F	AM60	None	Yes	Yes	0 to 1	N/A	750 μm
H	AM50	None	Yes	Yes	0	N/A	105 μm
I	AM50	AA, 5μm	No	Yes	1	N/A	N/A
J	AM50	AA, 5μm	Yes	Yes	0	N/A	30 μm
K	AM50	AA, 10μm	No	Yes	5	N/A	N/A
L	AM50	AA, 10μm	Yes	Yes	2	N/A	N/A
M	AM60	AC, 30s	Yes	Yes	0	N/A	N/A
N	AM60	AC, 50s	Yes	Yes	0	N/A	154 μm
O	AM60	AC, 2min	Yes	Yes	0	N/A	83 μm
P	AM60	AC, 4min	Yes	Yes	0	N/A	59 μm
Q	AM60	AC, 30s	No	No	N/A	N/A	49 μm
R	AM60	AC, 50s	No	No	N/A	N/A	97 μm
S	AM60	AC, 2min	No	No	N/A	N/A	86 μm
T	AM60	AC, 4min	No	No	N/A	N/A	77 μm
U	AM60	AC, 30s	Yes	No	N/A		1.3 mm
V	AM60	AC, 2min	Yes	No	N/A		1.47mm
W	AM60	AC, 4min	Yes	No	N/A		0.67 mm

AA: Anomag Anodization
AC: Alodine 5200 conversion coating

Table 3: Sample matrix and results of the galvanic corrosion study.

Sample ID	Metal combination	Treatment before assembly	Treatment after assembly	Visual inspecti on rating	Maximum depth of corrosion (mm)
A	Mg/Al/EG60	Sealed Anomag	Phosphate/E-coat	1	Not measured
B	Mg/EG60	Sealed Anomag	Phosphate/E-coat	3	0.79
C	Mg/tape/EG60	Sealed Anomag	Phosphate/E-coat	2	Not measured
D	Mg/Al/EG60	Alodine 5200	Phosphate/E-coat	2	Not measured
E	Mg/EG60	Alodine 5200	Phosphate/E-coat	4	0.34
F	Mg/Tape/EG60	Alodine 5200	Phosphate/E-coat	2	Not measured
G	Mg/CRS	Sealed Anomag	Phosphate/E-coat	10	2.3
H	Mg/EG60	Sealed Anomag /phosphate /E-coat	None	1	Not measured
I	Mg/EG60	Alodine 5200 /phosphate/E-coat	None	7	2.58
J	Mg	Sealed Anomag	Phosphate	10	1.7
K (no assembly)	Mg	Sealed Anomag	N/A	0	0
O	Mg/JS500 M8 J-clip	Sealed Anomag/ phosphate/E-coat	None	0	Not measured
**	Mg/Al/EG60	Alodine/powder coat	Phosphate/E-coat	1	Not measured
**	Mg/EG60	Alodine/powder coat	Phosphate/E-coat	2	Not measured.
**	Mg/CRS	Alodine/powder coat	Phosphate/E-coat	2	Not measured

**** Samples tested by Henkel Surface Technologies**
Al -- Aluminum 5183 alloy
EG60—Electrogalvanized steel, zinc coating weight 60g/m^2
CRS – AISI 1010 cold rolled steel
Tape – Polyester tape from 3M

2002-01-0082

Production and Die Casting of Semi-Solid Magnesium Alloy AZ91D

M. T. Shehata, V. Kao and E. Essadiqi
Materials Technology Laboratory/CANMET, Natural Resources Canada

C. A. Loong and C-Q Zheng
Industrial Materials Institute, National Research Council Canada

ABSTRACT

Semi-solid forming is a relatively new process whereby an alloy with thixotropic characteristics at a temperature between solidus and liquidus is cast into a component in a mould. This technology is based on research originally carried out at MIT in the 1970s on rheological properties of semi-solid metals subjected to mechanical stirring. By not starting with a super-heated melt as in conventional liquid pressure die casting, semi-solid processing offers distinct advantages such as low cycle time, less porosity because of non-turbulent flow, improved die life and significantly better mechanical properties.

The success or failure in the production of a component largely depends on the temperature uniformity and microstructural homogeneity of the semi-solid thixotropic feedstock prior to injection. This paper describes results of work related to the preparation of AZ91D feedstock by electro-magnetic stirring combined with superheat reduction, reheating of this material to the thixotropic state by induction and semi-solid die casting a number of complex box-like components. The reheating was carried out in a single induction coil programmed to provide variable power input such that a low temperature gradient and a minimal liquid metal segregation in the billet feedstock were realized. The microstructure was monitored throughout the processing to achieve the desired semi-solid microstructure for successful die casting. The optimal casting parameters (metal temperature, ram velocity and metal pressure) for die casting the box-like components are given.

INTRODUCTION

Semi-solid metal (SSM) casting is a relatively new process for producing components with high integrity using alloy billets reheated to a temperature between solidus and liquidus. These billets are typically cut from rheocast or grain-refined feedstock of a certain diameter. Upon reheating, the primary phase (often in the form of rosette particles) becomes globular and the resulting microstructure produces a thixotropic material having a viscosity low enough under shear to readily fill the die cavity of a pressure die casting machine. This technology is based on research originally conducted at MIT in the 1970s on rheological properties of semi-solid metals subjected to mechanical stirring (1).

Practically all high-volume SSM components produced today start with commercial billets in aluminum alloys A356 and A357 as feedstock materials (2-4). For magnesium components, the proprietary Thixomolding process (which involves preparation of the thixotropic slurry from magnesium alloy chips in an injection molding machine) is the most common production method used (5). Because fine-grained or rheocast magnesium alloy bars in different sizes are not available commercially, high-volume production components from billets has not been possible to date.

The initial feedstock material must be cast in a way that avoids the formation of dendrites or alternatively, a fine non-dendritic structure is obtained by shearing or breaking down the dendrites to small rosettes during solidification. The latter may be achieved through stirring the solidifying melt mechanically or by electromagnetic means, whereas the former can be achieved by casting at a low supper heat, or by the use of grain refiners. Much work has been done on thixotropic aluminium alloys, particularly A356 and A357. By comparison, very little has been done on magnesium alloys. Prior to semi-solid forming, the feedstock material should be reheated rapidly and held for sometime at the desired temperature (corresponding to the desired solid fraction) in order to complete the globularisation of the primary phase (6). There is an optimum temperature profile to achieve the best thixotropic behaviour (7), depending on the structure and the size of the feedstock.

This work presents the results for the casting of laboratory thixotropic AZ91D alloy feedstock material using a static mold casting method with electromagnetic stirring. It also presents the results for the optimized heating of this feedstock material in the form of 76mm diameter and 152mm long (3" dia. by 6" long) billets. In addition, microstructural changes in the as-cast thixotropic material as a result of heating to and holding at the desired semi-solid temperature are presented. Finally, semi-solid casting of the billets into a number of complex box-like components is covered, with attention paid to obtaining the optimal casting parameters and examining the morphological variations at different sections in the final part.

PRODUCTION OF THIXOTROPIC FEEDSTOCK

CASTING SIMULATION - AZ91D thixotropic feedstock billets were produced in a coated copper mould at the Materials Technology Laboratory / CANMET (Natural Resources Canada) using a resistance heated furnace with a capacity to melt 20kg of the alloy. The rosette or globular structure was obtained by incorporating an electromagnetic stirring unit manufactured by JME-Inverpower Canada Inc.(now ABB). A schematic drawing of the water-cooled copper mould is shown in Figure 1. Two thermocouples were inserted in the mould, one near the surface and the other at the center; both thermocouples were at mid-height of the mould to measure temperature variations during solidification. The chemical composition of the laboratory prepared material is that of a typical AZ91D alloy (8.7% Al, 0.89% Zn and 0.18% Mn). The casting is performed at a low super-heat, the melt pouring temperature being 620 °C.

Solidification microstructure of AZ91D at mid-radius of the billet without and with electromagnetic stirring are given in Figure 2. The structure is more equiaxed and rosette-like in the case of the stirred material compared to the non-stirred material, which contains dendrites. It will be seen below that the globularity of the alpha phase is achieved after heating the electromagnetic stirred material to the semi-solid temperature.

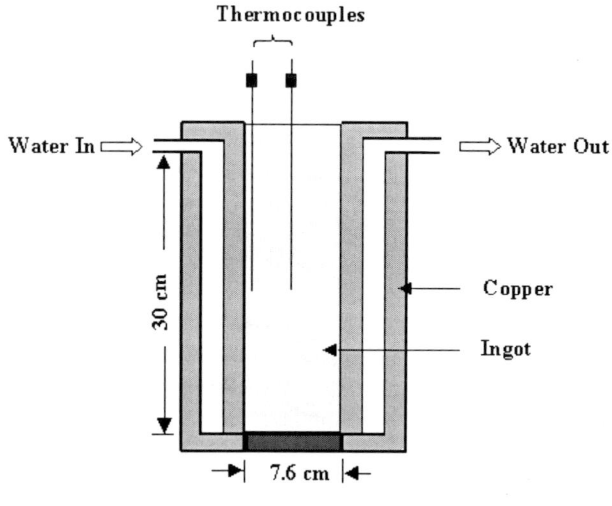

Figure 1. A schematic drawing showing details of the water-cooled copper mould.

REHEATING OF BILLETS BY INDUCTION

Inductive heat is generated in the billet by the thermal effect of a varying electromagnetic field when an a-c current is flowing in the coil. Electrical power is transformed into heat by induction in accordance with the well-known equation derived from Ohm's Law:

$$P=|I|^2 R/2$$

where P is the power, $|I|$ is the average value of sinusoidal current in the solenoid heating coil and R is the effective resistance of the workpiece.

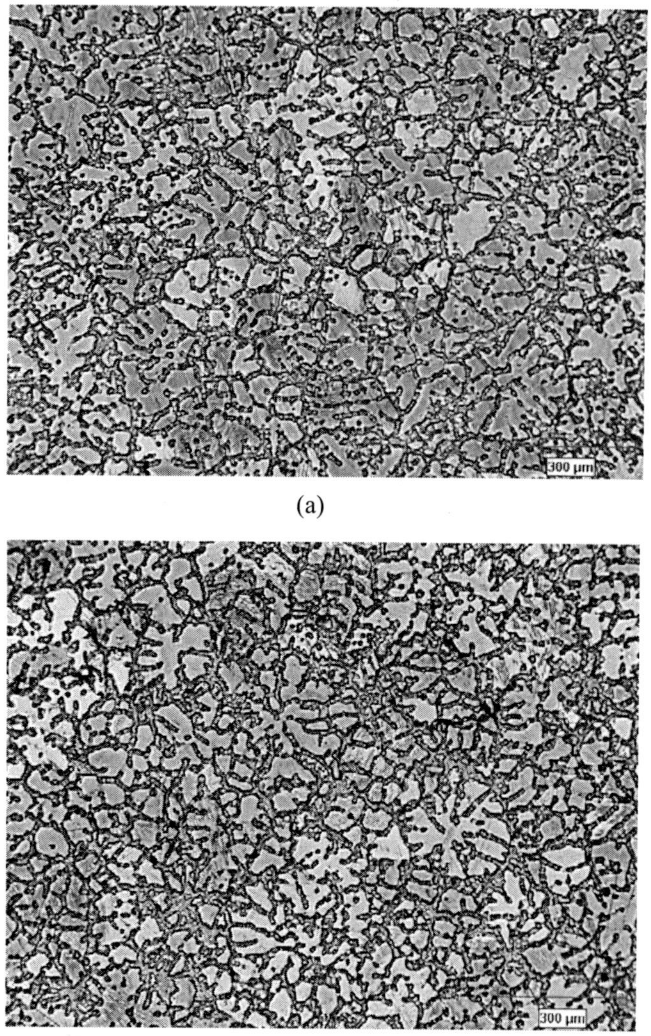

(b)

Figure 2. Microstructure of laboratory as-cast thixotropic feedstock AZ91D, (a) without and (b) with electro-magnetic stirring. The primary alpha phase is magnesium (with aluminium in solid solution) and the eutectic is mainly a mixture of alpha and beta (Mg17Al12)

A detailed treatment of the theories of induction heating is beyond the scope of this paper, however there are a few important factors that need be considered. First, commercial induction systems operate at frequencies that seldom exceed 1 kHz. The reason is that high frequency systems concentrate their energy on the surface and edges, thereby causing them to overheat. This results in a high surface temperature and an unacceptable amount of liquid metal draining from the surface. By operating the coil at a low frequency, the eddy current effect on the surface is reduced while the effective depth of penetration of the power applied is increased. This facilitates achieving a uniform temperature distribution between the core and the surface within a reasonable cycle time.

The effective resistance of a coil containing a billet is proportional to the square root of frequency for a good electrical conductor. Thus, the current required at a higher operating frequency is less than that at a lower operating frequency under the same heating power condition (9). Consequently, there is a clear advantage in using a high frequency system since the efficiency for such a system would be higher, but there is a need to overcome the shortfall of surface overheating.

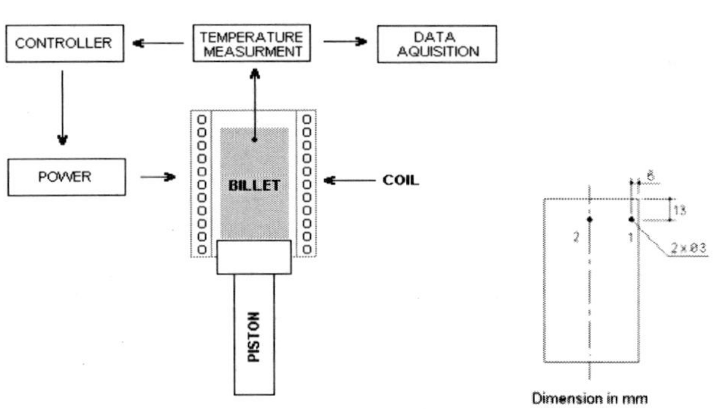

Figure 3. Schematic of set-up for reheating of billets; thermocouples were embedded at locations 1 and 2 as shown at the lower right.

In a series of experiments conducted to reheat the AZ91D billets, a single induction coil operating at 10 kHz and a power of 13.5 kW was used. The induction power unit was supplied by Norax Canada, a manufacturer of customized induction systems based on power transistors operating at high frequencies (9). The billet of 7.6 cm (3 in) diameter and 15.2 cm (6 in) in length was contained in a ceramic crucible and insulated at the top. It was introduced into the coil by means of a piston that could be raised and lowered pneumatically, Figure 3. The coil is fabricated from a 0.65 cm (0.25 in) diameter copper tubing and has a diameter of 12.7 cm (5 in) and a height of 20.3 cm (8 in). Power input into the coil is controlled by a programmable controller through a thermocouple inserted into a 0.32 cm (1/8 in) diameter by 1.28 cm (0.5 in) deep drilled hole 6 mm near the edge at the top surface. A second thermocouple is embedded in another drilled hole of the same dimensions at the center of the top surface so that temperature gradient between the two locations could be monitored.

In order to avoid over-heating of the surface in the workpiece, power in the coil was programmed to vary over the heating period of approximately 7 minutes as shown in Figure 4. Full power was turned on for the first 2 minutes until a temperature of 460 ° C (just below its solidus temperature) was attained, at which point it was reduced to 60% of its initial value. Additional reductions over the next two subsequent steps in the remaining 5 minutes brought the power input down even further. Towards the end of the heating cycle, the power required to achieve a final temperature of 575 - 580 ° C was 15% of the initial value.

A typical temperature difference plot between the two thermocouples as a function of time is shown in Figure 4. For over half the reheating cycle, there is a significant temperature difference of at least 10 ° C between location 1 and location 2. This difference becomes much narrower as the power is reduced such that towards the end of the reheating cycle, a difference in temperature of less than 5 ° C is realized.

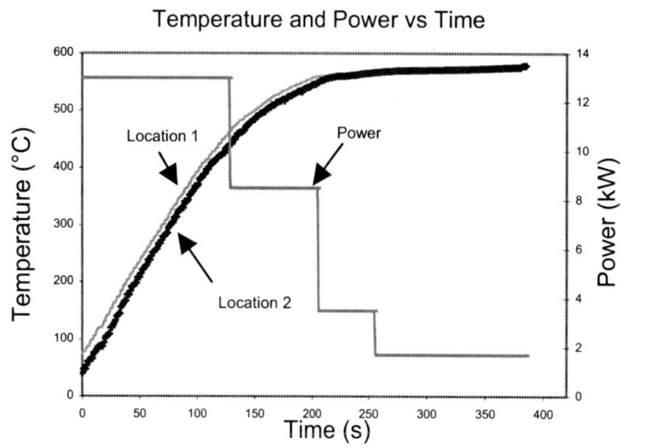

Figure 4. Temperature vs. time curves for thermocouples at locations 1 and 2 under different power levels.

The morphological changes of the primary solid phase, during reheating and holding at the semi-solid temperature, were studied by heating smaller samples (25x25x20 mm) and quenching from the semi-solid temperature. The quenching of samples was carried out in mineral oil. Metallographic samples were cut from the quenched samples, which were then subjected to conventional metallographic preparation (mounting, grinding, polishing and etching). The samples were etched with acetic-picral for 4-6 sec. The microstructures of the samples were then examined by optical microscopy. It was observed that the material becomes more globular by heating to 580 °C ; the longer the holding time at this temperature, the more globular was the resulting structure. At very prolonged holding times, grain growth of the primary phase is observed. This may not be desirable as it may lead to poorer mechanical properties in the final part after forming. It appears that an optimum holding time for globularisation of the primary phase grains without excessive grain growth is achieved at a holding time of about 90 seconds as shown in Figure 5.

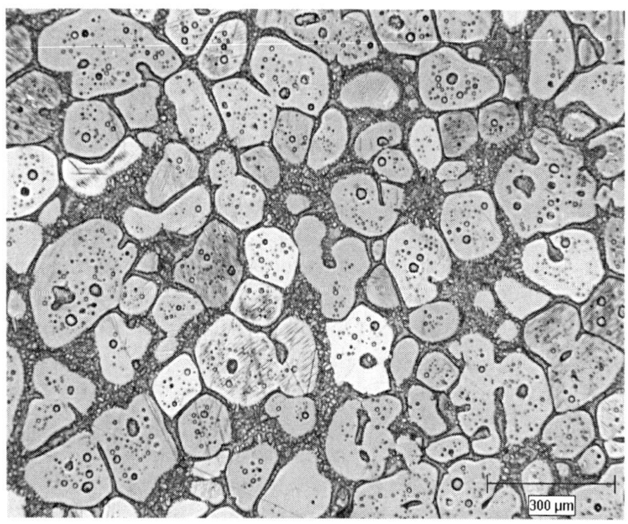

Figure 5. Microstructure of laboratory prepared AZ91D after heating to 580° C and holding for 90 seconds.

DIE CASTING OF SEMI-SOLID BILLETS

The reheated billets were cast into box-type components at the Industrial Materials Institute using a BÜhler 600 T capacity SC N/53 die casting machine. The die was designed to allow material from the central region of the billet to flow into the die cavity through a cylindrical gate, thereby leaving the oxide skin behind in the biscuit. Unlike conventional liquid metal casting, the sleeve has a wider and longer opening, which makes it easy to introduce the cylindrical billet into the shot well. The casting has a shot weight of 0.95 kg and a height of 6 cm. With a complex configuration, the casting wall thickness varies from 9 mm near the gate to 5 mm at the bottom edge, Figure 6.

The optimal parameters used for successfully producing high quality castings with a good surface finish may be outlined as follows:

BILLET TEMPERATURE - Billet temperature is critical in that fluidity of the semi-solid AZ91D alloy is dependant on viscosity and the volume fraction of solid. For consistency, it is vital to maintain the feedstock temperature as uniform as possible. It was found that a temperature range of 575 - 580 ºC was most suitable. This temperature range corresponds to a solid fraction of approximately 0.3 – 0.4 (10). Billets with a temperature lower than 570 º C could not be introduced readily into the shot sleeve.

DIE TEMPERATURE - As with conventional liquid metal die-casting, heating of the die prior to injection is required. The die temperature kept within a range of 230 – 260ºC by hot circulating oil was found to be suitable. However, no tests were carried out at a die temperature higher than this range.

RAM SPEED - The metal injection speed has a major influence on the nature of the flow front in the die cavity. In general, a slow speed should be employed as one of the principal reasons behind semi-solid casting is the ability to fill the die cavity in a laminar manner. A fast speed leads to flow turbulence, which negates this advantage and could result in an unacceptably high level of porosity. However, for complex components cast at a given metal pressure, it may not be possible to fill the die completely under a low ram speed. Tests conducted using the box die at ram speeds of 0.25 –1.25 m/s showed that best results were obtained when the ram speed reached 1.0 m/s, corresponding to a gate velocity of 9 m/s. A gate velocity higher than this level was found to increase gas entrapment in the cavity significantly.

METAL PRESSURE - In liquid pressure die casting, metal pressure often plays a very important role in expanding the castability window in the production of complex castings. Problems with castings not being able to meet quality specifications can sometimes be solved successfully by using a machine with a higher injection force. This is also the case with semi-solid die casting where minimal pressure requirement for most parts is at least two times greater than that of liquid metal die casting. This is because heat loss during injection results in a material that is increasingly viscous as it fills the cavity. Casting trials conducted at a metal pressure of 500 – 1200 bars (50–120 MPa), a billet temperature of 580 ºC and an injection speed of 1 m/s, showed that defects seen at a lower metal pressure could be minimized or even eliminated when a much higher pressure was applied. A pressure of 1200 bars (120 MPa) or higher was found to produce castings with high integrity.

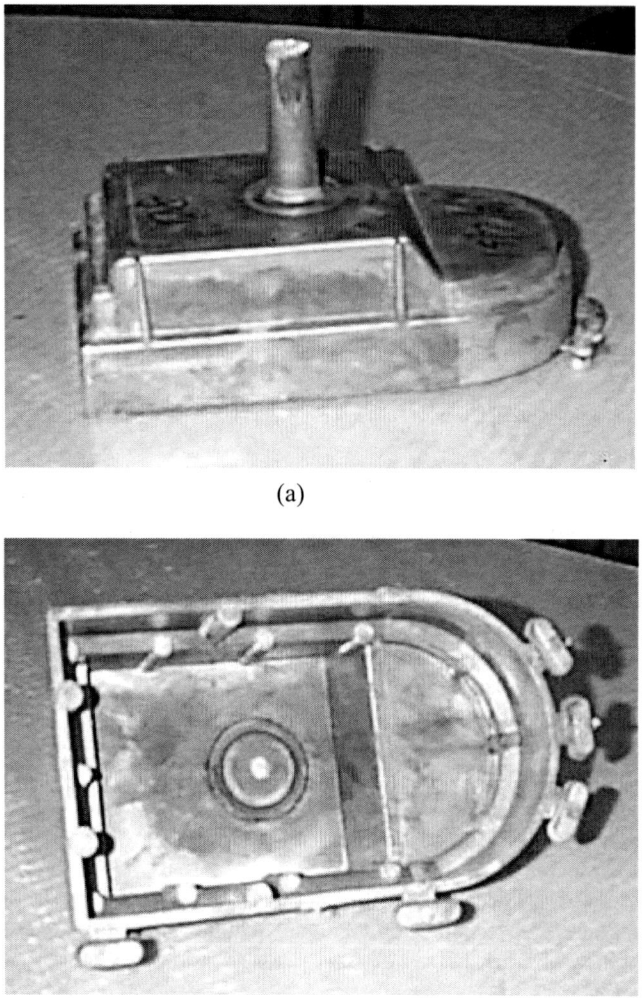

(a)

(b)

Figure 6. Box Casting from semi-solid billet, (a) external view with gate and (b) internal view.

A part produced using the optimal parameters was examined to determine if there was any significant liquid segregation during injection of the semi solid slurry into the die. For this purpose metallographic samples were cut from five different sections of this part. These locations do not have the same wall thickness and also contain corner areas where liquid segregation may be suspected. The different areas examined are shown schematically on Figure 7.

Some typical photomicrographs taken from the different areas are shown in Figures 8 and 9. No significant liquid segregation near the wall in all areas was observed, except possibly for a very small pocket at the very end of the thinnest wall (area 5 in Figure 7).

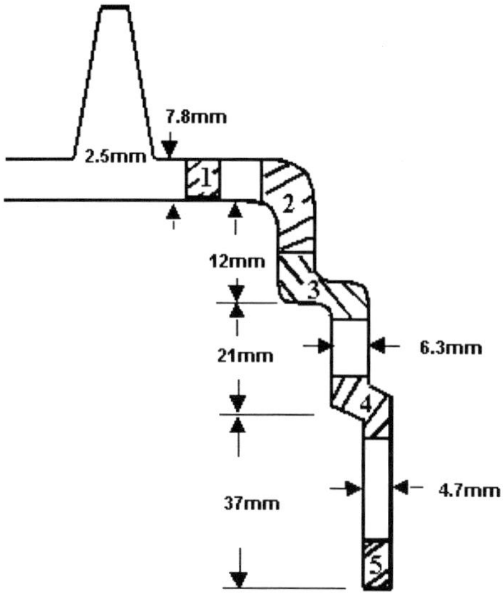

Figure 7. A schematic drawing representing the various areas where metallographic samples were cut from the diecast part shown in Figure 6.

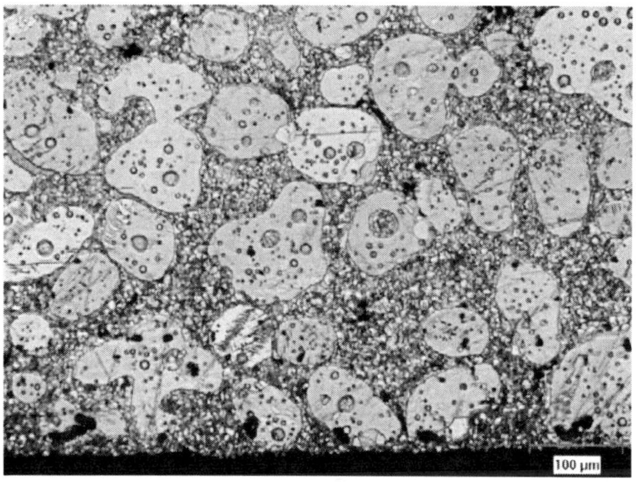

(a)

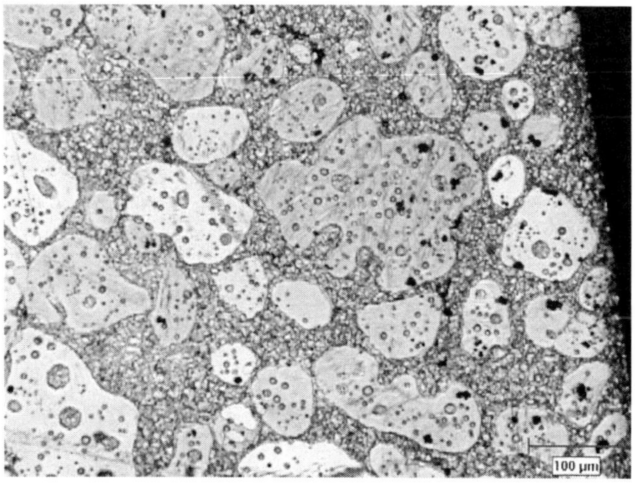

(b)

Figure 8. Photomicrographs taken near surface in, (a) area 1 and (b) area 2 of Figure 7.

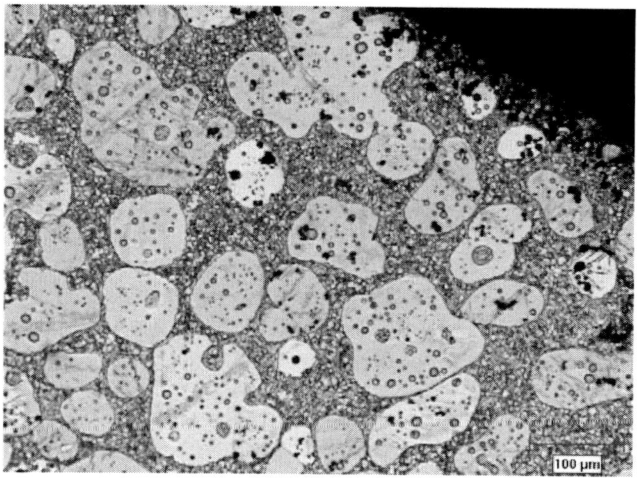

(a)

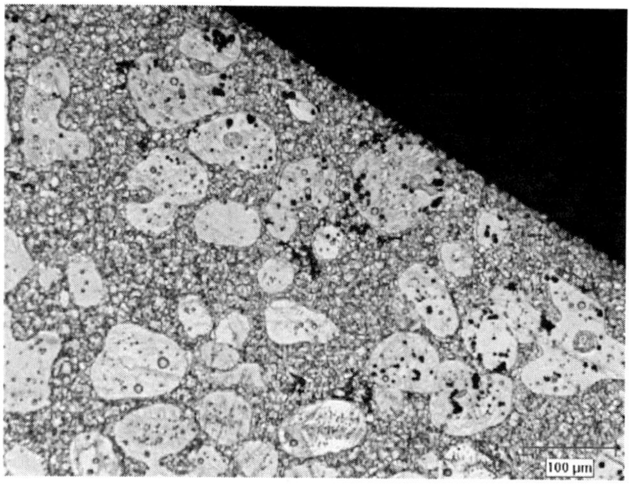

(b)

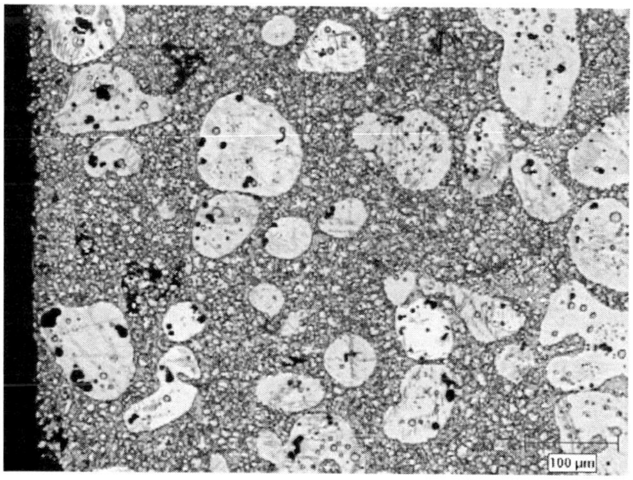

(c)

Figure 9. Photomicrographs taken near surface from various areas indicated in Figure 7, (a) from area 3 at corner, (b) area 4 and (c) area 5.

GENERAL DISCUSSION

It is evident that a combination of a low super-heat and electromagnetic stirring can produce AZ91D feedstock suitable for semi-solid casting. A morphology containing fine non-dendritic primary particles (with an average size of less than 150 micrometer) as starting material is important if a high degree of globularisation of the primary phase upon reheating by induction is the goal. It is the presence of a globular morphology that makes the alloy thixotropic and provides it with the low viscosity during die filling.

The temperature uniformity requirement for reheating alloy billets for semi-solid die casting makes it necessary for the induction system to have accurate control features that takes into account the "skin effect" produced by eddy currents on the billet surface. A single-coil high-frequency system programmed for power input would likely be more efficient compared with a low-frequency multi-station system used commercially. In terms of capital investment, a single coil reheating system would be less costly to set up. With the possibility of varying and controlling the power into the same coil, precision heating of the billet can be achieved with a number of heating stages in a fixed coil.

Results of the casting trials demonstrate that semi-solid die casting requires a much more precise control of the billet temperature and a well-defined window of castability for the alloy cast. Unlike liquid metal casting, semi-solid casting requires a higher metal pressure, a much slower gate velocity and a very narrow metal temperature range in the billet. On the other hand, by not starting with a super-heated melt as in conventional liquid pressure die casting, distinct advantages such as low cycle time, less porosity because of non-turbulent flow, improved die life and significantly better mechanical properties can be realized.

CONCLUSION

With the casting simulator of MTL/CANMET laboratory scale thixotropic feedstock of AZ91D (three inch diam.) was prepared through a combination of superheat reduction and electromagnetic stirring of the melt. The microstructure of feedstock consists of rosettes of primary alpha phase that becomes globular after reheating to 580° C.

Reheating of AZ91D billets for semi-solid die-casting can be done efficiently in a single-coil high-frequency induction system with controlled power input. Sound castings can be routinely produced with a high input power to heat the billet to the solidus temperature, followed by a rapid reduction of power, down to 15 % of its original value, prior to injection.

The quality of the casting is influenced by three main factors: the temperature of the billet, the speed of injection and the metal pressure used. At the optimum casting conditions very little or almost no liquid segregation is observed in the final die-cast part.

ACKNOWLEDGMENTS

The authors would like to acknowledge the financial support for this project from the Canadian Lightweight Materials Research Initiative (CLiMRI) of Natural Resources Canada. They also would like to acknowledge the technical help of A. Rivard and R. Canaj of MTL. Thanks are also extended to the many technical staff both at MTL/CANMET and IMI/NRC who contributed to this project.

REFERENCES

1. M.C. Flemings, "Behavior of Metal Alloys in the Semi-Solid State," Metallurgical Transaction, Volume A22, 952-981, May 1991.
2. K.P. Young and R. Fitze, " Semi-Solid Metal Cast Aluminum Automotive Components, " 3rd Int'l Conf. on Semi-Solid Processing of Alloys and Composites, pp. 155-177, 1994.
3. K.P. Young, "Recent Advances in Semi-Solid Metal (SSM) Cast Aluminium and Magnesium Components," 4th International Conference on Semi-Solid Processing of Alloys and Composites, pp. 229-233, 1996.
4. K. Young and P. Eisen, " SSM (Semi-Solid Metal) Technological Alternatives For Different Applications, " 6th Int'l Conf. on Semi-Solid Processing of Alloys and Composites, pp. 97-102, 2000.
5. D. Walukas, S. LeBeau, N. Prewitt and R. Decker. "Thixomolding Technology Opportunities and Practical Uses," 6th Int'l Conf. on Semi-Solid Processing of Alloys and Composites, pp. 109-120, 2000.
6. G. Hirt, R. Cremer, A. Winkelmann, T. Witulski, M. Zillgen, 3th Int. Conf. Semi-Solid Processing of Alloys and Composites, Tokyo, Japan, September (1994), pp. 107-116.
7. E. J. Zoqui, M. T. Shehata, M. Paes, V. Kao, E. Es-Sadiqi, Journal of Materials Science and Engineering A, Submitted, January 2001.
8. J-W. Liaw, T-F. Chen, C.A. Loong, C-Q. Zheng, P. Raducanu, K.T. Nguyen and C.K. Jen, " Theoretical Modeling and Experimental Verification of Induction Heating of Semi-Solid Billets," 6th Int'l Conf. on Semi-Solid Processing of Alloys and Composites, pp. 571-577, 2000.
9. H. Shahani and M. Shahani, "Resonance Switching Power Supply,"US Patent 5715155, 1998.
10. F.C Bennett, T.E. Leontis, and S.L. Couling, Proceeding International Magnesium Association, 34th Annual Meeting, Columbus, Ohio, May 1977, pp. 23-29.

Elevated Temperature Forming of Sheet Magnesium Alloys

Paul E. Krajewski
General Motors Research and Development Center

ABSTRACT

The use of sheet magnesium for automobile body applications is limited, in part, due to its low room temperature formability. Elevated temperature forming of magnesium sheet could enable the manufacture of automobile body closure and structural panels to meet vehicle mass targets. The effect of temperature in improving the formability of sheet magnesium has been known since the 1940's; however, automobile applications for sheet magnesium still have been very limited. The present work characterizes the elevated temperature mechanical behavior of commercially available magnesium sheet alloys at temperatures between 300°C and 500°C. The materials are then evaluated using both warm forming and superplastic forming technologies.

INTRODUCTION

Magnesium is the lightest structural material, having a density of 1.74 g/cm^3. Because of its low mass, magnesium usage in automobiles has steadily increased since the 1980's and additional utilization is expected [1]. Virtually all current magnesium applications are die castings. Wrought magnesium production applications are very limited, as is ongoing research and development on wrought magnesium alloys.

Sheet magnesium has been used as a structural material in the transportation industry since World War II [2]. The B36 bomber contained 9000 lbs of magnesium sheet, while jeeps and parachute delivery platforms were also made from sheet magnesium during the war. The U.S. transportation industry began to utilize sheet magnesium in the early 1950's, especially in freight trucks. Between 1955 and 1965, 6200 Metro-lite trucks were manufactured featuring magnesium sheet panels as well as structures made from magnesium plate and extrusions [2]. These trucks had increased payload capacity (for the same net truck weight) and showed excellent durability. The truck applications were perfect for magnesium because they did not require extreme formability, only structural integrity and low mass.

While sheet magnesium has not been used in high volume production automobiles, prototype components have been made using sheet magnesium. General Motors made prototype hoods for the Buick

Figure 1 – 1952 Buick LeSabre "car of the future" featuring sheet magnesium hood. (photo from GM achives)

Figure 2 – 1961 Chevrolet Corvette. Prototype hoods for this vehicle were made using magnesium sheet. (photo from GM achives)

LeSabre in 1952 and the Chevy Corvette in 1961. While both of the prototype parts were successful, neither was used in production. Recently, Volkswagen evaluated a sheet magnesium inner panel for its Lupo liftgate; however, they opted for a die cast magnesium panel [3].

The absence of sheet magnesium from mainstream production programs is predominantly due to the high material cost and magnesium's limited room temperature formability. The inferior room temperature formability of sheet magnesium compared to aluminum or steel is the result of its hexagonal close packed (HCP) structure, which limits the number of available slip systems for deformation. However, raising the temperature of magnesium increases the number of active slip systems, which enables significant improvements in plasticity. The increase in ductility with temperature is shown clearly in Figure 3.

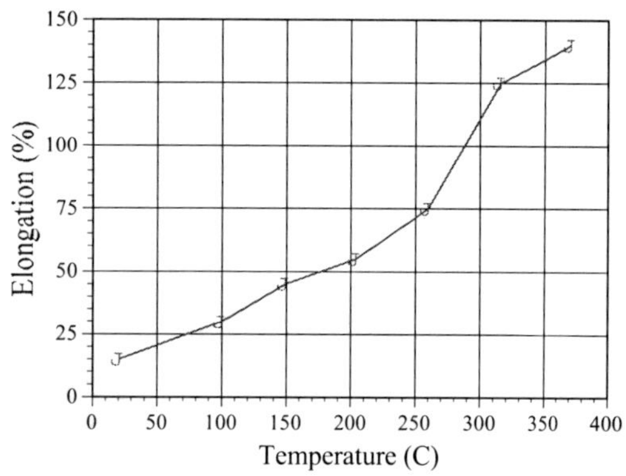

Figure 3 – The effect of temperature on total elongation in AZ31B-H24 (data taken from [4]).

While ductility values approaching 100% will result in significant formability improvements compared to room temperature, even higher values (>500%) have been achieved with superplastic (SP) magnesium sheet alloys [5-10]. Unfortunately, the methods used to manufacture SP magnesium such as rapid solidification [5,6], alloying with lithium [7,8], equi-channel angular extrusion [9], and compaction of machining chips [10], are either too expensive or impractical for high volume production.

The goal of the present work is to measure the elevated temperature performance of commercially available magnesium sheet materials and to evaluate their performance in elevated temperature metal forming processes.

EXPERIMENTAL PROCEDURE

MATERIALS - Three wrought magnesium alloys were evaluated: AZ31B-O (1.5 mm gage), AZ31B-H24 (2.6 mm gage), and AZ61A-O (1.4 mm gage). Mark Metals, Whittier, CA provided the AZ31B-H24 material and Spectrulite Consortium, Madison, IL supplied the AZ31B-O and AZ61A-O materials. The composition and typical mechanical properties of these materials are shown in Tables 1 and 2, respectively.

Table 1. Composition of the three magnesium alloys in the present study.

Actual Composition (wt. %)								
Alloy		Al	Zn	Mn	Si	Fe	Mg	
AZ31B-O		2.9	1.00	0.45	<0.1	<0.005	Bal.	
AZ31B-H24		3.0	0.88	0.42	<0.1	<0.005	Bal.	
AZ61A-O		5.9	0.56	0.24	<0.1	<0.005	Bal.	

Table 2. Typical room temperature mechanical properties of the three magnesium alloys in the present study. (taken from [4])

Material	U.T.S (MPa)	Y.S. (MPa)	Elong (%)
AZ31B-O	260	200	23
AZ31B-H24	290	220	15
AZ61A-O	310	230	16

TENSILE TESTING - Elevated temperature tensile testing was performed with a screw-driven machine, Instron model 5568, with an Instron 3119-007 high temperature chamber and Merlin data acquisition system. Tests were conducted at a constant crosshead speed. Tensile samples were 25.4 mm wide x 63.5 mm long with a 6.4 mm x 25.4 mm gage section, and machined with the rolling direction parallel to the gage length. The samples were heated for 15 minutes prior to testing and water quenched immediately after completion of the test. The reported results are the average of three tests at each condition with an approximate standard deviation of 20% elongation.

STAMPING - All stamping was performed on a 200-ton Niagara, single-action, mechanical press. Binder pressure was controlled with nitrogen cylinders positioned under each corner of the binder ring. No draw beads were used to control metal flow. A 305 mm x 381 mm pan die geometry with a depth variable up to 160 mm was used for all stampings.

Warm forming trials were conducted with the same die and press as for the room temperature stamping trials. Blanks for the warm forming trials were heated to between 200°C and 400°C in a furnace and rapidly transferred to the forming press. The forming temperature was typically 50°C cooler than the preheat temperature due to heat losses.

Blanks stamped at room temperature were lubricated with Alcan's Al070 prelube wax. Warm formed blanks were coated prior to preheating with Krytox liquid Teflon lubricant produced by Dupont. The binder and punch in the die were also sprayed with the Teflon lubricant for the warm forming trials.

SUPERPLASTIC FORMING (SPF) – SPF parts were formed on a 1200 ton press at the GM R&D Center. Steel dies were heated by conduction through platens on the press. Magnesium blanks were heated in the die for 5 minutes prior to forming to insure that they had reached the proper forming temperature. Air pressure was applied up to a peak of 500 psi in a controlled manner. Panels were coated with ZYP Lubricoat ™ (boron nitride) prior to forming.

RESULTS AND DISCUSSION

TENSILE TESTING –Elevated temperature tensile testing on all three magnesium sheet alloys was performed at initial strain rates of 0.001/sec and 0.01/sec. Stress-strain curves for the AZ31B-H24 material are shown in Figure 4. At 300°C, peak load was reached almost immediately, followed by rapid softening. At higher temperatures, samples actually exhibited "hardening". For example, at 500°C, the peak stress was not reached until 60 or 70% strain. After the peak was reached, the load decreased gradually as the material was strained further.

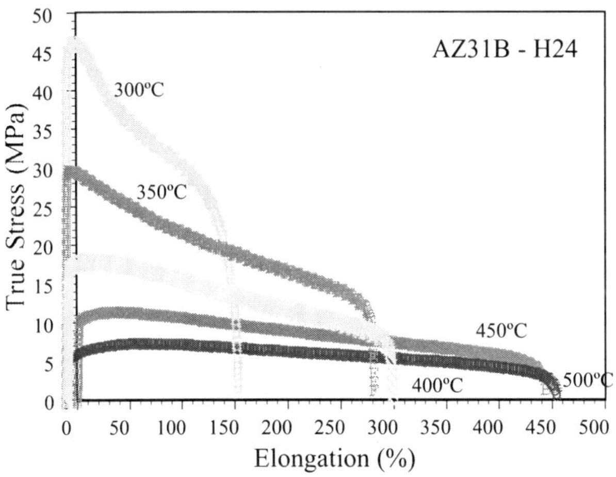

Figure 4 – The effect of temperature on tensile behavior of magnesium sheet alloy AZ31B-H24 *at an initial strain rate of 0.001.sec.*

Increasing the temperature from 300°C to 500°C produced a significant increase in ductility from 150% to over 400% for the AZ31B-H24 material. The effect of temperature on ductility for initial strain rates of 0.001/sec and 0.01/sec is shown for all three materials in Figures 5 and 6, respectively. At the slower initial strain rate (0.001/sec), AZ31B-H24 gave the best overall ductility, reaching over 400% elongation at 450°C and 500°C. The only temperature where AZ31B-H24 did not exhibit the highest ductility was at 300°C where AZ31B-O exhibited slightly higher ductility. AZ61 exhibited the lowest ductility at all temperatures at an initial strain rate of 0.001/sec.

For higher initial strain rates (e.g. 0.01/sec), the AZ31B-O material exhibited the highest ductility at both 450°C and 500°C. AZ61A-O exhibited a ductility of almost 250% at both temperatures, whereas the AZ31 materials exhibited less than 200% elongation. At temperatures below 450°C, the AZ31 materials exhibited better ductility than AZ61A-O.

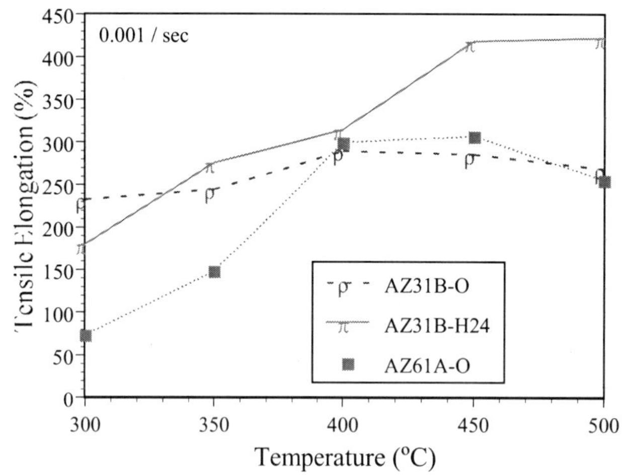

Figure 5 – The effect of temperature on total elongation in three commercially available magnesium alloys at an initial strain rate of 0.001/sec.

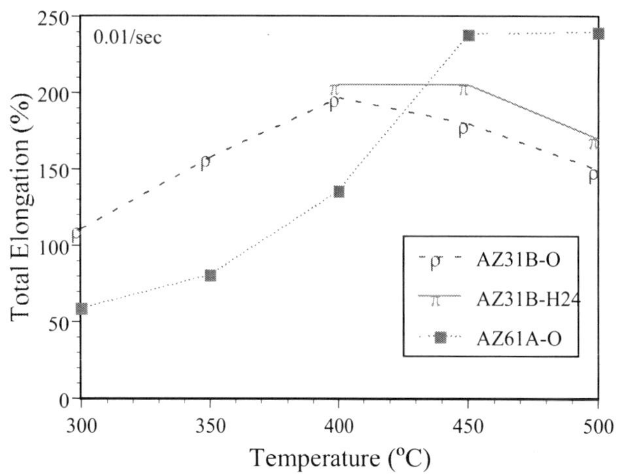

Figure 6 – The effect of temperature on total elongation in three commercially available magnesium alloys at an initial strain rate of 0.01/sec.

Material behavior at final failure was also affected by temperature. For example, the AZ31B-H24, samples tested at 450°C and 0.001/sec showed very rapid final failures. Once the material began to fail, the load dropped to zero very quickly. However, at 500°C and 0.001/sec, final fracture was more gradual, occurring over a range of about 25% strain. This gradual failure indicates that AZ31B-H24 could accommodate damage better and sustain more load carrying capability at the higher temperature. The change in failure behavior is shown clearly by the change in the fracture surface as temperature is increased (see Figure 7).

The AZ31B-H24 sample tested at higher temperatures exhibited necking down to a point whereas the sample tested at lower temperatures failed by a cavitation dominated mechanism. Correlation of the fracture behavior with tensile elongation indicated that the optimum ductility occurred where the material underwent a transition from cavitation dominated failure to necking dominated failure.

Figure 7 – SEM micrograph of failed tensile samples of AZ31B-H24 tested at 450°C and 500°C using an initial strain rate of 0.001/sec.

The ductility values exhibited by the commercial magnesium sheet alloys in the present study, especially the AZ31B-H24, are very comparable to those exhibited by SPF-grade aluminum alloys such as AA 5083 under similar conditions [11]. The comparable elevated temperature ductility suggests that commercially available sheet magnesium could be used to make parts that have been successfully made with 5083-type aluminum alloys.

The fact that these magnesium sheet materials were not given special thermomechanical processing and yet exhibited SPF-like behavior is very promising. The extra processing required to produce SPF-grade material typically increases the cost of the alloy significantly. It also is difficult to introduce high amounts of cold work into magnesium sheet alloys due to their low formability at room temperature. The tensile results from the present study indicate that commercially available magnesium could be used in a warm forming or superplastic forming process without additional thermomechanical processing.

FORMING EVALUATION –

Conventional Stamping (Baseline Comparison) – The room temperature formability of AZ31B-O magnesium sheet was evaluated by stamping rectangular pans. These trials were performed to establish a baseline for how far magnesium sheet can be formed at room temperature. The room temperature stamping trials were very unsuccessful. The magnesium sheet split at draw depths greater than 12 mm. An example of the failed AZ31B-O stamped pans is shown in Figure 8. The pan failed with almost no draw-in. For comparison, similar gage sheet of aluminum alloy AA5754 was stamped to draw depths of greater than 100 mm using the same die. Varying the binder pressure, blank size, or lubricant did not improve the draw depth of the AZ31B-O magnesium sheet.

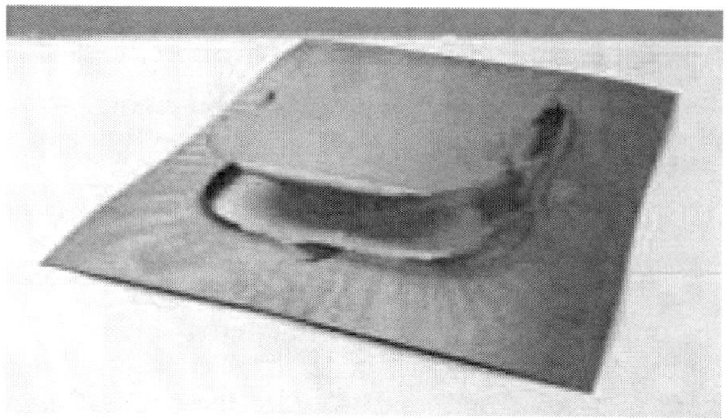

Figure 8 – Conventionally stamped pan formed using 1.4 mm AZ31B-O. Failed at 12mm draw depth.

The limited ductility of magnesium sheet at room temperature is also demonstrated by examining the fracture surface from the failed stampings, as seen in Figure 9. The fracture surface showed a cleavage-type failure with minimal local plasticity.

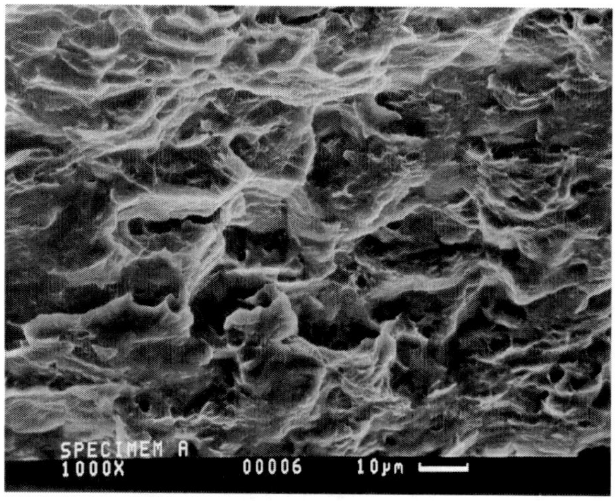

Figure 9 – Fracture surface of failed AZ31B-O pan conventionally stamped at room temperature.

The lack of formability at room temperature would prohibit the use of conventional stamping with sheet magnesium for anything except very simple shapes. In addition, secondary forming operations such as flanging, hemming, or clinching would be very difficult at room temperature. These results also suggest that the crash performance of magnesium sheet components also needs to be evaluated.

Warm Forming – The application of heat significantly improved the formability of AZ31B-O sheet. The draw depth of the rectangular pan increased from 12 mm to over 80 mm. The optimum conditions identified for forming this pan geometry occurred at approximately 350°C. A photograph of the deepest pan that was warm formed is shown in Figure 10. At larger forming depths, the panels exhibited splitting in the corners of the draw wall.

Figure 10 – Warm-formed pan drawn at 350°C to a depth of 83 mm with AZ31B-O.

The improvement in plasticity during warm forming compared to room temperature stamping can be seen in Figure 11, which shows the fracture surface of a pan formed at 250°C. The fracture surface exhibited characteristics more consistent with ductile fracture and local plasticity than the room temperature fracture surface shown in Figure 9.

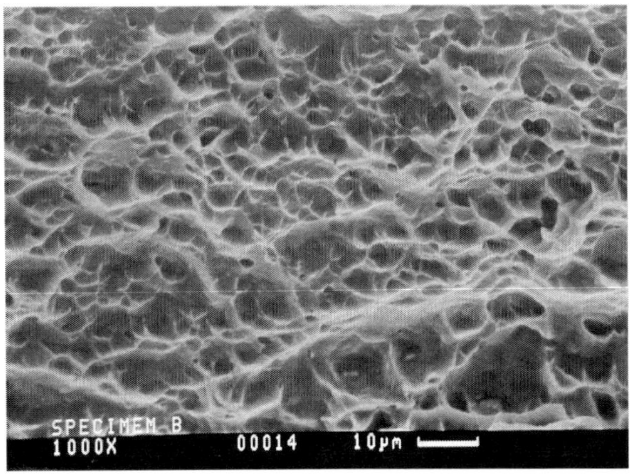

Figure 11 – Fracture surface of warm formed pan that was warm formed at 250°C.

Warm forming (200°C to 400°C) has been the standard process for making complex shapes from sheet and plate magnesium. The process has been used to make a variety of products from aerospace components to luggage. Warm forming guidelines for draw depths, stretching limits, and bending radii, have been summarized by Emley [13]. Spinning, hammer forming, and rubber pad forming have also been used at "warm" temperatures. Warm stamping of magnesium can enable forming a wide range of moderately complex automobile components. The issues limiting high volume application of warm forming for sheet magnesium are similar to those which currently prevent warm forming from being used with aluminum: blank preheating, thermal control in the dies, and lubrication among others.

Superplastic Forming – The superplastic formability of the commercial magnesium alloys was evaluated by forming two prototype components: a door inner and a reinforcement panel. The components were selected as representative automobile components that could be successfully formed using SPF aluminum alloys such as AA 5083 [12]. Both components were successfully formed with the AZ31B-H24 material at a temperature of 475°C. AZ31B-H24 was chosen because it exhibited the highest ductility in the elevated temperature tensile tests. 475°C was chosen because the hot tensile data indicated that peak elongation was observed at this temperature Examples of the magnesium panels are shown in Figures 12 and 13.

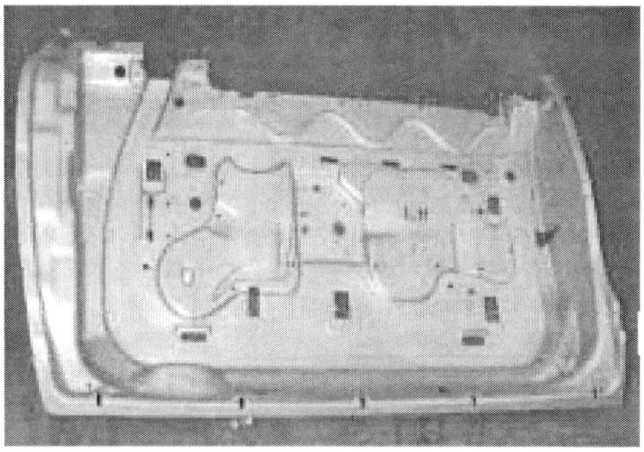

Figure 12 - Door inner panel superplastically formed with AZ31B-H24.

Figure 13 - Reinforcement panel superplastically formed with AZ31B-H24.

SUMMARY

The present study demonstrated that commercially available magnesium sheet alloys could exhibit significantly improved ductility at temperatures between 300°C and 500°C. The ductility enhancement leads to improved formability for both warm forming and superplastic forming technologies. This should enable the manufacture of complex automobile components such as door inners from magnesium sheet alloys. The ability of commercially available magnesium sheet alloys to exhibit SPF-like ductility is very important as it implies that special processing may not be required to produce SPF-grade magnesium sheet, which would eliminate the cost premium associated with SPF-grade materials.

ACKNOWLEDGMENTS

The contributions of Roy Sexton (tensile testing), Ron Lanzi (stamping), Jacob Lyjak (SPF forming), Bob Cuneo (forming) and Jerry Koralewski (forming) from the GM R&D Center are acknowledged.

REFERENCES

1. A. Luo, "Materials Comparison and Potential Applications of Magnesium in Automobiles", H.I. Kaplan, J.N. Hryn, B.B. Clow, Eds., Magnesium Technology 2000, Nashville, TN (Minerals, Metals, and Materials Society/AIME, PA, USA, 2000), pp. 89-98.
2. L.T. Barnes, "Rolled Magnesium Products: What Goes Around, Comes Around", Proceedings of the International Magnesium Association, 1992, Chicago, IL, pp. 29-43.
3. W. Diem, "Magnesium in Different Applications", *AutoTechnology*, vol. 1 (2001) pp.40-41.
4. "Selection and Application of Magnesium and Magnesium Alloys", *Metals Handbook*, 10th Ed., Vol. 2, 1990, pp. 455-479.
5. C.F. Chang and S.K. Das, "Rapidly Solidified Magnesium Base Alloy Sheet," U.S. Patent # 5,078,807, January 7, 1992.
6. J.K. Solberg, J. Torklep, O. Bauger, and H. Gjestland, *Mat. Sci. Eng.*, A134, 1991, pp. 1201-1203.
7. C.C. Kedenburg, "Superplasticity of Magnesium Base Alloys," Report within the C1 Project SFB 390 Magnesium Technology, Institut Fur Schweibtechnik und Trennende Fertigungsverfahren, Technische Universistat Clausthal, 1999.
8. P. Metenier, G. Gonzalez-Doncel, O.A. Ruano, J. Wolfenstine, and O.D. Sherby, *Mat. Sci. Eng.*, A125, 1990, pp. 195-202.
9. M. Mabuchi, K. Kubota, and K. Higashi, *J. Mat. Sci. Let.*, 12, 1993, pp. 1831-1832.
10. M. Mabuchi, K. Ameyama, H. Iwasaki, and K. Higashi, *Acta Mater.*, 47#7, 1999, pp. 2047-2057.
11. R. Verma, GM R&D, unpublished research, 1998.
12. GM R&D, unpublished research, 1998.
13. E.F. Emley, Principles of Magnesium Technology, Pergamon Press, New York, NY, 1966, pp. 584-603.

CONTACT

Paul E. Krajewski, PhD Staff Research Scientist,
General Motors Research and Development Center
Mail Code 480-106-212,
30500 Mound Rd, Warren, MI 48090-9055

paul.e.krajewski@gm.com.

2001-01-0422

Development of Creep-Resistant Magnesium Alloys for Powertrain Applications: Part 1 of 2

Bob R. Powell, Alan A. Luo, Vadim Rezhets, John J. Bommarito, and Basant L. Tiwari
General Motors Research and Development Center

ABSTRACT

A family of low-cost, creep-resistant magnesium alloys has been developed. These alloys, containing aluminum, calcium, and strontium are designated as "ACX" alloys. Developed for engine blocks and transmissions, the "ACX" alloys have at least 40% greater tensile and 25% greater compressive creep resistance than AE42, and corrosion resistance as good as AZ91D (GMPG 9540P/B corrosion test). These alloys are estimated to cost only slightly more than AZ91D and have as good castability. Creep data up to 200°C, tensile properties at room temperature and 175°C, corrosion results and microstructure analysis are presented and discussed. These alloys have the potential to enable the extension of the substantial weight reduction benefits of magnesium to powertrain components.

INTRODUCTION

The use of magnesium to reduce weight in automobiles has grown rapidly since the early 1990's. This growth has generally been limited to interior-component applications using the common die casting alloys AZ91D, AM50A and AM60B, which have good strength and/or ductility at ambient temperatures. With few exceptions such as the recently introduced Honda oil pan [1] and the Audi manual and CVT transmission cases [2], the only magnesium powertrain components currently in production are non-structural and low-temperature [3]. If the advantages of magnesium are to be extended to engines and automatic transmission components, several objectives will have to be achieved. These include greater creep strength at high temperature, lower alloy cost, and demonstrated corrosion resistance.

The aforementioned die casting alloys (AZ91D, AM60, and AM50) are limited to near-room-temperature applications because their mechanic properties decrease at powertrain operating temperatures and they are susceptible to creep. AE42, a rare earth element-containing magnesium die casting alloy, has creep resistance sufficient for automatic transmission operating temperatures (up to 150°C), but not engine temperatures (above 150°C) [4]. The additional cost of this alloy due to the presence of rare earth increases an already major barrier to the use of magnesium for powertrain components. Corrosion resistance is also a concern for magnesium components that will be exposed to road conditions and salt spray. Corrosion has been overcome in the low-temperature alloys because their purity is carefully controlled and fastening techniques to prevent galvanic coupling have been established. Any powertrain alloy will need to demonstrate this same level of corrosion resistance.

This paper describes the development of a family of Mg-Al-Ca-Sr alloys (referred to hence as ACX alloys) that have achieved the following characteristics:

- creep resistance – at least 25% greater than AE42 at 150°C
- cost and corrosion resistance – equivalent to AZ91D

Castability of the alloys is also promising with successful castings being made in specimens and parts weighing up to 10 kg. These alloys have outstanding potential for powertrain components.

BACKGROUND

The air-cooled engines of the VWs of the 1970's used Mg alloys. The first engines used AS 41 (Mg with 4% Al and 1% Si). This alloy was succeeded by AS21, which had better high-temperature properties than AS41. Both alloys relied on the reduced Al content (reduced presence of low-melting $Mg_{17}Al_{12}$) and the formation of Mg_2Si to impart creep strength to the alloy. The use of Mg for engines by VW ended because of several factors [5] including increasing alloy cost, higher engine operating temperature, and the change to water-cooling which brought corrosion concerns back.

Dow developed the rare earth-containing alloy AE42 for high temperature applications [6]. It was developed from non-Al Mg alloys in which rare earths were shown to increase creep resistance by forming fine Mg_9E precipitates in the grain boundaries [7]. Aluminum was added to improve castability and room temperature mechanical properties. Furthermore, the Al formed $Al_{11}E_3$

(also called Al_4E) under die casting conditions and greatly improved the creep resistance. However, the rare earth increases the cost of the alloy and AE42 has been reported to be susceptible to hot-cracking [8]. Furthermore, recent work has shown that the compressive creep resistance of AE42 decreases abruptly above 150°C [4].

It has been shown [9] that the breakdown in creep resistance of AE42 about 150°C is accompanied by a phase change in the microstructure of this alloy, specifically the decomposition of $Al_{11}E_3$ and the formation of Al_2E and $Mg_{17}Al_{12}$. $Mg_{17}Al_{12}$ is the low-melting-temperature phase that is present in AZ91D, AM60, and AM50 and to which is attributed the poor creep behavior of these alloys. It follows that increasing the thermal stability of $Al_{11}E_3$ might be a means for extending the creep resistance of AE42 to above 150°C. It also suggests the possibility of developing lower cost, creep-resistant alloys by replacing the rare earth in AE42 with a less expensive element that also forms an $Al_{11}E_3$ - type phase.

$Al_{11}E_3$ - type phases have been reported in Al - alkaline earth (Ca, Sr, and Ba) compounds [10]. Of the three alkaline earths, calcium is the least expensive on a cost per pound basis. It also has the lowest density and atomic weight, such that the "cost per atom of Ca" is significantly less than that of Sr or Ba. Previous work has shown that Ca indeed does impart creep resistance to Mg-Al alloys [5,11,12], but it has also been reported that the resulting alloys have poor castability; they are prone to die sticking and hot-cracking [5,13]. Gray prevented die sticking and hot-cracking by limiting the Ca level to below 0.5% [14]. These casting problems were also reduced by the addition of zinc [15], but the resulting alloy achieved the satisfactory creep resistance only up to 150°C. Honda's oil pan alloy contains Ca and rare earth [1]. It's tensile creep strength is greater than that of AE42, but is still expensive due to the 2.6% rare earth content.

The present work revisits Ca additions for increasing the creep-resistance of die cast Mg-Al alloys. As the previous studies were limited to less than 2% Ca, it was decided in this work, to extend the Ca to higher levels. Strontium and silicon were included as possible fourth-element additions for modifying precipitates and further improving the alloy characteristics.

EXPERIMENTAL PROCEDURE

COMPOSITION RANGES AND MELT PREPARATION - The alloys were cold chamber die cast at Lexington Die Casting in Lakewood, NY. The compositions cast are shown in Table 1. Alloying in the melt was done at the die casting site using procedures developed at the GM R&D Center. The metals used in alloying were AM50, Mg, Al, Ca, Sr (as Sr10-Al), and Si (as AS41 alloy containing about 1% Si). Recovery was greater than 95%. Melting and alloying was done under protective CO_2/SF_6 cover gas. The Ca-containing alloys were so unreactive that the SF_6 may have been unnecessary. The use of Ca to increase the ignition temperature of Mg melts has been reported [16].

DIE DESIGN AND CASTING CONDITIONS - The die insert has three cavities: one 12 mm-diameter tensile bar, one 6 mm-diameter tensile bar, and one 38 mm-square and 12 mm-thick compressive stress retention (CSR) coupon. The tensile bars were end-gated. The Lexington die cast unit (a 700 ton Lester machine) was equipped with QPC Prince shot profile control and recording. The melt temperature was controlled at 1250°F (677°C) plus/minus 5°F and the die surface temperature was maintained at about 660°F (350°C).

Table 1. Magnesium Alloy Compositions (weight percent)

Alloy	Designation	Chemical Composition (wt.%)			
		Al	Ca	Si	Sr
A	AM50	4.7	-	-	-
B	AC52	4.5	1.9	-	-
C	AC53	4.5	3.0	-	-
D	AC53+0.3%Si	4.5	2.9	0.26	-
E	AC53+0.3%Si+0.1%Sr	5.4	2.9	0.27	0.11
F	AC53+0.3%Si+0.15%Sr	5.7	3.0	0.28	0.15
G	AC53+0.03%Sr	4.7	3.1	-	0.03
H	AC53+0.07%Sr	5.0	3.1	-	0.07
I	AC53+0.15%Sr	5.7	3.1	-	0.15

SPECIMEN ANALYSIS - Sample chemistries were measured for each casting composition using inductively coupled plasma/atomic emission spectroscopy (ICP/AES). X-ray diffraction (XRD) was used to identify phases in the microstructure. The lattice parameters and weight percent of α-Mg were calculated using the Rietveld method. Additional microstructural analysis was done using analytical transmission electron microscopy (TEM) with energy dispersive spectroscopy and electron diffraction. The TEM samples were prepared by ion milling.

CREEP TESTING - Tensile creep testing was done at 150, 175, and 200°C. Samples for each test were selected on the basis of casting quality as determined by X-ray inspection. Threads were machined into the grip regions of the 6 mm-diameter tensile bars so that they could be held in the test fixtures. Tensile-creep was done under constant-load, constant-temperature conditions. Total creep extension in 100 h at the test temperature was recorded as were the primary and secondary regions of the creep curves.

Compressive creep was characterized by compressive stress retention (CSR) measurement at 150 and 175°C following the procedures reported by GM Powertrain [4]. CSR simulates the bolt load retention performance of the alloy and is a critical functional test for a powertrain component with regard to the integrity of the parts that are bolted to the component. CSR testing was done at KEI Engineering in Ann Arbor, MI.

TENSILE PROPERTIES - The tensile yield strength, ultimate tensile strength, and elongation were measured on the 6 mm-diameter bars at room temperature and at 175°C. Measurement procedures followed ASTM Standard B 557M-94.

CORROSION BEHAVIOR - CSR samples were evaluated using the GM9540P/B corrosion test [17] which was conducted by the Corrosion Engineering Group at the GM Proving Ground, Milford, MI. This accelerated laboratory corrosion test uses a combination of cyclic conditions (salt solution, various temperatures, humidity, and ambient environment) to simulate the equivalent of ten years corrosion exposure for some metal systems.

RESULTS AND DISCUSSION

TENSILE CREEP BEHAVIOR – The tensile creep behavior of each ACX alloy is shown in Table 2. The 100 h creep extensions of the ACX alloys range from 40 to 80% less than that of AE42 at 150°C. It is noted that because of the high stress and temperature conditions used in this study, all the ACX alloy samples entered the secondary creep region within 100 hours. The 0.1% creep strength of AE42 at this temperature has been reported, 9.4 ksi [8]; i.e., the total creep extension of AE42 at a load of 9.4 ksi and at 150°C will be approximately 0.1% in 100 hrs. This is in good agreement with the rest results for AE42 in this study. At 12 ksi (28% greater load) the creep strain of the ACX alloys averages 0.05%, less than half that of AE42 specimens. At 175°C, the ACX alloys are nearly 50% better than AE42. There is an indication in the creep data that microalloying with Sr further improves the creep resistant but the effect is very small. The limited data obtained for Si shows no effect.

COMPRESSIVE CREEP BEHAVIOR - Compressive creep resistance is an important criterion for the transmission and block material because it is a measure of how tight the bolts remain in the assembled unit. As measured by compressive stress retention (CSR), the ACX alloys are much better than AE42 (see Figures 1 and 2). In these figures CSR is presented as the percent of load remaining in the bolted sample as a function of the time of exposure up to 750 hrs at the indicated temperature. The previously published CSR behavior of AZ91D and aluminum A380 [4] is included in the figures for comparison.

Table 2. Total 100-Hour Creep Extension (%) of Magnesium Alloys.

Alloy	Designation	150°C, 12 ksi	175°C, 10 ksi	200°C, 8 ksi
	AE42	0.11	0.12	-
A	AM50	2.3[1]	4.5[2]	-
B	AC52	0.05	0.06	0.26
C	AC53	0.07	0.09	0.28
D	AC53+0.3%Si	0.06	0.07	0.25
E	AC53+0.3%Si+0.1%Sr	0.03	0.07	0.18
F	AC53+0.3%Si+0.15%Sr	0.05	0.06	0.14
G	AC53+0.03%Sr	0.06	0.08	0.28
H	AC53+0.07%Sr	0.05	0.06	0.20
I	AC53+0.15%Sr	0.04	0.08	0.16

[1]Sample broke in 2 hours
[2]Sample broke in 3 hours

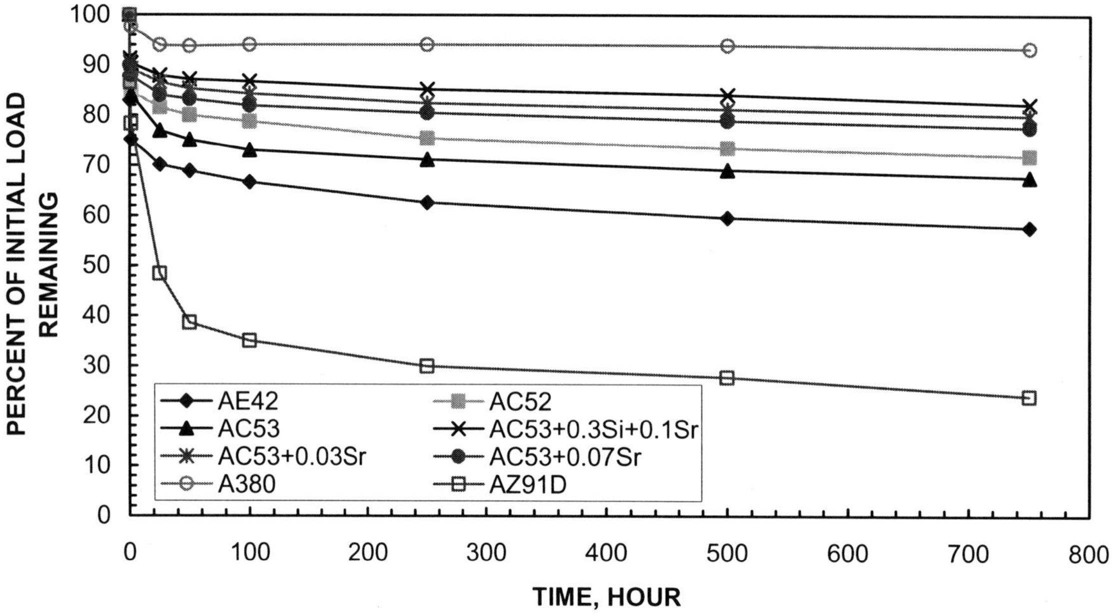

Figure 1. The compressive stress retention of AE42, aluminum A380, AZ91D and the ACX alloys at 150°C for times up to 750 hrs.

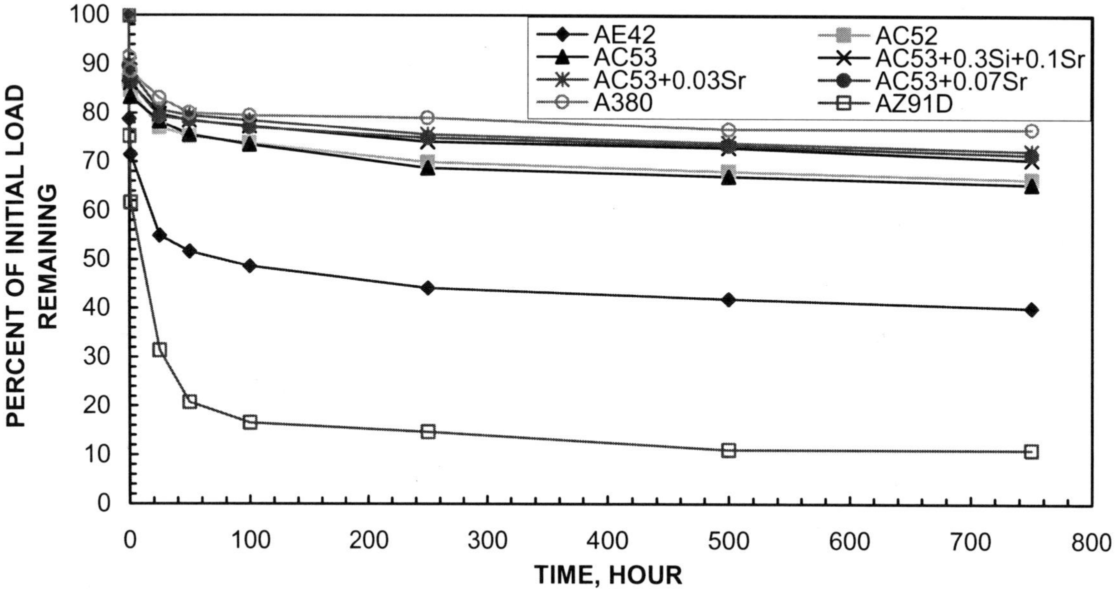

Figure 2. The compressive stress retention of AE42, aluminum A380, AZ91D and the ACX alloys at 175°C for times up to 750 hrs.

At 150°C and 750 hrs, AE42 retained 58% of the initial load while the ACX alloys CSR ranged from 68 to 82%, all better than AE42. At 175°C, the CSR of AE42 dropped considerably, to 40%, as was expected from the earlier study [4]. This has been explained [9] as being due to the decomposition of $Al_{11}E_3$ with the subsequent formation of $Mg_{17}Al_{12}$. The ACX alloys do not demonstrate the same deterioration with increasing temperature. They retain nearly as much load as they did at 150°C, 65 vs. 72%. As with the tensile creep results, the addition of Sr appears to further improve creep resistance but the effect is much more evident in the CSR results. In fact, the Sr-microalloyed AC53 samples performed almost as well as aluminum A380.

CORROSION BEHAVIOR - The benchmark for the corrosion resistance of the ACX alloys is that of AZ91D in the GM9540P ten-year equivalent accelerated corrosion test. Table 3 shows that the ACX alloys microalloyed with Sr perform as well as AZ91D. Over two independent test series, the AZ91D averaged 0.5% weight loss. AM50 was included in the second test set and lost 0.7% of its weight. The ACX alloys with X ranging from 0.05 to 0.1% Sr also achieved this level of corrosion resistance. The data shows that increasing Sr levels improved the corrosion resistance and the Si appeared to be detrimental. The effect of 2 vs. 3% Ca is not clear because there was more scatter in the individual results. Each reported value in each series was the average of three samples.

Also shown in Table 3 are corrosion results for AC33 and AC53 samples that were cast at a different site. These results are included because they show that Fe content is just as critical to corrosion rates in the ACX alloys as it is in AZ91D. Chemical analysis of the samples from these samples showed that the Fe level was 90 ppm, about twice the allowable level per the ASTM specifications for Mg alloys. The Ni and Cu levels were within specifications. The Fe levels in the Lexington Die Cast samples were 40 ppm. This resulted in nearly ten-fold lower mass loss in the GM 9540P test. A set of Fe-contaminated ACX samples was included in the second test set and the results show the reproducibility of the GM9540P test.

Table 3. The Percent Weight Loss of Mg Test Coupons in the GM9540P Cyclic Salt Spray Corrosion Test.

Alloy Composition	1st Set (%)	2nd Set (%)
AZ91D	0.6	0.4
AM50		0.7
AC52		1.5
AC53		2.1
AC53 + 0.3 Si		1.6
AC53 + 0.3 Si + 0.10 Sr		1.0
AC53 + 0.3 Si + 0.14 Sr		1.0
AC53 + 0.02 Sr		0.8
AC53 + 0.05 Sr		0.6
AC53 + 0.10 Sr		0.5
AC33 (high Fe)	15.6	17.1
AC53 (high Fe)	11.7	12.3

TENSILE PROPERTIES - Figures 3-5 show the tensile properties of the ACX alloys at room temperature and at 175°C. The tensile properties of AM50, also cast at Lexington, are included in the figures. All of the ACX alloys have higher yield and tensile strength than AM50 due to the calcium addition; AC53 is stronger than AC52 at both room temperature and 175°C. The effects of microalloying with Si and/or Sr are small. The room-temperature elongation data for the ACX alloys are shown in Figure 5. Calcium additions reduce total elongation while Si and Sr additions to the ACX alloys have no additional effect. The fracture surfaces of the tensile samples were free from metal inclusions. This in part explains the high total elongation of 17% measured for the AM50 alloy, which is much higher than the value of 10% reported in the literature [18]. Literature values tend to be conservative so as to simulate properties of cast parts rather than cast tensile bars. Whether the measured elongation values are due to the excellent casting conditions at Lexington (very little porosity was observed by X-ray inspection of the samples), or other factor remains to be studied.

MICROSTRUCTURE AND CASTING CHARACTERIZATION - In an early phase of this study, the Mg-Al-Ca ternary was surveyed for microstructural features by drawing pin samples from a Mg-4%Al melt after successive additions of Ca to the melt. Pin samples were collected by vacuum suctioning from the melt into a 5 mm. diameter glass tube. Below 1% Ca, only α-Mg was identified in the XRD pattern, see Table 4. At and above 1% Ca, a second phase was also identified, Mg_2Ca, the amount increasing as the Ca level in the melt was increased. Lattice parameter shifts are consistent with substitution of Al on Mg sites, $(Mg_{1-x}Al_x)_2Ca$, in this phase. The detailed characterization and analysis are reported in the accompanying paper [19]. As the Ca content of the melt increased, the lattice parameter shifted in the direction of lower substitution; i.e., less Al in the phase. However, at the same time, the amount of this phase increased from zero to nearly 20%. This would result in a shifting of Al from the primary Mg to the Mg-Al-Ca ternary.

As the Ca content increased and the amount of α-Mg decreased from 100 to 80%, the Mg phase also underwent a change in its lattice parameters that corresponded to the removal of Al from solution in the phase [20].

No other phases were identified, and no evidence of Al_4Ca or $Mg_{17}Al_{12}$ was detected, However, these results were based on the analysis of pin samples which, as noted, were used only to simulate die casting solidification rates. Subsequent TEM analysis (Figure 6) of the die cast AC53 clearly revealed the ternary lamella in the eutectic regions of the alloys. These lamella had the hexagonal Mg_2Ca phase structure with approximately half of the Mg atoms replaced with Al. Thus, while Al_4Ca was not detected, nor was $Mg_{17}Al_{12}$. This and the absence of Al solid solution in α-Mg indicates that the Ca is still performing its role of functionally removing Al from the alloy and preventing the formation of $Mg_{17}Al_{12}$.

CASTABILITY AND CASTING QUALITY - The ACX alloys have excellent creep resistance, corrosion resistance, and tensile properties. In our die casting experience to date, with the ACX alloys, they have shown excellent castability. Even on small samples, hot-cracking can be indicated by the surface condition of the parts. All samples showed smooth surfaces and no evidence of cracking. On the basis of these experiments we conclude that the castability of these alloys for small parts is excellent, at least as good as that of AZ91D.

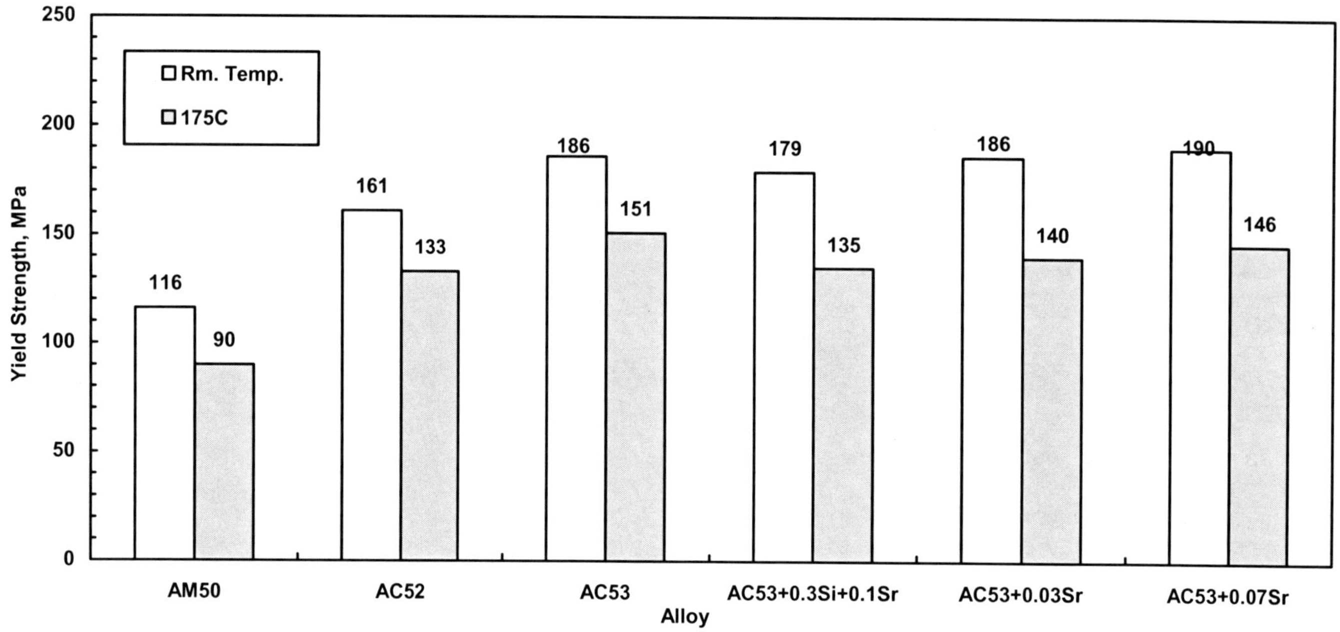

Figure 3. The tensile yield strength AM50 and the ACX alloys at room temperature and 175°C.

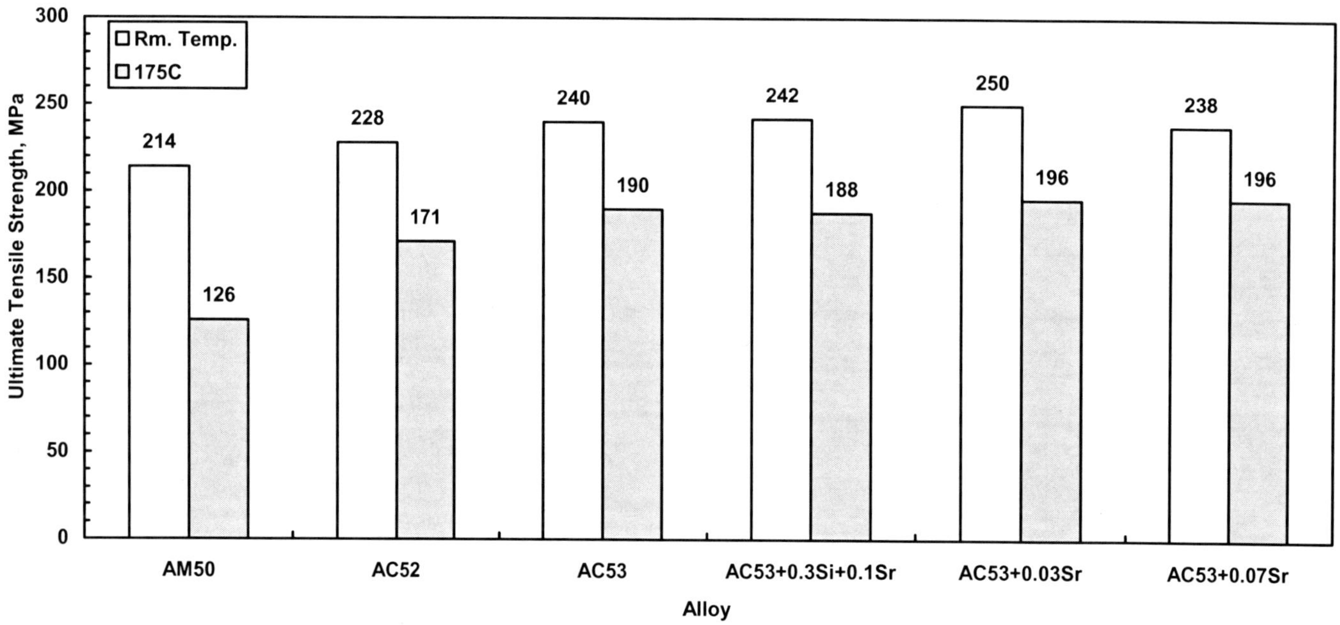

Figure 4. The ultimate tensile strength of AM50 and the ACX alloys at room temperature and 175°C.

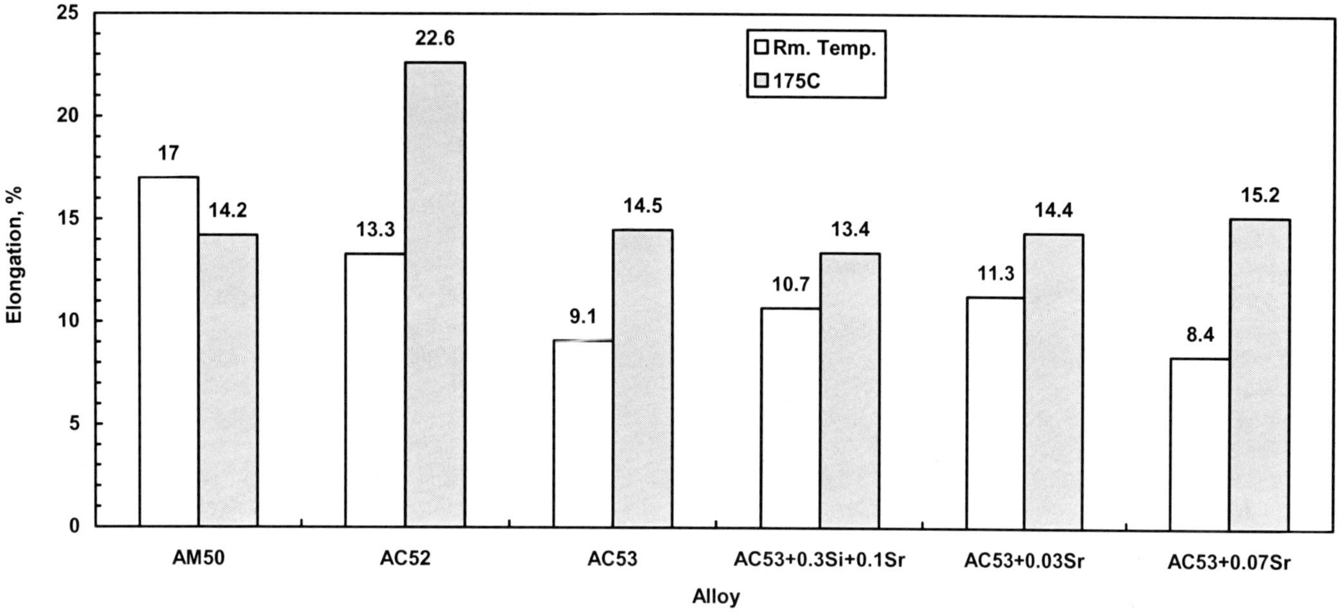

Figure 5. The tensile elongation of AM50 and the ACX alloys at room temperature and 175°C.

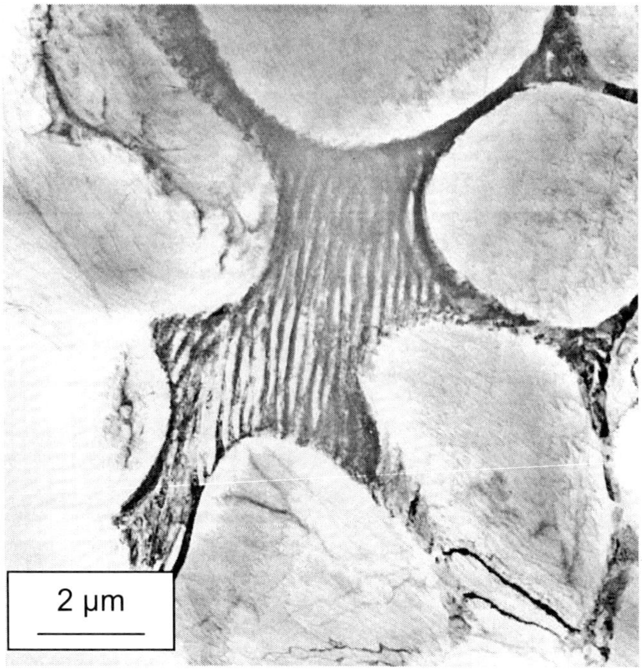

Figure 6. Analytical electron micrograph of AC53 alloy showing primary Mg grains and a grain boundary eutectic structure comprising $(Mg,Al)_2Ca$ and Mg. This structure is probably responsible for the improved creep resistance of the AC alloys.

CONCLUSIONS

Magnesium alloys containing aluminum, calcium, and fourth-element additions (ACX alloys) have been developed for use in structural powertrain components that operate at high temperature. The properties and characteristics of these alloys are summarized as follows:

1. Compared with the benchmark creep-resistant die casting alloy AE42. The new ACX alloys show at least 40% reduction in tensile creep extension.

2. The compressive creep behavior of the ACX alloys is 25% better than that of AE42 at 150°C. At 175°C the compressive load retention is nearly twice that of AE42 and rivals the behavior of aluminum A380.

3. It is estimated that the ACX alloys will cost no more than AZ91D because they do not require rare earth additions.

4. GM Proving Ground tests (GM9540P/B cyclic salt spray) show that the ACX (X = Sr) alloys performed as well as AZ91D.

5. The castability (fluidity) of these alloys for small parts and simple shapes is excellent.

6. The greater resistance to ignition of ACX alloys (compared to AZ91D and AE42) during melting and casting could reduce the need to use sulfur hexafluoride as a protective gas, thus providing an

additional advantage because sulfur hexafluoride is both costly and a powerful greenhouse gas.

ACKNOWLEDGMENTS

The authors appreciate the experimental support of Mike Balogh, Debra F. Esch, and George Kilbertus. Larry S. Thompson conducted the corrosion experiments at the GM Proving ground. We sincerely appreciate the help and professionalism of the management and staff at Lexington Die Casting.

CONTACT

Dr. Bob R. Powell
Staff Research Scientist
Materials & Processes Lab/MC 480-106-212
General Motors Research & Development Center
Warren, MI 48090-9055
Tel: 810-986-1293
Fax: 810-986-9204
e-mail: bob.r.Powell@gm.com

REFERENCES

1. Koike, S. et al., "Development of Lightweight Oil Pans Made of a Heat-Resistant Magnesium Alloy for Hybrid Engines," SAE Technical Paper 2000-01-1117, (2000).

2. "Most of Audi's Big Cars to be Equipped with CVTs," Automotive News, 74 [5849] 38 (1999).

3. Luo, A. "Materials Comparison and Potential Applications of Magnesium in Automobles," Magnesium Technology 2000, eds. H.I. Kaplan, J.N. Hryn, and B.B. Clow, The Minerals, Metals, & Materials Society (TMS), Warrendale, PA, pp. 89 (2000).

4. Sieracki, E.G., Velazquez, J.J., and Kabiri, K., "Compressive Stress Retention Characteristics of High Pressure Die Casting Magnesium Alloys," SAE Technical Publication No. 960421 (1996).

5. Hollrigl-Rosta, F., Just, E., Kohler, J., and Melzer, H.-J.: "Magnesium in Volkswagen", L. Met. Age, August 1980, pp. 22-29.

6. Nelson, K.E., "Magnesium Die Casting Alloys," SDCE Transactions, Paper no. 13 (1970).

7. Wei, L.Y. and Dunlop, G.L., "Precipitation Hardening in a Cast Mg-Rare Earth Alloy," in Magnesium Alloys and Their Applications, DGM Informationsgesell-schaft, Verlag, Germany, p. 335 (1992).

8. Mercer, W.E., "Magnesium Die Cast Alloys for Elevated Temperature Applications," SAE Technical Paper 900788, (1990).

9. Powell, B.R., Rezhets, V., Balogh, M., and Waldo, R., "The Relationship Between Microstructure and Creep Behavior in AE42 Magnesium Die Casting Alloy," The Metallurgical Society Annual Meeting , New Orleans, LA, February 11, 2001.

10. Buschow, K.H.J. and van Vucht, J.H.N., "Systematic Arrangement of the Binary Rare-Earth-Aluminum Systems," Philips Res. Reports, 22 233-45 (1967).

11. Raynor, G.V., The Physical Metallurgy of Mg and Its Alloys, Pergamon Press, Oxford, p. 103 (1959).

12. Gjestland, H., et al., "Stress Relaxation and Creep Behavior of Rapidly Solidified Magnesium Alloys," Mat. Sci. Eng., A134, 1197-1200 (1991).

13. Riopelle, L., personal communication, unpublished Hydro Magnesium research in the 1960's.

14. International Conference on Strontium-Containing Compounds, ed. by T.J. Gray, NRC, Ottawa, Canada, p. 232 (1973).

15. Luo, A. and Shinoda, T., "Development of a Creep-Resistant Magnesium Alloy for Die Casting Applications," in Magnesium Alloys and Their Applications, ed. by B.L. Mordike and K.U. Kainer, Werkstoff-Informationsgesellschaft, Frankfurt, Germany, p. 151 (1998). A US Patent was issued in 1998 covering the zinc-containing alloys.

16. Chang, S.-Y., Matsushita, M., Tezuka, H., and Kamio, A., "Ignition Prevention of Magnesium by Simultaneous addition of Calcium and Zirconium," Int. J. Cast. Metals. Res., 10 [6] 345-51 (1998).

17. "Accelerated Corrosion Test GM9540P," GM Engineering Standards/ Materials and Processes - Procedures, July 1991.

18. Hydro Magnesium data sheet (1996).

19. Luo, A., Balogh, M.P., and Powell, B.R., "Tensile Creep and Microstructure of Mg-Al-Ca Based ACX Alloys for Powertrain Applications: Part 2 of 2," SAE 2001, Detroit, MI (2001).

20. Murray, J.L., "Aluminum-Magnesium," in Phase Diagrams of Binary Magnesium Alloys, ed. by A.A. Nayen-Hashemi and J.B. Clark, published by ASM International, Metals Park, Ohio, pp. 17-34 (1988).

2001-01-0423

Tensile Creep and Microstructure of Magnesium-Aluminum-Calcium Based Alloys for Powertrain Applications – Part 2 of 2

Alan A. Luo, Michael P. Balogh and Bob R. Powell
General Motors Research & Development Center

ABSTRACT

This paper describes the tensile creep and microstructure of Mg-Al-Ca-based ACX magnesium alloys being developed for powertrain applications. Important creep parameters, i.e., secondary creep rate and creep strength, for the new alloys are reported. Tensile creep properties of the newly developed ACX alloys are significantly better than those of AE42 alloy, which is the benchmark creep-resistant magnesium die casting alloy. Creep mechanisms for different temperature/stress regimes are proposed. A new intermetallic phase, $(Mg,Al)_2Ca$, was identified in the microstructure of the ACX alloys, and is proposed to be responsible for the improved creep resistance of the alloys.

INTRODUCTION

In an accompanying paper [1], a new series of magnesium alloys containing 4-6% aluminum, 1.5-3.5% calcium, and small additions of strontium and/or silicon (designated as ACX alloys) was developed at the GM R&D Center. The new ACX alloys offer excellent creep resistance and low cost that meet the materials requirements for automotive powertrain applications. The previous paper [1] provides a good summary of the tensile, creep, corrosion, castability and microstructure of the ACX alloys. This paper presents additional creep test results and analyzes the data in a format suitable for design engineers. An in-depth analysis was also conducted to provide a further understanding of the microstructure and the creep mechanisms of the new alloys, which is extremely important in their optimization and commercialization in powertrain applications.

EXPERIMENTAL PROCEDURE

ALLOY PREPARATION AND DIE CASTING

Die casting of the ACX alloys was carried out at Lexington Die Casting, Lakewood, NY. Ten alloy compositions based on the Mg-Al-Ca system with microalloying additions of Si and/or Sr were prepared at Lexington in a steel crucible with a melt capacity of 650 lb. A 700-ton Lester cold-chamber die casting machine was used to cast the metal into a three-cavity die producing the following test specimens:

- one tensile bar 6mm in diameter and 25mm gage length;
- one tensile bar 12mm in diameter and 50mm gage length; and
- one 40x40x10mm block coupon with an 11mm-diameter core hole.

For the entire casting trial, the melt temperature was controlled at around 677°C, while the die surface temperature was maintained at about 350°C. For each alloy, at least 50 shots were made. Chemical analysis using ICP/AES (Inductively-Coupled Plasma/Atomic Emission Spectroscopy) techniques was performed on each alloy, and the results are listed in Table 1. The quality of the die cast test specimens was determined by X-ray radiography. The 6-mm tensile bars were generally sound castings and, therefore, were used for tensile creep testing.

TENSILE CREEP TESTING

Creep is the time-dependent deformation of a material at a stress which is often below the yield strength of the material. In this study, tensile creep testing was performed on as-cast 6-mm tensile bars under constant-load and constant-temperature conditions according to ASTM E139 procedures. Each alloy was tested under three temperature/load conditions, i.e., 150°C/83MPa, 175°C/70MPa and 200°C/56MPa, using two Satec and one Arcweld creep test machines. These conditions were selected based on creep performance requirements for various powertrain applications. For each specimen, the test temperature was monitored in three locations (top, middle and bottom) to maintain a temperature control of ±1°C. Each test was continued for at least 100 hours. Tensile displacement was measured using extensometers, and all temperature-strain-time data were collected using a Labview computer data-acquisition system. For each alloy at each test condition, the creep results were based on the average of at least two and, in most cases, three specimens.

Table 1. Chemical Compositions of the Experimental Alloys Cast at Lexington

Alloy	Designation	Chemical Composition (in wt.%)							
		Al	Ca	Si	Sr	Mn	Fe	Ni	Cu
B	AC52	4.5	1.9	-	-	0.27	0.003	<0.002	0.002
C	AC53	4.5	3.0	-	-	0.27	0.003	<0.002	0.003
D	AC53+0.3Si	4.5	2.9	0.26	-	0.28	0.003	<0.002	0.003
E	AC53+0.3Si+0.1Sr	5.4	2.9	0.27	0.11	0.26	0.005	<0.002	0.003
F	AC53+0.3Si+0.15Sr	5.7	3.0	0.28	0.15	0.26	0.004	<0.002	0.003
G	AC53+0.03Sr	4.7	3.1	-	0.03	0.30	0.003	<0.002	<0.002
H	AC53+0.07Sr	5.0	3.1	-	0.07	0.28	0.003	<0.002	<0.002
I	AC53+0.15Sr	5.7	3.1	-	0.15	0.27	0.003	<0.002	<0.002
K	AC52+0.1Sr	5.4	1.9	-	0.10	0.30	0.004	<0.002	<0.003
L	AC62+0.2Sr	5.9	1.8		0.17	0.29	0.005	<0.002	<0.003

MICROSTRUCTURAL ANALYSIS

The microstructure of the die cast alloy specimens was analyzed using optical microscopy, X-ray diffraction (XRD) and transmission electron microscopy (TEM). Specimens for optical and TEM were sectioned from the center of the 6-mm tensile bars, and then polished and etched according to standard metallographic procedures. Optical metallography was carried out using a Nikon Epiphot microscope. Specimens for XRD and TEM analyses were taken from the center of the 6-mm tensile bars of AC53 alloy. XRD data were collected from thin wafers using a Siemens D5000 powder X-ray diffractometer equipped with a Cu X-ray tube. Sample preparation for TEM consisted of mechanical thinning and final thinning by ion-milling. The thin foils were characterized in a Philips 430T TEM operated at 300 kV. The TEM was equipped with a X-ray energy dispersive spectroscopy (EDS) system for composition analysis of the phases.

RESULTS AND DISCUSSION

ANALYSIS OF CREEP DATA

Fig. 1 is a typical creep strain *vs.* time curve obtained from the constant-load and constant-temperature test for alloy AC52. As shown in Fig. 1, ***total creep extension*** (ε_t) measures the total time-dependent strain (creep strain) of a material under a constant load at a given temperature for a specific time period, and is the most frequently used parameter for reporting creep properties of magnesium alloys. Fig. 1 also shows that AC52 alloy, as most other metals and alloys, exhibits two stages of creep, i.e., primary or transient creep, and secondary or steady-state creep. As demonstrated for die cast AZ91D magnesium alloys such as AZ91 [2], AS41 and AE42 [3], the primary and secondary creep strains (ε_1 and ε_2, respectively) for ACX alloys can be described by the following equations:

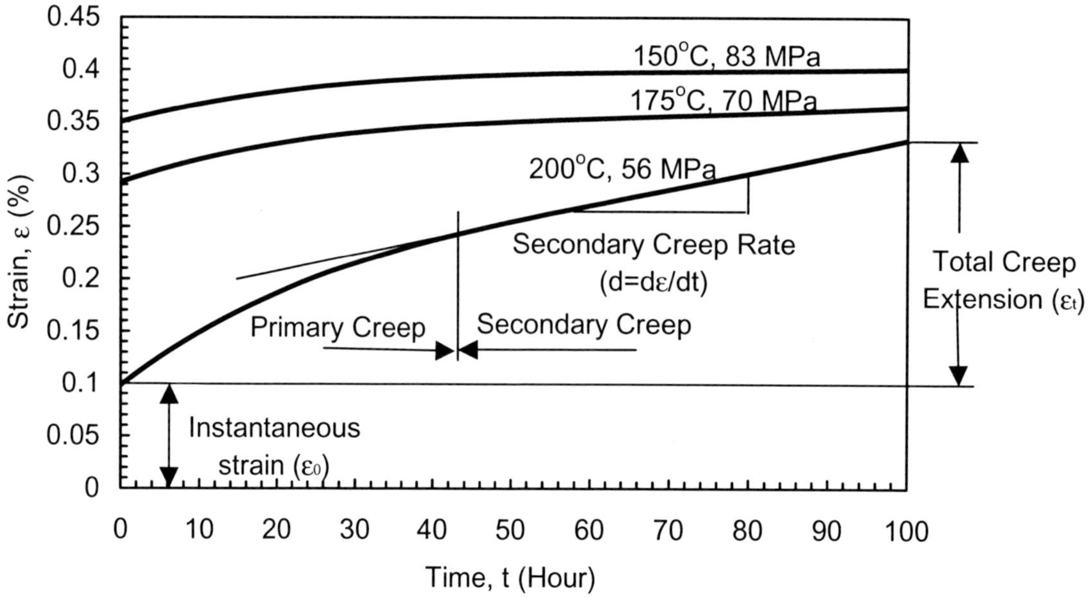

Fig. 1. Typical creep curves for AC52 alloy.

$$\varepsilon_1 = a\,t^b \quad (1)$$

$$\varepsilon_2 = c + d\,t \quad (2)$$

where, t is time; and a, b, c and d are constants. Among these four constants, d represents the ***secondary creep rate*** and is the most important design parameter derived from the creep curves [4, 5]. In this report, therefore, both ε_t and d data are reported for all alloy specimens tested.

Creep strength, defined as the stress required to produce a certain amount of creep at a specific time and a given temperature, is another creep parameter often required by design engineers for evaluating the load-carrying ability of a material in prolonged time periods. It is a common practice in light metals industry to report creep strength as the stress for producing 0.1% total creep extension in 100 hours at a given temperature. For magnesium alloys, Mercer [6] has proposed a linear relationship between creep stress (σ) and log creep strain (ε_t), based on which the creep strength of a given alloy at a given temperature can be interpolated using the σ vs. logε_t data obtained from two or three tests at different stress levels. Such a practice proved accurate for numerous magnesium and aluminum alloys and, hence, was used in this study to determine the creep strength at 175°C for several ACX alloys.

The secondary creep rate (d) varies with temperature, stress and material characteristics. For most metals and alloys including magnesium and its alloys [7-12], the temperature (T) and stress (σ) dependence of d can be expressed as [5]:

$$d = A\,\sigma^n \exp(-Q/RT) \quad (3)$$

where A is a material-related constant, R is the gas constant, Q is the ***apparent activation energy*** for creep and n is known as the ***stress exponent***. Based on Equation (3), the slope of the log d vs. log σ plot at a given temperature is the stress exponent n, and an Arrhenius plot (ln d vs. $1/T$) at a specific stress level will yield the apparent activation energy (Q) value. The n and Q parameters can be used to infer the dominant creep mechanisms for a material in specific ranges of stress and temperature. For AC53 as a typical ACX alloy, n and Q values were also obtained in this report to reveal the possible creep mechanisms involved.

TOTAL CREEP EXTENSION AND CREEP RATE

Table 2 summarizes the total creep extension (ε_t) and secondary creep rate (d) data for ACX alloys in comparison with AE42 magnesium alloy. For AE42 alloy, the total creep extension of 0.12% obtained at 175°C/70MPa in this study agrees reasonably well with the value of 0.10% measured by Chen et al [10] for the same alloy at 175°C/60MPa, given the slight difference in applied stress. The secondary creep rate data for AE42 alloy also show a good agreement with the values reported by Dargusch and Dunlop [8].

It can be seen from Table 2 that all ACX alloys show significant improvements over AE42 at both 150°C and 175°C test conditions: 50-79% reduction in total creep extension at 150°C and 40-67% reduction at 175°C. A closer examination of the data suggests the following:

- AC52 has slightly better creep resistance than AC53.
- The addition of 0.3% Si alone (Alloy D) seems to have no significant effect on the creep performance of AC53; however, the combination of 0.3% Si and 0.1-0.15% Sr additions to AC53 (Alloys E and F) provides an additional improvement in total creep extension, especially at 200°C (a 36-50% reduction).
- The additions of 0.03-0.15% Sr to AC53 (Alloys G-I) and 0.1% Sr to AC52 (Alloy K) further improve the creep resistance of the Mg-Al-Ca alloys. The data also indicates that the creep extension decreases with increasing Sr content.

Table 2. Tensile Creep Data for AE42 and ACX Alloys

Alloy	Designation	Total Creep Extension, ε_t (%)			Secondary Creep Rate, d (x 10^{-9} s^{-1})		
		150°C 83 MPa	175°C 70 MPa	200°C 56 MPa	150°C 83 MPa	175°C 70 MPa	200°C 56 MPa
	AE42	**0.11**	**0.12**	**-**	**0.985**	**1.452**	**-**
B	AC52	0.05	0.06	0.26	0.486	0.695	3.430
C	AC53	0.07	0.09	0.28	0.694	0.864	5.640
D	AC53+0.3Si	0.06	0.07	0.25	0.694	1.388	3.328
E	**AC53+0.3Si+0.1Sr**	**0.03**	**0.07**	**0.18**	**0.463**	**0.694**	**2.224**
F	AC53+0.3Si+0.15Sr	0.05	0.06	0.14	0.729	0.990	1.890
G	AC53+0.03Sr	0.06	0.08	0.28	0.926	1.235	5.449
H	**AC53+0.07Sr**	**0.05**	**0.06**	**0.20**	**0.579**	**0.926**	**1.853**
I	**AC53+0.15Sr**	**0.04**	**0.08**	**0.16**	**0.370**	**0.556**	**1.111**
K	AC52+0.1Sr	0.04	0.05	0.21	0.694	0.750	2.864
L	AC62+0.2Sr	0.06	0.08	0.19	0.728	1.042	3.472

The secondary creep rate data in Table 2 generally supports the above findings from the total creep extension data. The following three alloys (bolded in Table 2) are most creep-resistant among the ACX alloys tested when both total creep extension and secondary creep rate are considered:

- AC53+0.3%Si+0.1%Sr;
- AC53+0.07%Sr; and
- AC53+0.15%Sr.

The tensile creep data for AE42 and selected ACX alloys at 175°C/70MPa are plotted in Fig. 2, which clearly shows the significant improvement of ACX alloys over AE42 in both total creep extension and secondary creep rate.

CREEP STRENGTH

Fig. 3 compares the 175°C creep strength (stress to produce 0.1% creep extension in 100 hours) for several ACX alloys and conventional magnesium and aluminum die casting alloys. It is evident from Fig. 3 that AE42 alloy has the highest creep strength at 175°C among the currently available magnesium alloys. The creep strength of the new ACX alloys is 48-66% higher than that of AE42 alloy, which is in an excellent agreement with the percentage improvement in total creep extension (50-69%). However, it is still 10-20% less than that of aluminum die casting alloy A380, the leading competitive material for powertrain casting applications.

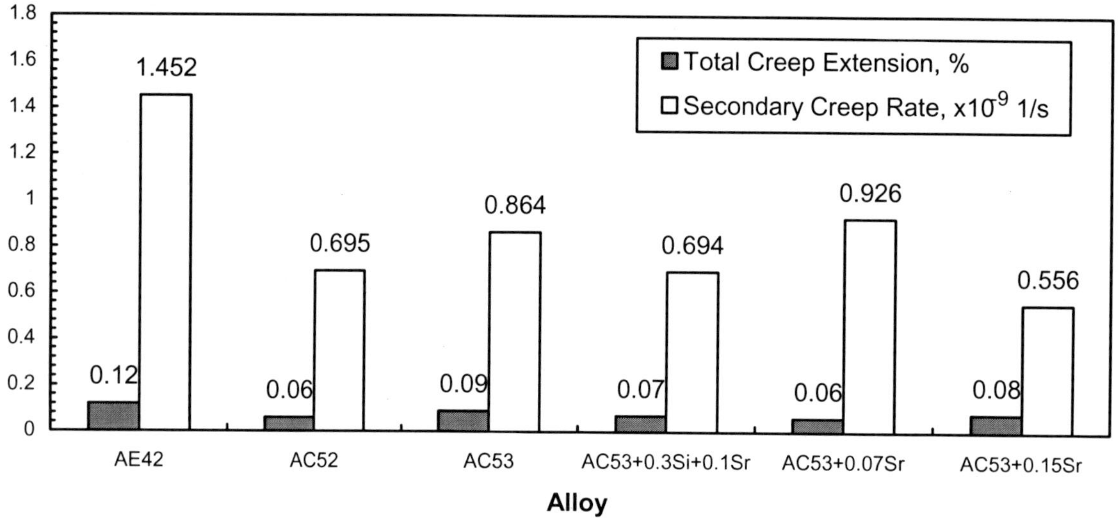

Fig. 2. Tensile creep data for AE42 and selected ACX alloys at 175°C/70MPa.

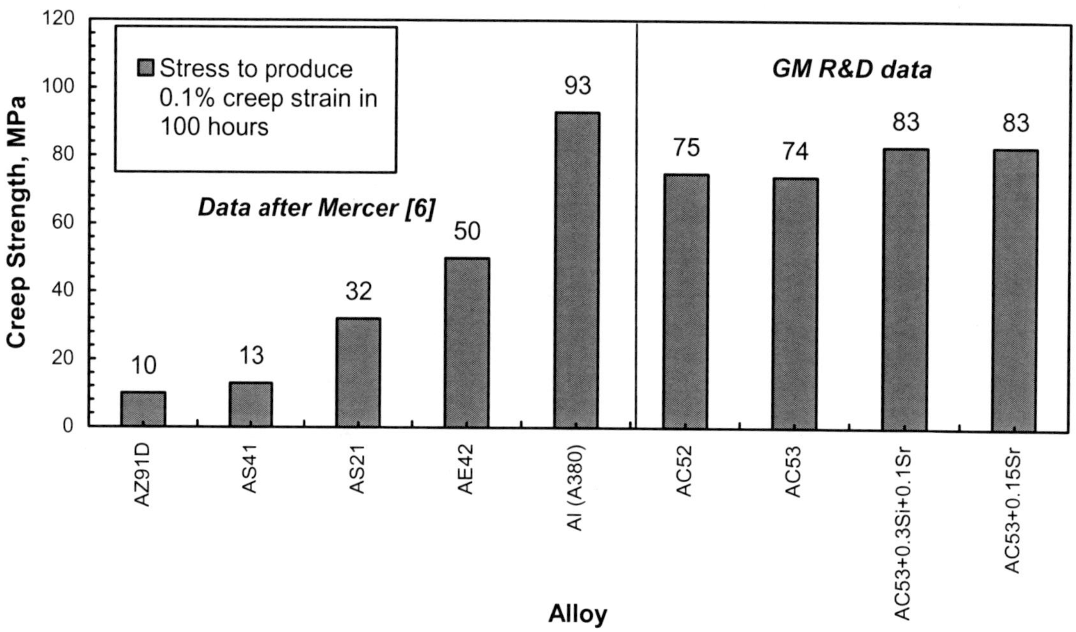

Fig. 3. Creep strength at 175°C.

CREEP MECHANISMS

Creep Mechanisms at 175°C The stress dependence of the secondary creep rate of AC53 alloy tested at 175°C is plotted in Fig. 4, which shows a two-stage behavior with *n*=1.5 for low applied stresses (42-70 MPa) and a much higher value of *n*=8.5 at high stress levels (70-97 MPa). The stress exponent value (*n*=1.5) of AC53 alloy at low stresses is close to 2 reported for other Mg-Al based die casting alloys tested at low-stresses [8]. The *n*-value of approximately 2 has been related to grain boundary sliding as the prominent creep mechanism in Mg-Al based alloys and is supported by strong experimental evidence [8]. At higher stresses, the *n*-value of 8.5 is significantly higher than *n*~5 reported for other die cast alloys such as AZ91D and AS21 [8]; but is closer to *n*=7 reported for more creep-resistant magnesium alloy QE22 (Mg-2%Ag-2%Nd) [7], in which dislocation glide was proposed as the rate-controlling mechanism. A recent study [11] on high-stress (75-115 MPa) creep of AZ91D alloy ingot showed *n*=11 and a creep rate one order of magnitude lower than that of the die cast AZ91D (*n*=6.9 at 150°C and n=5.4 at 180°C [12]) due to a larger grain size in the ingot microstructure. In both cases (ingot and die casting), creep mechanisms based on dislocation motions (on basal and non-basal planes) were proposed [11, 12]. In this study, the stress exponent values for AC53 alloy indicate that, at 175°C, the alloy experiences a grain-boundary-sliding-type of creep at a low applied stress (<70 MPa), and a dislocation-glide-controlled creep at high stresses (>70 MPa).

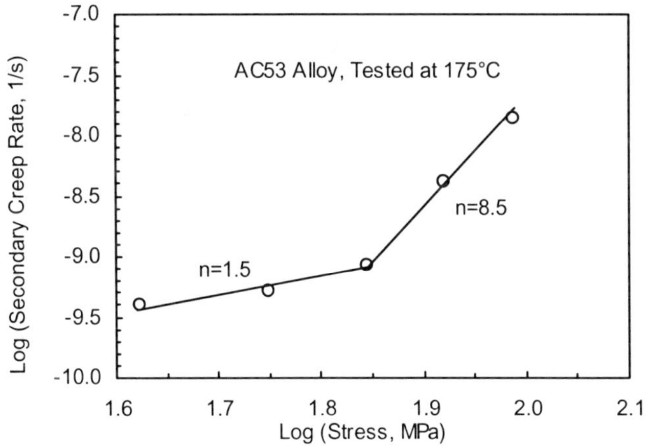

Fig. 4. Applied stress dependence of secondary creep rate for AC53 alloy at 175°C.

Creep Mechanisms at 83 MPa Fig. 5 is an Arrhenius plot which gives the temperature dependence of the secondary creep rate of AC53 alloy under an applied stress of 83 MPa. Similarly, it shows a two-stage linear relationship: an activation energy of *Q*=40 kJ/mol for the temperature range of 100-150°C; and *Q*=120 kJ/mol for temperatures of 150-200°C. The *Q*-value of 40 kJ/mol for low-temperature creep is in excellent agreement with other researchers [8], who reported 36-44 kJ/mol for creep of AZ91D and AS21 and related it to the grain boundary sliding mechanism promoted by discontinuous precipitation of $Mg_{17}Al_{12}$ for which the activation energy is 30 kJ/mol [13]. For high-temperature (150-200°C) creep of AC53 alloy, the activation energy value of 120 kJ/mol is very close to 125 kJ/mol for high-temperature creep of pure magnesium [5], and is in the range of 92-135 kJ/mol reported for the activation energy for dislocation glide in basal planes in pure magnesium which is also the self-diffusion energy in magnesium lattice. These results suggest that, for AC53 alloy under an applied stress of 83 MPa, there are two creep mechanisms operating in parallel: grain boundary sliding for which the activation energy is 40 kJ/mol; and dislocation glide creep with a activation energy of 120 kJ/mol. At low temperatures (<150°C), grain boundary sliding with a lower activation energy is in control; and at high temperatures (>150°C), the dislocation glide creep takes the control.

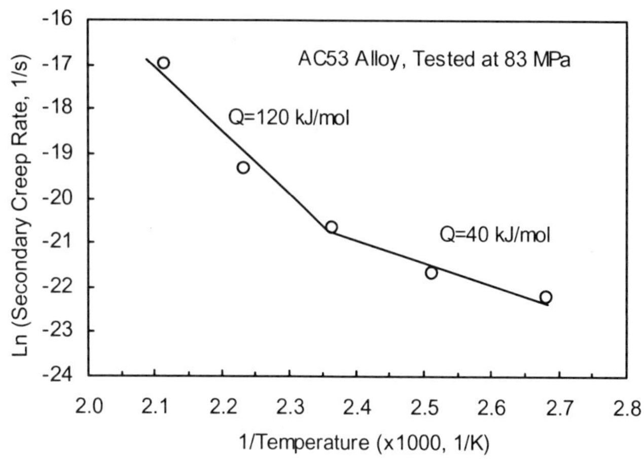

Fig. 5. Temperature dependence of secondary creep rate for AC53 alloy at 83 MPa.

OPTICAL MICROSTRUCTURE

Fig. 6 is a set of optical micrographs showing the as-cast microstructure of the creep test specimens (6-mm tensile bars) for several ACX alloys. Generally, the alloys show extremely fine α-Mg grains (~10μm) surrounded by a grain boundary eutectic phase. For Alloy D containing 0.3%Si, large particles of Mg_2Si are evident in the microstructure as marked in Fig. 6(c). Unfortunately, the non-uniform distribution and the low volume fraction of the Mg_2Si particles do not provide any considerable improvement in creep resistance of the alloy. However, when 0.1% Sr is added to the alloy, such as the case of Alloy E shown in Fig. 6(d), Mg_2Si particles are somewhat modified and more evenly dispersed. Whether the significantly improved creep resistance observed in Alloys E and F is due to the modification of Mg_2Si particles or the effect of Sr itself is not clear at this point. For Sr-containing alloys (Alloys G and I), no appreciable differences can be observed in the optical microstructure as shown in Figs. 6(e) and 6(f).

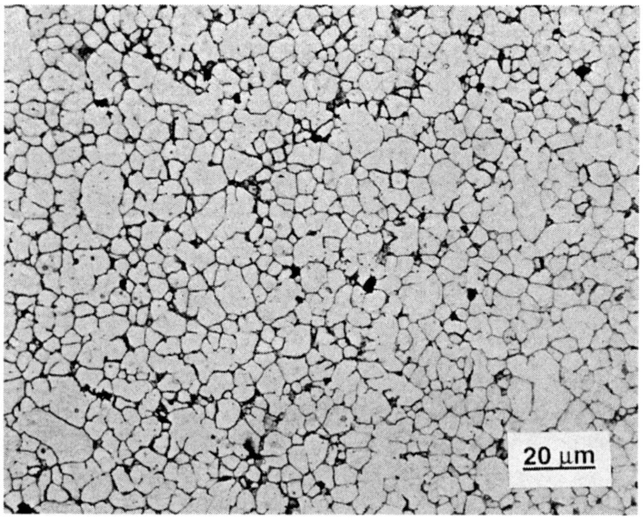

(a) Alloy B (AC52)

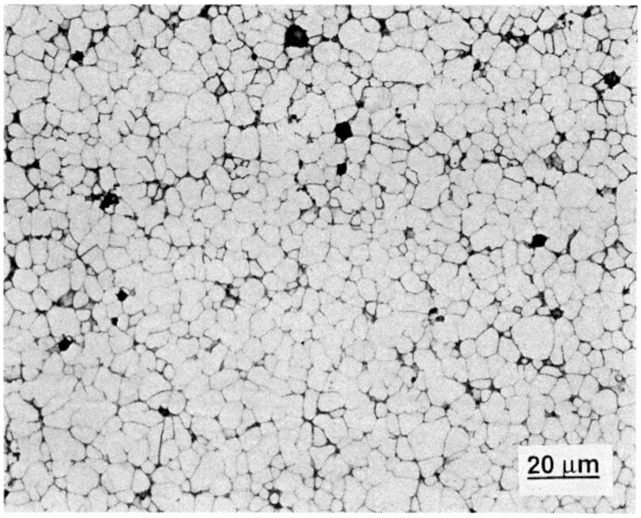

(b) Alloy C (AC53)

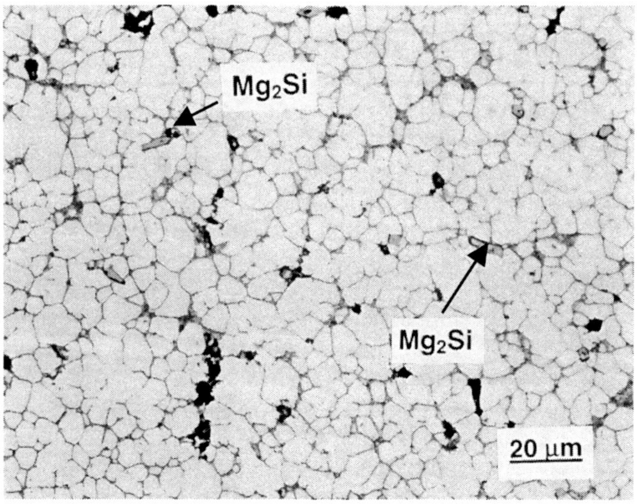

(c) Alloy D (AC53+0.3Si)

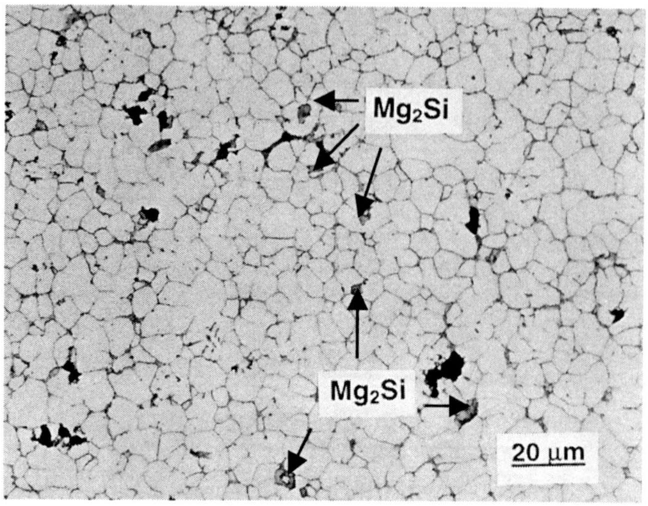

(d) Alloy E (AC53+0.3Si+0.1Sr)

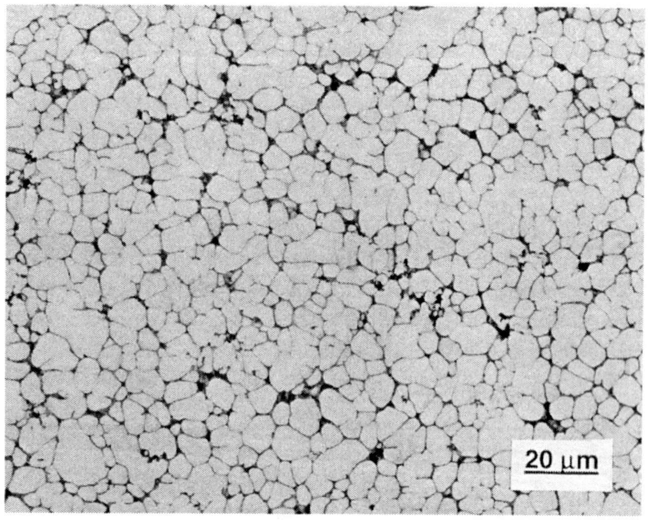

(e) Alloy G (AC53+0.03Sr)

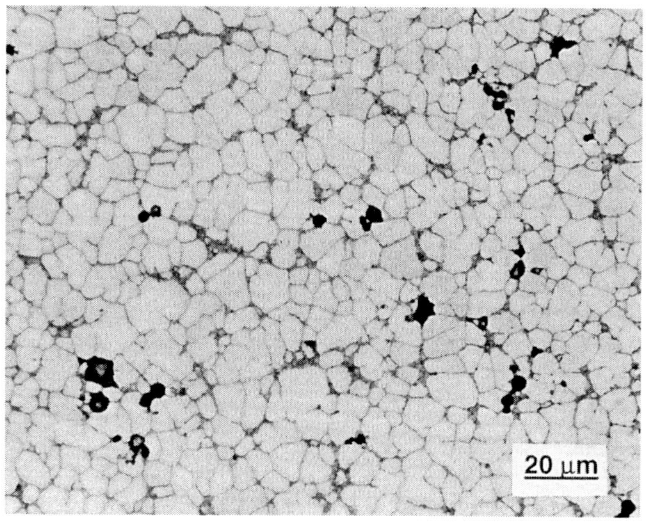

(f) Alloy I (AC53+0.15Sr)

Fig. 6. Optical micrographs showing the as-cast microstructure of several ACX alloys.

XRD AND TEM RESULTS

The XRD results on alloy AC53 reveal that the major diffraction peaks match the reference data for hexagonal α-Mg. Several smaller peaks are also present but do not match any reference data very well. However, the positions of the non-magnesium peaks do coincide with those of a hexagonal lattice. Lattice parameters for this phase calculated by Rietveld analysis match those of hexagonal Mg_2Ca with certain shifts in both *a* and *c* parameters (Table 3). Such large changes in *a* and *c* parameters for Mg_2Ca phase can be attributed to the incorporation of aluminum atoms in the hexagonal structure, leading to the formation of a $(Mg,Al)_2Ca$ phase.

The formation of $(Mg,Al)_2Ca$ phase in AC53 alloy is confirmed by TEM results. Fig. 7 shows that the grain boundary eutectic phase in the microstructure of AC53 alloy has a lamellae structure consisting of alternating layers of magnesium and a Mg-Al-Ca intermetallic phase. The structure of the intermetallic phase and its interface with magnesium are more clearly shown in TEM micrographs with higher magnifications (Fig. 8). The selected area diffraction (SAD) patterns in Fig. 9 confirm that the ternary phase has a hexagonal structure with lattice parameters of approximately

$$a = 5.8\ \text{Å},\ \text{and}$$
$$c = 9.6\ \text{Å}$$

Table 3. Lattice Parameters of Major Phases Identified in AC53 Alloy Obtained from XRD analysis.

Phase	Structure	Lattice Parameters (Å)	
		a	c
Primary Phase	hexagonal	3.2042	5.2034
Ref. for α-Mg	hexagonal	3.2094	5.2112
Secondary Phase	hexagonal	5.7480	9.6290
Ref. for Mg_2Ca	hexagonal	6.2386	10.146

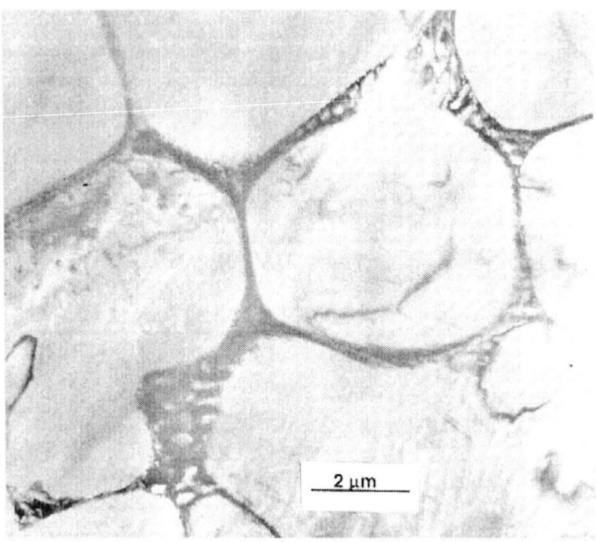

Fig. 7. TEM bright field image showing the formation of the grain boundary phase, $(Mg,Al)_2Ca$, in AC53 alloy.

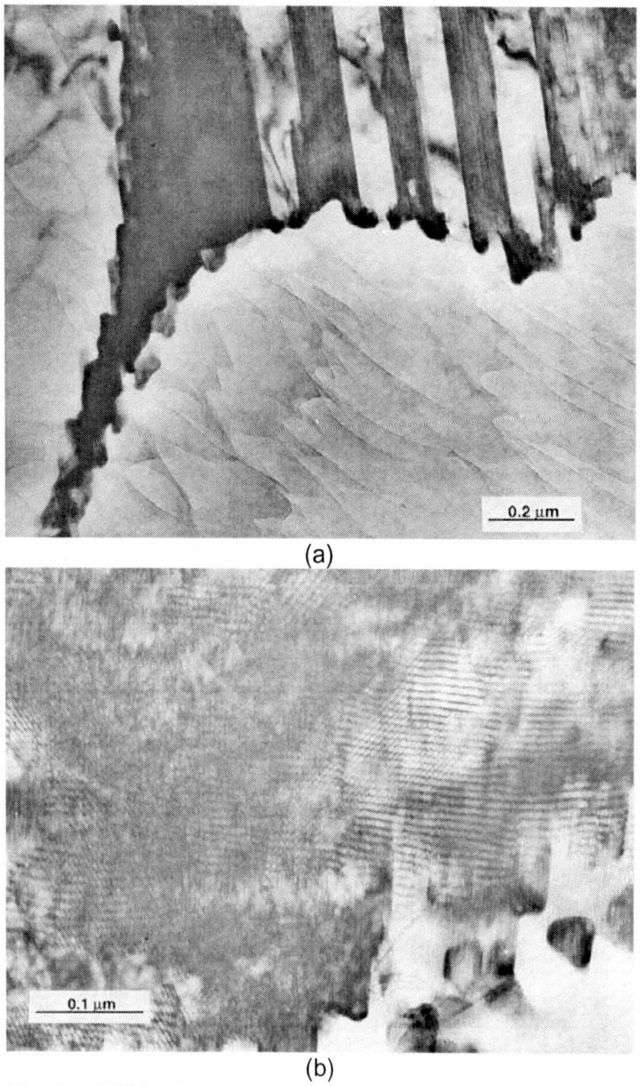

Fig. 8. TEM micrographs showing the morpholgy of the grain boundary phase, $(Mg,Al)_2Ca$, in AC53 alloy.

The SAD patterns also suggest the following orientation relationship of the eutectic magnesium and the $(Mg,Al)_2Ca$ phase:

$$(001)Mg\ //\ (001)(Mg,Al)_2Ca;\ \text{and}$$
$$[010]Mg\ //\ [110](Mg,Al)_2Ca$$

Moire fringes observed in Fig. 8(b) indicate a coherent interface between $(Mg,Al)_2Ca$ and the eutectic magnesium, which is also supported by the relatively small lattice mismatch calculated as 3.6% along the basal plane and 7.5% for the perpendicular plane. Evidence of stacking faults can be seen in the image of the $(Mg,Al)_2Ca$ intermetallic phase as shown in Fig. 8(a). The interface between the magnesium phases (both primary α-Mg grains and the eutectic magnesium) and the $(Mg,Al)_2Ca$ has a step-like appearance with one face of the step parallel to the basal plane and the other perpendicular to the basal plane.

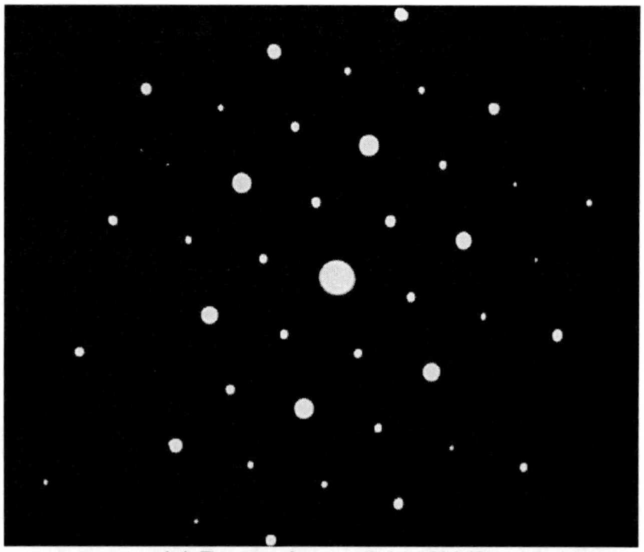

(a) Zone axis parallel to [001]

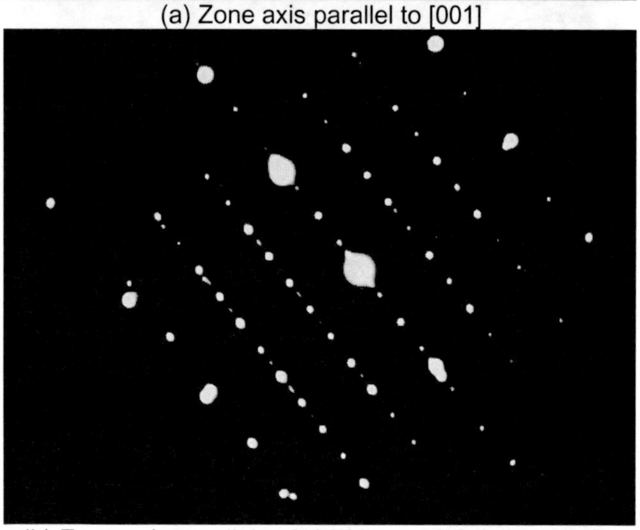

(b) Zone axis parallel to [120]Mg and [001] $(Mg,Al)_2Ca$

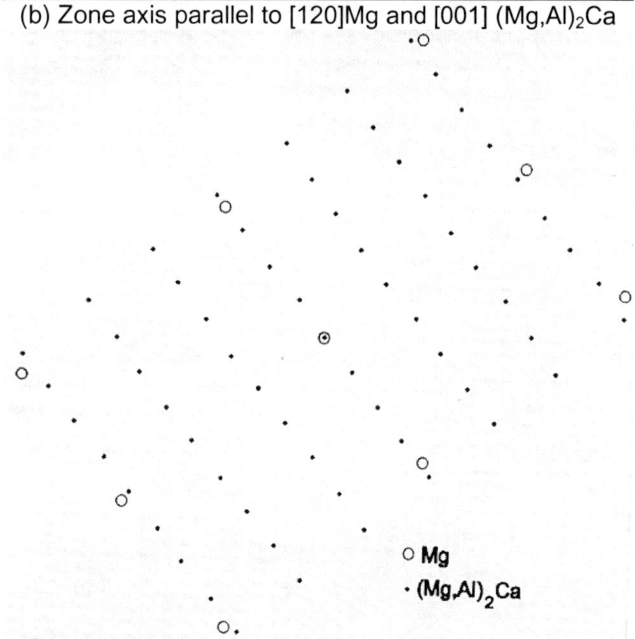

(c) Schematic illustration of the SAD shown in (b)

Fig. 9. Selected area diffraction patterns for the grain boundary phase, $(Mg,Al)_2Ca$, in AC53 alloy.

The thermal stability of the $(Mg,Al)_2Ca$ intermetallic phase can be inferred from the relatively high melting point and the eutectic temperature of its parent phase (Mg_2Ca), i.e., 715°C and 445°C, respectively [14]. The presence of the thermally stable $(Mg,Al)_2Ca$ phase at the grain boundary and its coherent interface with magnesium provide effective pinning to resist grain boundary sliding. The grain boundary pinning effect is further enhanced by step-like morphology at the α-Mg/$(Mg,Al)_2Ca$ interface as shown in Fig. 8(a). As discussed in the preceding section, grain boundary sliding is proposed as the dominant creep mechanism in ACX alloys the temperature and stress ranges that are critical to powertrain applications. While grain boundary sliding does not contribute significantly to the total creep at excessively high temperatures and/or high stresses when dislocation glide creep becomes dominant, it has been shown that grain boundary sliding must be present to maintain grain contiguity during diffusion-dependent dislocation creep [5]. Therefore, any restriction of grain boundary sliding can improve the overall creep resistance over different temperature and stress conditions.

CONCLUSIONS

1. Tensile creep properties (total creep extension, secondary creep rate and creep strength) of the newly developed Mg-Al-Ca-based ACX alloys are significantly better than those of AE42 alloy, which is the benchmark creep-resistant magnesium die casting alloy.

2. The stress dependency of the secondary creep rate of ACX alloys at 175°C suggests that the alloys experience a grain-boundary-sliding-type of creep at low applied stresses (<70 MPa), and a dislocation-controlled creep at high stresses (>70 MPa).

3. The temperature dependency of the secondary creep rate of ACX alloys at 85MPa indicates that grain boundary sliding with a lower activation energy is in control at low temperatures (<150°C); while the dislocation creep takes control at high temperatures (>150°C).

4. The new intermetallic phase, $(Mg,Al)_2Ca$, identified in the microstructure of the ACX alloys, has hexagonal crystal structure similar to magnesium, and forms a coherent interface with the magnesium matrix. Both the thermal stability and the interfacial coherency of the $(Mg,Al)_2Ca$ phase provide the pinning effect at the magnesium grain boundary, thereby, improve the overall creep resistance of these alloys.

ACKNOWLEDGMENTS

The authors wish to acknowledge Vadim Rezhets, Debra Esch and John Bommarito for experimental support, and Basant Tiwari, Anil Sachdev, Paul Krajewski and Raja Mishra for many helpful discussions. General Motors Corporation is acknowledged for permission of publishing this paper and Lexington Die Casting for assisting the die casting trials.

CONTACT

Dr. Alan Luo
Materials & Processes Lab
General Motors Research & Development Center
MC 480-106-212
Warren, MI 49098-9055
Tel: 810-986-8303
Fax: 810-986-9204
e-mail: alan.luo@gm.com

REFERENCES

1. B.R. Powell, A. Luo, V. Rezhets, J.J. Bommarito and B.L. Tiwari: "Development of Creep-Resistant Magnesium Alloys for Powertrain Applications: Part 1 of 2", SAE2001, Detroit, MI, 2001.

2. W.K. Miller: "Creep of Die Cast AZ91 Magnesium Alloy at Room Temperature and Low Stress", *Metallurgical Transactions A*, vol. 22A, 1991, pp. 873-877.

3. T.Kr. Aune and T.J. Ruden: "High Temperature Properties of Magnesium Die Casting Alloys", SAE Paper No. 920070, SAE, Warrendale, PA, U.S.A., 1992.

4. R.W. Evans and B. Wilshire: *Creep of Metals and Alloys*, The Institute of Metals, London, England, 1985, pp. 69-84.

5. G.E. Dieter: *Mechanical Metallurgy*, McGraw-Hall, New York, NY, U.S.A., 1986, pp. 432-457.

6. W.E. Mercer II: "Magnesium Die Cast Alloys for Elevated Temperature Applications", SAE Paper No. 900788, SAE, Warrendale, PA, U.S.A., 1990.

7. R.L. Mordike and P. Lukac: "Creep Behavior of Magnesium Alloys Produced by Different Techniques", in *Proceedings of the 3rd International Magnesium Conference*, The Institute of Metals, London, England, 1997, pp. 419-429.

8. M.S. Dargusch and G.L. Dunlop: "Elevated Temperature Creep and Microstructure of Die Cast Mg-Al Alloys", in *Magnesium Alloys and Their Applications*, ed., B.L. Mordike and K.U. Kainer, Werkstoff-Informationsgesellschaft, Frankfurt, Germany, 1998, pp. 277-282.

9. S.R. Agnew, K.C. Liu, E.A. Kenik and S. Viswanathan: "Tensile and Compressive Creep Behavior of Die Cast Magnesium Alloy AM60B", *Magnesium Technology 2000*, eds. H.I. Kaplan, J.N. Hryn and B.B. Clow, The Minerals, Metals & Materials Society (TMS), Warrendale, PA, 2000, pp. 285-290.

10. F.C. Chen, J.W. Jones, T.A. McGinn, J.E. Kearns, A.J. Nielsen and J.E. Allison: "Bolt-Load Retention and Creep of Die-Cast Magnesium Alloys", SAE Paper No. 970325, SAE, Warrendale, PA, U.S.A., 1997.

11. M. Regev, E. Aghion, A. Rosen and M. Bamberger: "Creep Studies of Coarse-Grained AZ91D Magnesium Castings", *Materials Science and Engineering A*, Vol. A252, 1998, pp. 6-16.

12. M. Regev, E. Aghion and A. Rosen: "Creep Studies of AZ91D Pressure Die Casting", *Materials Science and Engineering A*, Vol. A234-236, 1997, pp. 123-126.

13. H. Uchida and T. Shinya: "Estimation of Creep Deformation Behavior in Mg-Al Alloys by Using θ Projection Method", *J. Jpn. Inst. Light Metals*, Vol. 45, No. 10, 1995, p. 572.

14. T.B. Massalski: "Binary Alloy Phase Diagram", Vol. 1, 1986, p. 619.

2001-01-0425

Bolt-Load Retention Behavior of a Die Cast Magnesium-Rare Earth Alloy

Ian P. Moreno, Keun Yong Sohn* and J. Wayne Jones
University of Michigan

John E. Allison
Materials Science Department, Ford Research Laboratory, Ford Motor Co.

ABSTRACT

The need for improved understanding of new magnesium alloys for the automotive industry continues to grow as the application for these lightweight alloys expands to more demanding environments, particularly in drivetrain components. Their use at elevated temperatures, such as in transmission cases, presents a challenge because magnesium alloys generally have lower creep resistance than aluminum alloys currently employed for such applications. In this study, a new die cast magnesium alloy, MEZ, containing rare earth (RE) elements and zinc as principal alloying constituents, was examined for its bolt-load retention (BLR) properties. Preloads varied from 14 to 28 kN and test temperatures ranged from 125 to 175°C. At all test temperatures and preloads, MEZ retained the greatest fraction of the initial imposed preload when compared to the magnesium alloys AZ91D, AE42, AM50, and the AM50+Ca series alloys. The BLR behavior of MEZ did not show significant sensitivity to temperature within the range examined, whereas the other alloys displayed a clear decrease in bolt-load retention with increased temperature at a given preload. Retained bolt-load decreased for MEZ with increasing preload in a manner similar to the behavior of other alloys. The higher BLR can be attributed to the greater resistance to creep and arises mainly from the Mg-RE phases present at cell and grain boundaries and the relatively high solidus temperature (T_s) of MEZ. Additional means of improving BLR by varying geometrical dimensions in the bolted assembly for AZ91D and AM50 was investigated and no significant improvement were observed in the limited studies that were performed.

*Now with Korea Institute of Machinery and Materials (KIMM)

INTRODUCTION

Weight reduction remains an important goal in the automotive industry and the replacement of current materials with magnesium die cast alloys that can meet the targeted properties offers significant potential. The conventional high volume production magnesium alloys, such as high pressure die cast (HPDC) AZ91D and AM50, are limited to operating temperatures well below 150°C because of their poor elevated temperature creep properties. These alloys cannot be utilized in powertrain components such as automatic transmission cases where typical operating conditions can exceed this temperature.

Among the several magnesium alloys identified as potential replacements for aluminum-based alloys used in elevated temperature components, AE42 has been shown to be the most promising because of its superior creep resistance compared to AZ91D [1,2,3,4]. Although the cost of producing rare earth-containing alloys is a limiting factor and is reflected in their restricted use, it has been determined that the operating costs for casting and trimming AE42 would be similar to that of AZ91D if manufactured in comparable volumes [3]. The elevated temperature creep strength of AE42 has been attributed to rare earth additions that increase the volume fraction of higher melting, stable Al-RE phases [4,5]. AE42 offers creep resistance superior to other Mg-Al alloys but still displays marginal performance at 175°C [6]. It has been stated or inferred by some [1,7,8] that substantial improvement in creep resistance may come only with the development of a magnesium alloy outside of the Mg-Al system.

Magnesium Elektron, Inc. (Newcastle, UK) has developed a new alloy, designated Elektron MEZ, and has reported creep results superior to other Mg alloys considered for high temperature automotive applications, including AE42 [1,9]. Die castability was demonstrated in the production of a filter housing, gearbox, and oil pan. Corrosion resistance was similar to that of high purity Mg alloys. This paper reviews the bolt-load retention behavior of MEZ and material properties that contribute to its resistance to creep.

As an alternative to investing in more costly alloy systems, design changes in fastener configuration may hold promise where BLR is critical. For example, Pettersen and Fairchild reported substantial improvement in the bolt-load retention of AZ91D by varying the thickness and diameter of washers in their test assembly [10]. In the present study similar tests were conducted on AZ91D and AM50 using different fastener configurations. The results are described in this paper.

EXPERIMENTAL PROCEDURE

MEZ, with a nominal composition of Mg-2.5%RE-0.35%Zn-0.3%Mn, was pressure die cast at Mag-tec Casting Corporation (Jackson, MI) in a four-cavity die. Components used in the BLR test configuration, cylindrical tensile/creep and fatigue blanks were produced. Melt and die temperatures were 700-710°C and 177-232°C, respectively. Gate velocities were 32.9, 30.5, 28.3, and 28.3 m/s for case, flange, tensile/creep, and fatigue samples, respectively. The actual composition of as-cast MEZ obtained by inductively coupled plasma (ICP)-atomic emission spectroscopy and mass spectrometer on a bulk flange specimen, is given in Table 1, along with compositions of materials from other studies [4,11].

Cast components of the bolt-load retention assembly consisted of a flange and case as shown in Figure 1. The flange and case have a 40x40mm contact surface with a centered 11mm diameter hole. The wall of the threaded case hole is 4mm thick and is supported by four 4mm thick ribs. The BLR assembly is shown in Figure 2, where a standard M10 bolt, machined just below the head to accommodate a bonded four-strain gage assembly in a 90° T rosette layout, passes through the flange and is threaded into the tapped BLR case. The assembly, simulating the actual fastener and boss geometry in a typical automatic transmission housing, was machined to a 1.6μm finish on the mating surfaces.

Transmission bolts are commonly torqued to preloads of 75-90% of the bolt proof strength. Specimens were torqued slightly above a nominal preload level of 14, 21, or 28kN at room temperature and held for approximately one hour before being submersed in a silicon oil bath held constant to within ±0.1°C of 125, 150, and 175°C. This waiting period minimizes the effects of short-term relaxation following tightening of bolted joints, as described in [12]. Tests were conducted at least 100 hours, and data was amplified and scaled using the Measurements Group System 2210 signal conditioner

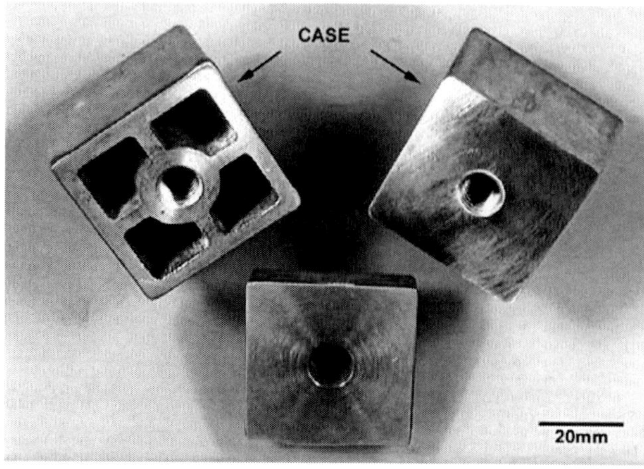

Figure 1. Photograph of machined BLR case and flange.

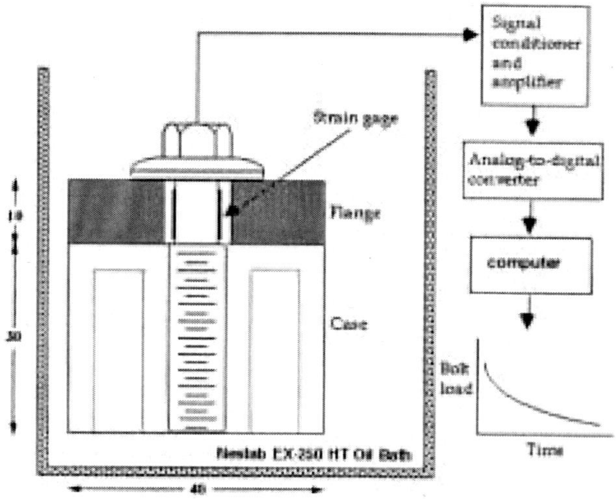

Figure 2. Schematic of the BLR test and data recording assembly; not to scale (dimensions are in mm).

and continuously acquired using the Measurements Group System 2000 analog-to-digital converter. Duplicate tests show repeatable results to within 3%. Slight differences in preload level, torque rate/method, and individual specimen properties (final machined state) are presumed to cause the modest variation that was observed.

AZ91D and AM50 specimens were also tested in this same manner but with the addition of up to three flat aluminum washers added to the assembly just beneath the bolt head. Aluminum washers had dimensions of approximately 10.6mm ID, 30.3mm OD, and 2.1mm thick. A single 1010 steel washer, of approximately the same dimensions (to within ±0.2mm) was placed under the bolt head in separate BLR tests.

Porosity levels of MEZ were evaluated using a precision Archimedes density method. Porosity levels of 2.8 and 3% for the BLR case and flange were measured. Microstructure was investigated using scanning electron microscopy (SEM) after etching in an ethylene glycol

Table 1. Chemical composition of HPDC Magnesium alloys MEZ, AE42, and AZ91D

Alloys	Composition (wt%)									
	Al	Zn	Mn	Si	Fe	Cu	Ni	RE total	Sn	Mg
MEZ	-	0.33	0.26	-	0.01	0.003	<0.001	1.92	-	Bal.
AE42	3.7	0.005	0.21	-	<0.005	0.003	0.005	2.69	-	Bal.
AZ91D	8.9	0.72	0.21	0.01	<0.005	0.004	<0.005	-	-	Bal.

* Rare earth (RE) consists of approximately 53%Ce, 25%La, 17%Nd, and 5%Pr.

solution, transmission electron microscopy (TEM) after jet-polishing in a perchloric acid solution and ion-milling, and electron microprobe analysis (EMPA) after final polishing. Chemical composition of selected areas in the microstructure was analyzed by energy-dispersive spectrometry (EDS) and wavelength-dispersive spectrometry (WDS). Melting temperatures were investigated using differential scanning calorimetry (DSC). Duplicate tensile tests, carried out by Westmoreland Mechanical Testing and Research Inc., were performed at ambient and elevated temperatures on machined specimens of MEZ. Yield strength was determined by the 0.2% offset method. Strain rate was 0.005 mm/mm/min until yield and 0.05 mm/mm/min until failure according to ASTM E8-99 and E21-92.

RESULTS AND DISCUSSION

BOLT-LOAD RETENTION (BLR) BEHAVIOR - The decrease in bolt load of the torqued BLR assembly for MEZ is shown in Figure 3a. Here the instantaneous bolt load is normalized by the applied preload and plotted as a function of time at the indicated temperature. As shown in Figure 3b, MEZ demonstrates bolt-load retention behavior that is superior to other magnesium alloys tested in the same manner. It is notably better than AE42. For example, after an initial nominal preload of 28kN and exposure for 100h at 175°C, MEZ retained 78.5% of the initial load while AE42 retained only 55.5%.

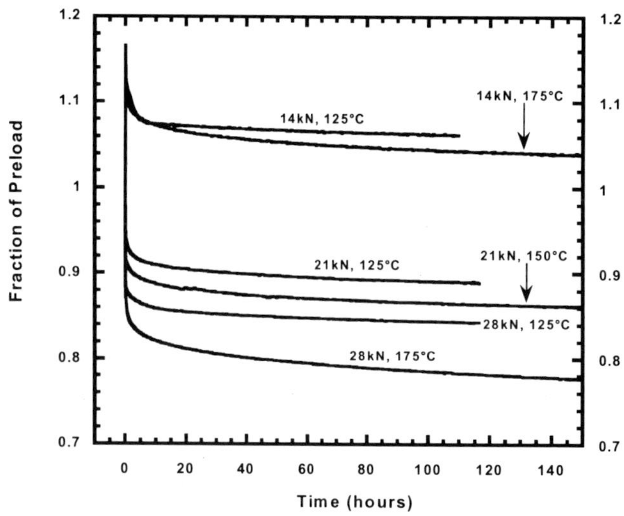

(a)

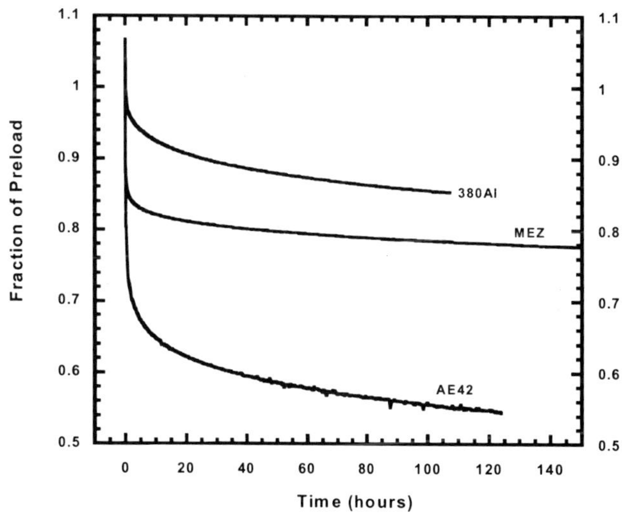

(b)

Figure 3. BLR curves of (a) MEZ at various temperatures and preloads and (b) MEZ, AE42, and 380Al BLR curves compared at 175°C, 28kN.

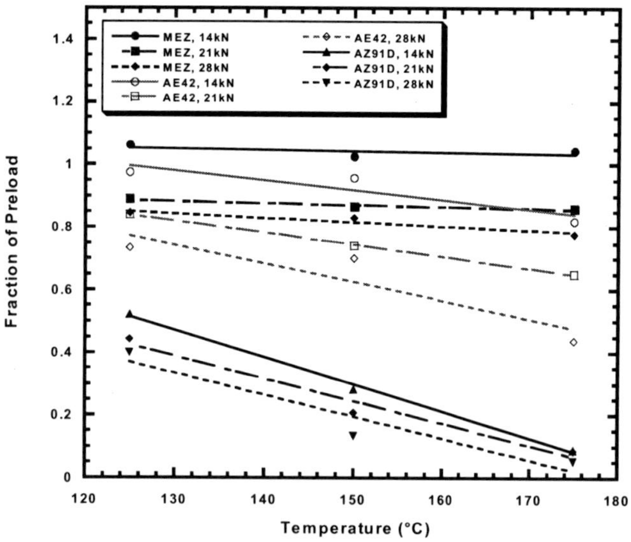

Figure 4. Comparison of BLR sensitivity to temperature (data represents retained load after 100 hours). AE42 and AZ91D data obtained from previous work [4].

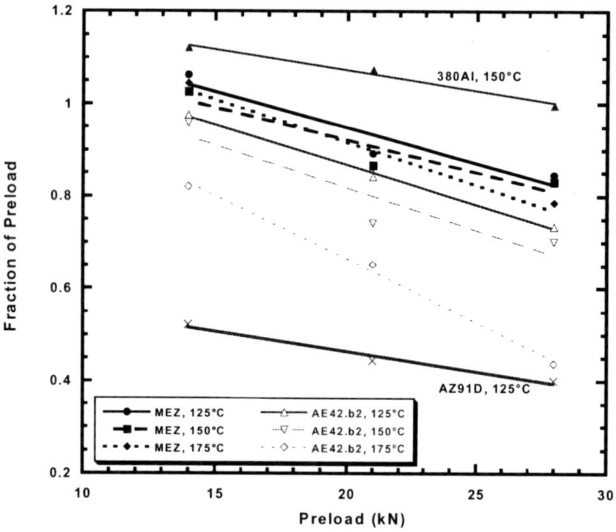

Figure 5. Comparison of BLR sensitivity to preload (data represents retained load after 100 hours). AE42 and AZ91D data obtained from previous work [4].

The dependence of BLR behavior on temperature and preload are shown for MEZ, AE42, and AZ91D in Figures 4 and 5. Unlike AZ91D and AE42, for MEZ there is little, if any, dependence of BLR on temperature for the preloads examined. As is observed for other alloys, the fraction of preload retained in MEZ decreases as the level of the preload increases, although its sensitivity to preload is not as great as for AE42, especially at 175°C.

The bolt-load retention behavior at very short exposure times is shown in Figure 6. Similar to AZ91D and AE42 [ref. 4,10], MEZ experiences an initial period of load increase caused by the thermal expansion mismatch between the magnesium case/flange and the steel bolt which is eventually countered by the onset of creep in the case/flange assembly. During this initial period, the steel bolt, having a lower thermal expansion coefficient, restricts the magnesium from expanding freely and imposes a greater compressive stress in the assembly. As expected, the initial load increase is greater at higher temperatures (greater thermal expansion effect) and lower preloads (lower stresses in the case/flange), with the preload level again showing a more pronounced effect than test temperature. At higher preloads, the initial loss of load is increased due to the higher compressive stresses in the assembly that result in an increase in creep deformation of the magnesium parts.

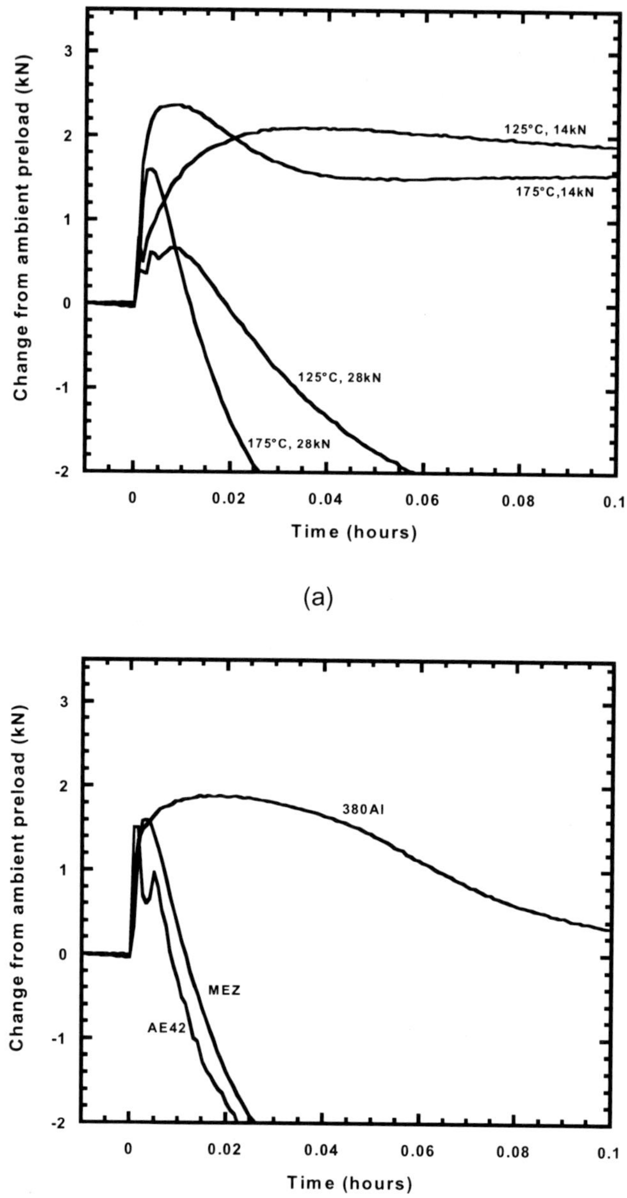

(b)

Figure 6. Initial transient response in bolt-load after submersion in oil bath (a) of MEZ at various temperatures and preload levels and (b) comparison of MEZ, AE42, and 380Al at 175°C, 28kN.

A bolt-load relaxation limit, F_r, above the lowest nominal preload of 14kN was displayed by MEZ at all temperatures and is extrapolated from the data presented in Figure 7. This value, representing the preload below which no stress relaxation occurs (bolt load drop of zero), is another representation of the superior creep resistance MEZ exhibits over other alloys. The initial thermal expansion of the alloy increases the bolt-load, as described earlier, and the alloy subsequently retains most of this load. MEZ exhibits this behavior even at 175°C, where its relaxation limit is 15.2kN in contrast to AE42 with F_r equal to 8.7kN.

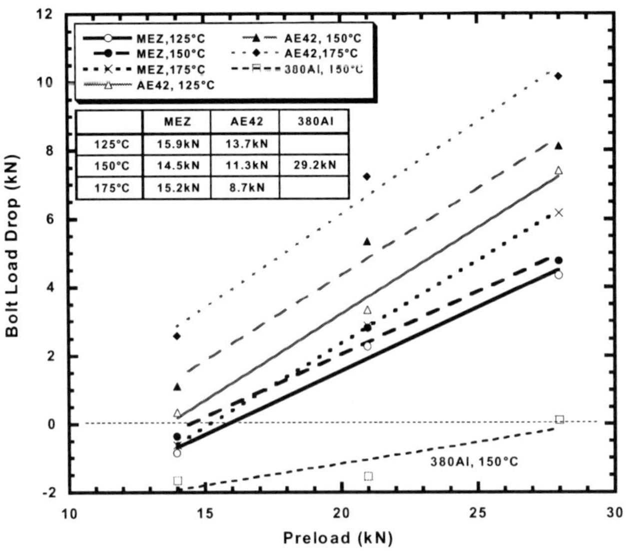

Figure 7. Drop from ambient preload after 100 hours at temperature as a function of ambient preload for MEZ, AE42, and 380Al. The bolt-load relaxation limit, F_r, which is extrapolated from this data, is included in the inserted table.

The relaxation limit, like the data shown in Figures 4 and 5, represents retained load after only 100 hours, but it is a good indicator of the behavior that a similar transmission case bolted-joint might display over its service life, and this is especially so for MEZ. Testing for over 200 hours at 175°C shows that the retained bolt-load was less than 2% lower (21.5kN) than at 100 hours (21.9kN). MEZ reaches a very shallow third stage stress relaxation period where further drop in bolt-load over time is minimal.

Similar to the BLR characteristics of other magnesium die casting alloys, MEZ displays an initial load increase followed by a considerable decrease in load (although this decrease is not as sharp as exhibited by other alloys), after which the load decreases at a fairly steady but gradual rate. This final, more gradual, state of load relaxation occurs when the stresses surrounding the bolt become less localized (in the upper end of the bolt) and more evenly distributed along the length of the threaded case. This has been analyzed by finite element modeling [16] and described in earlier reports on zinc and other magnesium castings [14,17]. Figure 8 shows the bolt-load MEZ retained after over 100 hours of exposure at different temperatures after a preload of 14, 21 and 28kN. A power law equation of the form $F(t)=ct^n$ can be used to describe the retained load in MEZ, with constants listed in Table 2. Except for the very early stage of the bolt-load test where there is a jump in the load as described in Figure 6, this power law depicts the BLR behavior of MEZ fairly well. For example, for 175°C and 28kN preload this equation would predict the retained load to be 21.6kN after 200h. The actual test resulted in a retained load of 21.5kN (< 5% error).

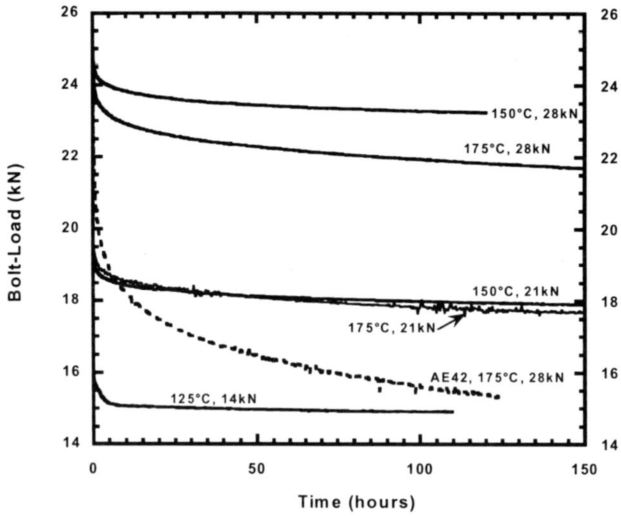

(a)

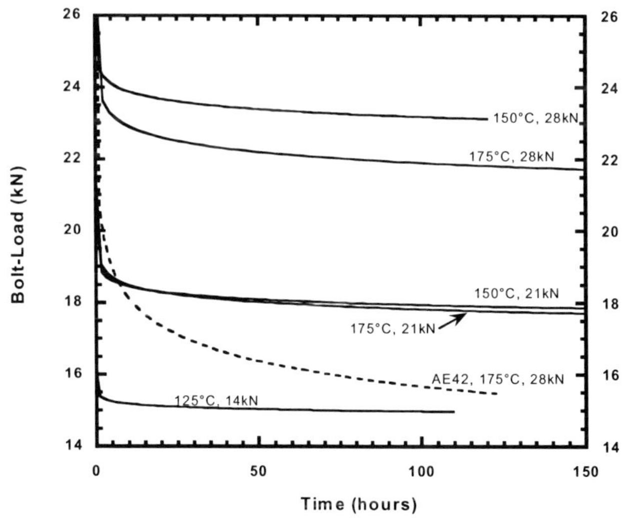

(b)

Figure 8. Retained bolt-load curves for MEZ at indicated preloads and temperatures (a) from experimental results and (b) described by the power law relationship, $F(t)=ct^n$. The AE42 BLR curve is added for comparison.

Table 2. Constants of the power law equation, $F(t)=ct^n$ for MEZ (AE42 data for 175°C, 28kN)

	125°C		150°C		175°C	
	c	n	c	n	c	n
14kN	15.39	-0.006	14.81	-0.004	15.45	-0.008
21kN	19.55	-0.013	18.98	-0.012	19.19	-0.016
28kN	24.89	-0.014	24.62	-0.013	23.98	-0.020
AE42					20.96	-0.063

TENSILE AND CREEP PROPERTIES – Results from duplicate tensile tests conducted on MEZ at RT, 125, 150, and 175°C are listed in Table 3 and compared to AE42 in Figure 9.

Figure 9 shows the variation of these properties with temperature compared to die cast AE42 tested previously in our laboratory. As shown, the yield and tensile strength for MEZ are lower than those measured for AE42 and decrease with increasing temperature. Elongation also increases with temperature to 150°C but decreases slightly at 175°C.

Table 3. Mechanical properties of die cast MEZ.

Temp. (°C)	YS (MPa)	UTS (MPa)	Elong. (%)
RT	97.6	134.8	2.9
125	83.6	115.9	6.8
150	78.0	110.0	7.5
175	73.5	99.7	5.0

Creep is known to occur at temperatures above approximately 0.4 of the absolute melting temperature of most metals. In alloys the melting range varies significantly depending on composition and the presence of higher melting phases. AZ91D, for example, begins to melt (solidus temperature, T_S) at approximately 475°C [18]. The onset of melting (T_S) for MEZ was determined to be 593°C by differential scanning calorimetry (DSC) This higher melting temperature is compared on an absolute basis with other alloys in Table 4, where it is evident that for the test temperatures of interest that MEZ is operating at a lower fraction of its absolute melting temperature.

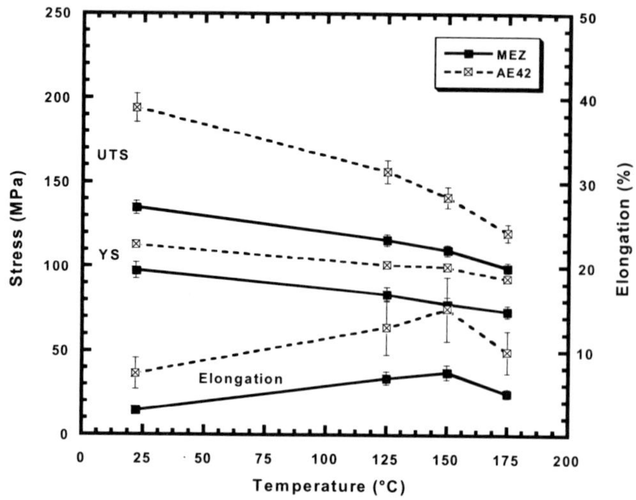

Figure 9. Mechanical properties of die cast MEZ and AE42 at various temperatures. (AE42-175°C data from previous work [11] obtained by crosshead displacement measurement)

Table 4. Homologous temperature values for die cast alloys. AE42, AZ91D, and 380Al data were obtained from previous work [18].

Test Temp. (°C)	MEZ T_S=593°C	AE42 T_S=613°C	AZ91D T_S=480°C	380Al T_S=540°C
125	0.46	0.45	0.53	0.49
150	0.49	0.48	0.56	0.52
175	0.52	0.51	0.59	0.55

MICROSTRUCTURE – With an absence of aluminum, MEZ lacks any Mg-Al or Al-RE phases. A Mg-RE phase exists and is partly responsible for the superior BLR behavior of the alloy. In a traditional sense, it can be expected that since grain boundary sliding commonly occurs during the creep of metals, a more coarsely grained microstructure is desired. However, with the introduction of stable precipitates along grain boundaries, grain boundary sliding at high temperatures can be impeded. WDS and EDS investigations using EMP and TEM, respectively, indicates that Mn is present within these grains of primary magnesium, and the rare earth elements and zinc are present along the grain and cell boundaries. A back-scattered SEM micrograph is shown in Figure 10, where the rare earth elements are shown in brighter contrast compared to the Mg matrix due to their higher atomic number.

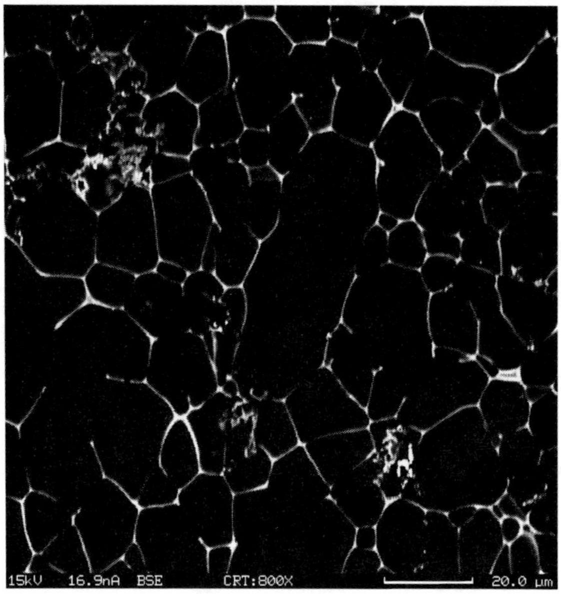

Figure 10. The brighter grain boundry regions reflect the presence of the rare-earth elements in this phase.

The rare earth elements were introduced in this alloy as a mischmetal (MM) (with nominally 50% cerium and the remainder consisting of lanthanum, praseodymium, and neodymium) rather than separated rare earth metals since it is less costly but still effective as an alloying method [7, 8]. It is estimated that the presence of one or more stable Mg-RE phases along grain and cell boundaries in MEZ contributes greatly to its superior bolt-load retention behavior at elevated temperatures through intergranular and transgranular strengthening.

ROLE OF FASTENER GEOMETRY ON BLR – The geometry of the BLR test assembly was varied to determine the merits of an alternative means to an improved bolt-load assembly, through the addition of aluminum washers. Up to two aluminum washers were added to the BLR assembly of AZ91D and AM50, and 100-hour tests were conducted at different temperatures and preloads. These washers were placed under the M10 bolt head, in an effort to reduce the localized stress and, thus, reduce the creep occurring in the flange. Figures 11 and 12 offer a graphical representation of these results. As can be seen, the benefits of adding aluminum washers were normally very minimal, if not detrimental. At a low temperature (125°C) AZ91D showed a decrease in retained bolt-load with the addition of an aluminum washer and increased slightly with added washers, however the retained load was still below the retained load with no washer added. The expected variance is approximately 3% from test to test, so these differences are not considered significant. It is concluded that adding washers had no significant effect for these test conditions.

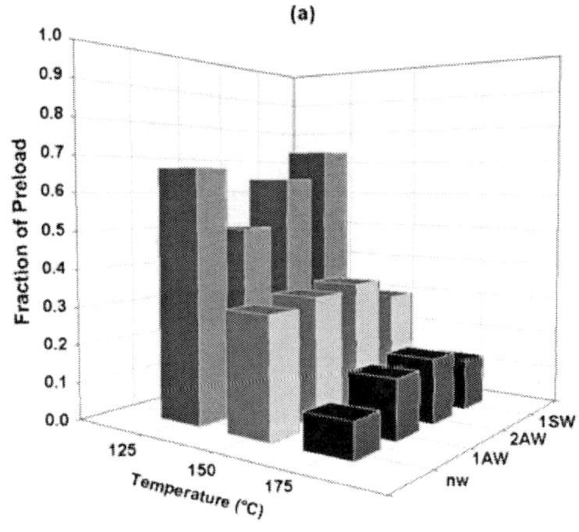

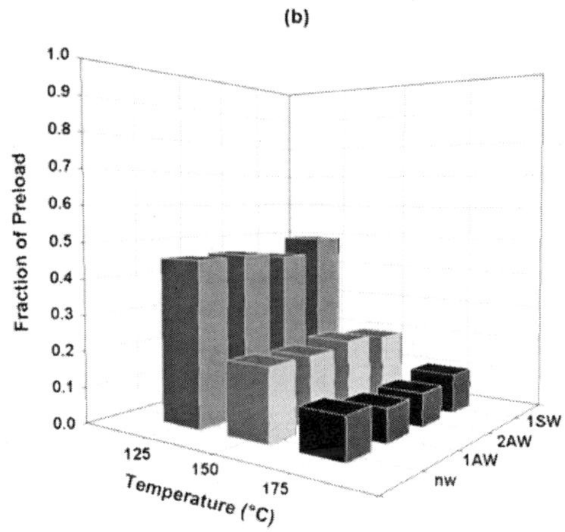

Figure 11. Bolt-load retention and the effect of adding washers to the BLR assembly of AZ91D at (a) 14kN and (b) 28kN. (nw = no washer, 1AW = 1 aluminum washer, 2AW = 2 aluminum washers, 1SW = 1 steel washer)

AM50 similarly displayed no significant improvement in retained bolt-load with the addition of aluminum washers. At a low preload (14kN) and temperature (125°C) adding aluminum washers increase the retained load only slightly. It is evident that at all other temperatures and preloads, the retained bolt-load is essentially unaffected by the addition of aluminum washers.

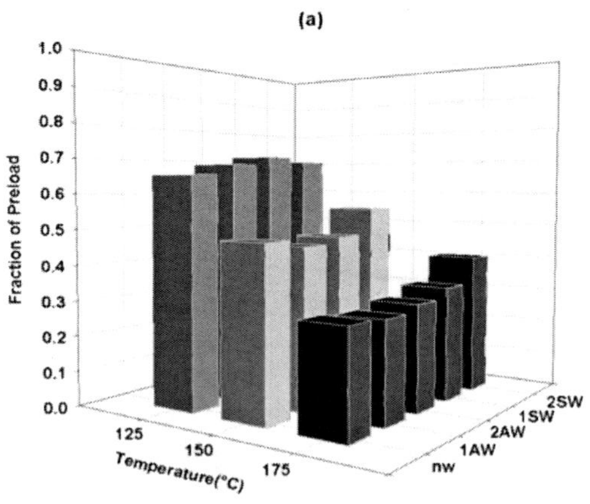

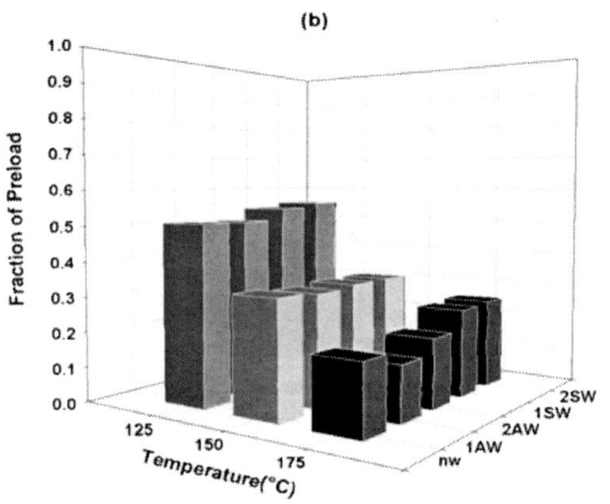

Figure 12. Bolt-load retention and the effect of adding washers to the BLR assembly of AM50 at (a) 14kN and (b) 28kN. (nw = no washer, 1AW = 1 aluminum washer, 2AW = 2 aluminum washers, 1SW = 1 steel washer, 2SW = 2 steel washers)

The aluminum washers may have crept slightly during the test, as washer deformation was visible after the tests, particularly at high preloads. With this deformation, the BLR assembly relaxes, as before. A single steel washer (up to two steel washers in the AM50 assembly) was added, separately, to the assembly to see if improved bolt-load retention prevailed. Deformation was not evident in the steel washers, however the BLR data obtained showed only a slight improvement by using a steel washer. These results did not appear to be statistically significant.

With the addition of washers, the effective length of bolt engagement decreases. This results in a greater induced load on the contact threads of the case, and thus increases the creep stresses. With the addition of every washer, approximately 1.5 contact threads were lost, and although the creep in the flange is further reduced as the stress on the flange face is more distributed, the loss of these threads works against the improvement of the stress-relaxation behavior. The amount of creep that occurs in the flange is not normally as significant as that which occurs in the case. Chen conducted studies on this with AZ91D and AE42 by separately replacing the magnesium flange or case with a stainless steel one. The relaxation limit for the stainless steel flange/magnesium case was normally lower than for the magnesium flange/stainless steel case, showing that, in most cases, more creep occurs in the threaded region of the case than at the flange/bolt-head interface [15]. Su, through the use of a finite element model, concluded that there is an uneven distribution of load at the early stages of the bolt-load retention test, with most of the stress initially localized in the upper threaded region [16]. However, this stress is more evenly distributed with time and the full thread contact area collectively contributes to opposing creep in that region [19].

Studies on the effect that increased bolt length has on the load-retention of zinc alloy pressure die castings were carried out by Murphy [17]. Results indicated that small increases in thread engagement length substantially improved the retained load. Through simulation, it was concluded that changing the contact area between the boss and the casting would have only a small effect on preload relaxation. Pettersen and Fairchild found substantial improvements in load retention by increasing the washer diameter and by increasing the thickness of the plate or the washer in the assembly [10]. However, they used a closed assembly, such that a nut was tightened at the end of the bolt. With increased thickness in the plate or washer, the effective bolt-length is increased, creating a combined effect that adds to the creep resistance of the clamping assembly. Such a closed assembly is not achievable in many drive train components, particularly in a typical automatic transmission case.

CONCLUSION

High pressure die cast MEZ displays far superior bolt-load retention behavior when compared to other magnesium alloys. The BLR characteristics of MEZ are nearly as good as those of 380Al even to temperatures as high as 175°C. The relatively high solidus temperature of MEZ and a stable microstructure containing higher melting rare earth phases contribute to its excellent creep resistance. Changes to the geometry of this particular bolt-load retention assembly by the use of washers did not improve the stress relaxation properties significantly as the thread engagement length of the case/flange and bolt was effectively decreased, resulting in greater creep in the BLR assembly. This last result indicates the importance of thread contact length for assemblies fabricated from magnesium alloys.

ACKNOWLEDGMENTS

I. Moreno thanks Carl Hendersen and Chris Palenik of the University of Michigan Electron Microscope Analysis Laboratory with help using the WDS and EDS on the EMP and TEM.

REFERENCES

1. J.F. King, "Development of Magnesium Diecasting Alloys". Paper presented at the Magnesium Alloys and their Applications conference, Wolfsburg, Germany. April 28-30, 1998, pp 37-47.
2. I. Nakatsugawa, S. Kamado, Y. Kojima, R. Ninomiya, and K. Kubota, "Corrosion of magnesium alloys containing rare earth elements." Corrosion Reviews, vol. 16, no. 1-2, 1998, pp 139-157.
3. J. Berkmortel, H. Hu, J.E. Kearns and J.E. Allison, "Die Castability Assessment of Magnesium Alloys for High Temperature Applications: Part 1 of 2", SAE Technical Paper 2000-01-1119, 2000.
4. K.Y. Sohn, J.A. Yurko, J.W. Jones, J.E. Kearns, and J.E. Allison, "Bolt-Load Retention Behavior of Die-Cast AZ91D and AE42 Magnesium", SAE Technical Paper 980090, 1998.
5. T.K. Aune and H. Westengen, "Property Update on Magnesium Die Casting Alloys", SAE Technical Paper 950424, 1995.
6. E.G. Sieracki, J.J. Velazquez, and K. Kabiri, "Compressive Stress Retention Characteristics of High Pressure Die Casting Magnesium Alloys", SAE Technical Paper 960421, 1996.
7. C.S. Roberts. Magnesium and Its Alloys. New York, John Wiley & Sons, Inc. 1960.
8. G.A. Mellor and R.W. Ridley, "Creep at 250°C and 300°C of Some Magnesium Alloys Containing Cerium", Journal of the Institute of Metals, vol. 81, 1952-53, pp 245-253.
9. P. Lyon, J.F. King, and K. Nuttall, "A New Magnesium HPDC Alloy for Elevated Temperature Use", Paper presented at the Proceedings of the 3rd International Magnesium Conference, Manchester, UK, April 1995, pp 99-108.
10. K. Pettersen and S. Fairchild, "Stress Relaxation in Bolted Joints of Die Cast Magnesium Components", SAE Technical Paper 970326, 1997.
11. K.Y. Sohn, J.W. Jones, and J.E. Allison, "Creep Mechanism in Die-cast Magnesium Alloys AZ91D and AE42", presentation made at the Automotive Alloys III, 1999 TMS Annual Meeting, San Diego, California, February 27-March 4, 1999.
12. J.H. Bickford. An Introduction to the Design and Behavior of Bolted Joints, 3rd Ed., New York, Marcel Dekker, Inc., 1995.
13. K.Y. Sohn, J.W. Jones, and J.E. Allison, "The Effect of Calcium on Creep and Bolt Load Retention Behavior of Die-Cast AM50 Alloy", Paper presented at the Magnesium Technology 2000, Nashville, Tn, March 12-16, 2000, pp 271-278.
14. K.Y. Sohn, J.W. Jones, J. Berkmortel, H. Hu, and J.E. Allison, "Creep and Bolt Load Retention Behavior of Die Cast Magnesium Alloys for High Temperature Applications: Part 2 of 2", SAE Technical Paper 2000-01-1120, 2000.
15. F.C. Chen, J.W. Jones, T.A. McGinn, J.E. Kearns, A.J. Nielson, and J.E. Allison, "Bolt-Load Retention and Creep of Die-Cast Magnesium Alloys", SAE Technical Paper 970325, 1997.
16. X. Su, K.Y. Sohn, D. Dewhirst, and J.E. Allison, "Finite Element Modeling of Bolt Load Retention of Die-Cast Magnesium", SAE Technical Paper 2000-01-1121, 2000.
17. S. Murphy and F.E. Goodwin, "Preload Relaxation of Steel Fasteners in Zinc Alloy Pressure Die Castings-Some Engineering Solutions", SAE Technical Paper 960763, 1996.
18. K.Y. Sohn, J.A. Yurko, F.C. Chen, J.W. Jones, and J.E. Allison, "Creep and Bolt-Load Retention Behavior of Die-Cast Magnesium Alloys", in Automotive Alloys II, San Antonio, Tx, TMS, February 16-19, 1998, pp 81-90.
19. X. Su, Personal communication.

About the Editor

Thomas Ruden is the market development manager for Lunt Manufacturing, a global leader in magnesium die casting company, headquartered in Schaumburg, IL. Mr. Ruden's career has spanned the automotive industry for more than 25 years, with 17 of those years directly involved with the magnesium and light metal industries.

Currently, Mr. Ruden conducts workshops and seminars regarding magnesium and works closely with engineers and product planners to convert applications from alternate materials into magnesium. He holds a B.Sc. in mechanical engineering from Lawrence Institute of Technology and an MBA from the University of Detroit.

In addition to application and market development work, he has authored more than 30 technical papers and writes about the advancements of magnesium. He holds several U.S. patents. Tom has been a member of the Society of Automotive Engineers for 22 years, an organizer for their Magnesium Technology Sessions, and formerly the Chair of SAE's Non-ferrous committee.